EXPERIMENTAL NUCLEAR PHYSICS IN EUROPE

EXPERIMENTAL NUCLEAR PHYSICS IN EUROPE

ENPE 99
Facing the Next Millennium

Sevilla, Spain   June 1999

EDITORS
Berta Rubio
CSIC, Valencia, Spain

Manuel Lozano
University of Sevilla, Spain

William Gelletly
University of Surrey, U.K.

American Institute of Physics

AIP CONFERENCE
PROCEEDINGS 495

Melville, New York

Editors:

Berta Rubio
Instituto de Fisica Corpuscular
CSIC-Valencia
Edificio Institutos de Paterna
Apartado de Correos 2085
46071 Valencia
SPAIN

E-mail: berta.rubio@ific.uv.es

Manuel Lozano
Departamento de Fisica Atomica, Molecular y Nuclear
Facultad de Fisica
Aptdo. 1065
Universidad de Sevilla
41080 Sevilla
SPAIN

E-mail: lozano@cica.es

William Gelletly
School of Physics and Chemistry
University of Surrey
Guildford GU2 5XH
Surrey
UNITED KINGDOM

E-mail: w.gelletly@surrey.ac.uk

L.C. Catalog Card No. 99-067684
ISBN 1-56396-907-6
ISSN 0094-243X
Printed in the United States of America

CONTENTS

EXOTIC NUCLEI

REACTIONS AND RADIOACTIVE NUCLEAR BEAMS

HIGH ENERGY COLLISIONS

APPLICATIONS

CLOSING REMARKS

PREFACE

Dear colleagues, dear friends

In this present year, the year of our lord 1999, a considerable number of nuclear physicists gathered together in the city of Sevilla (Spain), from the 21st to the 26th of June, to discuss the state of knowledge in their subject.

Some five centuries ago Sevilla was undoubtedly the most important trading centre in Europe. It was here that both goods and ideas were exchanged between the New World and Old Europe, then still emerging from the Middle Ages into the full enlightenment of the Renaissance. There was no lack of wise men at that time, who knew much about philosophy, art, navigation, botany, astronomy etc. Although strongly influenced by religious ideas, the best of them could absorb the high points of Arabic, Jewish and Christian culture and ideas. Such people were poised, with eyes open, ready to exploit the new discoveries.

From then until now, we have advanced enormously in terms of our knowledge and understanding of the world we live in. Some hints of the changes to come were just visible at the time but some could not even be glimpsed by the most percipient minds. Amongst the most dramatic changes is our knowledge of the most intimate nature of matter. Without the discovery of Radioactivity, 100 years ago, and the later technical developments which allowed us to ionise atoms and accelerate them to produce beams of atomic nuclei, we would never have begun to understand how the subatomic part of our world is constructed and how it behaves.

It is one of the fundamental constituents of matter, the atomic nucleus, which was the focus of discussion during the month of June in Sevilla. Just as five centuries ago Sevilla was the meeting point for people from many different countries so it was now. They came from the shores of the Mediterranean, from the North, East and West of Europe, from the American continent, the Middle and Far East. Now, as then, they were greeted with open arms by this generous and hospitable city (now with the advantage of air conditioning!)

In this learned gathering we were to hear about the most recent and interesting discoveries; everything from the most exotic nuclei, extremely difficult to synthesise in a terrestrial laboratory and whose mere existence proves a challenge to our nuclear models, to rare and exotic effects in those nuclei more accessible to our experiments. In our field advances are the result of a common, collective effort in terms not only of the exchange of ideas but also in shared efforts to create technological improvements. It is here that we encounter another of the major goals of the Sevilla Conference.

When a group of people works at the cutting edge of experimental technique, they may realise that they have opened the way to some completely orthogonal application. Just as in the Sevilla of five centuries ago some of the adventurous in spirit will change direction and start a new line of research which will improve the quality of life for all of us. In our meeting we saw some beautiful examples; how to determine the composition of particular materials, pollutants in our environment, metals in prehistoric artefacts, the abundance of the elements in the stars immediately spring to mind. In some applications it even allows us to eliminate things which may cause us harm, such as the cells of a tumour or long-lived radioactive residues.

All these and many other applications of nuclear physics, were presented beautifully at the conference, and were also broadcast to the general public through articles in the press and on television and radio. Of course behind all that was the idea of convincing our scientific authorities that in Nuclear Physics, as in any other branch of science, research brings knowledge, knowledge brings progress and progress brings wealth for all of us, not only in economic terms but also in terms of the quality of life, particularly for the country where this progress is made. Science is driven by ideas, and ideas and the people who have them face no national frontiers. There are many, very good, young Spanish nuclear physicists who are welcome and working all around the world. However, as happens elsewhere, our country should make an effort to absorb these highly qualified people and to give the Spanish nuclear physics community the tools to contribute to progress in our field. Given the tools we will be happy to do this job and for a reasonable investment our country will benefit enormously.

Let me turn finally to you, the participants in the conference, whose contributions were the key ingredients in making it a success. It was the quality of your presentations and participation in the discussions which led to the excellent atmosphere inside and outside the conference hall. Let me thank you from the depth of my heart for your carefully written contributions to the book. The spoken word may go with the wind, but if the words are written down they stay. Let me thank the people who made this event possible and the Institutions which supported it. In particular I give my warmest thanks to Rocio Peinado, our Conference Secretary, without whom this conference would never have been possible.

This conference is but a small grain of sand on the heap to try to make our field and our people more successful. As Seneca, a famous Andalusian, said many years ago:

"If one does not know to which port one is sailing, no wind is favourable"

I think we know our port of call.

Berta Rubio
IFIC, Valencia, 20th September 1999

CONFERENCE CHAIRPERSONS

B. Rubio (Valencia)

M. Lozano (Sevilla)

LOCAL ORGANIZING COMMITTEE

J. M. Arias (Sevilla)

J. Diaz (Valencia)

J.L. Egido (Madrid)

J.L. Ferrero (Valencia)

M.J.G. Borge (Madrid)

A. Lallena (Granada)

E. Moya de Guerra (Madrid)

INTERNATIONAL ADVISORY COMMITTEE:

U. Amaldi (Italy)

G. Amsel (France)

P. Armbruster (Germany)

P. Butler (United Kingdom)

G. de Angelis (Italy)

S. Gales (France)

W. Gelletly (United Kingdom)

P. G. Hansen (USA)

M.N. Harakeh (The Netherlands)

R. Julin (Finland)

J.S. Lilley (United Kingdom)

A. Mueller (France)

A. Richter (Germany)

D. Schardt (Germany)

J.C. Soares (Portugal)

J. Vervier (Belgium)

ACKNOWLEDGMENTS

- Universidad de Sevilla

- Consejo Superior de Investigaciones Científicas

- Instituto de Física Corpuscular (CSIC-Univ. Valencia)

- Centro de Investigaciones Científicas Isla de la Cartuja.

- Centro Informático y Científico de Andalucía

- Ministerio de Educación y Ciencia, Dirección General de Enseñanza Superior e Investigación Científica.

- Junta de Andalucía
  Dirección General de Universidades e Investigación.

- Iberia L.A.E

- R.E.N.F.E.

- Nuclear Ibérica

- EDP Sciences

- PROCONSUR, S.A.

EXOTIC NUCLEI

Light Dripline Nuclei

Björn Jonson

Experimental physics, Chalmers University of Technology and Göteborg University
S-412 96 Göteborg, Sweden

Abstract. This paper reviews the recent developments in the experimental studies of nuclei at or beyond the driplines.

INTRODUCTION

The isobaric lines across the nuclear chart are terminated at both the neutron-deficient and neutron-rich side of stability where the nuclei become unstable against emission of nucleons. The two lines connecting these last bound members of the isobars are referred to as the *driplines*. For more than a decade there have been intense experimental and theoretical activities concerning light nuclei in the dripline region. The reason for the interest is that light dripline nuclei have shown novel structural features that make them scientifically very attractive. This is illustrated in Fig. 1 for the two isobars A=11 and 17. One of the most striking discoveries was observation that dripline nuclei could develop dilute, spatially very extended nuclear matter distributions referred to as nuclear halo states. The nucleus ^{11}Be is one example of a dripline nucleus where the last bound neutron (S_n=504 keV) forms a halo. The prime example of a nucleus with a halo state as its ground state is ^{11}Li. Its halo belongs to the so called Borromean halo states, where the three-body system (^{9}Li+n +n) is bound while the binary subsystems (^{10}Li, ^{2}n) are unbound. The Borromean nuclei are of great current interest since they are situated at the threshold that separates the discrete and the continuous spectrum. They are held together by forces that in neighboring nuclei give rise to scattering states, which leaves strong imprints on those states in form of final-state interactions. It is therefore important to understand the continuum well. There are also indications of the disappearance of the shell closures at N=8, 20, 28 at the driplines, and there are changes in the ordering of the shell-model states. An example of the latter is the $1s_{1/2}$ intruder state, which forms the ground state of ^{11}Be.

There have been several reviews on the physics of dripline nuclei, which describes the main experimental and theoretical progress over the past 15 years (1,2,3,4). In this talk I shall point to the many new results that have come from studies performed at different laboratories worldwide during the past few years.

CP495, *Experimental Nuclear Physics in Europe*, edited by B. Rubio et al.
© 1999 American Institute of Physics 1-56396-907-6/99/$15.00

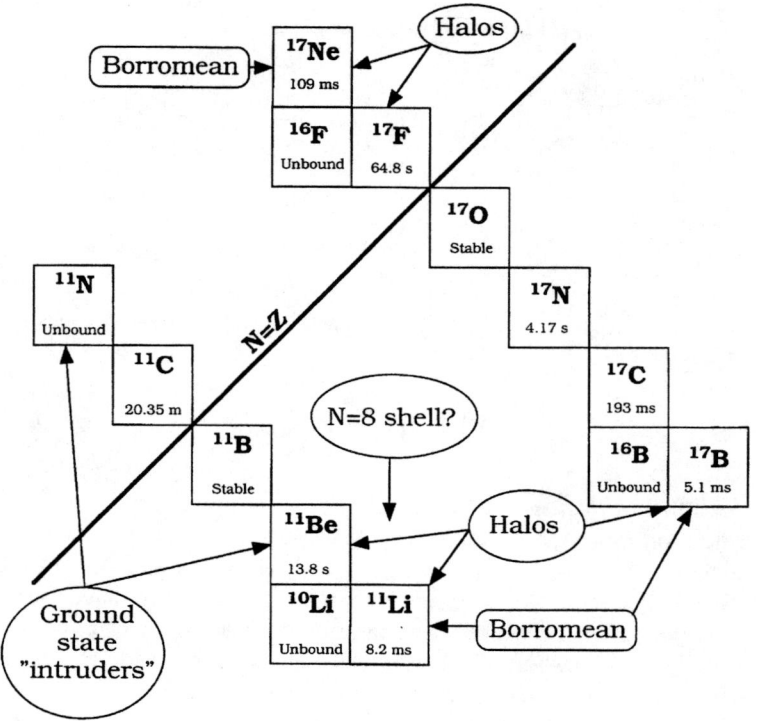

FIGURE 1. Many, partly interconnected phenomena occur for light dripline nuclei. This is here illustrated for the mass 11 and 17 isobars.

EXPERIMENTS ON DRIPLINE NUCLEI.

The progress in the study of dripline nuclei depends very much on the availability of good production methods of the exotic nuclei. We have heard about the progress and plans for the production and acceleration of radioactive beams from Daniel Guerreau and Gottfired Münzenberg during this conference and there are several investigations going on how to create the "ideal" accelerator facility to serve the physics community.

There has also been a strong development of new methods to investigate exotic nuclei. Here I may mention:

- Laser spectroscopic methods to determine spins and nuclear moments. (5)
- Determinations of charge radii from reaction experiments.

- Mass measurements in traps (6) and with the MISTRAL spectrometer at ISOLDE (7) and at the ESR at GSI (8)
- Investigations of beta-delayed particles with multi-detector arrangements (9).
- Reaction studies with radioactive beams in complete kinematics experiments.

In the following pages I give some examples on recent experiments on nuclei in the dripline region where the methods mentioned above have been utilized. I organize the material as a function of mass number.

Experiments on $^{6-7}$He.

The first dripline nucleus to be discovered was ^{6}He, which was synthesized as early as in 1936 (10) for the first time. Still, there are many new experimental results concerning ^{6}He and it has served as an important test case since its main nuclear properties are well known.

- In experiments at GSI one-neutron knockout reactions were studied using a 240 MeV/u beam of ^{6}He. The shape of the (α+n) invariant mass spectrum showed that the dominating reaction mechanism was a two-step process with one neutron knocked out followed by the decay of the ^{5}He resonance (11).
- It was found, in the same experiment, that the knockout reaction leads to a spin alignment of ^{5}He in a plane perpendicular to the ^{5}He-momentum vector (12).
- The correlation function was found to be $W(\vartheta_{\alpha n}) \propto 1 + A\cos^2(\vartheta_{\alpha n})$ with A=1.5(3). This result was used (13) to determine that the contribution of the $(0p_{1/2})^2$ component in the ^{6}He ground state wave function is 7%.
- In experiments at Dubna, a beam of 151 MeV/u ^{6}He was used to study two-neutron exchange reactions in a gaseous helium target (14). The results revealed large cross sections in backward angles, which can be interpreted as due to the di-neutron component of the ^{6}He wave function.
- At GSI the three-body breakup ^{6}He→α+n+n was used to study continuum excitations in ^{6}He. The comparison between the *E1* strength and the cluster sum rules give a direct insight in the geometry of the ground-state wavefunction (15).
- In an experiment at RIKEN a search for excitations of the ^{7}He resonance was performed (16). A new resonance, at an energy 3.3 MeV above the ^{7}He ground state ($\Gamma \approx 2.2$ MeV), was observed and interpreted as due to a state with $I^{\pi} = 5/2^-$ and the proposed structure $[p_{1/2} \otimes ^6$He*(2$^+$)].

Experiments in the A=11 Region.

The two halo nuclei ^{11}Be and ^{11}Li have been studied in a large number of different experiments and new data continues to become available. Some of the most recent results for these nuclei are:

- Peripheral fragmentation of a 287 MeV/u beam incident on a carbon target was studied at GSI in a complete kinematics experiment (17). The angular correlation, between the knocked out neutron and the decay neutron from the ^{10}Li resonance, shows a skew distribution (see Fig. 2). This provides a model-independent indication of an admixture of states with different parity in the ground state of ^{11}Li.

- In a recent experiment at MSU ^{10}Li was studied and the relative velocity spectrum of the neutron and the ^{9}Li fragment was measured using the method of sequential neutron decay spectroscopy (18). The results show evidence for low-lying s-wave strength, confirming earlier GSI data (19) and theoretical predictions (20). The s-wave scattering length was $a_s < -20$ fm, corresponding to a peak energy of <50 keV.

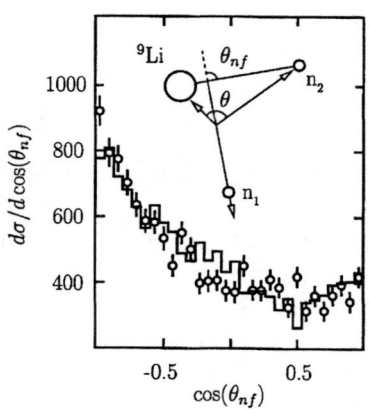

FIGURE2. Distribution of the decay neutron from ^{10}Li formed in ^{11}Li neutron knockout reactions.

- The momentum distribution of the ^{10}Li fragments gives direct access to the momentum of the removed neutron. The shape of the distribution, measured at GSI, was used to determine that the $(1s_{1/2})^2$ component in the ground-state wave function is 45±10 %.

- In an experiment at GANIL core breakup reactions of ^{11}Be and ^{11}Li were studied (21,22). The measured neutron multiplicity for ^{11}Be gave a direct manifestation of the shadowing (23) of the neutron in the core-breakup reactions.

- In experiments at MSU coincidences between ^{10}Be and gamma rays after breakup of ^{11}Be were studied. The results gave the contribution of the $[(0d_{5/2}) \otimes {}^{10}Be(2^+)]$ in the ^{11}Be ground state. The data also showed coincidences with gamma rays depopulating the 1⁻ and 2⁻ states in ^{10}Be. They stem from the configurations $[(0p_{3/2}^{-1}) \otimes (1s_{3/2})]$ after core-neutron removal (24).

- Isotopes of Be have recently been available for studies at ISOLDE where the Be atoms were ionized in a two-step ionization scheme with a subsequent excitation to an auto-ionizing state (25). The magnetic moment of ^{11}Be was measured with a β-NMR technique. The result, $\mu(^{11}Be) = -1.6814(13)$ μ_N, which is close to the Schmidt limit of -1.91 μ_N, seems to favor a relatively pure $(1s_{1/2})$ nature of the 1/2⁺ ground state (26).

Investigations of dripline nuclei beyond A=11.

I shall end this paper by listing some additional, new results on dripline nuclei:

- In experiments at GANIL one studied (27) the breakup of a 31.5 MeV/u beam of ^{12}Be, inelastically excited in carbon and $(CH_2)_n$ targets, into ^{6}He+^{6}He and ^{4}He+^{8}He. Rotational states in the energy interval 10 to 25 MeV were observed and the inferred moment of inertia was consistent with a cluster decay of an exotic α-4n-α structure (28). (See also the contribution by H.G. Bohlen to this conference.)

- The beta-delayed neutron emission probability of the two-neutron halo nucleus ^{14}Be was measured at ISOLDE with a time-correlation technique (29). The results show that most of the beta decays are followed by one-neutron emission, while the multi-neutron branches are consistent with zero.

- In experiments at GANIL, with the DEMON array, dissociation of ^{14}Be was studied. In the analysis of these data the technique of intensity interferometry was employed. A new iterative method was applied and correlation functions were extracted leading to root-mean-square n-n separations in good agreement with predictions from three-body models (30).

- Longitudinal momentum distributions from the breakup of ^{15,17,19}C have been studied at MSU and the results are discussed in (27).

- At the FRS at GSI (31) the longitudinal momentum distributions for ^{17,19}C were found to have momentum widths of 141 MeV/c and 69 MeV/c, respectively (32,33)

- Coulomb dissociation of 67 MeV/u ^{19}C in a carbon target has been studied (34) at RIKEN. They used their data for an indirect determination of the neutron separation energy in ^{19}C, based on an analysis of the center of mass angular distribution of ^{18}C+n. They arrived at a result of S_n=530±130 keV, which is a number preferred by several theoretical investigations.

FUTURE PERSPECTIVES

As a final comment in my talk I shall mention a few directions that I believe will be of interest for the continuation of our field. Higher intensity, by brute force or by elegance, would be most welcomed in most of the experiments at the driplines since our field, to some extent, is suffering from too low data rates. New beams in the driver accelerators, as for example ^{48}Ca, should also be employed. The post acceleration of radioactive beams is becoming more and more into the focus of interest. We shall soon see SPIRAL at GANIL and REX ISOLDE at CERN in operation. Several international committees are working on more or less exotic schemes for the FACILITY of the next millennium. On the European scene there shall be a NuPECC report by the end of this year where the first steps towards a joint European proposal will be sketched (35). The work towards the proposal will also address detector and spectrometer developments. In my opinion it is clear that our field needs exotic radioactive nuclei available both as thin radioactive samples *and* as beams from the lowest energy up to the GeV/u region. If this can be done at one geographic location in Europe is not yet clear but a strong intellectual network joining our forces has to be set up soon.

On the physics side it is clear that we want to reach the dripline for much heavier neutron-rich nuclei than today. Maybe one may come close to the r-process path and in some regions of the nuclear chart we might envisage physics beyond nuclear astrophysics. It seems as if the heavy element search now has reached the impressive $Z=118$ region. One may speculate if our technique will be able to find (possible) exotica far beyond the driplines, as for example ^{13}Li. The next millennium will tell us....

REFERENCES

1. M.V. Zhukov et al., *Phys. Rep.* **231**(1993) 151
2. P.G. Hansen, A.S. Jensen, B. Jonson, *Rev.Nucl.Part.Sci.* **45** (19959 591-634.
3. I. Tanihata, *J. Phys.* **G22** (1996) 157-
4. B. Jonson, K. Riisager, *Phil. Trans. R. Soc. Lond.* **A356** (1998) 2063-2081.
5. W. Geitner et al., *The ISOLDE Lab. Portrait*, to be published in *Hyperfine Interactions*.
6. G. Bollen, *these proceedings*.
7. C. Monsanglant, *these proceedings*.
8. H. Geissel, *these proceedings*.
9. O. Tengblad, *these proceedings*.
10. T. Bjerge, K.J. Broström, *Nature (London)* **138** (1936) 400-402.
11. D. Aleksandrov et al., *Nucl. Phys.* **A633** (1998) 234 -246.
12. L.V. Chulkov et al., *Phys. Rev. Lett.* **79** (1997) 201-204.
13. L.V. Chulkov, G. Schrieder, *Z. Phys.* **A359** (1997) 231-234.
14. G.M. Ter-Akopian et al., *Phys. Lett.* **B426** (1998) 251-256.
15. T. Aumann et al., *Phys. Rev.* **C59** (1999) 1252-1262.
16. A.A. Korshenninikov et al., *Phys. Rev. Lett.* **82** (1999) 3581-3584.
17. H. Simon et al., *Phys. Rev. Lett.* **83** (1999) 496-499.
18. M. Thoennesen et al., *Phys. Rev.* **59** (1999) 111-117.
19. M. Zinser et al., *Phys. Rev. Lett.* **75** (1995) 1719-1722.
20. I.J. Thompson, M.V. Zhukov, *Phys. Rev.* **C49** (1994) 1904-1907.
21. H. Grevy et al., *Nucl. Phys.* **A650** (1999) 47 - 61.
22. L. Axelsson et al., *these proceedings*.
23. P.G. Hansen, *Phys. Rev. Lett.* **77** (1996) 1016-1019.
24. A. Navin et al., *these proceedings*.
25. U. Georg et al., *these proceedings*.
26. W. Geithner et al., CERN-EP/99-73, submitted to *Phys. Rev. Lett.*
27. M. Freer et al., *Phys. Rev. Lett.* **82** (1999)1383-1386.
28. W. von Oertsen, Z. Phys. A357 (1997) 355- 365.
29. U. Bergmann et al., *Nucl. Phys. in press.* See also these proceedings.
30. F.M. Marqués et al., *Phys. Rev. Lett.* In press.
31. D. Cortina-Gil *et al, these proceedings*.
32. T. Baumann et al., Phys. Lett. B439 (1998) 256- 261.
33. K. Markenroth, et al., *these proceedings*.
34. T. Nakamura et al., *Phys. Rev. Lett.* **83** (1999) 1112-1115.
35. Report from the *NuPECC Study Group of the Future European Radioactive Beam Facility.*

Elastic 2n-transfer in the ^{4}He(^{6}He,^{6}He)^{4}He scattering

R. Raabe*, A. Piechaczek*[1], A. Andreyev*[2], D. Baye†,
W. Bradfield-Smith‡, S. Cherubini*, T. Davinson‡,
P. Descouvemont†, A. Di Pietro‡, W. Galster*, M. Huyse*,
A.M. Laird‡, J. McKenzie‡, W.F. Mueller*, A. Ostrowski‡,
A. Shotter‡, P. Van Duppen* and A. Wöhr*[3]

*Instituut voor Kern- en Stralingsfysica, University of Leuven
B-3001 Leuven, Belgium
†Physique Nucléaire Théorique et Physique Mathématique, C.P. 229
Université Libre de Bruxelles, B-1050 Brussels, Belgium
‡Department of Physics & Astronomy, University of Edinburgh
Edinburgh EH9 3JZ, United Kingdom
*Institut de Physique Nucléaire, Université Catholique de Louvain
B-1348 Louvain-la-Neuve, Belgium

Abstract. The elastic scattering ^{4}He(^{6}He,^{6}He)^{4}He has been investigated at center-of-mass energies of 11.6 and 15.9 MeV. Differential cross sections are determined using a post-accelerated ^{6}He ($T_{1/2} = 0.807$ s) beam in the center-of-mass angular range between 50 and 140 degrees. The comparison of the measured data with calculations using a double folding potential shows evidence for the 2n-transfer process in the ^{4}He(^{6}He,^{6}He)^{4}He elastic scattering.

Great interest has been devoted to halo nuclei since their first discovery [1]. The halo-nucleus ^{6}He [2] is a favorable case to study because of its simple structure: it may be described as an inert ^{4}He core surrounded by two weakly bound valence neutrons. Furthermore, the ^{4}He-n and n-n interactions are well known. The main properties of ^{6}He have been reproduced using either macroscopic three-body models [3] or microscopic three-cluster models [4]. Experiments performed so far focus mainly on the extended tail in the matter distribution of ^{6}He. However, little

[1] Present address: Department of Physics and Astronomy, Louisiana State University, Baton Rouge, LA 70803, U.S.A.
[2] Permanent address: FLNP, Joint Institute for Nuclear Research, 141980 Dubna, Moscow Region, Russia
[3] Present address: Department of Physics, University of Oxford, Oxford OX1 3PU, United Kingdom

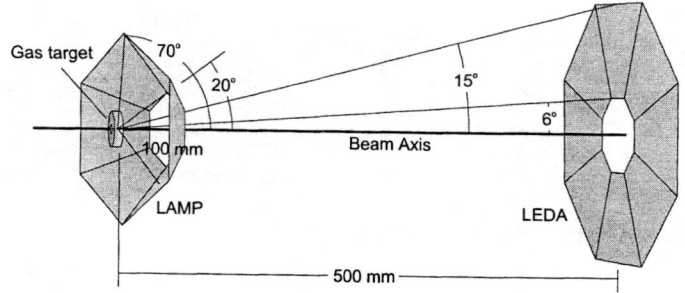

FIGURE 1. Arrangement of LEDA and LAMP silicon detectors with respect to the ^{4}He gas target.

is known about the details of the neutron halo structure.

In order to reveal information on a possible 2-neutron component in the ^{6}He wavefunction, we investigated the elastic scattering ^{4}He(^{6}He,^{6}He)^{4}He at beam energies of 29.6 and 40 MeV, corresponding to center-of-mass energies of 11.6 and 15.9 MeV, respectively. Assuming the cluster-structure of ^{6}He, elastic two-neutron transfer is expected to occur between the two α-cores [5]. The experiments were performed at the ARENAS^3 [6] Radioactive Ion Beam Facility in Louvain-La-Neuve, Belgium. There, a post-accelerated low-energy ^{6}He beam is available with intensity and purity not obtainable elsewhere. At the detection point, a gas target of 1 cm thickness was placed, containing ^{4}He at 500 mbar pressure. Reaction products were registered using a segmented silicon detector system [7], covering the laboratory scattering angles between 6 and 15 degrees and between 20 and 70 degrees. The detector arrangement with respect to the target is shown in Fig. 1. Eight segments were placed at a distance of 500 mm (LEDA); six more segments were inclined with respect to the beam axis to form a lamp-shade shaped structure (LAMP). Each segment is divided into 16 annular strips. ^{4}He(^{6}He,^{6}He)^{4}He elastic scattering events were searched for among all multiplicity-2 events detected. An effective separation of the ^{6}He–^{4}He pairs from elastic and inelastic background processes was possible by applying kinematic conditions. The center-of-mass scattering angle of each event was determined from the deposited energies. Fig. 2 shows the experimental data, compared with theoretical calculations that are discussed below. The obtained values have an overall uncertainty of ±20% due to tolerances in the integrated beam current, target thickness and efficiency determination. The statistical errors are shown in the figures.

Analysis of the data was performed using a double-folding interaction potential suggested by Baye *et al.* [4], based on a microscopic three-cluster model which is used to calculate the matter distribution of ^{6}He. The potential is described analytically by a gaussian part and a Woods-Saxon part; the latter is included to account for the extended matter distribution in ^{6}He [4]. To investigate effects related with the elastic transfer of the two neutrons, we included a parity-dependent

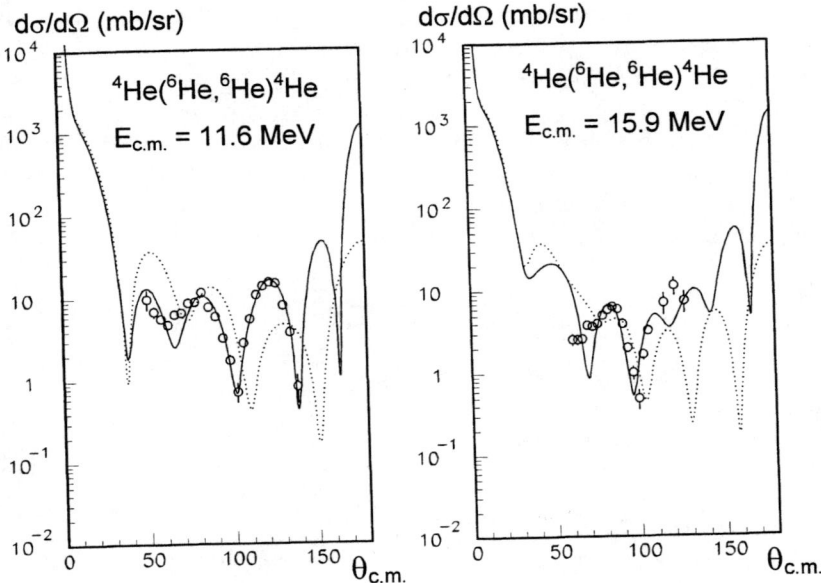

FIGURE 2. Measured ^{4}He(^{6}He,^{6}He)^{4}He differential cross sections. The full curves are best fits using the double folding potential [4], including a parity-dependent term. The dotted curves are best fits obtained without the parity-dependent term.

potential V_P which has the Yukawa asymptotic form $\exp(-kR)/R$ [4,5]. W. von Oertzen and H.G. Bohlen [8] showed that the sign of V_P depends upon the number of particles involved in the exchange process, their angular momentum, and the spin of the core particle. The coefficient k of the Yukawa asymptotic form can be related to the effective binding energy and to the reduced mass of the exchanged particle or cluster with respect to the core [5]. However, deriving a value for k in our case is not obvious because the di-neutron is unbound; moreover, the halo of ^{6}He has a much more complicated structure than a di-neutron cluster coupled to a ^{4}He core.

We performed a best fit on the two sets of data simultaneously. The results are shown in Fig. 2 (solid lines). The sign of V_P obtained from the fit is opposite to theoretical expectations. However, W. von Oertzen and H.G. Bohlen [8] pointed out that the sign is well-determined only when a strong minimum or maximum is present at 90 degrees. This is not the case for our data: at both energies a maximum is present at 83 degrees and a minimum at 100 degrees. A difference in the sign of V_P is the signature of a more complex process than the exchange of a single 2n-cluster. Contributions from other effects in potential scattering can also be present: these could come from the extended matter distribution in ^{6}He, reflected in the Woods-Saxon term of the double-folding potential. Still, the contribution of the parity-dependent term is essential in order to reproduce the data. The dotted curves in Fig. 2, calculated from the previous ones omitting the parity-dependent term, differ significantly from the experimental data.

In conclusion, the need for a parity-dependent term to reproduce the data shows that the elastic 2n-transfer process is present in the ^{4}He(^{6}He,^{6}He)^{4}He scattering at $E_{c.m.} = 11.6$ MeV and 15.9 MeV. The absence of a strong minimum or maximum at 90 degrees, suggests a more complicated process than the transfer of a 2n-cluster. This confirms that the internal structure of ^{6}He is more complicated than a di-neutron coupled to the ^{4}He core, and is better described by a microscopic model.

REFERENCES

1. I. Tanihata et al., Phys. Lett. 160B (1985) 380.
2. I. Tanihata et al., Phys. Lett. B 289 (1992) 261.
3. M.V. Zhukov et al., Phys. Rep. 231 (1993) 151.
4. D. Baye, L. Desorgher, D. Guillain, and D. Herschkowitz, Phys. Rev. C 54 (1996) 2563.
5. W. von Oertzen, Nucl. Phys. A148 (1970) 529.
6. M. Gaelens, M. Huyse, P. Van Duppen, M. Loiselet and G. Ryckewaert, Nucl. Instr. Meth. B 126 (1997) 125.
7. P.J. Sellin et al., Nucl. Instr. Meth. A 311 (1992) 217.
8. W. von Oertzen and H.G. Bohlen, Phys. Rep. 19 (1975) 1.

Low lying states in ^{6}He by (p,p') scattering.

F. Auger, A. Lagoyannis, A. Musumarra, <u>E. C. Pollacco</u>, N. Alamanos, F. Braga,
A. Drouart, G. Fioni, A. Gillibert, V. Lapoux, S. Ottini-Hustache,
DAPNIA/SPhN, CEA-Saclay, F-91191 Gif-sur-Yvette, Cedex, France
Y. Blumenfeld, E. Khan, D. Santonocito, J. A. Scarpaci, T. Suomijarvi,
Inst. de Physique Nucléaire, IN2P3-CNRS, F-91406 Orsay, France
W. Mittig, P. Roussel-Chomaz,
GANIL, BP 5027, F-14021, Caen, France
M. La Commara, D. Pierroutsakou, M. Romoli, M. Sandoli.
INFN, Sezione di Napoli, I-80125 Napoli, Italy

Over the last three years we have been developing an experimental program at GANIL to study the nuclear structure of neutron and proton rich light nuclei from elastic, inelastic and transfer reactions using inverse kinematics on proton or deuterated targets. In this short communication we present the (p,p') data for ^{6}He. ^{6}He has generated a considerable interest. It is Borromean. It is considered as a halo nucleus. Predictions of "di-neutron" or "cigar-like" configurations have been made[1]. Predictions have also been made on soft dipole modes[2]. Many structure calculations are available in the literature[3]. In contrast data is limited.

The experiment was performed using a primary ^{13}C beam from GANIL. The resulting ^{6}He at 40.9$A \cdot MeV$ traversed two CATS[4] time-position detectors. Target of $(CH_2)_3$ were used. The beam on target had a rate of ~10^5 Hz and the ions were tracked event-by-event by employing CATS which gave a position resolution of 0.3mm (FWHM) in both x- and y-direction. Proton recoils were detected in 8 MUST[5] telescopes. The energy threshold for protons was 0.7MeV. Energy and time resolutions offered a good mass and charge selection over the dynamic range of interest. The scattered helium ions were detected using a plastic scintillator hodoscope, LIGHT. The beam was analysed in a separate small scintillator. The collected light output v.s. TOF(CATS-Scintillator) allowed a good separation between the beam contaminants, ^{6}He and ^{6}He decay ($\rightarrow$ ^{4}He + 2n). Data was analysed event wise by setting software gates on both the proton (MUST) and ^{4,6}He (LIGHT) and by computing the incident and exit trajectories. The data was normalised by the target thickness, average incident beam characteristics and angular acceptance of MUST.

In fig. 1 the reconstructed proton energy spectrum is given and shows the good separation between the g.s. and the first excited state at 1.8MeV. Fig. 2 shows the extracted angular distributions for the two states. The angular distribution for the 1.8MeV level is presented for the first time. With the data points given in fig. 2 we plot the calculations by S. Karataglidid et al.[6]. No description of the calculation is given here, but we underline that it is *shell model based* (no core, 4$\hbar\omega$ space, G-matrix interaction of Zheng et al.[7]) *with no free parameters.* Given that that <u>no</u> normalisation factor are introduced for the comparison (solid curves), the description is excellent. The calculations reconfirms the $J^\pi = 2^+$ [8]. Further, the calculated reaction cross-section of 406mb reproduces the measured value of 415±21mb[9]. To demonstrate the sensitivity of

CP495, *Experimental Nuclear Physics in Europe*, edited by B. Rubio et al.
© 1999 American Institute of Physics 1-56396-907-6/99/$15.00

this comparison the calculation was redone with an increase in the valency neutron binding energy (dashed curves). With higher neutron binding the model shows that the halo is transformed into a neutron skin of thickness ~ 0.5fm. The elastic scattering is modified principally at backward angle beyond 65° where our data is limited but in agreement with the solid curve. In contrast for the 2^+ we note an interesting modification in the shape over the whole of the measured angular range. Hence we conclude that our data is consistent with a halo description. From this comparison we hold that elastic and inelastic high quality measurements and an extended theoretical study will yield a means to probe low density matter distributions and disclose the underlying spectroscopy.

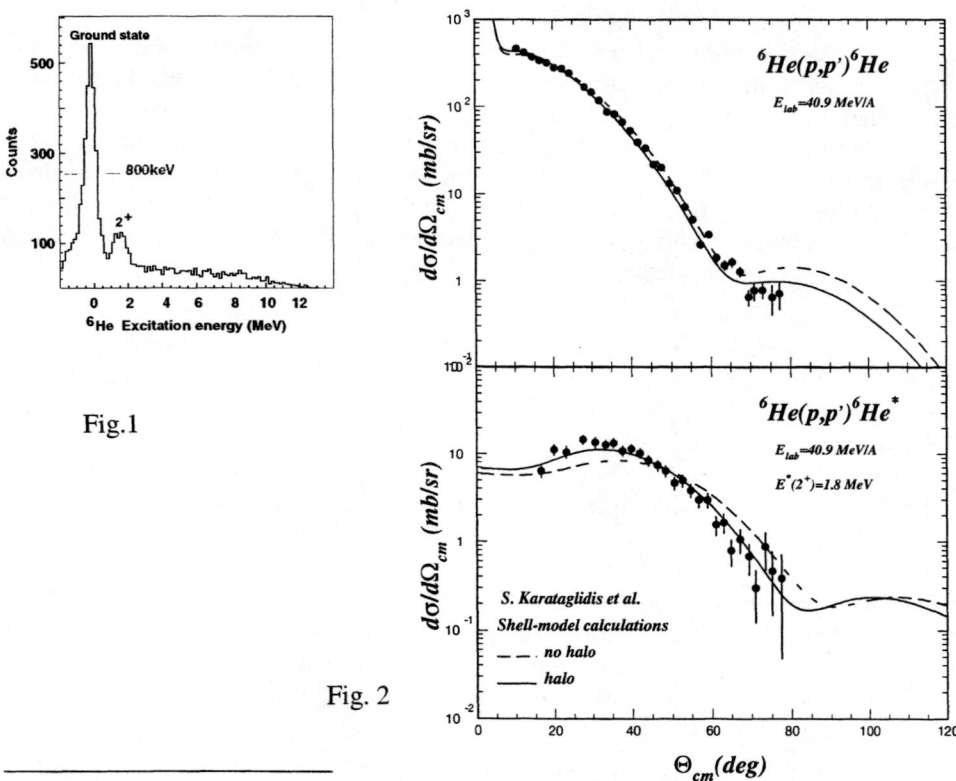

Fig.1

Fig. 2

REFERENCES

1 M. V. Zhukov et al., Phys. Rept. **231**(93)151

2 B. V. Danilin eta al., PRC **55**(1997)R577.

3 S. Funada et al., Nucl. Phys. **A575**, 93 (1994), K. Arai et al., PRC **59**(1999)1432, , ref. therein.

4 S. Ottini et al., NIM **A431**(1999)476.

5 Y. Blumenfeld et al., NIM **A421**(1999)471.

6 S. Karataglidid et al., private communication., P.J. Dortmans et al., PRC **58**(1998)2249.

7 D.C. Zheng et al., PRC **52**(1995)2488

8 F. Ajzenberg-Selove, Nucl. Phys. **A490**, 1 (1988).

9 P. Roussel-Chomaz et al., to be published.

Probing the Halo
with Intensity Interferometry

F. Miguel MARQUÉS MORENO

Laboratoire de Physique Corpusculaire,
IN2P3-CNRS, ISMRA et Université de Caen, F-14050 Caen cedex, France

Abstract. The technique of intensity interferometry is presented as a probe of n-n haloes. After exploring the sensitivity of interferometry to the n-n configuration, it is demonstrated that, owing to the specific characteristics of the halo (principally the low momentum content), this technique can only be employed using a new iterative method which provides access to the intrinsic n-n correlation function. The method has been applied to a simultaneous measurement of the dissociation of the three established two-neutron halo systems, ^{6}He, ^{11}Li and ^{14}Be, and the n-n correlation functions and corresponding source sizes extracted for the first time.

INTRODUCTION

For some nuclei in the vicinity of the neutron drip-line, the very weak binding of the valence neutrons leads to the formation of the halo — a low-density distribution extending far beyond the core of the nucleus. The recent experimental and theoretical efforts aim to the understanding of these new nuclear structures, as well as of the implications in nuclear reactions where they might be involved [1]. Of particular interest are the Borromean halo nuclei [2], three-body systems (core-n-n) in which the two-body subsystems are unbound. These unique objects represent a testing-ground for three-body effects and the n-n interaction at low densities.

The different experimental approaches fail, however, to probe the structure of the n-n subsystem as they are only sensitive to the total size of the system [3] or neglect the distorting effects of the reaction [4]. We have proposed a new probe for the core+n+n dissociation channel, intensity interferometry, aiming to determine the spatial configuration of the halo neutrons within the original system.

INTENSITY INTERFEROMETRY

Intensity interferometry has evolved from astronomy in the 50s towards particle physics in the 60s, and it has been extensively applied in both nuclear and particle physics since [5]. Basically it extracts from the emission probability of two identical

CP495, *Experimental Nuclear Physics in Europe,* edited by B. Rubio et al.
© 1999 American Institute of Physics 1-56396-907-6/99/$15.00

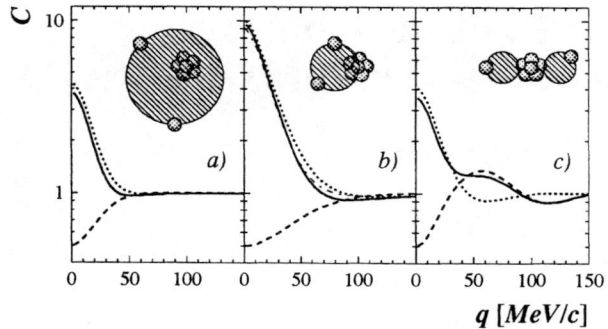

FIGURE 1. Correlation functions (solid lines) calculated for simultaneous emission from Gaussian sources (hatched areas) [6]. The individual contributions from QSS and FSI are indicated by the dashed and dotted lines, respectively; in b) the dot-dashed line corresponds to a emission time distribution of variance $\tau_0 = 50$ fm/c.

particles with momenta p_1 and p_2 the effects of their final-state interaction (FSI) and quantum statistical symmetries (QSS), among *all* other physical/experimental contributions:

$$C(X) = \frac{d^2 n / dp_1 dp_2}{(dn/dp_1)(dn/dp_2)} \quad , \quad X = f(p_1, p_2) \tag{1}$$

These two effects are governed by the space-time characteristics of the source, and their contribution to the correlation function defined in Eq. (1) is displayed in Fig. 1 for different source distributions. The parameter X commonly used to project the correlation function is the relative momentum $q = |\vec{p}_1 - \vec{p}_2|$.

RESIDUAL CORRELATIONS

The main difficulty relies on the determination of the independent single-particle distributions[1] of the denominator in Eq. (1), as they must contain all the effects in the measured two-particle probability *but* the FSI and QSS. As all the dissociation events consist of two neutrons in the final state, the procedure usually adopted is one by which "virtual" single-particle distributions are constructed by mixing particles from different events [7]. Unfortunately, this does not provide for fully independent distributions, as every event involves the emission of two neutrons and thus some "memory" of the initial correlation is retained.

These residual correlations are related to the degree of correlation in the whole data set, i.e. the overlap in q between the magnitude of the correlation function

---

[1] The one a particle would exhibit if it was not influenced by the other.

itself and the distribution of pairs. This overlap becomes extremely important for the *simultaneous* emission of *halo* neutrons, as their low momentum leads to a distribution in q peaked at low values and there the correlation function is much higher than 1 (Fig. 1). This effect leads to an overestimation of the source size, but it can be eliminated through an iterative method which compensates for the memory of the initial correlation in the denominator till convergence in C is obtained [6].

EXPERIMENT AND RESULTS

We have applied the technique to the simultaneous measurement, at GANIL, of the dissociation of ^{6}He, ^{11}Li and ^{14}Be secondary beams (50, 30 and 35 MeV/N, respectively) on a Pb target. The core was identified using a Si-CsI detector telescope and the neutrons detected using 99 elements of the DEMON array. Details about the setup and other results of the experiment can be found in Ref. [6,8].

The experimental correlation functions are displayed in Fig. 2. The suppression of residual correlations is clearly observed in the evolution of the source size with the iteration step, till convergence towards the intrinsic value is obtained. This correction is stronger for ^{11}Li as it exhibits a narrower single-neutron momentum distribution, and then lower q values and higher overlap with C (higher residual correlations).

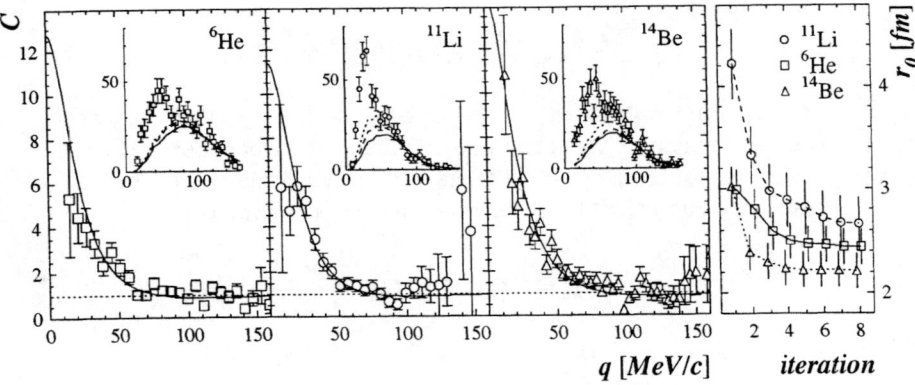

FIGURE 2. Correlation functions constructed for the dissociation of the three nuclei. The solid lines correspond to the fit assuming a Gaussian source. In the insets, the measured numerator (symbols) and the successively reconstructed denominators (dotted, dashed and solid lines for $i = 1, 2, 8$, respectively) of C are shown. The right-most panel displays the evolution of r_0 with the number of iterations [6].

At first sight the spatial configuration of the halo neutrons in ^{14}Be seems more compact than in the other two systems (higher interference signal). The values extracted for the root-mean-square n-n separations are 5.9 ± 1.2, 6.6 ± 1.5, and 5.4 ± 1.0 fm for ^{6}He, ^{11}Li and ^{14}Be, respectively, which represent the first measurement of the spatial configuration of n-n haloes and are in good agreement with the values predicted by various three-body models [6].

CONCLUSION AND PERSPECTIVES

We have demonstrated how FSI, which up to now have been considered as an unwanted and distorting effect on the halo structural information, can be exploited among all other effects by applying the intensity interferometry technique to core+n+n dissociation events. We have extracted the n-n correlation function for the three established two-neutron haloes, and obtained the rms distance between the neutrons for the first time. This technique will also open up new possibilities for the study of multi-neutron halo structures along the drip-lines, such as ^{8}He, as well as of the final state of β-n-n decays in neutron-rich systems.

We have also started a new programme aiming to eliminate FSI: proton radiative-capture onto core-n-n systems. A first experiment has been undertaken [9] for the reaction ^{6}He$(p, \gamma)\{d, t,^{5,6,7}Li\}$ at 40 MeV/N with the Château-de-Cristal BaF$_2$ array, and very preliminar results show clear evidence of capture of the proton onto subsystems in ^{6}He. As an example, events from ^{5}He$(p, \gamma)^6$Li provide undistorted information (no FSI for the γ) about the momentum of the spectator halo-neutron and about the probability of finding ^{5}He+n structures in ^{6}He [10].

ACKNOWLEDGMENTS

I am grateful to my colleagues at LPC, specially to Nigel Orr who is the main responsible of the experimental programme briefly described here, Marc Labiche for the data analysis of the DEMON experiment, and Emmanuel Sauvan for the first results of the (only two months ago) proton-capture experiment.

REFERENCES

1. See, for example, R. Morlock et al., Phys. Rev. Lett. **79**, 3837 (1997).
2. M.V. Zhukov et al., Phys. Rep. **231**, 151 (1993).
3. J.S. Al-Khalili, J.A. Tostevin, Phys. Rev. Lett. **76**, 3903 (1996).
4. N.A. Orr, Nucl. Phys. A **616**, 155c (1997) and references therein.
5. F.M. Marqués et al., Phys. Rep. **284**, 91 (1997).
6. F.M. Marqués, M. Labiche, N.A. Orr et al., submitted to Phys. Rev. Lett. (1999).
7. G.I. Kopylov, Phys. Lett. **B50**, 472 (1974).
8. M. Labiche et al., in preparation.
9. F.M. Marqués, H.W. Wilschut et al., proposal E302 to GANIL (1997 and 1998).
10. E. Sauvan, Thesis Université de Caen, in preparation.

Beta decay asymmetry in mirror nuclei: $A = 9$.

Olof Tengblad[a], L. Axelsson[d], U. Bergmann[b], M.J.G. Borge[a],
L.M. Fraile[a], H.O.U. Fynbo[b], P. Hornshøj[b], Y. Jading[c], B. Jonson[d],
T. Nilsson[c], G. Nyman[d], K. Markenroth[d], I. Martel[c], I. Mukha[b],
K. Riisager[b], F. Wenander[d], K. Wilhelmsen Rolander[e] and
the ISOLDE collaboration[c]

[a] *Instituto de Estructura de la Materia, CSIC, Serrano 113, E-28006 Madrid, Spain*
[b] *Institut for Fysik og Astronomi, Århus Univ.,DK-8000 Århus, Denmark*
[c] *PPE Division, CERN, CH-1211 Geneva 23, Switzerland*
[d] *Department of Physics, Chalmers Univ. of Technology, S-41296 Göteborg, Sweden*
[e] *Department of Physics, Univ. of Stockholm, S-11385 Stockholm, Sweden*

Abstract. Investigations of light nuclei close to the drip lines have revealed new and intriguing features of the nuclear structure. The occurrence of halo structures in loosely bound systems has had a great impact on the nuclear physics research in the last years, as intriguing, but not yet solved, is the nature of transitions with very large beta strength. We report here on the investigation of this latter feature by an accurate measurement of the beta decay asymmetry between the mirror nuclei in the $A = 9$ mass chain.

INTRODUCTION

More specifically the aim of this experiment is to study asymmetries in beta decay rates of mirror nuclei. First some comments on why we have chosen to look at ^{9}C: There are indications that the two mirror decays ^{9}C $\overset{\beta}{\to}$ ^{9}B and ^{9}Li $\overset{\beta}{\to}$ ^{9}Be are very relevant. An asymmetry has already been observed [1,2] in the decays to the 5/2$^-$ states at excitation energy ∼2.4 MeV, see the decay scheme in figure 1. The asymmetry parameter $\delta = (ft)^+/(ft)^- - 1$ is determined to be 1.2 ± 0.5 (and possibly larger) which is one of the largest observed asymmetries of mirror states [1]. The large uncertainty in the δ value mainly arises from incomplete knowledge of the ^{9}C decay. The 5/2$^-$ states are relatively narrow and have branching ratios of 15–30 % and can thus be picked out rather easily experimentally.

A potentially larger asymmetry — and a much more interesting one — seems to exist in the decays to highly excited states at ∼12 MeV excitation energy fed with

CP495, *Experimental Nuclear Physics in Europe*, edited by B. Rubio et al.
© 1999 American Institute of Physics 1-56396-907-6/99/$15.00

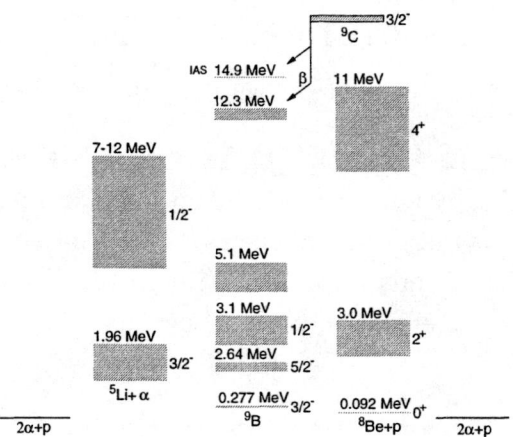

FIGURE 1. Decay scheme for ^{9}C with the resonances in the binary subsystems ^{8}Be and ^{5}Li indicated. Energy values are relative to 2αp. States with known spin-parity other than $1/2^-$, $3/2^-$ or $5/2^-$ are not shown.

branching ratios of 2–3 %. The transition has been investigated in detail for ^{9}Li and has [2] $B_{GT} = 5.6 \pm 1.2$. Only a lower limit on B_{GT} value is given in [1] for the corresponding transition from ^{9}C.

The possible asymmetry for the decay to the states around 12 MeV is interesting not only due to the fact that the individual B_{GT} values are large (with large overlap in wave-functions, an unambiguous interpretation is much easier made), but also due to the special role played by this transition for the ^{9}Li decay. It seems to belong to a class of high-B_{GT} transitions observed [3] at the neutron drip line and has been suggested to be due either to a lowering of the giant Gamow-Teller resonance [4] or to the occurrence of "two-neutron $\rightarrow$ deuteron" transitions [3,5]. Knowing whether the mirror transition on the proton rich side has a similar strength would help greatly in identifying what causes the large transition strengths. This type of "superallowed" transitions has been observed in other light nuclei as ^{6}He, ^{8}He and ^{11}Li but their mirror partners ^{6}Be, ^{8}C and ^{11}O respectively are particle unbound. This makes the decay ^{9}C $\xrightarrow{\beta}$ ^{9}B* the unique case to study the preservation of superallowed GT-transition in mirror nuclei.

THE EXPERIMENTAL SET-UP

A MgO target was used for the production of the ^{9}C nuclei, which were extracted via the $A = 25$ CO$^+$ sideband. The beam of ^{9}C ions was passed through a hole in an annular detector and implanted in a thin carbon foil. When the ^{9}C nucleus decays, the excited daughter nucleus ^{9}B break into three charged particles: a proton and two α-particles. These three particles were measured in coincidence using the

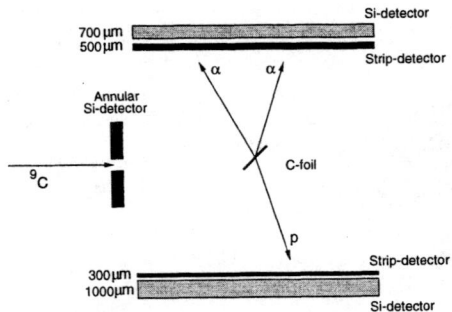

FIGURE 2. Detector setup employed for the measurement of the β-decay of ^{9}C. The ^{9}C nuclei are stopped in a Carbon foil. The foil is viewed by two 16×16 Si strip detectors, two thick Si-detectors placed behind the strip detectors and a thick annular Si-detector. A decay event is shown in the figure.

setup shown in Fig. 2. The sum of the three measured energies gives directly the excitation energy in ^{9}B. The two ΔE detectors were double sided strip detectors with 16x16 strips and total dimension 5x5 cm^2, that can register two or more particles in coincidence. To detect the high energy protons the ΔE detectors were complemented by a 700 μm thick 2000 mm^2 E-detector and a 1 mm thick 5x5 cm^2 Si-PAD detector respectively. The total thickness of the ΔE-E telescopes is 1.3 mm for a perpendicular impact, which suffices to stop the protons from decays of the 12 MeV state, as well as to stop protons from the so-far undetected narrow isobaric analog state (IAS) of ^{9}C in ^{9}B (E*=14.66 MeV, T=3/2).

The energies and angles of all fragments are thus measured in order to obtain the proton+alpha+alpha correlations. We can distinguish protons and alpha particles kinematically due to the complete detection.

DATA PREVIEW

As the analysis of our data for ^{9}C is still in progress the results shown in this section should not be taken as final results, but only as a taste of what will come.

Figure 3 gives an overview of our data for the β-decay of ^{9}C. The left part is a contour plot of the correlation between the summed energy and the individual energies for events with two α-particles and a proton, the right part gives the projection onto the ordinate. In the contour plot the diagonal lines corresponding to the ground states of ^{8}Be and ^{5}Li can both be identified, whereas there is an indication of a contribution from the first excited state of ^{8}Be. The projection onto the ordinate reflects a combination of the detection efficiency and the distribution of β-strength over the resonances in ^{9}B. We identify a strong narrow peak close to 2 MeV, a continuum stretching between this peak and a broad peak at 12 MeV, and the IAS at roughly 14.5 MeV. The narrow peak at 2 MeV sits on top of a

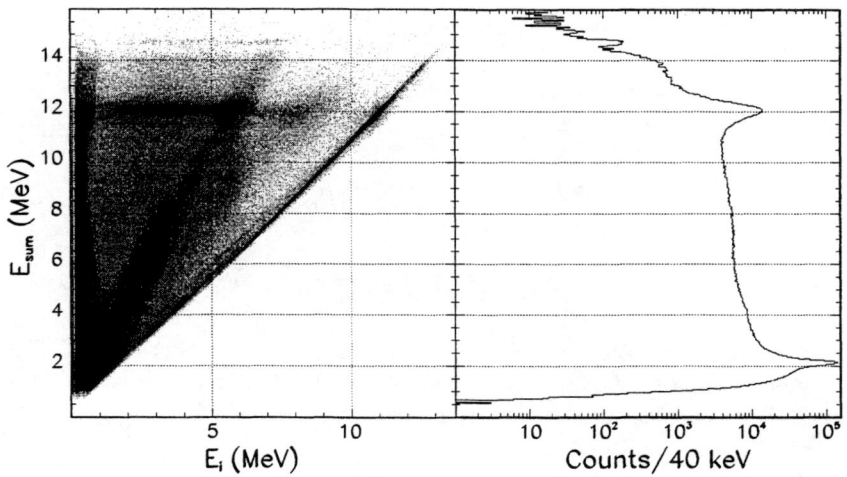

FIGURE 3. The figure gives an overview of the multiplicity-three events from the full experiment (log scale). To the left is shown the sum energy of the three particles against the energy E_i of the individual particles. Each event is represented by three points lying on the same horizontal line. The right part is the projection onto the ordinate.

broad peak. This spectrum is consistent with that observed in [1], but obtained here with much improved statistics.

At this early stage of the analysis it is possible to conclude that our data will provide a significant improvement of our knowledge of the decay of the states in ^{9}B. As an example we have seen that for the 12 MeV state there are contributions from both ^{8}Be and ^{5}Li channels. It is hoped that our data will inspire theoretical studies of the breakup of the resonances in ^{9}B. A comparison to the shell model in [1] will provide strong tests of the interactions used.

Our present data set suffers from detection thresholds which are known to be higher than particle energies from low energy states in ^{9}B, e.g. the strong ground state transition is not observable in our data. In order to be able to obtain absolute branching ratios and ft-values a remeasurement with improved threshold conditions will be carried out in the fall of 1999.

REFERENCES

1. Mikolas D. *et al.*, Phys. Rev. **C37** (1988) 766
2. Nyman G. *et al.*, Nucl. Phys. **A510** (1990) 189
3. Borge M.J.G. *et al.*, *Z. Phys.* **A340** 255 (1991).
4. Sagawa H., Hamamoto I. and Ishihara M., *Phys. Lett.* **303B** 215 (1993).
5. Poves A., Retamosa J., Borge M.J.G. and Tengblad O., *Z. Phys.* **A347** 227 (1994).

Momentum distributions from core-breakup reactions of ^{11}Be and ^{11}Li

Leif Axelsson

Chalmers, Sweden

S. Grévy, J. C. Angélique, R. Anne, D. Guillemaud-Mueller, P. Hornshøj, B. Jonson,
M. Lewitowicz, A. C. Mueller, T. Nilsson, G. Nyman, N. A. Orr, F. Pougheon,
K. Riisager, M.-G. Saint-Laurent, M. Smedberg and O. Sorlin

Abstract.
The experiment was performed at GANIL and the secondary beams of the halo nuclei ^{11}Be and ^{11}Li impinged onto a Be target with energies of 38.5 MeV/u and 29.9 MeV/u, respectively. The distribution of neutrons in coincidence with fragments from the core-breakup reactions were studied. This reaction is interesting as the halo neutrons are expected to be released without any major distortion due to the reaction.

Even though ^{11}Be and ^{11}Li are the most extensively studied halo nuclei, there are still important pieces missing for a complete understanding of their structure. A continuation of the earlier reaction studies is the core-breakup reaction which has been investigated in this work [1,2]. The reason to use such violent reactions to study the halo phenomenon is based on the idea of a core and a loosely bound halo; for reactions taking place on a sufficiently short time scale, i.e. large beam velocities compared with the intrinsic momentum of the halo neutrons, the halo can be thought to be only slightly perturbed in the reaction. Three other experiments have used core-breakup reactions in the study of neutron halo nuclei ^{11}Be, ^{11}Li and ^{19}C [3–5]. In these experiments no proper subtraction of the core component was done and this influence was thought to be negligible.

The aim of this experiment was to measure the distribution of neutrons from core-breakup reactions from both the halo nucleus and the core to be able to subtract the neutrons arising from the core. The setup of the experiment is described in [1] and the results from this measurement can be found in [1,2].

The results from the angular distributions of the neutrons and a measure of the shadow effect is published in [1]. The angular distribution of neutrons from ^{11}Li with a width $\Gamma = 42 \pm 6$ show good agreement with previous measurements. The widths of $\Gamma = 42 \pm 4$ and $\Gamma = 32 \pm 4$ were obtained for the neutrons from ^{11}Be in coincidence with Li and He fragments, respectively. This is in good agreement with

CP495, *Experimental Nuclear Physics in Europe*, edited by B. Rubio et al.
© 1999 American Institute of Physics 1-56396-907-6/99/$15.00

the longitudinal distribution of the core ^{10}Be when including a decrease of 10% of the width caused by shadowing. In [1] it is seen that the spatial extension of the halo nuclei ^{11}Be and ^{11}Li is limited. This is a result from the low neutron multiplicities obtained which tells us that a large portion of the neutrons are absorbed in the reaction. The shadowing will affect the measured momentum distribution and several theoretical articles have included this effect in their calculations [6–8].

The halo-neutron wavefunction is not a "pure simple functional shape", compare with the assumption of Lorentzian for a halo neutron in an s wave. It is therefore important to investigate the shape of the momentum distribution of halo neutrons in a model-independent approach. We have studied the $p_\parallel$ distribution as a function of $p_\perp$ (actually θ). This requires a double-differential measurement and as we are studying nuclei close to the driplines the statistics will be limited. In this case the right statistical methods should be chosen to extract as much information as possible from the data. The mathematical tools introduced to this analysis and the results are shown in [2]. The dependence of the shape on the angle of the halo-neutron longitudinal momentum distribution follows a Lorentzian for small angles but for larger angles it is something in between a Gaussian and a Lorentzian. The distribution of the neutrons stemming from the core on the other hand follows the shape of a Gaussian as is expected from the Goldhaber model.

To conclude with the core-breakup reactions have been shown to give important clues about the neutron halo structure. When studying this reaction it is necessary to make a proper subtraction of the core component and also to be as selective as possible in the exit channel to isolate any possible final state interactions.

REFERENCES

1. S. Grévy et al., *Nucl. Phys A* **650**(1999)47.
2. L. Axelsson et al., *In progress*.
3. R. Anne et al., *Nucl. Phys. A***575**(1994)125.
4. T. Nilsson et al., *Europhys. Lett.* **30**(1995)19.
5. F. M. Marqués et al., *Phys. Lett. B* **381**(1996)407.
6. P. G. Hansen *Phys. Rev. Lett.* **77**(1996)1016.
7. G. F. Bertsch, K. Hencken and H. Esbensen, *Phys. Rev. C***57**(1998)1366.
8. E. Garrido, D. V. Fedorov and A. S. Jensen, *Phys. Rev. C***59**(1999)1272.

High-spin states in ^{9,10,11}Be isotopes

Pierre Descouvemont

Physique Nucléaire Théorique et Physique Mathématique, CP229
Université Libre de Bruxelles - B1050 Bruxelles - Belgium

Abstract. The ^{9,10,11}Be isotopes are studied in a microscopic multicluster model involving two α particles with surrounding neutrons. Spectra and E2 transition probabilities are calculated, and a band structure is suggested.

Interest for neutron-rich Be isotopes has been recently revived by the observation of highly excited states in ^{11}Be [1] and ^{12}Be [2]. These states are interpreted as "molecular states", i.e. presenting a cluster structure. Clustering effects are known in ^{7}Be and ^{8}Be for a long time. More recently the existence of cluster states in heavier Be isotopes has been suggested by von Oertzen [3] and by Kanada-En'yo et al. [4] from theoretical approaches.

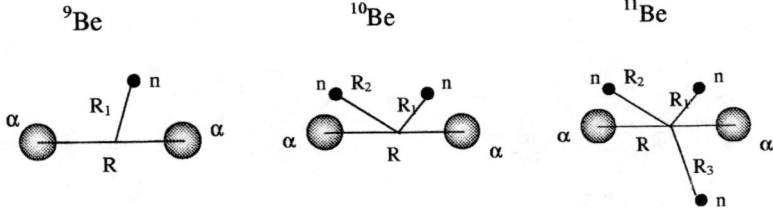

FIGURE 1. Cluster structure.

In the present work, we investigate the ^{9,10,11}Be isotopes in a microscopic multicluster model [5,6]. The wave functions are defined in the Generator Coordinate Method with a cluster structure involving two α particles, and one, two or three external neutrons for ^{9}Be, ^{10}Be or ^{11}Be respectively (see figure 1). In such a model, the physics of the problem is determined from a nucleon-nucleon interaction only, without any further parameter. Antisymmetrization between all the nucleons is exactly taken into account in order to satisfy the Pauli principle. The wave functions are projected on good quantum numbers, such as spin and parity. This model can be applied both to the spectroscopy of light nuclei, and to low-energy reactions (see ref. [5] and references therein).

CP495, *Experimental Nuclear Physics in Europe,* edited by B. Rubio et al.
© 1999 American Institute of Physics 1-56396-907-6/99/$15.00

First, we investigate the adiabatic energies of these systems, obtained for a fixed distance R between the α clusters, and a full set of generator coordinates for the external neutrons. These adiabatic energy curves present a minimum near 3 fm for the three nuclei, which shows that α clustering is important in neutron-rich Be isotopes, as it is well known in ^{7}Be and ^{8}Be. A minimum is obtained even for high angular momenta, and suggests the existence of long-lived high-spin states. We illustrate ^{9}Be in figure 2 (left panel).

In a second step, the spectra are studied with a full basis, without freezing the $\alpha - \alpha$ interdistance. For low-spin states, the present results are consistent with the available experimental data. From the analysis of E2 transition probabilities, we suggest the existence of molecular bands with α clustering [7]. These bands involve states up to $J = 13/2^+$ in ^{9}Be (see figure 2) and $J = 7^-$ in ^{10}Be. For each isotope, the band structure depends on the parity. In the parity corresponding to the $0\hbar\omega$ configuration, the lowest bands are limited to a given J value ($9/2^-$, 4^+ and $7/2^-$ for ^{9}Be, ^{10}Be, ^{11}Be respectively). Calculations on ^{11}Be are in progress [7].

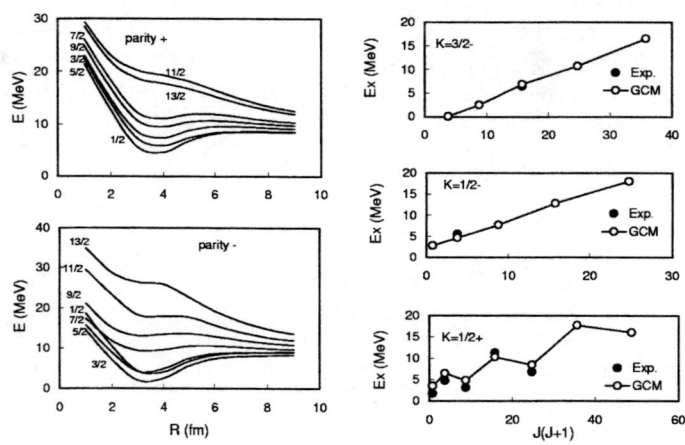

FIGURE 2. Adiabatic energies (left panel) and spectra (right panel) of ^{9}Be

REFERENCES

1. Bohlen H.G. *et al.*, *Prog. Part. Nucl. Phys.* **42**, 17 (1999).
2. Freer M. *et al.*, *Phys. Rev. Lett.* **82**, 1383 (1999).
3. von Oertzen W., *Z. Phys.* **A357**, 355 (1997).
4. Kanada-En'yo Y. *et al.*, *J. Phys.* **G24**, 1499 (1998).
5. Descouvemont P. , *Nucl. Phys.* **A626**, 647 (1997).
6. Dufour M. and Descouvemont P. , *Nucl. Phys.* **A605**, 160 (1996).
7. Descouvemont P. , to be published.

β-delayed neutron emission from ^{12,14}Be

U.C. Bergmann[1], L. Axelsson[2], M.J.G. Borge[3], V.N. Fedoseyev[4],
C. Forssén[2], H.O.U. Fynbo[1], S. Grévy[5], P. Hornshøj[1], Y. Jading[6],
B. Jonson[2], U. Köster[7], K. Markenroth[2], F.M. Marqués[5],
V.I. Mishin[4], T. Nilsson[6], G. Nyman[2], A. Oberstedt[8], H.L. Ravn[6],
K. Riisager[1], G. Schrieder[9], V. Sebastian[10], H. Simon[6],
O. Tengblad[3], F. Wenander[2], K. Wilhelmsen Rolander[11] and the
ISOLDE collaboration[6]

[1] *Aarhus Universitet*, [2] *Chalmers Göteborg*, [3] *CSIC Madrid*, [4] *Russian Academy of Sciences Troitsk*, [5] *LPC Caen*, [6] *EP CERN*, [7] *TU München*, [8] *Örebro Universitet*, [9] *TH Darmstadt*, [10] *Johannes Gutenberg Universität Mainz*, [11] *Stockholms Universitet*

Abstract. The existence of β-delayed particle emission allows one to study the β-strength distribution, and thereby the ground-state structure, of exotic drip-line nuclei. A time-correlation method has been applied to recent data on the heaviest Be-isotopes to determine their neutron branches. Improved results were obtained for ^{12}Be and new results were found for the very short-lived neutron-emitter, ^{14}Be.

EXPERIMENT, ANALYSIS AND RESULTS

The neutron branches from ^{14}Be have been studied previously in several experiments [1–3]. In [1] the authors found that 14 ± 3 % of the β-decays proceed to particle-bound states in the daughter ^{14}B. Shell-model calculations seem to disagree with this result [3], predicting close to 100% feeding to a single state around the one-neutron threshold. The validity of the measurement could not be verified in two later time-of-flight measurements [2,3], due to an uncertain detection efficiency for low-energy neutrons.

We can now report on new results for the one-, two- and three-neutron branches from ^{14}Be, obtained with a setup which thermalized the neutrons before detection, giving an essentially constant detection efficiency down to a few keV. More details about the experiment, its setup, results and in particular its analysis based on a time-correlation technique, can be found in [4].

Pulsed radioactive beams of Be were produced at the CERN-ISOLDE PS-Booster with high selectivity using the resonance laser ion source. Average yields of 16000 and 17 atoms/pulse (pulse separation 2.4 seconds) were obtained for ^{12}Be and ^{14}Be, respectively. Even better yields of ^{14}Be are expected in future experiments due to new target techniques and a proton beam-energy upgrade [5].

CP495, *Experimental Nuclear Physics in Europe,* edited by B. Rubio et al.
© 1999 American Institute of Physics 1-56396-907-6/99/$15.00

TABLE 1. Measured total neutron emission probability $P_n = \sum_i i P_{in}$ and individual i-neutron branches P_{in}, given in %. Literature values taken from [a] [6], [b] [1] and [c] [7].

Precursor	This Work			Literature			
	P_n	P_{0n}	$P_{2n}+3P_{3n}$	P_n	P_{0n}	P_{2n}	$P_{2n}+3P_{3n}$
^{12}Be[a]	0.50(3)	-	-	0.48(11)	-	-	-
^{14}Be[b]	101(4)	<4	0.8(8)	91(4)	14(3)	5(2)	-
^{11}Li[c]	106(5)	-	11.5(16)	101.7(13)	6.3(6)	4.2(4)	10.0(7)

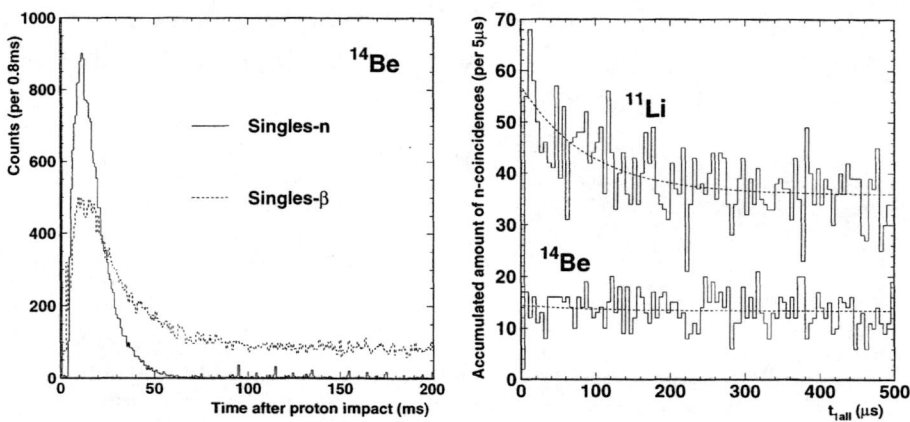

FIGURE 1. Left: ^{14}Be neutron and beta arrival-times measured relative to proton-pulse impact. Right: n-n time differences, comparison between ^{14}Be and ^{11}Li.

The value of $P_n = \sum_i^3 i P_{in}$ was determined from the neutron and beta yields (left histogram of Fig. 1). The sum $P_{2n}+3P_{3n}$ was determined from the exponential component in the spectrum of n-n time separations (right histogram of Fig. 1). The results are gathered in table 1. The data show that ^{14}Be has essentially only a single-neutron branch of intensity close to 100%. Having the spectrocopy results from [3] in mind, it is likely that nearly all of this intensity goes to a single state at 287 keV above the one-n threshold. There is no indication of any feeding to bound states, in contrast to previous results [1]. The data and the analysis methods were checked with a beam of ^{11}Li, a nucleus with well known particle decay properties.

REFERENCES

1. Dufour, J.P. *et al.*, *Phys. Lett.* **B206**, 195 (1988).
2. Belbot, M.D. *et al.*, *Phys. Lett.* **C51**, 2372 (1995).
3. Belbot, M.D. *et al.*, *Phys. Lett.* **C56**, 3038 (1997).
4. Bergmann, U.C. *et al.*, to be published in Nuclear Physics A
5. Georg, U. *et al.*, these proceedings.
6. Köster, U. *et al.*, p. 989 in AIP Conf. Proc. 455 from ENAM98 (Bellaire, 1998).
7. Borge, M.J.G. *et al.*, *Phys. Rev.* **C55**, R8 (1997).

One-Nucleon Removal Reactions at the FRS

T. Baumann ^{a}, D. Cortina-Gil ^{a}, H. Geissel ^{a}, L. Axelsson ^{b},
U. Bergmann ^{c}, M.J.G Borge ^{d}, L. Fraile ^{d} M. Hellström ^{a},
M. Ivanov ^{e}, N.Iwasa ^{g},B. Jonson ^{b}, H. Lenske ^{f},
K. Markenroth ^{b}, G. Münzenberg ^{a}, F. Nickel ^{a}, T. Nilsson ^{b},
A. Ozawa ^{g}, K. Riisager ^{c}, C. Scheidenberger ^{a}, W. Schwab ^{a},
G. Schrieder ^{h}, H. Simon ^{h}, B. Sitar ^{e}, M. Smedberg ^{b},
P. Strmen ^{e}, K. Sümmerer ^{a}, T.Suzuki ^{g}, M. Winkler ^{a}

^{a} GSI Darmstadt, ^{b} Chalmers Tekniska Högskola Göteborg, ^{c} Institut for Fysik og Astronomi, Aarhus Universitet, ^{d} IEM, CSIC Madrid, ^{e} Comenius University Bratislava, ^{f} Institut für Theoretische Physik I Giessen, ^{g} RIKEN Saitama, ^{h} Institut für Kernphysik, TU Darmstadt

Abstract. The fragment separator FRS at GSI was used as an energy-loss spectrometer to measure the longitudinal momentum distributions of ^{11,16,18}C fragments after one-neutron removal, and these of ^{7,9}Be after one-proton removal in a C target at relativistic energies. The one-nucleon removal cross-sections (σ_{-1n}, σ_{-1p})were also measured and compared with existing results for stable neighboring nucleus. The width of the momentum distribution obtained for ^{18}C and ^{7}Be fragments was smaller compared with the same measured quantities for ^{12}C and ^{8}B. At the same time a significant increase in the cross-section of the one-proton removal channel($\sigma(^{19}C)_{-1n}$ and $\sigma(^{8}B)_{-1p}$) was observed. Those results, support a neutron halo structure in ^{19}C, and the one -proton halo of ^{8}B.

The neutron-rich nucleus ^{19}C has been recently the object of several investigations at MSU [1] and GANIL [2]. The momentum distributions of ^{18}C fragments and neutrons from ^{19}C breakup reactions were reported to be very narrow compared to its neighbors and interpreted as indication for a one-neutron halo structure in ^{19}C. Parallel investigations using the same experimental techniques, have also been carried out for the proton drip-line nucleus ^{8}B, at GSI, MSU and GANIL [3–5] and have shown narrow momentum distribution widths of ^{7}Be. Those experiments, proved that momentum distribution of fragments after breakup reactions are a good method to obtain nuclear structure information. However, they need to be combined with other quantities to deduce more complete structure information. We have studied at the FRS [6] momentum distributions of ^{11,16,18}C fragments from ^{12,17,19}C break-up and one-neutron removal cross-sections for secondary ^{17,19}C beams. The same combined measurements of momentum distributions of ^{7,9}Be

CP495, *Experimental Nuclear Physics in Europe*, edited by B. Rubio et al.
© 1999 American Institute of Physics 1-56396-907-6/99/$15.00

fragments from ^{8,10}B secondary beams and one-proton removal cross-sections were also performed.

TABLE 1. One-nucleon removal cross section for C and B projectiles in a C target at mid-target energies. The value corresponding for ^{12}C was reported by Kidd et al. [9] and is shown here for comparison with the neutron rich nuclei.

Isot.	E(MeV/u)	σ_{-1n}(mb)	Isot.	E(MeV/u)	σ_{-1p}(mb)
^{12}C	1050	44.7± 2.8	^{8}B	1440	94±6
^{17}C	904	129± 22	^{10}B	1450	17±2
^{19}C	910	233±51			

Primary beams of ^{40}Ar at 1 GeV/u and ^{12}C at 1.5 GeV/u were delivered by the heavy-ion synchrotron SIS. The production of the secondary beams(C and B isotopes) was possible by nuclear fragmentation on a thick Be target placed at the entrance of the magnetic spectrometer FRS. The first stage of the FRS was set to select the secondary beam. Behind the breakup target (C 4.4 g/cm^2) placed at the dispersive mid- plane, the magnetic fields of the dipoles were set to select the one-nucleon removal products. A schematic view of the experimental set-up and details about the experimental method can be found in [10,12]. Breakup fragments arriving at the final focus were identified by measuring the time-of-flight between two scintillators , by determining the magnetic rigidity from the position measurement in position sensitive detectors , and by a coincident energy-deposition measurement in an ionization chambers. All the different isotopes were well separated in charge and mass.

The one-nucleon removal cross-sections were deduced from the ratio of the number of projectiles in front of the breakup target at the intermediate focus and the number of fragments arising after the one-nucleon removal and detected at the final focus by using the method described previously, corrected for the acquisition dead time, secondary reactions and transmission losses calculated by applying the ion-optical ray tracing code MOCADI [7] and using as input information the measured momentum widths of the fragments [8]. The obtained cross-sections are summarized in Table 1 for C and B isotopes. The presence of contaminants coming from other reaction channels has been studied and accounts less than 2%.

The longitudinal momentum distribution, p_z, of the core fragments was directly deduced from the position measurements with two position sensitive detectors and the experimental measured dispersion at the final focus. The momentum distributions obtained were corrected for the ion-optical aberrations, atomic energy straggling, angular straggling and location straggling in the relatively thick target. These corrections represents 2-3 % of the measured momentum widths.

The main results of the experiment are summarized in Figure 1.

In Figure 1a the longitudinal momentum distributions of the fragments have been normalized in their maxima for comparison. The Full Width at Half Maximun (FWHM), deduced from fitting amounts to 220 $\pm$ 12 MeV/c for ^{12}C, 143 $\pm$ 5

TABLE 2. Full width at half maximum for the longitudinal momentum distributions of ^{11,16,18}C fragments after one-neutron removal of ^{12,17,19}C in a C target at mid-target energies and of ^{7,9}Be fragments after one-proton removal of ^{8,10}B in a C target at mid-target energies.

Isot.	E(MeV/u)	FWHM(MeV/c)	Isot.	E(MeV/u)	FWHM(MeV/c)
^{12}C	1500	220± 12	^{8}B	1440	95±5
^{17}C	904	143±5	^{10}B	1450	165±8
^{19}C	910	68±3			

MeV/c for ^{17}C, and 68 ± 3 MeV/c for ^{19}C [10]. These results are summarized in Table 2. The width of the ^{12}C distribution agrees nicely with the prediction of the Goldhaber model (230 MeV/c) [11]. We interpret this agreement as a confirmation of our experimental method. However, this statistical model fails in the description of the loosely bound systems ^{17,19}C. In Figure 1b the momentum distributions obtained for the different fragments have been weighted using the measured σ_{-1n}. The increase of the cross-section and the small width of the fragments's momentum distribution of loosely bound nuclei are both signatures for an extended nuclear structure, specially important for ^{19}C.

Several theoretical approaches(RPA by H. Lenske and many-body theory by Ridikas et al.) attempt to describe the measured width of longitudinal momentum distributions after one-neutron removal reactions [10], yielding to the common conclusion of the big importance of core-polarization for this reaction (the ^{18}C core is already very neutron-rich and in consequence very easy to excite to its first 2^+ excited state). This 2^+ state contributes to the ^{19}C nuclear structure up to 60%.

For B isotopes, the obtained momentum distributions are shown in Table 2. For ^{8}B, the presence of Coulomb barrier and the contribution of the angular momentum add some difficulties to the valence proton to tunnel out the potential. Still under these considerations the same effects described for the neutron-rich isotopes were observed in our experiment for the proton-rich case.

The same theoretical approaches used for the ^{19}C case were used to describe the width of the momentum distribution of ^{7}Be fragments after one-proton removal [12,13]. Both approaches show a very good agreement in this case, yielding to the very similar core+proton configurations with very closed calculated spectroscopic factors.

As conclusion we can say that the combination of momentum distribution of the fragments after one-nucleon removal with the measured one-nucleon removal cross-sections yield complementary nuclear structure information. This kind of measurement performed on an isotopic chain shows nicely the evolution of the nuclear structure from the tightly bound isotopes to the very exotic ones. A second generation of experiments measuring these quantities in coincidence with γ rays

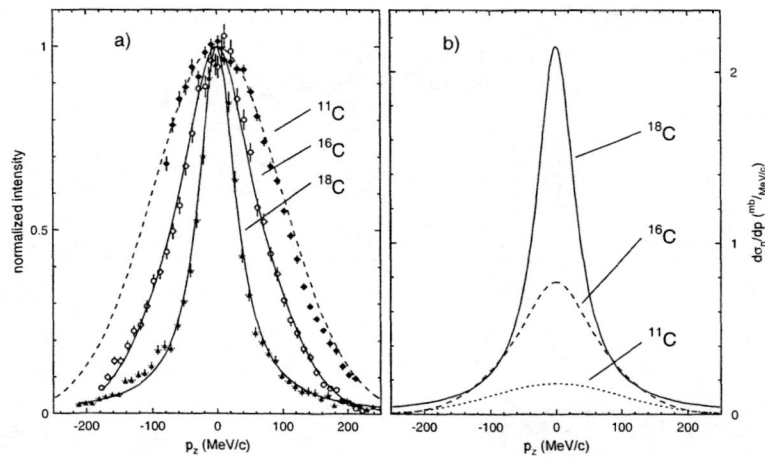

FIGURE 1. a)Longitudinal momentum distributions (p_z) of ^{11}C (closed circles), ^{16}C (open circles) and ^{18}C fragments (closed triangles) after one-neutron removal of ^{12,17,19}C normalized to their maxima. The statistical model of Goldhaber [11] (dashed line) describes only the ^{11}C distribution. The solid lines represent a simple fit to the ^{17,19}C data to guide the eyes. b) Longitudinal momentum distributions weighted by the measured σ_{-1n}

seems promising. The coincidence of the fragments with the in-flight de-excitations will provide very valuable information on the spectroscopic factors to be assigned to the different core+nucleon orbital occupations, allowing a real improvement in our understanding of the nuclear structure. The first experiment done in the FRS with this novel technique is presently under analysis.

REFERENCES

1. D. Bazin et al., Phys. Rev. C. 57 (1998) 2156.
2. F.M. Marqués et al., Phys. Lett. B381 (1996) 407.
3. W. Schwab et al., Z. Phys. A350 (1995) 283.
4. J.H. Kelly et al., Phys. Rev. Lett. 74 (1995) 30.
5. F. Negoita et al., Phys . Rev C54(1996)1787
6. H. Geissel et al., Nucl. Instr. and Meth. B 70 (1992) 286.
7. N. Iwasa et al., Nucl. Instrum. Meth. B 126 (1997) 284.
8. D. Cortina-Gil et al., under preparation
9. J.M. Kidd et al. Phys. Rev. C 37 (1988)
10. T. Baumann et al., Phys. Lett. B 439 (1998)
11. A.S. Goldhaber, Phys. Lett. B 53 (1974) 306.
12. M.H. Smedberger et al., Phys Lett. B452(1999) 1-7
13. H. Lenske et al., private comunication.

Studies of Halo Nuclei Using Longitudinal Momentum Distributions

Karin Markenroth

for the GSI-Chalmers-AArhus-Madrid collaboration[†]

Chalmers, Sweden

[†] T. Aumann, L. Axelsson, T. Baumann, U. Bergmann, M. J. G. Borge, D. Cortina-Gil, L. Fraile, H. Geissel, M. Hellström, M. Ivanov, N. Iwasa, R. Janik, B. Jonson, H. Lenske, K. Markenroth, G. Münzenberg, F. Nickel, T. Nilsson, A. Ozawa, A. Richter, K. Riisager, C. Scheidenberger, G. Schrieder, W. Schwab, H. Simon, B. Sitar, M. H. Smedberg, P. Strmen, K. Sümmerer, T. Suzuki, M. Winkler, H. Wollnik, M. V. Zhukov

Abstract. The one-nucleon removal reactions from loosely bound carbon isotopes have been studied at FRS, GSI, using beam energies around 920 MeV/u. In particular, the longitudinal momentum distributions have been investigated [1,2], but cross sections have also been extracted [3].

Measurements of the longitudinal momentum distributions of fragments from halo-nucleon removal reactions have proven to give essential information regarding the structure of these drip line systems. Recent points of interest are the loosely bound, odd-A carbon isotopes ^{17,19}C. Both theoretical and experimental results indicate that in particular ^{19}C could be the second case of a one-neutron halo system. The one-neutron separation energy for ^{17}C is 729±18 keV [4], while it is still not determined with great accuracy for ^{19}C. The $S_n(^{19}$C) is often taken to be 242±95, but recent data suggest that the value could be twice as large [5]. The halo size, and thus the width of the $p_{//}$ distribution, is very sensitive to this value. The ground states of ^{17,19}C are not known, and theoretical calculations give practically degenerate $1/2^+$, $3/2^+$ and $5/2^+$ states.

The experiment was performed at the fragment separator FRS at GSI during 1997. As primary beam, ^{40}Ar at 1 GeV/u with intensity $\approx 10^9$ ions/second was used. The secondary beams were produced in fragmentation reactions in a beryllium target. For comparison, and as a test of the experimental procedure as well as analysis methods, the $p_{//}$ distributions of the fragments ^{39}Cl and ^{39}Ar, the resulting products after one-proton and one-neutron removal from ^{40}Ar, were also measured [6]. These distributions have Gaussian shapes and large widths, as is expected for stable nuclei with one-nucleon separation energies about 10 MeV [7], see table 1. In fig. 1, the experimental $p_{//}$ distributions for the carbon isotopes are shown together with Lorentzian fits. It is clearly seen that a Lorentzian curve describes the ^{18}C data well, but that the tails of the ^{16}C distribution falls of much faster than this curve would suggest. Both ^{16}C and ^{18}C have low lying 2^+ excited

CP495, *Experimental Nuclear Physics in Europe*, edited by B. Rubio et al.
© 1999 American Institute of Physics 1-56396-907-6/99/$15.00

Secondary beam	Detected fragment	FWHM $p_{//}$ (MeV/c)
^{40}Ar	^{39}Ar	207±7
^{40}Ar	^{39}Cl	280±8
^{19}C	^{18}C	69±3
^{17}C	^{17}C	133±6

TABLE 1. The measured widths of the longitudinal momentum distributions, given as FWHM. The results presented here are for a carbon target.

states (1.77 MeV and 1.62 MeV, respectively), suggesting that the ground states of ^{17,19}C should be mixtures of $s_{1/2}$ and $d_{3/2,5/2}$ halo neutrons copuled to the core 0^+ ground state as well as to the 2^+ excited state of the core. The deduced widths are listed in table 1. The large $p_{//}(^{16}$C) width indicates that the valence neutron is to a large extent a d wave. The result for ^{19}C might seem surprisingly large at first, but theoretical models which include the ^{18}C excited 2^+ state describe the data very well. This means that the ground state of ^{19}C has an appreciable amount of relative s motion coupled to the core excited 2^+ state, see [1] and references therein.

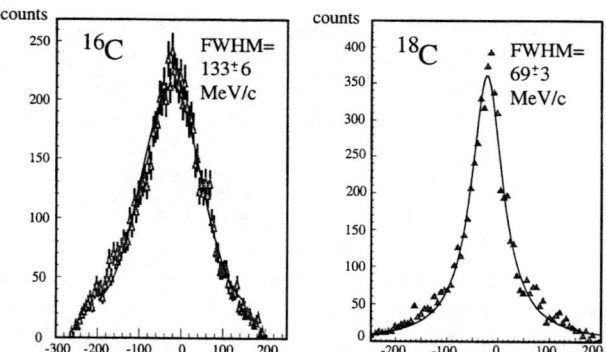

FIGURE 1. Longitudinal momentum distributions of the core fragments ^{16,18}C after breakup of the secondary beams ^{17,19}C. The solid curves are Lorentzian fits.

REFERENCES

1. T. Baumann *et al.* Phys. Lett. **B439** (1998) 256
2. M. H. Smedberg *et al.* Phys. Lett. **B452** (1999) 1
3. D. Cortina-Gil *et al.* in preparation
4. G. Audi and A. H. Wapstra, Nucl. Phys. **A565** (1993) 1
5. T. Nakamura *et al.*, Phys. Rev. Lett. **83** (1999) 1112
6. K. Markenroth, Licenciate Thesis (1998), Dept. of Physics, Chalmers (unpublished) University of Technology and Göteborg University (unpublished).
7. A. S. Goldhaber, Phys. Lett. **B53** (1974) 306

Systematics of Coulomb-Energy Differences of Mirror Nuclides in the $0d_{5/2}$ Subshell

F. Everling*

IPN at the University of Kiel, Germany

Abstract. Coulomb-energy differences of mirror nuclides for $T_z=\pm1/2$ to ±3 are shown to support a hypothesis regarding the spacial distribution of the $0d_{5/2}$ subshell.

The spacial nucleon distribution is investigated by using the Coulomb repulsion of the protons.

Fig. 1 shows the binding-energy differences of mirror nuclides plotted up to $T_z=\pm3$. They represent the Coulomb energies between 1, 2, 3, and 4 excess protons and all others, together with their mutual Coulomb-interaction energies. All excess protons (and neutrons) belong to the $0d_{5/2}$ subshell; $5/2^+$ excited states are used for ^{19}F,Ne, ^{21}Ne,Na and ^{23}Na,Mg (therefore the star in ΔB^*); at $A=20$, 22, and 24, the 4^+ states are used according to known coupling rules.

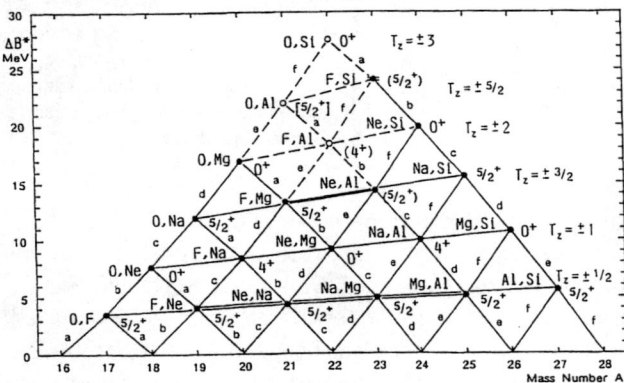

FIGURE 1. Coulomb-energy differences of mirror nuclides. Dots are used for ground states and the pairs of excited states mentioned in the text, circles indicate predictions. Uncertainties are ≤50 keV unless indicated. Brackets denote an expected spin. For letters „a" to „f", see text.

The **hypothesis** is adopted that the ^{16}O core does not change its spacial distribution during the build-up process and that the spacial extension of the subshell does not change either, independent of how many nucleons occupy it. Thus every value of the diagram is characterized by the **number of interactions with the core**, the **number of pair interactions,** and the **number of non-pair interactions.**

If the hypothesis is correct without exceptions, there are certain equalities of the lengths of connecting lines (or their vertical projections) indicated by letters „a" to „f".

---

*Address: Ringheide 24 f, D-21149 Hamburg, Germany; e-mail: F.Everling@t-online.de

CP495, *Experimental Nuclear Physics in Europe*, edited by B. Rubio et al.
© 1999 American Institute of Physics 1-56396-907-6/99/$15.00

There are 21 points in Fig.1, three of which are predictions (open circles). Another three points, ^{17}O,F, ^{18}O,Ne, and ^{19}O,Na, could not be used because of the step phenomenon (1) explained below. The point ^{19}F,Ne was also omitted because of an as yet unexplained exceptional shift. The resulting system of 14 linear equations for the three types of Coulomb interactions leaves 11 independent equations for checking the hypothesis; they are identical with special cases of Garvey-Kelson relations. They fit within **about 50 keV**, thus **confirming the hypothesis above** $A=19$ within this precision. The hypothesis was equally successful in the $0p_{3/2}$ and $0d_{3/2}$ subshells (2).

Fig. 2 shows the negative values of the **average binding energies** for the mirror nuclides of which the **differences** were shown in Fig. 1. A term linear in A is added to compensate the steep decrease, and the nuclides ^{16}O, ^{20}Ne, ^{24}Mg, and ^{28}Si are included.

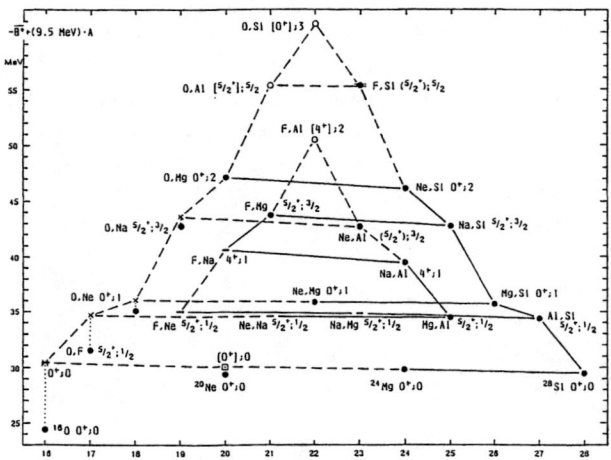

FIGURE 2. Averages of the binding energies used. Excited states are indicated by dashes here. After the element symbols, „$J^\pi;T_z$" is given. The 0^+ states for $T_z=1$ and 2 are connected directly, not via the 4^+ states. Uncertainties are ≤50 keV unless indicated. The square at ^{20}Ne 0.65 MeV indicates the position at which a very early paper (3) reported a level with insufficient evidence which would contradict the rule of an energy gap.

The step phenonenon (1), which occurs at magic numbers N or $Z = 2$, 8, 20, and possibly 40, is interpreted as a rearrangement before the next subshell can be accomodated. At ^{16}O, the rearrangement energy is 6.05 MeV or close to this value, depending on whether this 0^+ level is the rearranged state with an empty $0d_{5/2}$ subshell or - if this state does not exist - ^{12}C with a filled $1s_{1/2}$ subshell. Based on the empirical fact of nearly straight trends within subshells, the extrapolations from the right to the left side, leading to the crosses, indicate smaller steps at $A=17$, 18, and 19 which suggest that the rearrangement of the core has not yet been completed.

The mass excesses of ^{21}Al, ^{22}Al, and ^{22}Si are predicted as 26.08±0.05, 18.01±0.05, and 32.11±0.08 MeV, if the spins of the first two mirror pairs are $5/2^+$ and 4^+.

REFERENCES

1. Everling, F., Nucl. Phys. **40**, 670-689 (1963).
2. Everling, F., in ENAM 98, Exotic Nuclei and Atomic Masses, Bellaire, MI, USA, June 1998, ed. Sherrill, B.M., Morrissey, D.J., and Davids, C.N., Conf. Proc. 455, Am. Inst. Phys., Woodbury, NY, 1998, p. 298-301.
3. Morita, S., and Takeshita, K., J. Phys. Soc. Japan **13**, 1241-1247 (1958).

Research on Exotic Nuclei – Experiments at the Limits of the Nuclear Landscape

Juha Äystö

Department of Physics, Accelerator Laboratory, University of Jyväskylä, Finland

Abstract Experimental progress in studies of exotic nuclei and decay modes near the proton drip line and near the boundary of known neutron-rich nuclei is reviewed with emphasis on recent developments in detector systems and separation techniques.

INTRODUCTION

During the last few years the development of accelerators and instrumentation have led to significant improvements in our possibilities to produce and study atomic nuclei far from the valley of stability under extreme conditions with respect to excitation energy, collective motions and/or neutron-to-proton number ratio (1). Four well advanced areas of research covered here are the spectroscopic studies near the proton drip line from the sd-shell up to the Pb region, recent highlights from the recoil-decay tagging experiments on heavy elements, studies of the shell and collective structures of neutron-rich nuclei near N=20 and 28 as well as between N=50 and 82 and the newest developments in radioactive ion beam techniques. Two important topics not covered are neutron halos and superheavy elements, which are covered in two other presentations in this conference by Jonson and Hofmann, respectively.

SPECTROSCOPY NEAR THE PROTON DRIP LINE

Due to Coulomb repulsion, nuclei with increasing number of protons become weaker bound until they finally reach the proton-drip line, where the last proton or a proton pair can no longer be bound to the nucleus. The proton-drip line is located on the proton-rich side of the Z=N line up to ^{100}Sn, beyond which it is expected to occur for Z<N nuclei only.

CP495, *Experimental Nuclear Physics in Europe,* edited by B. Rubio et al.
© 1999 American Institute of Physics 1-56396-907-6/99/$15.00

Decay modes of $T_z=-5/2$ nuclei

Most proton-rich nuclei accessible to experiments are found among the $2s_{1/2}, 1d_{5/2}$, $1d_{3/2}$ and $1f_{7/2}$ shells between Mg and Ni (2). Experimental progress in these studies has been tremendous due to the developments in detectors and separators both at fragmentation and ISOL-type facilities. High-precision beta-decay studies in the sd-hell have recently progressed to the family of nuclei with $T_z=1/2(N-Z)=-5/2$. Two examples of such studies, one on the beta-delayed two-proton decay of ^{31}Ar performed at ISOLDE and another one on beta-decay study of ^{35}Ca studied at the LISE-facility of GANIL, are described.

Two-proton emission is of considerable interest due to its potential to provide information on nucleon correlations on the nuclear "surface". However, no ^{2}He emission from a well-defined quantum state has been detected so far. For studying this phenomenon a novel setup consisting of a hemisphere of 15 Si detectors and one 16 x 16 double-sided Si strip detector was used at ISOLDE with a rate of 3 ^{31}Ar atoms/s (3). High granularity in combination with efficiency of the order of 50 % for singles and several % for multiplicity-two events allowed for the first time detailed energy and angular correlation measurement for β2p decay of both isospin allowed as well as forbidden decays. The observed spectrum displays several 2p transitions, of which the three most energetic ones could be assigned as decays to the first three states in ^{29}S. Several transitions are also found from the levels below IAS, representing the first observation of the isospin allowed two-proton decay, which was not detected for ^{22}Al and ^{26}P. The energy and angular distributions of individual protons showed that both in isospin allowed and forbidden decays the emission mechanism is sequential. Again, no clear signature of ^{2}He emission is observed. The only observed decay channels from the IAS are 1p and 2p, which represent only two thirds of the strength expected for the Fermi transition. Three-proton decay is also energetically possible and had been reported earlier. However, no evidence for this decay mode was found with an upper limit of 1.1 10^{-3} (99% C.L.) for the 3p branch from the IAS in ^{31}Cl (4). The missing decay intensity could be due to other unobserved decay modes, especially an M1 γ-decay.

An other $T_z=-5/2$ nucleus ^{35}Ca was studied at GANIL using fragmentation of ^{40}Ca beam and the LISE spectrometer. Production rate of 0.3 atoms/s into the focal plane Si-detector system allowed detailed spectroscopy and determination of exact half-life of 25.7(2) ms and the beta feeding pattern providing information on beta strength distribution (5). Two-proton decay from the IAS in ^{35}K was observed, but only to the ground state of ^{33}Cl. The overall Gamow-Teller strength could be determined up to 8 MeV excitation energy indicating a quenching factor of 0.78(4) for the strength as compared to the standard shell model calculation. Earlier, the value of only 0.41(6) was observed for ^{31}Ar with the energy window extending up to 14.5 MeV (6). Such a large difference for these two nearby $T_z=-5/2$ nuclei calls for further studies of the interactions used in the shell model calculation. In both mentioned cases the same universal sd-shell Hamiltonian, obtained by a massive fit to known binding and excitation energies was used.

Proton-drip line in and above the fp-shell

The proton drip line has been experimentally reached throughout the sd-shell and most of the $f_{7/2}$ shell. It has been crossed for most if not all odd-Z nuclei up to Sn. For all these cases the lightest unbound isotopes have Z>N. Above Sn a large number of proton emitters have been discovered during the recent few years and the spectroscopy has become an important spectroscopic tool of nuclear structure.

The most proton-rich bound nuclei detected so far are the T_z=-7/2 nuclei ^{45}Fe and ^{49}Ni in the $f_{7/2}$ shell, whereas in the sd-shell it is ^{22}Si with T_z=-3. However, no direct ground state proton or two-proton radioactivity have been detected in these mass regions (7). Decay spectroscopy has been limited so far to the T_z=-2 family in the $f_{7/2}$ shell. Above Ni the proton drip line has been crossed for ^{69}Br and ^{73}Rb, as proven by the fragmentation experiments at the LISE spectrometer at GANIL (8) and in experiments at ISOLDE (9).

Above Sn there exists experimental data on over 20 proton decaying isotopes and isomers. These provide information not only on the decay rates and wave functions but also on binding energies far from stability and even beyond the proton drip line (10). Experiments on proton radioactivity have been carried out mostly using the FMA separator at Argonne and more recently also at Oak Ridge with a similar device. In these studies it has been found that spherical emitters can be described within a low-seniority shell model calculation. Similarly, the decay rates of highly deformed nuclei, such as ^{131}Eu and ^{141}Ho could be explained using a formalism for deformed proton radioactivity and assign Nilsson orbitals to the ground state (11). Newest experiments have also detected fine structure in proton spectra. (12).

Physics along the Z=N –line

Nuclei with equal numbers of protons and neutrons filling the same orbitals possess variety of exotic shapes, deformations and shape transitions, mainly due to mutual reinforcement of shell gaps. Similarly the pairing interaction between protons and neutrons is at its maximum for odd-odd Z=N nuclei. Around the Z=N symmetry line isospin breaking, particularly for the heaviest mirror nuclei is also of great interest.

Fusion evaporation reactions in combination with efficient γ-arrays and/or recoil separators have allowed studies of excited structures of Z=N nuclei up to A=84 with the most recent odd-odd system studied being ^{70}Br (13). Deacy studies of these nuclei have appeared difficult, although a series of experiments using fusion evaporation, fragmentation as well as spallation reactions have been conducted to measure, for example, the half-lifes of T_z=0 nuclei. Now all bound Z=N nuclei, with the exception of ^{98}In, have a measured half-life (14). Only a modest precision of these half-lives, especially above A=74, calls for more experiments in order to gain a full benefit from these measurements, for example in testing the CVC hypothesis in nuclear beta decay.

Only few beta-decay half-lives have been measured so far for nuclei with $T_z<0$ above Zn. A recent measurement of the half-life of 57(21) ms for ^{70}Kr, the heaviest known nucleus with $T_z=-1$, consistent with the pure Fermi decay, is of significance in the rp-process reaction flow calculations (15). The most proton-rich nuclei with a measured decay property are $T_z=-3/2$ isotopes ^{61}Ge, ^{65}Se, ^{69}Kr and ^{75}Sr. The latter two are known only with very modest accuracy.

NEUTRON-RICH NUCLEI

An important goal of nuclear structure studies is to investigate magic numbers far from the valley of beta-stability. Breakdown of magicity has been suggested for the N=20 shell closure for neutron-rich nuclei near Z=11 in various experiments on masses, spins and excited states. Moreover, the weakening of the N=28 shell closure has been detected in ^{43}S and ^{44}S by employing Coulomb excitation and fragmentation reactions at GANIL, MSU and RIKEN; for an example see ref. (16). Recently, the same physics has been approached via beta decay studies of ^{34}Al and ^{35}Al isotopes at ISOLDE leading to information on the evolution of single particle states across the major sd- and fp-shells as well as information on the coexistence of spherical and deformed shapes in ^{34}Si; see the contribution of P. Baumann to this conference. β-decay strengths observed both for ^{34}Al and ^{35}Al to levels at and above 7 MeV provide additional information for improving knowledge on interactions used in the shell model calculations.

Studies of structures of neutron-rich nuclei near and below the Z=28 and N=50 doubly magic shells have seen only small progress over the last years. This has been partly solved by the experiments utilizing transfer reactions and fragmentation produced isomeric beams. Another new approach under investigation to produce these nuclei is superasymmetric fission at intermediate energies (17). Using this approach and selective resonant ionisation decay properties of neutron-rich $^{68-74}$Ni and $^{66-70}$Co isotopes were studied at the laser ion guide facility in Louvain la Neuve (18). Information of these measurements provides new insight into the evolution of the s.p. structure from N=40 towards N=50, with particular reference to effects of filling the $vg_{9/2}$ neutron orbital. Similar experiments were carried recently out on n-rich Mn isotopes employing the high-energy proton induced fission at ISOLDE (19).

Experiments utilizing symmetric fission at the ion guide based separator at Jyväskylä have lead to discoveries of several new isotopes and beta decays of neutron-rich isotopes of Y, Zr, Nb, Mo, Tc, Ru, Rh and Pd [20]. Similarly, at ISOLDE, experiments near ^{132}Sn have produced a wealth of new information on shell structures as well as decay properties of nuclei of astrophysical interest, including the half-life of the r-process waiting point nucleus ^{129}Ag (21). For heavier fission products newest results reported by Ichikawa et al. in this conference included new isotopes of ^{162}Sm and ^{166}Gd penetrating into the region of n-rich nuclei, where very little information is obtained so far.

IN-BEAM SPECTROSCOPY OF HEAVY ELEMENTS

Tremendous progress has taken place recently in studies of excited structures of exotic isotopes far from the valley of beta-stability. High-transmission recoil separators in combination with a fusion evaporation reaction have been a key element in this development. As a good example of this in the region of heavy nuclei are the in-beam γ-ray experiments at the gas-filled recoil separator RITU (22) employed together with powerful germanium arrays. The RDT principle, for Recoil Decay Tagging, has been employed in these experiments. It involves the detection of prompt γ-rays in the target in delayed coincidence with recoils observed in the focal plane implantation detector followed by the correlated observation of their subsequent radioactive decays. This method has allowed to perform in-beam γ-ray experiments with cross sections as low as few 100 nbarns. A nice example of such experiments is the study of the excited states of ^{254}No both at Argonne and Jyväskylä using the ^{48}Ca+^{208}Pb reaction (23). These experiments could detect the ground state band up to the spin 16h, and based on global systematics and the extrapolated 2_1^+ energy quadrupole deformation of 0.27(3) was extracted for ^{254}No. It is interesting to note that no indication of the fission barrier causing a band termination or strong deviations from the rotational pattern were observed in this experiment.

NOVEL EXPERIMENTAL TECHNIQUES

Decreasing production rates when moving towards more exotic nuclei set stringent requirements for the experimental methods. High efficiency and high selectivity as well as universality of separators and detector systems is required. Newest developments are pure monoisotopic radioactive beams and sources produced by selective laser ionization, development of novel techniques for cooling, trapping, bunching and mass purification of rare isotope beams in broad energy range.

Recently, at ISOLDE selective laser ionization has been used in a variety of experiments on Be isotopes produced in high-energy p-induced reactions (24). These experiments have included the production of ^{7}Be for ^{7}Be(p,γ)^{8}B cross section measurement, the measurement of the nuclear magnetic moment of ^{11}Be leading to strong indication of single neutron halo and the decay study of ^{14}Be, a candidate for a two-neutron halo.

Another significant experimental development is related to ion cooling and trapping techniques. Excellent examples of such applications are given by high-precision mass measurements performed using the ISOLTRAP, as described by G. Bollen in this conference. On the European Scheme this is being performed within the EXOTRAPS collaboration (25), which includes the development of ion manipulation techniques for high-precision measurements at ISOLTRAP and MISTRAL mass spectrometer at ISOLDE, for heavy element research using the SHIPTRAP at GSI, in connection with the ion guide isotope separators at Jyväskylä and Louvain La Neuve, as well as in connection with the SPIRAL-project at GANIL.

REFERENCES

1. NuPECC Report "Nuclear Physics in Europe: Highlights and Opportunities", December 1997
2. Äystö J. and Cerny J., in Treatise on Heavy-Ion Science, Vol. 8, Bromley D. A. (ed.), Plenum Press 1989, p. 207
3. Axelsson L. et al, Nucl. Phys. A628, 345 (1998) and Fynbo H. et al., to be published.
4. Fynbo H. et al., Phys. Rev. C 59, 2275 (1999)
5. Trinder W. et al., Phys. Lett. B, in print.
6. Axelsson L. et al., Nucl. Phys. A634, 475 (1998)
7. Blank B. et al., Phys. Rev. Lett. 77, 2893 (1996)
8. Blank B. et al., Phys. Rev. Lett. 74, 4611 (1995)
9. Jokinen A. et al., Z. Physik A 355, 227 (1996)
10. Woods P. J. and Davids C. N. Annu. Rev. Nucl. Part. Sci 47, 541 (1997)
11. Davids C. et al., Phys. Rev. Lett. 80, 1849 (1998); K. Rykaczewski et al., Phys. Rev. C 60 (1999)
12. Davids C. et al., to be published
13. Borcan C. et al., Eur. Phys. J. A 5, 243 (1999)
14. Dessagne P. et al. and Wefers E. et al., contributions to these Proceedings
15. Oinonen M. et al., contribution to these Proceedings
16. Ibbotson R. W. et al., Phys. Rev. C 59, 642 (1999)
17. Huhta M. et al., Phys. Lett. B 405, 230 (1997)
18. Franchoo S. et al., Phys. Rev. Lett. 81, 3100 (1998)
19. Hannawald M. et al., Phys. Rev. Lett. 82, 1391 (1999)
20. Wang J. et al., Phys. Lett. B 454, 1 (1999)
21. Kratz K. L. et al., to be published
22. Leino M. et al., Nucl. Instr. Meth. B 99, 653 (1995)
23. P. Reiter et al., Phys. Rev. Lett. 82, 509 (1999) and M. Leino et al.,Eur. Phys. J., in print
24. ISOLDE proposals; see for example CERN/ISC 98-6, ISC/P99
25. http://www.jyu.fi/~armani/exotraps/frames.htm

Study of Nuclei Far from Stability at Intermediate Energies: Present Status and Future

Marek Lewitowicz

GANIL B.P. 5027, 14076 Caen Cedex 5, France

Abstract.
In the present talk, in order to illustrate general tendencies in studies of nuclear structure far from stability, several representative examples of recent experiments performed with intermediate energy beams are presented. The perspectives of new experiments in physics of exotic nuclei opened by high intensity radioactive nuclear beams combined with new 4π detection systems are shortly discussed.

In-flight Production of Exotic Nuclei at Intermediate Energies

Study of nuclei far from stability is one of the most important and fast developing branch of modern nuclear physics. Among the main goals of these research one should mention: study of the isospin-dependent term(s) in effective nucleon-nucleon interactions, links between the structure of nuclei and astrophysics and search for new phenomena at and beyond the drip-lines.

From the experimental point of view to reach these ambitious and difficult goals a large variety of methods developed for intermediate energy beams is currently used. Thanks to them and due to the constantly increasing beam intensities, the last decade was rich in spectacular results in studies of nuclei far from stability.

TABLE 1. Comparison of the intermediate energy and relativistic energy fragmentation used for production of RNB.

Property	Interm. energy fragm.	Relat. energy fragm.
Beam Energy (AMeV)	<150	>500
Total interaction probability in a thick target/detector (%)	1	30-40
The highest rates for	Z<40, A<80	Z>40, A>80

CP495, *Experimental Nuclear Physics in Europe*, edited by B. Rubio et al.
© 1999 American Institute of Physics 1-56396-907-6/99/$15.00

Among the main advantages of the production of RNB at intermediate energies one should mention:

1) Fragment separation independent of the chemical properties of selected ions,

2) Short separation time - typically 100 ns,

3) Unambiguous event-by-event A versus Z identification and clean fragment selection for Z<55 nuclei,

4) The use of nuclear reactions with RNB in inverse kinematics.

In the following the present trends in physics of intermediate energy RNBs are illustrated by three examples, representative for direct in-flight identification methods and decay studies. Other experimental approaches are discussed by F. Azaiez [1], B. Jonson [2] and G. Münzenberg [3].

Mass Measurements far from Stability

Precise measurements of the ground state properties as mass, quadrupole and magnetic moments and half-life and even identification of new nuclei in vicinity of the drip-lines allow for a sensitive test of the nuclear models. One of the first indication of changes in nuclear structure far from stability like existence of an onset of deformation is usually obtained via mass measurements. Recently, a direct mass

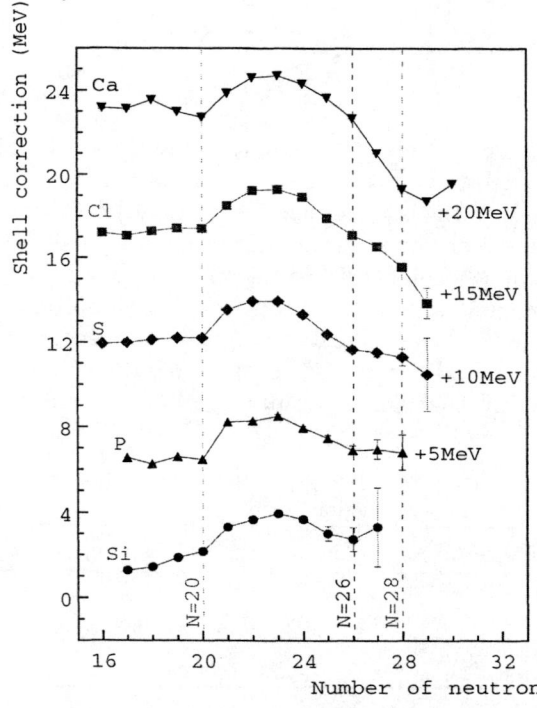

FIGURE 1. Shell corrections as a function of neutron number for the isotopes of Si,P,S,Cl and Ca. See text for details.

determination based on a precise time-of-flight measurement performed simultaneously for a wide range of neutron-rich isotopes from Si to Ca produced in the fragmentation of a 65 AMeV ^{48}Ca beam was performed at the SPEG spectrometer at GANIL. The technique, constantly improved during last ten years, in this case gave a first evidence of a disappearance of the N=28 shell gap. The results [4] are presented in figure 1 as a difference between the measured masses and the calculated macroscopic part of the Finite Range Liquid Drop Model [5]. The extracted in this way shell corrections for sulfur, potassium and silicon isotopes are stabilized at N=26 rather than at the "standard" closed shell N=28. These results together with the recent B(E2) values measured in the coulomb excitation of intermediate energy RNBs and important progress in the beta-gamma and isomer decay spectroscopy indicate dramatic changes in the shell structure of the fp neutron-rich nuclei.

Gamow-Teller and Fermi Strength in the Beta Decay of Drip-line Nuclei

The decay and in-beam spectroscopy of the exotic nuclei provides, often very detailed, information on deformation, shell closures and role of pairing far from stability. New experimental tools like silicon multi-strip detectors, 4π-germanium arrays and total gamma absorption spectrometers as well as a new generation neutron detectors allowed for the precise measurement of the Gamow-Teller (GT) and Fermi (F) strengths [6].

As an example of this kind of study may serve the recent measurement of the GT and F strengths for the β decay of ^{40}Ti. This measurement is related to the efficiency calibration of the ICARUS liquid argon neutrino detector, which is currently under construction in the Grand Sasso Laboratory [7]. The B(GT) value for the Gamow-Teller absorption reactions in ^{40}Ar can be deduced from the mirror β decays of ^{40}Ti. Under the assumption of isospin symmetry, the $B(GT)$ value for a ^{40}Ar neutrino capture transition to an excited state of ^{40}K is identical to the $B(GT)$ of the corresponding ^{40}Ti β decay to the mirror state of ^{40}Sc.

TABLE 2. Comparison of the results of the experiments performed at GANIL and GSI on the β decay of ^{40}Ti.

Quantity	GANIL[a]	GSI[b]
$T_{1/2}$, ms	52.7(1.5)	54(2)
B(GT)	5.52(20)	5.84(39)
B(F)	3.84(17)	4.01(31)
$\Sigma\sigma_\nu \times 10^{-43}$cm^2	14.0(3)	13.8(6)
Total of ^{40}Ti impl.	6x10^4	1.1x10^4
Yield ^{40}Ti/sec.	0.3	0.017

[a] ref. [8]
[b] ref. [9]

45

The summary of the experiments performed at GANIL with an intermediate energy beam of ^{50}Cr and at GSI with a relativistic ^{58}Ni beam are summarize in table 2. The results of both experiments are in a nice agreement and they imply that the ICARUS ^{40}Ar detector has an effective absorption cross section for ^{8}B solar neutrinos of $14.5(4) \times 10^{-43}$ cm^2; 73% of the total cross section arises from Gamow-Teller transitions. This contribution was neglected in early estimates of the ICARUS efficiency.

Shells far from stability: Short-lived Isomers Towards ^{78}Ni

An important and attractive field of research was opened recently by the production of secondary isomeric beams. They can be used for both: study of properties of the isomers themselves and production of other isomeric states with a higher spin and/or energy.

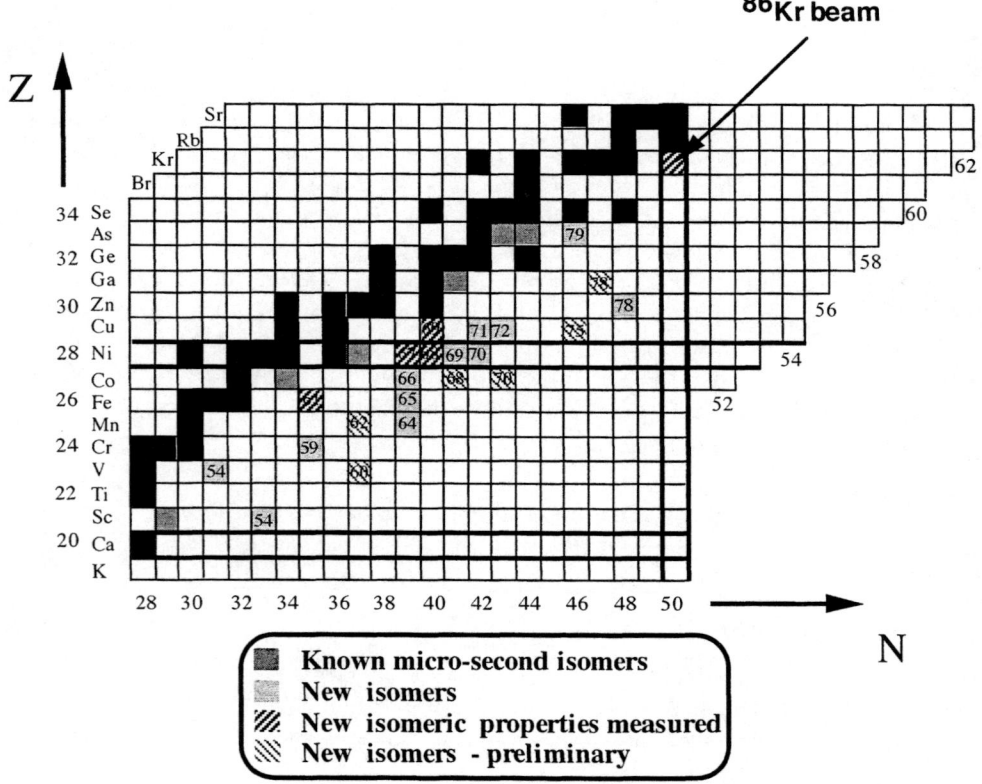

FIGURE 2. Chart of nuclei with isomeric states studied in the recent experiments at GANIL using a 60 AMeV ^{86}Kr beam.

Nucleus	$T_{1/2}$(ns)	E_γ(keV)
^{50}K	125(40)	171.4(5)
		127.4(5), 43.5(10)
		70.5(10), 101.3(10)
^{60}V	320(90)	98.9(5)
	13(3)	103.2(5)
^{62}Mn	95(2)	113.3(5)
^{68}Co	101(10)	48.4(5)
^{70}Co	54(10)	155.8(5), 161(1)
		164.1(5), 235.0(5)
		273.2(5)
^{75}Cu	195(15)	66.5(5)
	1500(1000)	61.8(5)
^{78}Zn	319(9)	144.7(5), 908.3(5)
		889.9(5), 729.6(5)
^{78}Ga	110(3)	(46), 453.3(5)
		157.5(5), 341.3(5)
		217.8(5), 281.0(5)

FIGURE 3. Half-lives and γ-transitions of the new isomeric states observed recently at GANIL using a 60 AMeV ^{86}Kr beam.

The large program of search and study of the short-lived isomeric states far from stability is currently going on at GANIL, GSI and MSU (see for example [10,11]).

Thanks to a very clean ion identification and relatively high production rates intermediate energy experiments are particularly successful in studies so far inaccessible neutron-rich light nuclei. Among the recent results one may mention a discovery of the 8$^+$ isomers in ^{70}Ni [12] and ^{78}Zn [13]. In figure 2 a chart of nuclei with the recently discovered and studied isomeric states is shown. Some characteristics of the new short-lived isomers found in the recent experiment at GANIL using a 60 AMeV ^{86}Kr beam are presented in figure 3. In many cases an interpretation of the observed isomeric states in the framework of the shell model will require important theoretical development.

Perspectives

It is well beyond a scope of this talk to cover all current directions in studies of nuclei far from stability at the intermediate energy accelerators, but let us mention few other fast developing experimental techniques.

The Coulomb excitation of the neutron-rich radioactive beams is extensively employed in mapping of low energy and low spin states in the vicinity of N=20, 28 and 40 magic or semi-magic numbers. Recently, a use of elastic and inelastic scattering as well as of direct reactions with radioactive nuclear beams (RNB) like (p,p') or (p,n) became a new tool in study of unbound nuclei and one- or two-nucleon halo systems. The transfer and/or break-up reactions allow to examined

47

in a detail the structure of light drip-line nuclei including their possible molecular states.

All these experiments obviously take or will take a benefit from the on going (GANIL, GSI) or planned in the next future (NSCL, RIKEN) increase of the primary beam intensity - typically by factor of 10 to 100. A spectacular upgrade of the fragment separators (SISSI device at GANIL, new spectrometers at NSCL and RIKEN) allows for further improvement of the RNB intensity and purity.

Last but not least, one observes extremely fast development of the detection devices like 4Π-γ arrays (for example EXOGAM at GANIL), high acceptance magnetic spectrometers (S800, VAMOS) or neutron detectors (TONNERRE and DEMON at GANIL and scintillator arrays at NSCL and RIKEN).

Thus, already in the near future one may expect that on the proton- rich side all drip-line nuclei up to ^{100}Sn will become accessible for the elastic and inelastic scattering and direct nuclear reactions studies. Similar, but for obvious reasons slower progress should be possible for neutron-rich nuclei up to ^{78}Ni.

REFERENCES

1. Azaiez, F., invited talk on this conference.
2. Jonson, B., invited talk on this conference.
3. Münzenberg, G., invited talk on this conference.
4. Sarazin, F. et al., Bormio Winter School, 1999.
5. Möller, P. and Nix, J. R., *ADNDT* **59**, 185 (1995).
6. Roeckl, E., contribution to this conference.
7. ICARUS Collaboration, Proposal, ICARUS II, *A second-generation proton decay experiment and neutrino observatory at Gran Sasso.* See also *http://www.aquila.infn. it:80/icarus/mail.html.*
8. Bhattacharya, M. et al.,*Phys. Rev.C* **58**, 3677 (1998).
9. Liu, W. et al.,*Phys. Rev.C* **58**, 2677 (1998).
10. ENAM 98, eds. B.M. Sherill, D.J. Morissey, C.N. Davids, AIP Conf. Proc. 455, Woodbury, NY, 430 (1998)
11. Pfützner et al., contribution to this conference.
12. Grzywacz, R.et al.,*Phys. Rev. Lett.* **81**, 766 (1998).
13. Daugas, J.M. et al.,submitted to *Phys. Lett. B*

In beam γ spectroscopy of neutron rich nuclei around N=20 using projectile fragmentation

M.J. López-Jiménez* and collaborators

*Grand Accélérateur National d'Ions Lourds, B. P. 5027, 14076 Caen Cedex 5, France

Abstract. The structure of nuclei far from stability around ^{32}Mg have been recently investigated by means of a novel method. In-beam γ-decay spectroscopy of a large number of exotic neutron-rich nuclei produced by a projectile fragmentation has been performed, using coincidences between the recoil fragments collected at the focal plane of SPEG and the γ-rays emitted at the target location. Preliminary results on both the population mechanism and the decay of excited states are presented.

Introduction

The typical cases of ^{32}Mg,^{31}Na (N=20) and ^{44}S (N=28), where a large collectivity has been found experimentally [1], [2] have brought some evidence for a shell-gap weakening at large neutron excess. The ^{32}Mg lies just at a neutron shell closure and therefore is expected to be spherical. Its large quadrupole excitation probability B(E2) [3] can only be understood in the frame of a larger configuration space, including f-d shell excitations for neutrons [4], [5].

Experimental technique

A ^{36}S beam, with a 15 nAe intensity was impinged onto a thin a 2.77 mg/cm^2 Be target at an energy of 77.4 MeV/u in order to produce the neutron rich nuclei. The target was located at the entrance of the SPEG spectrometer at GANIL. The "Château de cristal" array composed of 74 BaF$_2$ crystals was used around the Be target in order to detect the desexcitation γ-rays in coincidence with fragments. The large Doppler broadening of the measured γ-rays energies as well as the γ-rays background were been corrected.

Population of excited states

The study of the projectile fragmentation mechanism is essential in order to understand how and to what extend associated in-beam γ-spectroscopy could be used to study the structure of nuclei far from stability.

CP495, *Experimental Nuclear Physics in Europe*, edited by B. Rubio et al.
© 1999 American Institute of Physics 1-56396-907-6/99/$15.00

The excitation energy of fragments is given by measurements of the total energy by γ-emission (γ-energy sum), the spin by the measurement of the γ-multiplicity which is in turn deduced from the hit-pattern of the 74 BaF_2 detectors (γ-fold).

The figure 1 shows the dependence of the average energy sum (middle part) and the average raw fold (lower part) versus fragmentation products masses. One can see that both quantities slightly increase when going from fragments in the vicinity of the projectile to fragments well away from the projectile. These data are not corrected for the detector efficiency. A point which is not yet understood is the fact that for ^{11}Be, the fold and energy sum are so high (M = 2.4 and E_{sum} = 3.5 MeV). The E_{sum} value is in this case simi-

lar to that measured for other nuclei even if the one neutron separation energy of ^{11}Be (S_n = 508 keV) is extremely low (S_n plot on the upper part of picture). This means that not only the γ rays and γ continuum of projectile fragment are detected but also neutrons or other γ rays coming from nuclear reaction processes itself as target excitation etc. Neutron contribution could not been supressed because not enough resolution exits in the time of flight spectrum in order to separate γ and neutrons events. Furthermore, a odd-even staggerin is clearly seen, reflecting the pairing energy contribution to the reaction Q values.

More extended information about this experiment and results could be found in the proceeding of F. Azaiez in this conference and the reference [6]. Results obtained up to now prove the feasibility of such in beam γ-spectroscopy experiment using projectile fragmentation, a 4π multigamma detector and a high selectivity spectrometer.

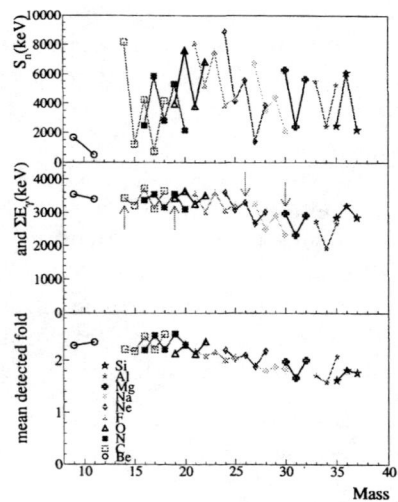

FIGURE 1. S_n, energy sum and fold versus the projectile mass

REFERENCES

1. C. Detraz et al., *Phys. Rev.* **C19** (1978) 171

2. N. A. Orr et al., *Phys. Lett.* **B258** (1991) 29

3. T. Motobayashi, *Phys. Lett.* **B346** (1995) 9

4. N. Fukunishi et al., *Phys. Lett.* **B296** (1992) 279

5. A. Poves et al., *Nucl. Phys.* **A571** (1994) 221

6. M.J. López-Jiménez et al., *Ricerca Scientifica ed Educazione Permanente*, **Vol 37** (1999), pp. 416-427

Intermediate-energy Coulomb excitation of ^{28,29,30,31}Na

B. Pritychenko[1,2], T. Glasmacher[1,2], P.D. Cottle[3], R.W. Ibbotson[1], K.W. Kemper[3], A. Sakharuk[1], H. Scheit[1,2] and V. G. Zelevinsky[1,2]

[1] *National Superconducting Cyclotron Laboratory, Michigan State University, East Lansing, Michigan 48824*
[2] *Department of Physics and Astronomy, Michigan State University, East Lansing, Michigan 48824*
[3] *Department of Physics, Florida State University, Tallahassee, Florida 32306*

Abstract. The neutron-rich radioactive isotopes ^{28,29,30,31}Na have been produced by nuclear fragmentation of ^{48}Ca and ^{40}Ar primary beams at intermediate energies. The energies and excitation cross sections to the lowest excited states were measured via intermediate-energy Coulomb excitation on a ^{197}Au target. Experimental results suggest a large degree of collectivity in ^{31}Na. Intrinsic ground state and transition quadrupole moments in ^{28,29,30,31}Na are compared.

First evidence about the existence of the so-called 'island of deformed nuclei' near the $N = 20$ shell closure was obtained in 1975 by Thibault et al. [1] from the mass measurements of sodium isotopes. It has been found that ^{31}Na and ^{32}Na are more tightly bound than expected from spherically symmetrical sd shell model. In the same year this phenomenon was explained by Campi et al. [2] via introduction of neutron $f_{7/2}$ intruder orbits for $Z < 14$ nuclei (i.e., an 'inversion' of the standard shell ordering) and Hartree-Fock calculations. For 20 years progress in this field was hampered by lack of experimental information until properties of ^{32}Mg were studied at Orsay and RIKEN. Motobayashi et al. [3] (using the technique of intermediate-energy Coulomb excitation) measured that the reduced transition probability to the first excited 2^+ state in ^{32}Mg was $B(E2; 0^+_{g.s.} \rightarrow 2^+_1) = 454(78)$ e^{2}fm^4. A similar result was recently obtained at the NSCL [4]. Recent calculations of Caurier et al. [5] predict that intruders dominate ground states in ^{30}Ne, ^{31}Na and ^{32}Mg. To improve our knowledge about the island of deformed nuclei we decided to study ^{28,29,30,31}Na. Our results represent a first measurement of transition energies and excitation cross sections in ^{28,29,30,31}Na.

The present experiment was performed at the National Superconducting Cyclotron Laboratory (NSCL) at Michigan State University. Primary beams of ^{48}Ca^{13+} with an energy of 80 MeV/nucleon and intensities as high as 8 particle-

CP495, *Experimental Nuclear Physics in Europe*, edited by B. Rubio et al.

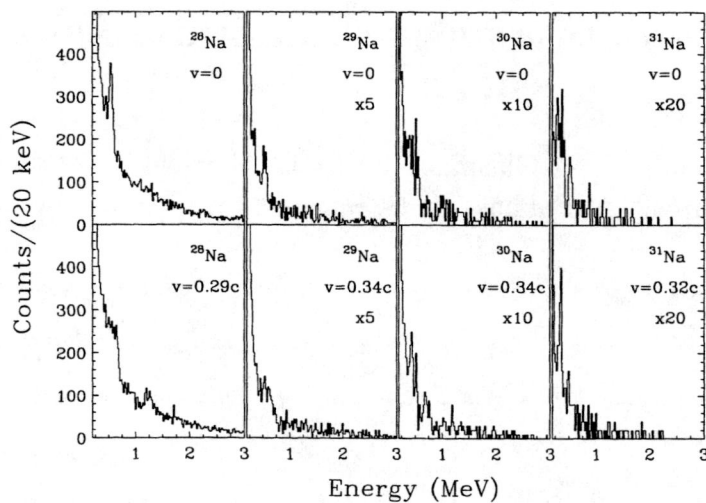

FIGURE 1. Upper panels contain background subtracted photon spectra in the laboratory frame. The 547.5 keV ($7/2^+ \rightarrow g.s.$) transition in the gold target is visible as a peak, while transitions in each projectile are very broad. Lower panels contain Doppler-corrected, background-subtracted γ-ray spectra.

nA, and ^{40}Ar^{12+} at 90 MeV/nucleon and 80 particle-nA were produced with the NSCL superconducting electron resonance ion source and the K1200 cyclotron. The secondary beams of sodium isotopes were obtained via fragmentation of the ^{48}Ca (^{40}Ar) primary beam in a 376 mg/cm^2 (564 mg/cm^2) thick ^{9}Be primary target located at the mid-acceptance target position of the A1200 fragment separator [6].

The position and direction of each fragment incident on the secondary gold target were measured with two parallel plate avalanche counters (PPAC). For small scattering angles ($\theta_{lab} \leq 4.0°$) Coulomb excitation dominates and we can neglect the nuclear contribution. The time of flight between a thin plastic scintillator located after the A1200 focal plane and plastic phoswich at 0° with respect to the beam located after the secondary target was recorded for each fragment and provided positive identification of the fragment after interaction in the target.

Photons were measured in coincidence with the scattered fragments with the NSCL NaI(Tl) array [7]. The energy resolution of a typical detector was 8% at 662 keV, and position resolution was 2 cm. This resulted in an angular resolution for the inner-ring detectors better than 10° for the emitted photon. The angular information was used to correct for the large Doppler shift of the photons emitted from the secondary beam particles. The time difference between the detection of the photon in the NaI(Tl) detectors and the detection of the scattered fragment in the phoswich detector was recorded for each event so that accidental coincidences could be excluded from the γ-spectrum. The time-gated Doppler-corrected γ-ray spectra obtained with scattered ^{28,29,30,31}Na nuclei are shown in Fig. 1. The photons emitted from fragments, which had velocities of $\approx 0.3c$, could be distinguished from

TABLE 1. Experimental parameters and results. The secondary fragments were positively identified on an event-by-event basis and only desired fragments were analyzed. The energy spread of the secondary beam was $\pm 2\%$.

Primary beam	Secondary beam	E_{beam}/A (MeV/A)	Total beam particles/10^6	E_γ (keV)	σ (mb)	θ_{lab}^{max}
^{40}Ar	^{28}Na	43.11	82.44	1240.0(11.0)	26(6)	3.96
^{48}Ca	^{29}Na	59.97	12.96	700.8(20.0)	26(21)	2.80
^{48}Ca	^{30}Na	55.56	3.30	432.8(16.0)	42(14)	2.80
^{48}Ca	^{31}Na	51.54	1.28	350.0(20.0)	119(35)	2.80

target excitations by their Doppler shifts.

Comparing the cross sections for ^{28,29,30}Na and ^{31}Na one may conclude that excitations in ^{31}Na are much more collective than in ^{28,29,30}Na, see Table 1. That is in good agreement with theoretical predictions of Caurier et al. [5], who predicted that $3/2^+$ ground state of the ^{31}Na is dominated by intruders and the first excited state should have spin and parity $5/2^+$ and energy ~ 200 keV. Preliminary estimates show that nuclear excitations [8] can account for 10-15% of the excitation cross sections. It is known experimentally that spin and parity for the ground state of ^{28,29,30}Na are 1^+, $3/2$ and 2^+, respectively [9]. We aren't aware of any theoretical calculations of the nuclear structure of ^{28,29,30}Na. Assuming static deformation and a rotational nature of low-lying excitations we can suggest 2^+ spin and parity assignments for the first excited state in ^{28}Na, $5/2^+$ in ^{29}Na and 3^+ in ^{30}Na. Two peaks are observed in the Doppler corrected spectrum of ^{30}Na. Our data analysis indicates that only the 432.8(16.0) keV transition belongs to ^{30}Na, and the 700.8(20.0) keV line originates from the neutron-stripping reaction and further deexcitation of ^{29}Na. In fact, a weak peak at this energy can be seen in the ^{29}Na data. However, the current experiment has little statistics for this isotope. Observation of a neutron-stripping reaction in ^{30}Na is consistent with the one-neutron separation energies for ^{28,29,30,31}Na, which are 3520(80) keV, 4420(120) keV, 2100(130) keV and 4000(190) keV [9], respectively. In the ^{28}Na data, the 1240(11) keV transition is present.

Assuming that all cross section can be atributed to excitations to the first excited state, we compare our data with the measured values of the ground state electric quadrupole moments [10]. The corresponding intrinsic quadrupole moments can be extracted using the formalism described in [11] and the experimental knowledge of the ground state spins. The same quantity for the transition moments can be obtained from our experimental cross sections (assuming pure E2 excitations) using the standard formalism [12]. The intrinsic quadrupole moments for sodium isotopes are presented in Figure 2. The results for ^{28,29,30}Na suggest that these nuclei are statically deformed. At the same time the observed difference between intrinsic quadrupole moments for the ground and excited states in ^{31}Na indicates dynamic deformation.

This work was supported by the National Science Foundation under Grants PHY-

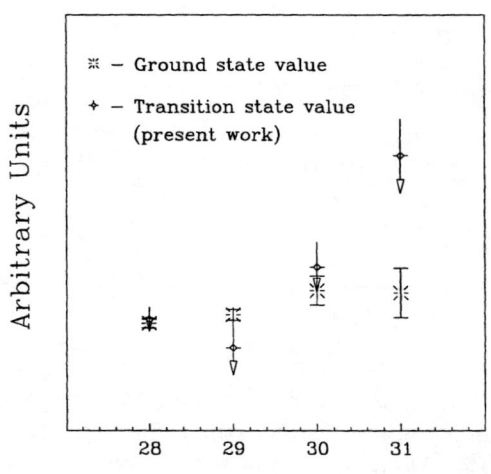

FIGURE 2. The intrinsic quadrupole moments for the ground and transition states in sodium isotopes (ignoring possible feeding from higher lying states).

9528844, PHY-9523974, and PHY-9605207.

REFERENCES

1. C. Thibault et al., Phys. Rev. C 12 (1975) 644.
2. X. Campi et al., Nucl. Phys. A 251 (1975) 193.
3. T. Motobayashi et al., Phys. Lett. B 346 (1995) 9.
4. B. V. Pritychenko et al., Phys. Lett. B (1999), in print.
5. E. Caurier et al., Phys. Rev. C 58 (1998) 2033.
6. B. M. Sherill et al., Nucl. Instr. Meth. B 56 (1991) 1106.
7. H. Scheit et al., Nucl. Instr. Meth. A 422 (1999) 124.
8. J. Raynal, Coupled channel code ECIS97, unpublished.
9. Table of Isotopes, edited by R. B. Firestone and V. S. Shirley (John Wiley and Sons, Inc.,1996) Volume I.
10. M. Keim, Proc. ENAM98: Exotic Nuclei and Atomic Masses, Bellaire, Michigan (1998) 50.
11. A. Bohr and B.R. Mottelson, *Nuclear Structure, Vol. 1 and 2*, (World Scientific, 1998).
12. P. Ring and P. Schuck, *The Nuclear Many-Body Problem*, (Springer-Verlag, 1980).

The ^{34,35}Al beta decay to ^{34,35}Si

S. Nummela*, P. Baumann‡, E. Caurier‡, Ph. Dessagne‡, A. Jokinen*, A. Knipper‡, Ch. Miehé‡, F. Nowacki‡, M. Oinonen†, Z. Radivojevic*, M. Ramdhane♮, G. Walter‡ and J. Äystö*

*Department of Physics, University of Jyväskylä, P. O. Box 35, Jyväskylä, Finland
‡Institut de Recherches Subatomiques, BP28, F67037, Strasbourg, France and the ISOLDE Collaboration
†PPE division, CERN, CH-1211 Genève 23, Suisse
♮Université de Constantine, Constantine, Algérie

Abstract. The ^{34,35}Al beta decay was studied at the CERN on-line mass-separator ISOLDE by β-γ, β-γ-γ and β-n measurements. In the ^{34}Al decay, the new data confirm the previous results and complement the level scheme in ^{34}Si and ^{33}Si. An intruder (2p-2h) 0_2^+ state is proposed at 2.1 MeV in ^{34}Si in agreement with different shell-model predictions but its decay is still unobserved. A ^{35}Al β-decay scheme to ^{35}Si bound and unbound states is established. The ^{35}Al half life (41.6(22) ms) was measured with a good precision. The first ^{35}Si level scheme including three levels at 911, 974 and 2169 keV with respectively $J^\pi = (3/2)^-$, $(3/2)^+$ and $(5/2)^+$ was obtained. The spin and parity values are consistent with a $(5/2)^+$ assignment for the ^{35}Al ground state. The value P_n=0.41(13) was also deduced. New information on the proton-neutron interaction in the (sd-fp) space is obtained.

INTRODUCTION

Studies [1,2] of very neutron-rich light nuclei have given evidence for an "island" of deformed nuclei near the N=20 shell closure ($Z < 14$). We have shown in beta-decay studies [3] that ^{34}Si (N=20, Z=14) lies at the edge of this island of deformation and behaves like a doubly-magic nucleus. Its ground-state was described as an 0p-0h state while the first excited 2^+ state at 3326 keV has a large fp-shell intruder 2p-2h component. Measurements of B(E2; $0_1^+ \rightarrow 2_1^+$) (note B(E2↑)) values were performed at MSU by Ibbotson et al. [4] in ^{32,34,36,38}Si. For ^{34}Si, a B(E2↑) value of 85(33) e^{2}fm^4 was determined which shows the quadrupole collectivity of the 2_1^+ state and reflects the 0p-0h and 2p-2h mixing. By shell-model calculations these authors found a value of 44 e^{2}fm^4. Their calculations also suggest the existence of an 0_2^+ intruder state at 2.02 MeV in ^{34}Si linked to the 2_1^+ state by a large B(E2↑) value : 215 e^{2}fm^4. Intruder states were also predicted by Heyde et al. [5] and by Caurier et al. [6]. The latter authors predict the 0_2^+ state at 3.0 MeV, linked

CP495, *Experimental Nuclear Physics in Europe*, edited by B. Rubio et al.
© 1999 American Institute of Physics 1-56396-907-6/99/$15.00

to the 2_1^+ state at 3.5 MeV by B(E2; $0_1^+ \rightarrow 2_1^+$) = 260 e^{2}fm^4. Having in mind those predictions, we carried out new ^{34}Al beta-decay experiments in order to improve the ^{34}Si level description and to locate the 0_2^+ intruder state.

Another interest was to determine the level structure of ^{35}Si (N=21, Z=14), namely to locate the p$_{3/2-}$ and d$_{3/2+}$ states and see their evolution in the N=21 isotones (from ^{41}Ca (Z=20) to ^{35}Si (Z=14)). The evolution of the $7/2^-$-$3/2^-$ splitting in the N=21 isotones gives indications on the N=28 shell closure on the one hand and, on the other hand, the location of the d$_{3/2+}$ intruder state complement information on the island of inversion and the gap between the 0p-0h and 2p-2h states at N=20.

EXPERIMENT

The ^{34,35}Al isotopes were produced at the ISOLDE facility at CERN by fragmentation of an uranium carbide target by the 1 GeV pulsed-proton beam of the PS/Booster. The reaction products were ionized in a tungsten surface-ionization source, mass selected in the ISOLDE separator and collected on a movable aluminized mylar tape in order to reduce the long-lived daughter activities. Standard spectroscopic measurements (β-γ, β-γ-γ, β-n, β-γ-n coincidences and multiscaling of β and γ emission) were performed using Ge counters, a 4π plastic counter surrounding the collection point for beta detection and 8 low threshold neutron detectors (threshold for the neutron detection $\sim$ 50 keV), placed at 45 cm from the collection point, for neutron time of flight measurements. The typical yield available for the experiment was 30 at/s and 10 at/s for ^{34}Al and ^{35}Al respectively.

RESULTS

The ^{34}Al β decay

We confirm the previously reported [3] ^{34}Si level structure. The ^{34}Al half-life was measured and yields to the value of 55.6(13) ms in agreement but much more precise than the published one (60(18) ms). Due to an improvement of the production yield, we observe in addition to the previous transitions, 7 new lines which can be assigned to the ^{34}Al decay. In ^{34}Si a new level was found at 4970 keV decaying by a 590 keV transition to the 4379 keV state. Also a new branch was found between the 4379 and 3326 keV levels. Two of the new transitions are placed in the ^{33}Si level scheme. A P$_n$ value of 0.26(4) was measured in agreement with our previous value (0.27(5)). A full report on the ^{34}Al decay scheme will be found in ref. [7].

As a candidate for 0_2^+ intruder state, we propose a level at 2133 keV linked to the 2_1^+ 2p-2h state at 3326 keV by a 1193 keV transition. We select this transition on the basis of its decay rate compatible with the ^{34}Al one and on its intensity (0.06) compatible with the intensity (0.94) of the $2_1^+ \rightarrow 0_1^+$ (g.s.) transition. The decay of the 2133 keV state is still unobserved. Taking into account the B(E2↑) value, 85(33) e^{2}fm^4 for the $0_1^+ \rightarrow 2_1^+$ transition [4] and the branching values, we found

$B(E2\uparrow) = 900(500)$ e^{2}fm^4 for the $0_2^+ \rightarrow 2_1^+$ transition. This value will be discussed further.

The ^{35}Al β decay

The ^{35}Al half-life was measured and yields to the value of 41.6(22) ms which differs from the two known values (30(4) ms [8] and 150(50) ms [9]). No β-delayed γ rays were reported before. We give in Fig. 1 the main properties we observed in the ^{35}Al decay. The first information on the ^{35}Si level structure is obtained. According to the β intensities and logft values we assign to the three levels 911, 974 and 2169 keV respectively $J^\pi = (3/2)^-$, $(3/2)^+$ and $(5/2)^+$. These values are consistent with $(5/2)^+$ for the ^{35}Al ground state. The P_n value (0.41(13)) is found larger than value (0.26(4)) from ref. [8]. In the β-delayed neutron spectrum we observed two maxima at 0.98 and 3 MeV. Based on the fact that the feeding (with l=0) of the 2$^+$ state in ^{34}Si is more probable than the 0$^+$ state's one (l=2 for the neutron angular momentum), excited states in ^{35}Si corresponding to the observed neutron emission are located around 6.81 and 8.88 MeV.

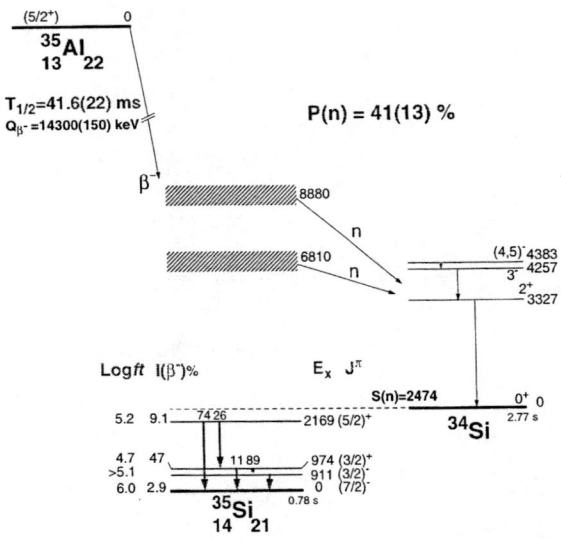

FIGURE 1. The ^{35}Al decay scheme.

57

DISCUSSION

The location of the 0_2^+ intruder state at 2133 keV in ^{34}Si is in good agreement with the predicted value of ref. [4] (2.0 MeV) but our extracted B(E2↑) = 900(500) e^2fm^4 value appears to be too large even if we consider the lowest limit (400 e^2fm^4) taking into account the error bar. The previous B(E2↑) measurement (85(33) e^2fm^4) [4] of the $0_1^+ \rightarrow 2_1^+$ transition shows the mixing 0p-0h and 2p-2h (in the limit of a zero mixing this transition does not exist). If that transition (0_1^+ (0p-0h) $\rightarrow 2_1^+$ (2p-2h)) would not exist, the maximum collectivity we can obtain for the ($0_2^+ \rightarrow 2_1^+$) in a pure 2p-2h space is around 450 e^2fm^4 similar to the collectivity observed in ^{32}Mg. The location of the 0_2^+ state is close to the predictions but the relative intensity of the E2 transitions from the 2_1^+ state is at variance with the theory and ask for more measurements.

The energy of the $3/2^-$ state in ^{35}Si gives new information on the proton-neutron interaction in the (sd-fp) space. One part of this information, already known, was extracted from the evolution of the $d_{3/2}$-$s_{1/2}$ splitting in the potassium isotopes [10,11]. Our new measurements affect mainly the cross-interaction $p_{3/2}/(s_{1/2}d_{3/2})$ and are taken into account in the residual interaction to reproduce exactly the behaviour of the N=21 isotones. One observes that the $f_{7/2}$-$p_{3/2}$ gap reduces when going from ^{41}Ca to ^{35}Si and this has as a consequence, the weakening of the shell closure at N=28 far from stability.

The ^{35}Al beta decay feeds the non-natural parity particle-hole states in ^{35}Si. Those states give a picture of the global sd-fp gap in ^{35}Si. This gap decays regularly from ^{41}Ca to ^{35}Si until it produces, according to the predictions, the inversion of the $f_{7/2}$ and $d_{3/2}$ shells for the ^{33}Mg ground state.

REFERENCES

1. Thibault C. et al., *Phys. Rev.* **C12**, 644 (1975).
2. Klotz G. et al., *Phys. Rev.* **C47**, 2502 (1993) and refs. therein.
3. Baumann P. et al., *Phys. Rev.* **B228**, 458 (1989).
4. Ibbotson R.W. et al., *Phys. Rev. Lett.* **80**, 2081 (1998).
5. Heyde K. and Wood J.L., *J. Phys.* **G17**, 135 (1991).
6. Caurier E. et al. , *Phys. Rev.* **C58**, 2033 (1998).
7. Nummela S. et al. , *Phys. Rev.* **C**, to be submitted (1999).
8. Reeder P.L. et al. , *Proc. Intern. Conf. on Exotic Nuclei and Atomic Masses*, Arles, France, June 19-23 (1995) p. 587.
9. Bazin D. et al. *AIP Conf. Proc.*, Vol. 164, ed. Towner I., (AIP, New York, 1987) p. 722.
10. Walter G. et al. , *Proc. of the Workshop on Nuclear Structure of Light Nuclei far from Stability*, Obernai, France, 1989, ed. Klotz G., (CRN Strasbourg, 1991) p. 71.
11. Retamosa J. et al. , *Phys. Rev.* **C55**, 1266 (1997).

First results using a new technology for measuring masses of very short-lived nuclides with very high accuracy: the MISTRAL* program at ISOLDE

C. Monsanglant[1], C. Toader[1,2], G. Audi[1], G. Bollen[3], C. Borcea[2],
G. Conreur[1], R. Cousin[1], H. Doubre[1], M. Duma[2], M. Jacotin[1], S. Henry[1],
J.-F. Képinski[1], H.-J. Kluge[4], G. Lebée[3], G. Le Scornet[1,3], D. Lunney[1],
M. de Saint Simon[1], C. Scheidenberger[3], C. Thibault[1]
and the ISOLDE collaboration[3]

[1]CSNSM-IN2P3-CNRS bât 108, F-91405 Orsay-campus, France
[2]Inst. Atomic Physics, Bucharest, Romania
[3]CERN, EP division, Geneva, Switzerland
[4]GSI, D-64291 Darmstadt, Germany

Abstract : MISTRAL is an experimental program to measure masses of very short-lived nuclides ($T_{1/2}$ down to a few ms), with a very high accuracy (a few 10^{-7}). There were three data taking periods with radioactive beams and 22 masses of isotopes of Ne, Na$^+$, Mg, Al$^+$, K, Ca, and Ti were measured. The systematic errors are now under control at the level of 8×10^{-7}, allowing to come close to the expected accuracy. Even for the very weakly produced ^{30}Na (1 ion at the detector per proton burst), the final accuracy is 7×10^{-7}.

Mistral is a new technology for measuring masses of very short-lived nuclides with very high accuracy. Accurate measurements of atomic masses particularly far from stability are crucial to put constraints on nuclear mass models and on their parameters, especially for the description of the nucleosynthesis r-process path, and to describe new structures and properties of nuclear matter in extreme isospin conditions. Given such motivations, MISTRAL was installed at ISOLDE in CERN during the summer 1997.

MISTRAL is a transmission radiofrequency mass spectrometer based on the principle defined by L.G. Smith (2). Mass-ratios are determined through cyclotron frequency-ratio measurements. When ions A and B (A=reference and B=unknown mass) are rotating in a given homogeneous magnetic field, the product of their mass

---

* Mass measurement at ISolde using a Transmission Radiofrequency spectrometer on-Line
\+ This work is part of C. Toader's thesis (1).

CP495, *Experimental Nuclear Physics in Europe*, edited by B. Rubio et al.
© 1999 American Institute of Physics 1-56396-907-6/99/$15.00

and their cyclotron frequency is constant: $m_A f_{c_A} = m_B f_{c_B}$. We therefore make relative measurements. f_c is determined with the help of two radiofrequency modulations of the ion kinetic energy. Ions may only reach the final detector if this radiofrequency f_{RF} is related to f_c through $f_{RF} = (n + 1/2) \cdot f_c$ (3,4,5,6). Accurate measurements require a very homogeneous magnetic field, stable in time and a good overlap of the two beam trajectories. The total transmission from the ISOLDE focal plane to the MISTRAL detector is about 1×10^{-4} and the resolving power is around 100 000.

A measurement from MISTRAL yields a spectrum of transmission versus radiofrequency. Shown in Figure 1 is a recorded peak for ^{32}Mg from the April 1999 run. The resonance frequency is derived from a triangular fit which is the theoretically expected lineshape (7).

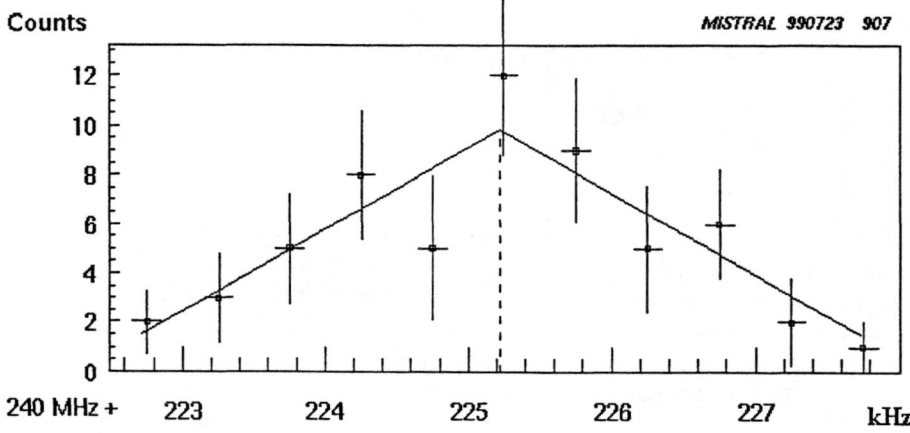

FIGURE 1. Transmission versus radiofrequency for ^{32}Mg from the April 1999 run. This nuclide is very weakly produced at ISOLDE and has a rather short half-life of only 120 ms. This measurement corresponds to 1 hour of data taking. The center accuracy is 9×10^{-7} and the resolving power is about 80 000.

During three data taking periods, the masses of isotopes of Ne, Na, Mg, Al, K, Ca, and Ti were measured. These measurements extend from the valley of stability to ^{30}Na and ^{32}Mg. Eight of them have a half-life under 1s, the shortest half-life being 31ms for ^{28}Na. Only the sodium results from the two 1998 data taking periods will be discussed here; they are part of C. Toader's thesis (1).

The uncertainties of the measurements obtained with MISTRAL are the quadratic sum of three components. First, there are the classical uncertainties due to statistics. They range from 1×10^{-7} to 9×10^{-7} for the very weakly produced ^{32}Mg. Second, the measured frequencies appear to be shifted proportionally to the mass jumps between reference (^{23}Na or ^{39}K) and unknown masses (probably due to imperfect trajectories coincidence and to insufficient homogeneity of the magnetic field). These shifts are calibrated with well-known masses. This correction introduces a calibration uncertainty of 4×10^{-7}. Third, it was also found that for different configurations of ISOLDE and MISTRAL, inconsistencies of the results exist at the level of 7×10^{-7}. The total precision for each single measurement ranges thus from 8×10^{-7} to 12×10^{-7}. Despite the systematic errors above, most masses measured with MISTRAL have been improved.

Figure 2 shows a comparison between the MISTRAL results and the Ame'95 mass table (8) for sodium isotopes and ^{27}Al. The well-known masses, used as calibrators, ^{23}Na, ^{24}Na, ^{25}Na and ^{27}Al measured against ^{23}Na and ^{39}K are in very good agreement with the table. The precision for the other sodium masses ranges from 10 keV to 22 keV for ^{30}Na. The agreement with the Ame'95 mass table is fair with the exception of ^{30}Na as discussed below. We improve the error on all these masses and the comparison of our results with the 1995 mass table has a consistency of 0.9 standard deviation which gives confidence in the MISTRAL measurements.

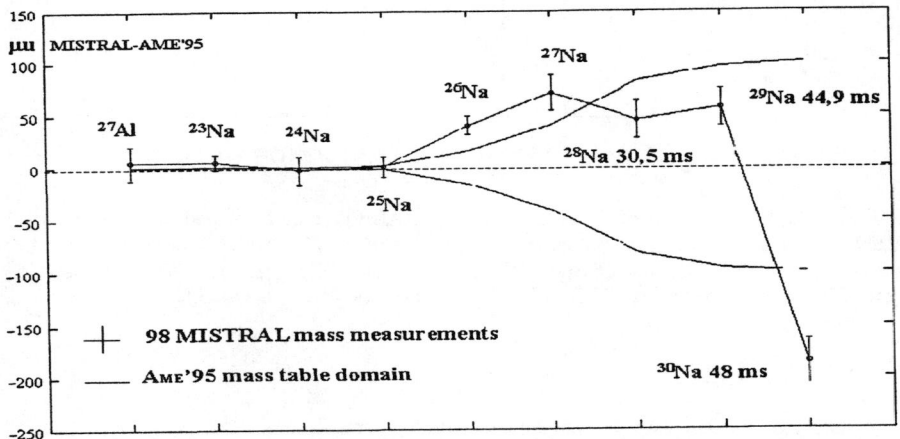

FIGURE 2. Comparison between the MISTRAL results and the AME 1995 mass table for sodium isotopes and ^{27}Al. The zero line is for the mass table values and the two symmetrical lines represent the limits of their error.

Prior to our work, six other measurements of ^{30}Na were performed using various methods (Fig. 3). MISTRAL has a very small error bar of 20 keV compared to 150 keV and up for the others. Thus almost one order of magnitude has been gained, but MISTRAL seems to have a shift of -200 keV compared to the Ame'95 table. Examining more closely the ^{30}Na results, it appears that the TOFI 91 measurement (9) had the most important contribution in the 1995 mass adjustment and is responsible for this disagreement. Apart from the two TOFI measurements (9,10), all data are in agreement with the MISTRAL result.

Using a RF mass spectrometrer, accurate determination of masses of very short-lived sodium nuclides were obtained. Other data are being analysed. Measurements on isotopes in the area of the next neutron shell closure at $N=28$ will complete this phase of the MISTRAL program at ISOLDE (11). Improvements are planned to increase sensitivity and to minimize the systematic frequency shift. The measurement program will then be focused on heavier, very short-lived, neutron-rich isotopes towards the stellar nucleosynthesis r-process path and also nuclides close to the proton dripline of interest for nuclear structure studies.

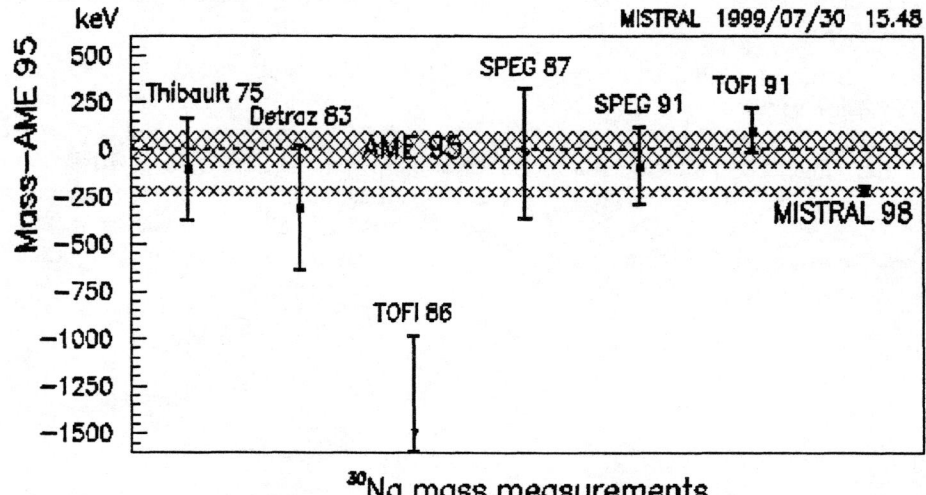

FIGURE 3. All measurements of ^{30}Na including MISTRAL value are compared to the AME'95 values (AME'95 dashed area). MISTRAL has an error bar of only 20 keV. Thibault75, mass spectrometry (12); Detraz83, Q_β (13); TOFI86, time of flight (10); SPEG87, time of flight (14); SPEG91, time of flight (15); TOFI91, time of flight (9); MISTRAL98, RF mass spectrometry (this work).

REFERENCES

1. Toader, C. F., *Ph.D.* Université Paris-Sud thesis (1999,unpublished)
2. Smith, L. G., *Phys. Rev.* **C4**, 22 (1971) and references therein.
3. http://csnwww.in2p3.fr/massatom/mistrpres_en.html
4. Lunney, D., et al., *Hyp. Int* **99**, 105-114 (1996).
5. Monsanglant., C., et al., *AIP* **447**, 175-182 (1998).
6. Lunney, D., et al., *AIP* **455**, 995-998 (1998).
7. Coc, A., et al., *Nucl. Inst. and Meth.* **A271**, 512-517 (1988).
8. Audi, G. and Wapstra, A. H., *Nucl. Phys. A* **595**, 409 (1995).
9. Zhou, X. G., et al., *Phys. Lett. B* **260**, 285 (1991).
10. Vieira, D. J., et al., *Phys. Rev. Lett.* **57**, 3253 (1986).
11. Lunney, D., et al., *CERN/ISC/P107* **99-3**, (1999).
12. Thibault, C., et al., *Phys. Rev. C* **12**, 644 (1975).
13. Détraz, C., et al., *Nucl. Phys. A* **394**, 378 (1983).
14. Gillibert, A., et al., *Phys. Lett. B* **192**, 39 (1987).
15. Orr, N. A., et al., *Phys. Lett. B* **258**, 29 (1991).

Recent beta-decay experiments on nuclei beyond ^{56}Ni

E. Roeckl[a], J. Äystö[b], R. Borcea[a], P. Dendooven[b], M. Gierlik[c],

M. Górska[a], H. Grawe[a], M. Hellström[d], A. Jokinen[b], M. Karny[c],

Z. Janas[c], R. Kirchner[a], M. La Commara[a], P. Mayet[a],

A. Niemenen[b], H. Pentillä[b], A. Płochocki[c], M. Rejmund[a],

M. Sawicka[c], C. Schlegel[a], K. Schmidt[a], R. Schwengner[e]

[a] Gesellschaft für Schwerionenforschung, D-64291 Darmstadt, Germany
[b] University of Jyväskylä, FIN-40351 Jyväskylä, Finland
[c] Institute of Experimental Physics, Warsaw University,
PL-00681 Warszawa, Poland
[d] Department of Physics, Lund University, S-221 00 Lund, Sweden
[e] Institut für Kern- und Hadronenphysik, FZ Rossendorf,
D-01314 Dresden, Germany

Abstract. By using heavy-ion induced fusion-evaporation reactions at the on-line mass separator of GSI, we investigated β-decay properties of the neutron-deficient isotopes ^{56}Cu, ^{57}Zn and ^{61}Ga. The results will be presented and discussed in comparison with shell-model predictions.

1. Introduction

The study of the β decay of very proton-rich fp-shell nuclei with $N \cong Z$ in general, and of nuclei between the double shell closure at ^{56}Ni and ^{100}Sn in particular, is of great current interest to nuclear physics and astrophysics. As far as nuclear physics is concerned, the high energy release in β decay allows one to measure the Gamow-Teller (GT) strength B(GT) for a large range of excitation energies in the daughter nucleus, and to thus stringently test theoretical B(GT) predictions (1). The related tasks are to determine the experimental distribution of B(GT) as a function of the excitation energy in the daughter nucleus *completely*, i. e. including (weak) decay branches to high-lying levels, and to investigate whether theoretical calculations are able to reproduce this distribution, with particular emphasis being put on the possible quenching of the calculated GT strength. The astrophysical interest is related to, e. g., the rp process (2) and to electron-capture cooling of supernovae (3).

CP495, *Experimental Nuclear Physics in Europe*, edited by B. Rubio et al.
© 1999 American Institute of Physics 1-56396-907-6/99/$15.00

Based on these motivations, there has been a recent upsurge of both theoretical and experimental work on β-decay of such nuclei. For example, by using fragmentation of high-energy heavy-ion beams, the superallowed $0^+ \rightarrow 0^+$ decays of ^{78}Y, ^{82}Nb and ^{86}Tc were observed at GANIL (4), and the half-lives of neighbouring N$\cong$Z nuclei were determined at GSI (5). Furthermore, the β decay of the proton-rich nuclei ^{58}Zn (6), ^{71}Kr (7), ^{72}Kr and ^{76}Sr (8) were studied at ISOLDE. By using the on-line mass separator of GSI, the β-decay properties of proton-rich nuclei near double shell closure at ^{56}Ni, i. e. ^{56}Cu ($T_z = -1$), ^{57}Zn ($T_z = -3/2$), and ^{61}Ga ($T_z = -1/2$), were investigated in experiments performed in July 1997 and October 1998. While the results from the earlier measurements have already been published for the case of ^{56}Cu (9) and ^{61}Ga (10), we shall sketch in this report the preliminary results obtained from the later experiment.

2. Experimental Technique

By using ^{28}Si(^{32}S,xpyn) and ^{28}Si(^{36}Ar,xpyn) fusion-evaporation reactions, radioactive sources of neutron-deficient copper-to-gallium isotopes were produced at the on-line mass separator of GSI. For the measurements of β–delayed protons, the mass-separated 60 keV beam was implanted into a thin carbon catcher which was viewed by a telescope. The latter comprised a gas detector and a silicon detector for the measurement of energy loss and energy of protons. The spectroscopy of β-delayed γ rays was accomplished by implanting the mass-separated beam into a tape that stayed at rest during a preselected time period and subsequently moved the collected activity away from the β-γ detection setup in recurring implantation-transport cycles. Positrons were recorded in a cylindrical NE102A plastic-scintillation detector that surrounded the implantation point, and γ rays were measured by means of two germanium detectors of the Euroball-Cluster (11) and Clover (12) type, mounted close to this point.

3. Results and Discussion

Concerning β-delayed γ rays from the decays of ^{56}Cu and ^{61}Ga, the statistics of the new experimental data is one to two orders of magnitude higher than that obtained in our earlier work (9,10) which represented the first observation of these decays. The β-delayed proton data obtained for ^{57}Zn, which probe excited states in the single-proton nucleus ^{57}Cu, were considerably improved, with respect to source purity and energy resolution, over those gained in the one and only previous measurement (13) of this decay.

Before going to a more detailed discussion of the ^{56}Cu data, we want to stress that we have any reason to believe that the current evaluation of the new ^{57}Zn and ^{61}Ga data will allow us to experimentally identify additional β-decay branches beyond those observed for these decays so far, and to thus deduce more complete experimental B(GT) distributions for a comparison with theoretical predictions. In the case of ^{61}Ga this would mean that further ^{61}Zn states with spin/parity assignments of $1/2^-$, $3/2^-$ or $5/2^-$ could be identified. So far, five B(GT) values have been observed (10) for this short-lived decay ($T_{1/2}$=140(70) ms). The new data will, in particular, be used for an attempt to improve the very limited accuracy that was previously reached (10) for the B(GT) value of the ground-state to ground-state decay of ^{61}Ga.

The experimental results obtained for ^{56}Cu are compiled in Table 1. Besides the γ transitions of 1225, 2506, 2701 and 2783 keV, that were already observed in our earlier work and used to identify four excited states of ^{56}Ni, we found additional γ lines of 951, 2189 and 3287 keV which were preliminarily assigned to 4^+,T=1→(4_2^+),T=0, (3_1^+),T=0→2_1^+,T=0, and (3_2^+),T=0→2_1^+,T=0 transitions, respectively. The experimental log ft values were determined by using the branching ratios from this work, a half-life of 78(15) ms from Ref. (9), and a Q_{EC} value of 15300(140) keV (15). As the latter was derived from systematical (Coulomb energy) trends, the resulting log ft values represent semi-empirical estimates, the Q_{EC}-related contribution to their uncertainties ranging from 7 to 9%.

TABLE 1. Beta-decay data for low-lying ^{56}Ni states. The experimental level energies E_{exp}, β-intensities I_β, and $logft_{exp}$ values, determined from a preliminary evaluation of results obtained in this work, are confronted to level energies E_{theor} and $logft_{theor}$ values obtained from shell-model calculations. The spin/partity/isospin assignments (I^π, T) were deduced from reaction data (14) except for the 4890, 5483 keV and 5988 levels whose configurations were tentatively deduced from a comparison with shell-model predictions.

E_{exp} (keV)	E_{theor} (keV) Ref. 9	E_{theor} (keV) Ref. 3	I^π, T	$I_{exp}^{(\beta)}$ (%)	$log ft_{exp}$	$log ft_{theor}$ Ref. 9	$log ft_{theor}$ Ref. 3
2701(1)	3219	–	2^+,0	–	–	–	–
3926(1)	4071	4495	4^+,0	27.9(7.1)	4.38(0.20)	4.64	4.38
4890(2)	4954	4902	(3^+),0	10.9(3.2)	4.59(0.20)	4.60	4.80
5483(2)	5643	5893	(4^+),0	8.4(1.8)	4.56(0.16)	3.92	4.33
5988(2)	6462	6335	(3^+),1	3.8(0.8)	4.79(0.17)	4.15	4.28
6432(2)	6180	6113	4^+,1	48.9(5.1)	3.56(0.12)	3.49	3.49

Table 1 lists also results obtained from large-scale shell-model calculations. The first one, called Model C in Ref. (9), used the FDP6* interaction to determine the strengths of GT transitions between the $[(f_{7/2})^{-1} \times p_{3/2}](4^+, T = 1)$ ground-state of ^{56}Cu and the $[(f_{7/2})^{-1} \times p_{3/2}](3^+, 4^+, 5^+; T = 0, 1)$ and $[(f_{7/2})^{-1} \times p_{1/2}](3^+, 4^+; T = 0, 1)$ excited states of ^{56}Ni. Up to three particles excited from the $f_{7/2}$ orbital to the $p_{3/2}$, $p_{1/2}$ and $f_{5/2}$ orbitals were taken into account. The second calculation (3) was based on the monopole-corrected KB3 interaction. In both calculations the global effective quenching of the GT matrix elemnets were included as determined by Martinez-Pinedo et al. (16). As can be seen from Table 1, qualitative agreement has been obtained between experimental and theoretical log ft values. However, the *total* B(GT) value measured for the ^{56}Cu decay amounts to 0.42(0.09) compared to the FDP6* result of 4.29. The FDP6* calculation predicts most of the GT strength to reside at higher ^{56}Ni excitation energies which are difficult to experimentally access due to their small phase space.

4. Summary

By using heavy-ion induced fusion-evaporation reactions at the on-line mass separator of GSI, we determined considerably improved data for the β-decays of ^{56}Cu, ^{57}Zn and ^{61}Ga. For the ^{56}Cu decay, qualitiative agreement between experimental B(GT) values for low-lying ^{56}Ni states and the corresponding theoretical predictions has been obtained, even though the measurement is far from yielding the *complete* B(GT) distribution. A detailed discussion of the nuclear-physics aspects of these results as well as of their astrophysical relevance can not be given within the scope of this report and will thus be presented elsewhere.

5. Acknowledgement

This work was supported in part by the European Community under Contract No. ERBFMGECT950083 and by the Polish Committee for Scientific Research under grant KBN 2 P03B 086 17.

REFERENCES

(1) J.L. Tain, contribution to this Conference
(2) H. Schatz et al., Phys. Rep. **294**, 167 (1998)
(3) E. Caurier et al., Nucl. Phys. **A 653**, 439 (1999)
(4) C. Longour et al., Phys. Rev. Lett. **81**, 3337 (1998)
(5) E. Wefers et al., contribution to this Conference
(6) M. Oinonen et al., Eur. Phys. J. **A 3**, 271 (1998)
(7) M. Oinonen et al., Phys. Rev. **C 56**, 745 (1997)
(8) Ch. Miehé et al., in "ENAM 98, Exotic Nuclei and Atomic Masses", AIP Conf. Proc. **455**, 789 (1998)
(9) M. Ramdhane et al., Phys. Lett. **B 342**, 22 (1998)
(10) M. Oinonen et al., Eur. Phys. J. **A 5**, 151 (1999)
(11) J. Eberth et al., Prog. Part. Nucl. Phys. **39**, 29 (1997)
(12) J. Gerl et al., in Proc. Conf. on Physics from Large γ-ray Detector Arrays, Berkeley, LBL 35687, CONF 940888, UC 413, p. 159 (1994)
(13) D.J. Vieira et al., Phys. Lett. **B 60**, 261 (1976)
(14) H. Junde, Nucl. Data Sheets **67**, 523 (1992)
(15) G. Audi et al., Nucl. Phys. **A 624**, 1 (1997)
(16) G. Martinez-Pinedo et al., Phys. Rev. **C 53**, R2602 (1996)

Core-Excited States in the Doubly Magic ^{68}Ni and its Neighbor ^{69}Cu

T. Ishii*, M. Asai*, A. Makishima†, I. Hossain‡, M. Ogawa‡,
J. Hasegawa‡, M. Matsuda*, and S. Ichikawa*

*Advanced Science Research Center, Japan Atomic Energy Research Institute,
Tokai, Ibaraki 319-1195, Japan
†Department of Liberal Arts and Sciences, National Defense Medical College,
Tokorozawa, Saitama 359-8513, Japan
‡Research Laboratory for Nuclear Reactors, Tokyo Institute of Technology,
Meguro, Tokyo 152-8550, Japan

Abstract. The $(\nu g_{9/2}^2 \nu p_{1/2}^{-2})_{8+}$ isomer with $T_{1/2} = 23(1)$ ns at 4208 keV in ^{68}Ni was found by deep-inelastic collisions of ^{70}Zn(8MeV/nucleon)+ ^{198}Pt and the $\nu g_{9/2}$ $E2$ effective charge was determined to be 1.5(1)e. In ^{69}Cu, the $(\pi p_{3/2}\nu g_{9/2}^2\nu p_{1/2}^{-2})_{19/2-}$ isomer with $T_{1/2} = 22(1)$ ns at 3691 keV was identified and its decay data were calculated quite accurately by a parameter-free shell model calculation using experimental level energies. Proton $2p$-$1h$ excitation, fed by another $T_{1/2} = 39(6)$ ns isomer, induces large collectivity in ^{69}Cu.

The doubly closed-shell nucleus $^{68}_{28}$Ni$_{40}$ gives valuable information to extend our knowledge of nuclear structure far from the β-stability line. In particular, the $(\nu g_{9/2}^2 \nu p_{1/2}^{-2})_{8+}$ isomer in ^{68}Ni provides neutron-neutron two-body residual interactions and a neutron effective charge in the $g_{9/2}$ orbital. The magicity of the neutron number 40 at $Z \neq 28$ is also an interesting subject. This property can be tested by excited states in $^{69}_{29}$Cu$_{40}$, especially by the proton two-particle one-hole $(2p$-$1h)$ excitation. Recently, experimental technique for studying neutron-rich nuclei near ^{68}Ni has made a remarkable progress. Grzywacz et al. [1] identified μs isomers around this nucleus produced in projectile fragmentation. Franchoo et al. [2] measured the β decay of $^{68-74}$Ni using an isotope separator with a laser ion-source. Broda et al. [3,4] measured in-beam γ rays of $^{64-68}$Ni produced in heavy-ion deep-inelastic collisions (DIC's) with a large array of γ detectors. We have also succeeded [5,6] in measuring in-beam γ rays from isomers, with $T_{1/2} > 1$ ns, produced in deep-inelastic collisions by an isomer-scope developed by ourselves [5].

In the present experiment, a ^{198}Pt foil, 4.3 mg/cm^2 in thickness, was bombarded with a ^{70}Zn beam of 566 MeV from the JAERI tandem booster. The γ rays from isomers were measured with an improved isomer-scope which detects projectile-like fragments (PLF's) with ΔE-E telescopes. Four Si ΔE detectors, each of diameter 20 mm and thickness 22 μm, were arranged symmetrically around the beam axis and were placed in front of a Si E detector of an annular shape, 100 mm in outer-diameter. Four Ge detectors surrounded the periphery of the Si E detector to observe the γ rays from the stopped fragments; these Ge detectors were placed in a cross geometry and each Ge detector was adjacent to each ΔE detector. A tungsten block shields these Ge detectors from the intense γ radiation from the target. Sorting the γ emitters by atomic numbers from the ΔE-E-$\gamma(\text{-}\gamma)$ coincidence data, we have greatly improved the sensitivity to detect the γ rays of interest. This geometry also allows us to measure in-plane to out-of-plane ratios of γ rays emitted by PLF's.

Gamma-ray spectra of nickel and copper isotopes are shown in Fig. 1(a) and (b). Figure 1(c) shows a $\gamma\gamma$ spectrum in ^{68}Ni coincident with the low-lying transitions identified previously [3]. In these spectra, we have found γ-ray from a new isomer at 4208 keV in ^{68}Ni and two new isomers at 3691 and 3827 keV in ^{69}Cu. The half-lives of these isomers were derived from the $t_{PLF-\gamma}$ coincidence data. The present results are summarized in the decay schemes shown in Fig. 2. According to a simple picture of DIC's, the angular momentum of the PLF is aligned perpendicular to the reaction plane defined by the beam axis and a ΔE detector. Thus, by measuring

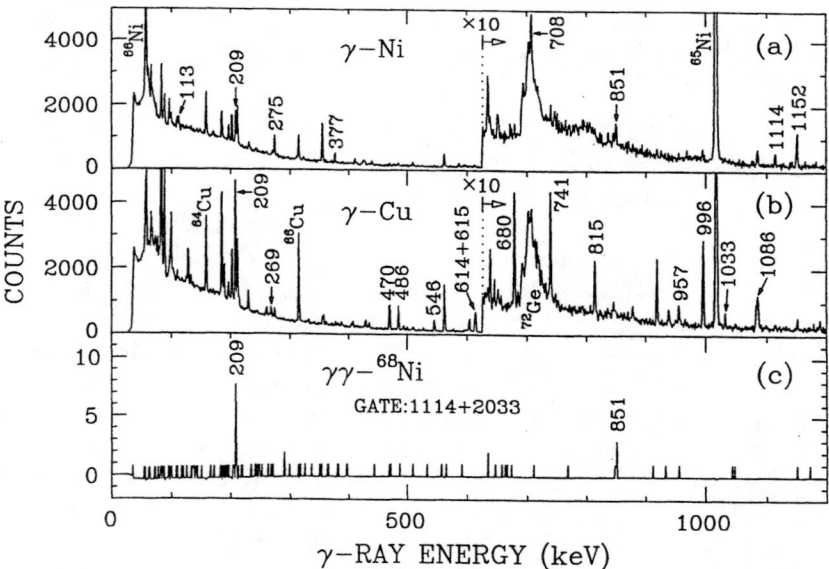

FIGURE 1. (a) A γ-ray spectrum of Ni isotopes. The γ-ray energies are depicted for ^{68}Ni. (b) A γ-ray spectrum of Cu isotopes. The γ-ray energies are depicted for ^{69}Cu. (c) A γ-γ spectrum in coincidence with the 1114 and 2033 keV γ rays in ^{68}Ni. These three spectra were obtained from the ΔE-E-$\gamma(\text{-}\gamma)$ coincidence data, by setting a $t_{PLF-\gamma}$ window of 20 − 100 ns and sorting by atomic numbers.

the in-plane to out-of-plane ratios, one can determine the multipolarities of γ-rays, if the nuclear orientation is well preserved. In the present experiment, the anisotropies of the γ-rays from the new isomers were observed enough to distinguish between a dipole and a quadrupole type. The spins shown in Fig. 2 were assigned mainly on the basis of the γ-ray multipolarities determined by this method.

The level spacings between the 8^+, 6^+, 4^+, and 2^+ states in ^{68}Ni give a clue to the configurations of these states. If these states had a pure $\nu g_{9/2}^2 \nu p_{1/2}^{-2}$ configuration, these level spacings in ^{68}Ni should be the same as those in ^{70}Ni because of the presumable conservation of seniority. The spacing between the 6^+ and 4^+ states in ^{68}Ni is much wider than that in ^{70}Ni, while the spacing between the 8^+ and 6^+ states is nearly the same. This fact suggests that the 4^+ and 2^+ states in ^{68}Ni have a significant admixture of other components, such as $\nu g_{9/2}^2 \nu f_{5/2}^{-2}$. On the other hand, the 8^+ and 6^+ states have a very pure $\nu g_{9/2}^2 \nu p_{1/2}^{-2}$ configuration. Thus, from the $B(E2; 8^+ \rightarrow 6^+)$ value of 26(4) e^2fm^4 in ^{68}Ni, the $E2$ effective charge for the ^{66}Ni core is determined to be $e_{\text{eff}}(\nu g_{9/2})/e = 1.5(1)$; here, $\langle g_{9/2} \mid r^2 \mid g_{9/2} \rangle = 22$ fm^2 is used.

The decay data of the $19/2^-$ isomer in ^{69}Cu can be calculated quite accurately by a shell model calculation with a minimum model space of $\pi p_{3/2} \nu g_{9/2}^2$ using

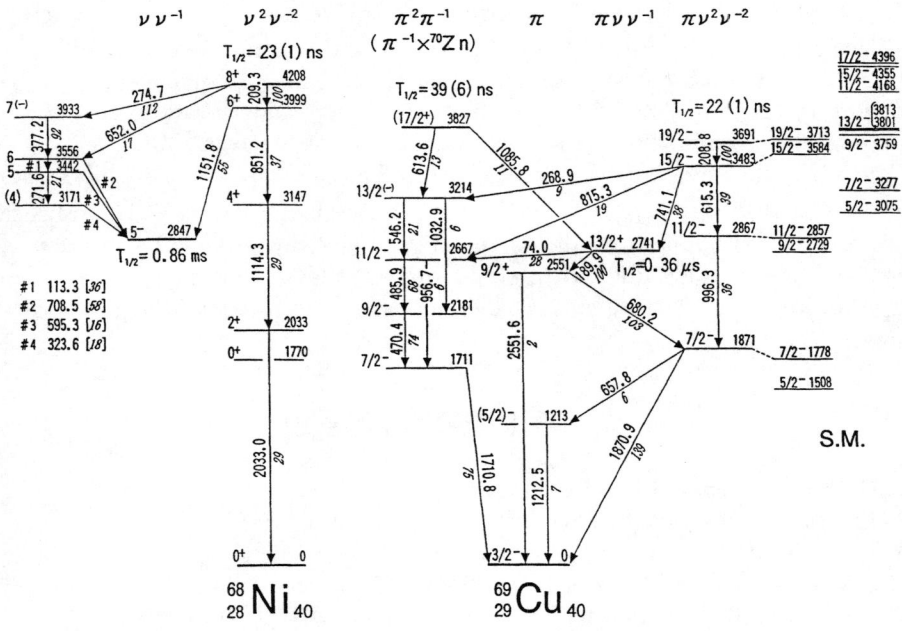

FIGURE 2. Decay schemes of the isomers in ^{68}Ni and ^{69}Cu. The relative γ-ray intensities are depicted in italics. The experimental levels in ^{69}Cu denoted by $\pi \nu^2 \nu^{-2}$ are compared to the shell model calculation (see text); the calculated yrast levels are shown next to the experimental ones.

experimental levels as residual interactions. The relative residual interactions of $(\nu g_{9/2}^2)_{2^+,4^+,6^+,8^+}$ are taken from the levels in ^{68}Ni and those of $(\pi p_{3/2}\nu g_{9/2})_{3^-,4^-,5^-,6^-}$ from the levels in ^{68}Cu [7]. Using the $\nu g_{9/2}$ single-particle energy in ^{67}Ni [4] and the relevant six ground state masses [8], we can carry out this calculation without free parameters. The agreement between calculation and experiment is excellent, as displayed on the right-hand side in Fig. 2. This result supports that the ^{68}Ni levels are good input-parameters for the calculation of the ^{69}Cu data, although the 2^+ and 4^+ states in ^{68}Ni contain other configurations as well as $\nu g_{9/2}^2$. This is probably because the interaction energies taken from experimental levels well absorbs the effect of the configuration mixing. The $B(E2; 19/2^- \rightarrow 15/2^-)$ value in ^{69}Cu, 63(3) e^2fm^4, can be also explained. Using $e_\nu = 1.5e$ obtained from the present work and $e_\pi = 2.0e$ of an assumed value, this shell model calculation gives 56 e^2fm^4, in good agreement with the experiment.

The $7/2^-$ to $13/2^{(-)}$ levels depicted on the left-hand side in the ^{69}Cu decay scheme, named the $\pi^2 \pi^{-1}$ band, are considered as proton $2p$-$1h(\pi f_{7/2}^{-1})$ states, because the $7/2^-$ level at 1711 keV has a large spectroscopic factor in the proton pick-up reactions [9,10]. In this band, the $\Delta I = 2$ transitions compete with the $\Delta I = 1$ ones. This competition indicates that the $E2$ strengths in this band are significantly large, because the $M1$ strengths dominated by the large g-factor of $\pi f_{7/2}^{-1}$ should have an order of magnitude of the Weisskopf estimate, namely, 1.8 μ_N^2. Using this $B(M1)$ value, the $B(E2)$ values in this band are comparable to the $B(E2; 2^+ \rightarrow 0^+)$ value in $^{70}_{30}$Zn$_{40}$. Furthermore, the $\Delta I = 2$ spacings in this band are close to those in ^{70}Zn. Thus, we have concluded that the states in the $\pi^2 \pi^{-1}$ band have large collectivity as the two-valence-proton nucleus ^{70}Zn.

The 3827 keV isomer in ^{69}Cu can be assigned as the $(\pi p_{3/2}\pi g_{9/2}\pi f_{7/2}^{-1})_{17/2^+}$ state, decaying to the $(\pi p_{3/2}\pi f_{5/2}\pi f_{7/2}^{-1})_{13/2^-}$ state through the 614 keV $M2$ transition. However, this assignment results in the 1086 keV γ ray as an $E2$ transition with large hindrance. This hindrance would suggest that this isomer has a different shape from the $13/2^+$ state at 2741 keV. Further investigation is required for this isomer.

REFERENCES

1. Grzywacz R. *et al.*, Phys. Rev. Lett. **81**, 766 (1998).
2. Franchoo S. *et al.*, Phys. Rev. Lett. **81**, 3100 (1998).
3. Broda R. *et al.*, Phys. Rev. Lett. **74**, 868 (1995).
4. Pawłat T. *et al.*, Nucl. Phys. **A574**, 623 (1994).
5. Ishii T. *et al.*, Nucl. Instrum. Methods Phys. Res. A **395**, 210 (1997).
6. Ishii T. *et al.*, Phys. Rev. Lett. **81**, 4100 (1998).
7. Sherman J.D. *et al.*, Phys. Lett. **67B**, 275 (1977).
8. Audi G. *et al.*, Nucl. Phys. **A624**, 1 (1997).
9. Zeidman B., and Nolen, Jr J.A., Phys. Rev. C **18**, 2122 (1978).
10. Ajzenberg-Selove A. *et al.*, Phys. Rev. C **24**, 1762 (1981).

Shell model analyses
of aligned j^2j' three-particle isomers.

P.Kleinheinz[a,b1], T.Ishii[c], M.Ogawa[a], A.Gadea[b2] and J.Blomqvist[d]

[a] *Research Institute for Nuclear Reactors, Tokyo Institute of Technology.*
[b] *Instituto de Física Corpuscular, Burjassot-Valencia.*
[c] *Advanced Science Research Center JAERI, Tokai.*
[d] *Physics Department, Royal Institute of Technology, Stockholm.*

Abstract.
  Minimum configuration space shell model calculations, with the residual interactions taken from the measured excitation energies of the respective two-nucleon multiplets, predict the j^2j'-type energy spectra of a number of three-particle nuclei outside doubly closed shells, including the $\nu g_{9/2}^2 \pi p_{3/2}$ configuration in the newly identified neutron-rich nuclide $^{71}_{29}\text{Cu}_{42}$. The agreement with experiment is excellent in all cases.

Recently a $19/2^-$ isomer at 2.756 MeV with $0.25_3 \mu s$ half life was identified [1,2] in ^{71}Cu, a neutron rich three-particle nucleus with two $g_{9/2}$ neutrons and one $p_{3/2}$ proton outside the ^{68}Ni doubly closed core. The isomer is assigned as $\nu g_{9/2}^2 \pi p_{3/2}$, and it de-excites through a cascade of four stretched E2 transitions to the $\pi p_{3/2}$ ground state, with spacings similar to those of the $\nu g_{9/2}^2$ (8^+ to 0_g^+) level sequence [3] in the ^{70}Cu neighbour. This is in conflict with the E4-E2-E2 cascade of the analogous $(\pi g_{9/2} \nu d_{5/2})_{21/2^+}$ isomer in the valence mirror nucleus ^{93}Mo, and this difference was interpreted [2] as the absence of residual proton-neutron interactions in ^{71}Cu. However, the $\nu g_{9/2}^2 \pi p_{3/2}$ energy spectrum of ^{71}Cu is predicted with high accuracy [1] in a shell model recoupling calculation in the minimum configuration space of three particles in the $\nu g_{9/2}$ and $\pi p_{3/2}$ orbitals using the two-nucleon residual interactions from the five $\nu g_{9/2}^2$ ($0^+, 2^+ .. 8^+$) states observed [3] in ^{70}Ni and the four $\nu g_{9/2} \pi p_{3/2}$ (3^- to 6^-) states [4] of ^{70}Cu. These nine experimental energies fully specify the configuration space of the calculation. The agreement with experiment is excellent, giving an average deviation $|\overline{\Delta E}|$ of 84 keV (Table 1), and theory readily predicts the electric quadrupole character of the isomeric transition.

---

1) Work partially supported by Dirección General de Enseñanza Superior, M.E.C. Spain
2) Fellowship supported by the EC under contract n° ERBFMBICT983127

TABLE 1. Shell Model analyses of $j^2 j'$ 3-particle configurations.

Nucleus	Configuration	I^π_{max} [a]	$E^*_{I_{max}}$ (keV)	$T_{1/2}$	# of states S.M.	# of states obs.	$E^*_{exp} - E_{SM}$ (keV) $E_{0,SM}$ [b]	$\lvert \Delta E \rvert$
^{43}Sc	$\nu f_{7/2}^2\, \pi f_{7/2}$	$19/2^-$ ○	3123	468ns	22	4	-411_2	156_2
^{43}Ti	$\pi f_{7/2}^2\, \nu f_{7/2}$	$19/2^-$ ○	3066	560ns	22	4	-349_9	141_9
^{51}Ti	$\pi f_{7/2}^2\, \nu p_{3/2}$	$15/2^-$ ○	2754	0.8ns	13	4	-73_{34}	77_{34}
^{53}Fe	$\pi f_{7/2}^2\, \nu f_{7/2}^{-1}$	$19/2^-$ ●	3040	2.58m	22	4	-17_{16}	122_{16}
^{69}Cu	$\nu g_{9/2}^2 j_0^{-2}\, \pi p_{3/2}$	$19/2^-$ ○	3691	22ns	17	4	[c]	65_{110}
^{71}Cu	$\nu g_{9/2}^2\, \pi p_{3/2}$	$19/2^-$ ○	2756	0.25μs	17	5	-133_{480}	84 [d]
^{91}Zr	$\pi g_{9/2}^2 j_0^{-2}\, \nu d_{5/2}$	$21/2^+$ ◇	3167	0.35μs	24	5	[c]	39_8
^{93}Mo	$\pi g_{9/2}^2\, \nu d_{5/2}$	$21/2^+$ ●	2425	6.9h	24	19	-3_{10}	42_{10}
^{147}Gd	$\pi h_{11/2}^2 j_0^{-2}\, \nu f_{7/2}$	$27/2^-$ ●	3582	27ns	38	1	[c]	94_{18}
^{149}Dy	$\pi h_{11/2}^2\, \nu f_{7/2}$	$27/2^-$ ●	2661	0.49s	38	4	85_{72}	16 [d]

[a] I_{max} and I_{max-2} states: ● inverted, ○ monotonic, ◇ inverted in S.M., not in experiment.

[b] $E_{0,SM}$ = excitation energy calculated for the ground state. The errors quoted derive from the pertinent 6 ground state masses used in the calculation.

[c] Configuration ground state not observed.

[d] Normalized at the ground state.

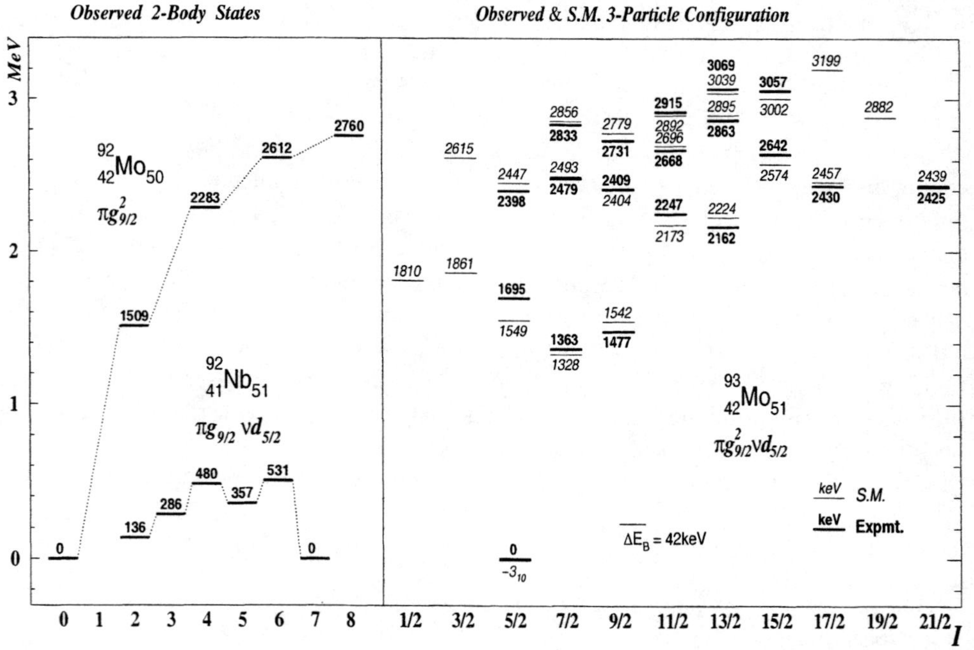

FIGURE 1. Observed 2-body states (left) and comparison between observed levels and shell model calculation for the 3-particle configuration (right).

We have then also recalculated the $\pi g_{9/2}^2 \, \nu d_{5/2}$ energies of ^{93}Mo, similar to the original calculation in the pioneer years of the nuclear shell model [5], but also considering the more complete and accurate present experimental data. Of the 24 levels expected for the configuration 19 have been observed, and theory predicts them within 42_{10} keV (cf. Fig.1 and Table 1). We should mention that, except for spin 1/2 and 3/2, the data provide quite clear configuration assignment, primarily from the extensive γ-ray results obtained in different type experiments. All states shown have unique I^π assignments; we reassign only one, at 2668 keV, from $13/2^+$ to $11/2^+$, which is equally compatible with the data.

In the table we compile the results of analogous calculations for other observed $j^2 j'$ three-particle configurations, where again in some cases earlier results exist but we have now taken into account the recent data. Sometimes similar $j^2 j'$ isomers have also been identified in the j'-*single* particle nucleus, were the j^2-component is a two-particle two-hole excitation of the core. These $j^2 j'$ isomers, with an additional 0^+ boson, are calculated with the j^2 sequence - or at least its highest members - observed in the core, and the respective proton-neutron states from the odd-odd j'-particle j-hole nucleus. The table includes three such cases, among them the newly [6] identified $\nu g_{9/2}^2 \, j_{0^+}^{-2} \, \pi d_{3/2}$ isomer of ^{69}Cu.

Agreement with experiment in general is excellent. It also should be emphasized that these elementary calculations are parameter free, and moreover they provide a straight-forward result for the $j_0^2 j'$ nuclear ground state in an absolute energy scale.

REFERENCES

1. T.Ishii et al. *Phys. Rev. Lett.* **81**, 4100 (1998).
2. R.Grzywacz et al. *Phys. Rev. Lett.* **81**, 766 (1998).
3. M.Pfützner et al. *Nucl. Phys.* **A626**, 259$_c$ (1997).
4. J.D.Sherman et al. *Phys. Lett.* **67B**, 275 (1977).
5. N.Auerbach and I.Talmi *Phys. Lett.* **9**, 153 (1964).
6. T.Ishii et al. *preceding contribution to this volume.*

Probing the N~Z line via β decay

Markku Oinonen

CERN
EP Division
CH-1211 Geneva 23
Switzerland

IS351, IS353 and the ISOLDE Collaboration

Abstract. This contribution reports several beta-decay studies performed at ISOLDE On-line Mass Separator at CERN recently for nuclei close to N = Z line. Beta decay of ^{58}Zn provides a possibility to compare Gamow-Teller strength extracted from complementary beta-decay studies and charge-exchange reactions. Measurement on beta-decay half-life of ^{70}Kr shows importance of experimental information in modelling the path of the astrophysical rp process. Decay of ^{71}Kr is an example of a mirror beta decay and extends the systematics of these particular decays towards highly deformed region close to A = 80.

Introduction

The ISOLDE On-line Mass Separator facility at PS-Booster, CERN [1] offers selective methods to produce nuclei close to N=Z line and beyond. Low production cross sections make the selectivity a key issue to distinguish the wanted products among high isobaric contamination. Typically when dealing with nuclei beyond the proton drip line the yields are of the order of few atoms/s or even less. In this contribution results for three nuclei close to proton drip line, ^{58}Zn, ^{70}Kr and ^{71}Kr, are presented.

Gamow-Teller strength in β decay of ^{58}Zn

Beta decay of ^{58}Zn provides a way of comparing the Gamow-Teller strength obtained from decay measurements and charge-exchange reactions. Discrepancies between results of these probes have been observed recently among in sd shell [2,3].In fp shell there are only few possibilities for this comparison due to lack of suitable target materials for charge-exchange reactions.

CP495, *Experimental Nuclear Physics in Europe*, edited by B. Rubio et al.
© 1999 American Institute of Physics 1-56396-907-6/99/$15.00

Ions of ^{58}Zn were produced from Nb-foil target using laser ionization. Transitions to first two excited states and the ground state in ^{58}Cu were observed. This allowed the half-life determination and extraction of Gamow-Teller strength [4].

The results can be directly compared to existing results of (p,n) and (^{3}He,t) reactions [5,6]. The Gamow-Teller strength for the transition to the second 1^+ state gave 0.54(26) which is comparable to the results from (p,n) and (^{3}He,t) reactions 0.74(14) and 0.46(10), respectively. Shell-model calculation resulted in a value of 1.122 which reproduces the well-known quenching factor of ~ 0.6. As a conclusion, no discrepancy between the decay and charge-exchange results was observed. This conclusion has gained support recently from the other β-decay experiment on ^{54}Ni [7].

β-decay half-life of ^{70}Kr

At astrophysical temperatures about 1.5 GK destruction rate of waiting-point nucleus ^{68}Se becomes proportional to β-decay half-life of ^{70}Kr due to (2p,γ) - (γ,2p) equilibrium [8]. Thus the β-decay half-life of ^{70}Kr affects the rate of rp process above A = 70. Kr isotopes were produced at ISOLDE from Nb foil target using hot plasma ion source equipped with water-cooled transfer line. Using high-energy β's the half-life for ^{70}Kr was determined to be 57(21) ms. This is lower than the value used in the recent rp-process flow calculation [8]: 390 ms. This discrepancy is due to Fermi decay contribution which was neglected in the QRPA-calculations [9] used in ref. [8]. At typical X-ray burst peak temperature T = 1.5 GK this reduces the effective half-life of the waiting point nucleus ^{68}Se from 35.5 s to 14.4 s [10]. Thus the possible rp-process flow above A = 70 during such an X-ray burst is accelerated producing possibly p-nuclei like ^{92}Mo and ^{96}Ru.

Mirror β-decay of ^{71}Kr

Easily identified mirror β transitions have been traditionally used for determining the Gamow-Teller quenching factor. These studies need accurate information for half-lives, decay energies and branching ratios. Observation of β-delayed γ's and protons in the decay of ^{71}Kr at ISOLDE allowed determination of $T_{1/2}$ and mirror transition branching ratio with accuracies of 2% and 3%, respectively [11]. Although the accuracy in the Q_{EC} value remained fairly poor, two goals were achieved: the discrepancy between the previous $T_{1/2}$ measurements [12,13] were solved and Gamow-Teller strength systematics of mirror transitions were extended to region of high deformation. This data combined with HF calculations have been later on used for determination of the ground-state spin of ^{71}Kr to be $3/2^-$ [14]. This would mean breaking in the mirror symmetry since the daughter ^{71}Br has been suggested to have $5/2^-$ for the ground state [?].

REFERENCES

1. E.Kugler et al., Nucl. Instr. and Meth. **B70**, 27 (1992).
2. B.Anderson et al., Phys. Rev. C **54**, 602 (1996).
3. W.Trinder et al., Nucl. Phys. **A620**, 191 (1997).
4. A.Jokinen et al., Eur. Phys. Journ. A **3**, 271 (1998).
5. J.Rapaport et al., Nucl. Phys. **A410**, 371 (1983).
6. Y.Fujita et al., Phys. Lett. B **365**, 29 (1996).
7. I.Reusen et al., Phys. Rev. C **59**, 2416 (1999).
8. H.Schatz et al., Phys. Rep. **294**, No.4 167 (1998).
9. P.Moller et al., At Data Nucl. Data Tables **66**, 131 (1997).
10. M.Oinonen et al., accepted for publication in Phys. Rev. C (1999).
11. M.Oinonen et al., Phys. Rev. C **56**, 745 (1997).
12. G.Ewan et al., Nucl. Phys. **A352**, 13 (1981).
13. B.Blank et al., Phys. Lett. B **364**, 8 (1995).
14. P.Urkedal and I.Hamamoto, Phys. Rev. C **58**, R1889 (1998).
15. J.Arrison et al., Phys. Lett. B **248**, 39 (1990).

Beta Decay Study of the N=Z nucleus ^{72}Kr

I. Piqueras[a], M.J.G. Borge[a], Ph. Dessagne[b], J. Giovinazzo[b],
C. Longour[b], Ch. Miehé[b], O. Tengblad[a,c] and the ISOLDE
Collaboration[c]

[a] Instituto de Estructura de la Materia, CSIC, Serrano 113 bis, E-28006 Madrid, Spain.
[b] IReS Strasbourg, Université Louis Pasteur, F-67037, Strasbourg Cedex 2, France.
[c] EP Division, CERN, CH-1211 Geneva 23, Switzerland.

Abstract. Beta decay of the N=Z even-even nucleus ^{72}Kr has been studied at the ISOLDE PSB facility at CERN. ^{72}Kr has been produced by fragmentation of a niobium target using the 1 GeV proton beam of the PS Booster. Delayed charged particles has been search for, X-ray, γ and γ-γ coincidences have been measured. Our results enrich the decay scheme in 15 unreported levels. Information on the Gamow-Teller strength distribution has been obtained for the region of the Q_{EC} window accesible to us.

INTRODUCTION

The mass region above A=70 is immersed in a zone of very surprising features: extreme deformation at low excitation energies, changes from strong prolate to large oblate deformation in neighbouring nuclei and in some cases shape coexistence. In this region of strong deformation, I. Hamamoto and coworkers [1,2] have made calculations for even-even N≈Z nuclei in the frame of Hartree-Fock formalism and the quasi-particle Tamm-Dancoff approximation. They found that the Gamow-Teller decay process, which is a powerful tool to investigate the nuclear structure in the isospin degree of freedom, may also assign unambiguously the nuclear deformation of the parent ground state.

Our collaboration decided to profit from this unique feature of the region besides the structural knowledge given by beta-decay and to investigate the β^+-EC decay of different proton rich nuclei along the N=Z line with A $\geq$ 70. Furthermore the decay of these nuclei is open to beta-delayed proton emission that allows for a sensitive mapping of the GT-strength at high excitation energies.

CP495, *Experimental Nuclear Physics in Europe*, edited by B. Rubio et al.
© 1999 American Institute of Physics 1-56396-907-6/99/$15.00

EXPERIMENTAL TECHNIQUE

The ^{72}Kr ion beam was produced at ISOLDE facility with average rate of 10^4 atoms/second. The experimental setup consisted of two independent but simultaneously operated measurement stations interconnected by a tape transport system. At the collection point two large volume HPGe detectors were used. Gamma data was gated by β^+-signals from a 4π plastic scintillator. The analysis of the $\gamma\gamma$ coincidences allowed to establish the low spin level scheme of ^{72}Br. At the measurement point intended mainly to study the β-delayed proton branch three different types of detectors were used: a gas-Si telescope sensitive to low energy protons, a X-ray detector and another large volume HPGe detector for coincidence purposes.

RESULTS AND CONCLUSIONS

The β^+-decay scheme has been established and a more precise value for the half-life, $T_{1/2} = 17.1(2)$s has been obtained. No protons were observed from this decay and an upper limit of 10^{-6} was set for the delayed proton branch. The experimental Gamow-Teller strength intensity is presented in Table 1 and compared with the previously mentioned theoretical calculations. From the results it is not possible to infer the sign of the deformation of this nucleus because the experimental results obtained for bound states are compatible with both, prolate and oblate deformations.

TABLE 1. Comparison of the total Gamow-Teller stregth $\sum$B(GT) obtained experimentally with theoretical calculations.

$\sum$B(GT)	prolate	oblate	experimental
Bound states	0.5	0.8	0.5(1)
Unbound states	0.7	1.4	$\lesssim 0.025$

In this work a high beta feeding from the 0^+ ^{72}Kr ground state to the ^{72}Br ground state has been determined. This feeding is not compatible with the 3^+ spin parity assignment given to the ^{72}Br ground state in in-beam studies [3]. Our work favours a 1^+ assignment for the spin and parity of the ^{72}Br ground state as it was done in earlier works [4].

REFERENCES

1. Hamamoto, I. and Zhang, X. Z., *Z. Phys.* **A353**, 145 (1995).
2. Frisk, F. et al., *Phys. Rev.* **C52**, 2468 (1995).
3. Collins, W. E. et al., *Phys. Rev.* **C9**, 1457 (1974).
4. Schmeing, H. et al., *Phys. Lett.* **44B**, 449 (1973).

Gamow-Teller[‡] and Fermi[♯] decay of N=Z nuclei above A=70

Ph. Dessagne[‡ ♯ a], D. Applebe[♯ b], L. Axelsson[♯ c], B. Blank[♯ d],
M.J.G. Borge[‡ e], A.M. Bruce[♯ f], W.N. Catford[♯ g], C. Chandler[♯ g],
R.M. Clark[♯ h], D.M. Cullen[♯ b], S. Czajkowski[♯ d], J.M. Daugas[♯ i],
A. Fleury[♯ d], L. Frankland[♯ f], J. Garcés Narro[♯ g], W. Gelletly[♯ g],
J. Giovinazzo[‡ ♯ d], B. Greenhalgh[♯ j], R. Grzywacz[♯ k], M. Harder[♯ f],
A. Huck[‡ a], A. Jokinen[‡ l], K.L. Jones[♯ g], N. Kelsall[♯ j], A. Knipper[‡ a],
T. Kszczot[♯ k], M. Lewitowicz[♯ i], C. Longour[‡ ♯ a], C. Miehé[‡ ♯ a],
R.D. Page[♯ b], C.J. Pearson[♯ g], I. Piqueras[‡ e], M. Ramdhane[‡ m],
V. Rauch[‡ a], A.T. Reed[♯ b], P.H. Regan[♯ g], O. Sorlin[♯ n],
O. Tengblad[‡ e], R. Wadsworth[♯ j]

[a] IReS Strasbourg, Université Louis Pasteur, F-67037 Strasbourg Cedex 2, France
[b] Oliver Lodge Laboratory, University of Liverpool, Liverpool, L69 3BX, UK
[c] Department of Physics, Chalmers University of Technology, S-412 96 Göteborg, Sweden
[d] CEN Bordeaux-Gradignan, Le Haut-Vigneau F-33175 Gradignan Cedex, France
[e] Instituto de Estructuro de la Materia, Serrano 113bis, E-28006 Madrid, Spain
[f] Cockcroft Building, University of Brighton, Brighton, BN2 4GJ, UK
[g] Department of Physics, University of Surrey, Guildford, GU2 5XH, UK
[h] Nuclear Science Division, LBNL, Berkeley, CA 94720, USA
[i] GANIL, BP 5027, F-14021, Caen Cedex, France
[j] Physics Department, University of York, Heslington, York, Y01 4DD
[k] Institut of Experimental Physics, Warsaw University, Pl-00681, Warsaw, Poland
[l] Department of Physics, Accelerator Laboratory, University of Jyväkylä, FIN-40351, Finland
[m] Institut of Physics, Department of Theoretical Physics, University of Constantine
25000 Constantine, Algeria
[n] Institut de Physique Nucléaire, 91406 Orsay, France

Abstract. The study of the β decay of some N=Z nuclei in the A=70 mass range have been performed using the GANIL and CERN/ISOLDE facilities. The fast decays of three T_z=0 odd-odd nuclei have been measured for the first time which extend the number of superallowed transitions. Information has been obtained on the Gamow Teller strength distribution in the β^+-EC decay of two even-even N=Z isotopes ^{76}Sr and ^{72}Kr. Results are compared with theoretical predictions giving indication about the shape of the ground state of the parent nucleus.

CP495, *Experimental Nuclear Physics in Europe*, edited by B. Rubio et al.
© 1999 American Institute of Physics 1-56396-907-6/99/$15.00

In the A=70 mass region along the N=Z line a lot of experimental [1], [2] and theoretical [3] investigations have been carried out to search for deformation signature and shape coexistence. In our work we have taken advantage of the β decay which is a major tool to study the nuclear structure and plays an important role to obtain a better understanding in astrophysical processes. Of special interest is the determination of the Gamow-Teller strength distribution which is expected to show a strong dependence on nuclear deformation, residual interaction and pairing [4], [5], [6]. In the $T_z = 0$ nuclei the interplay of neutron and proton is favoured and they are the best candidates to study the neutron-proton pairing which is maximum for odd-odd species where the competition between T=0 and T=1 pairing determines the ground state properties [7]. The study of the structure of these nuclei may provide also a test of the charge independence of nuclear forces in terms of isospin mixing for low lying states and this effect is important in the context of the V_{ud} matrix element evaluation. The beta decay processes are governed by simple operators and for Gamow Teller(GT) transition a good and complete description of the ground state parent nucleus and of the levels populated in the daugther isotope should provide a good value for the GT strength and its distribution over a wide energy range in the Q_{EC} window, after renormalization (see the contribution of A. Poves).

To investigate the mass region of interest and more particularly the odd-odd species, the projectile fragmentation reaction with associated in flight separation techniques have been used. We have observed the fast decays of the three isotopes ^{78}Y, ^{82}Nb and ^{86}Tc produced by fragmentation of a primary ^{92}Mo beam on a nickel target at the GANIL facility. These fragments were collected and separated with the LISE3 spectrometer [8]. At the final focus they were stopped in a three elements silicon telescope one of which was a 12 strips detector. The results presented here are part of our investigation from which the isotopes ^{77}Y, ^{79}Zr and ^{83}Mo were observed for the first time and the limit of stability was reached for nobium element (Z=41) [9], isomeric decays have been identified in ^{74}Kr [10], ^{80}Y and ^{84}Nb [11]. To determine the half-life of the three nuclei under study, only the β particles from the silicon strip where the isotope is implanted have been taken into account, in order to reduce the background and radioactivity build up influence. The obtained spectra and the halflives are presented in Figure 1. For ^{74}Rb the value is compatible with the more precise determination of D'Auria et al. [12] and provide a croos check of our results. For these fast decays, taking into account the Q_{EC} quantities from Audi and Wapstra [13] mass evaluation, and assuming a branching ratio of 100% and a ground state to ground state transition, the resulting $log\ ft$ values are compatible with 3.5. They are indicative of superallowed Fermi transitions as spin and parity of the ground state of the daugther nuclei are 0^+ and isospin is 1. A question arise which concern the nature of the ground state of the parent nucleus, it is a T=0 or a T=1 state? From the systematics and from the theoretical approach of A. Poves et al. [14] we can expect a T=1 configuration for the ground state of the three isotopes of interest which concern the $g_{9/2}$ shell. To be taken into account in the

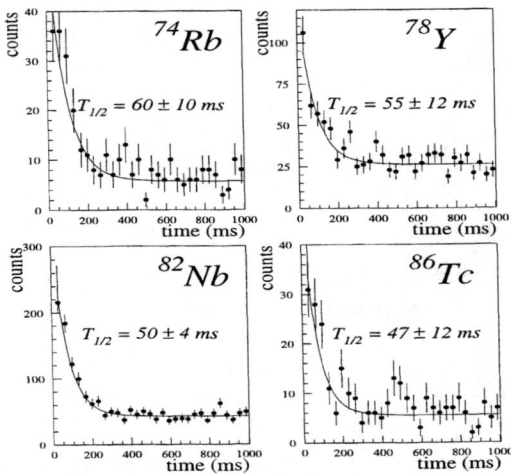

FIGURE 1. The time spectra for the odd-odd N = Z nuclei as measured in the present work.

context of the CVC test [15], accuracies to be reached for these three superallowed Fermi transitions, are 0.05% for $T_{1/2}$ and Q_{EC} (for example see [16], [17]).

In this mass region another field of investigation concern the nuclear ground state deformation of even-even T_z=0 nuclei. A pioneer theoretical work in this frame has been performed by I. Hamamoto et al. [5]. It reveals that a large part of the GTGR is accessible by $\beta^+ - EC$ decay and has shown for the nuclei near the N=Z line a strong dependence of the GT strength on nuclear shape. The interest of the experimental investigations is reinforced by the results of the extensive calculations presently performed by P. Sarriguren et al. [6]. In this spirit we have studied the $\beta^+ - EC$ decay of two even-even N=Z nuclei ^{72}Kr, ^{76}Sr at the CERN/ISOLDE facility. The results on ^{72}Kr are presented by I. Piqueras in this conference. For ^{76}Sr, the established decay scheme yielded the GT strength distribution reported in Figure 2. Above 4.5 MeV excitation energy in ^{76}Rb the strength comes from the delayed proton emission and the gap observed around 4 MeV is due to experimental limitation on γ detection as the β feeding is fragmented at high excitation energy, the deexcitation proceed through many different paths and the primary γ rays are of high energy. Therefore a great part of the GT feeding at high excitation energy is not observed. In spite of these limitations a comparison between theory [5] and experiment can be attempted (table 1). An agreement is found with prolate shape for bound states after renormalization and for unbound states after renormalization and an estimate of the missed γ decays (more than 95%) based on a statistical approach. Considering the importance of the theoretical investigations and predictions and taking into account the limitation on the experimental side it is of interest to obtain complete GT strength distribution in the A=80 mass region. In conclusion to study the properties of the N=Z nuclei we have used two complementary experimental techniques which are promising. Concerning the odd-odd isotopes we have observed for the first time the fast decay of three species which

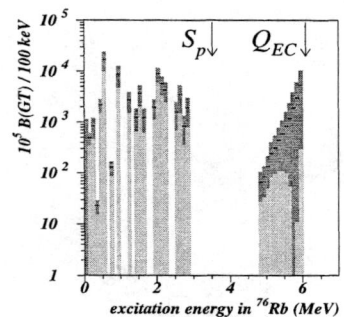

FIGURE 2. Experimental Gamow Teller strength distribution for ^{76}Sr.

TABLE 1. Experimental and theoretical Gamow–Teller strength for ^{76}Sr.

ΣB(GT)	experiment	theory	
		oblate	prolate
unbound states	0.07 - 0.22	0.4	4.5
bound states	0.6 - 0.8	2.9	1.4

extends towards higher masses the known superallowed transitions. For even-even nuclei, using the ISOL technique, indications have been obtained about the shape of the parent nucleus. However to overcome the experimental limitations we will in a further step use a Total Absorption gamma Spectrometer.

REFERENCES

1. Gelletly, W. et al., Phys. Lett. **B 253** (1991) 287.
2. Lievens, P. et al., CERN report, CERN-PPE/95-160.
3. Nazarewicz, W. et al. Nucl. Phys. **A435** (1985) 397.
4. Hamamoto, I. et al., Phys. Rev. **C 48** (1993) 2960.
5. Hamamoto, I. et al., Z. Phys. **A353** (1995) 145.
6. Sarriguren, P. et al., Nucl. Phys. **A 635** (1998) 55. Contribution to this conference.
7. Van Isacker, P. et al., Phys. Rev. Lett.**78** (1997) 3266.
8. Mueller, A.C. and Anne, R. Nucl. Ins. Methods Phys. Res., sect.**B56/57** (1991) 559.
9. Janas, Z. et al., Phys. Rev. Lett. **82** (1999) 295.
10. Chandler, C. et al., Phys. Rev. **C56** (1997) R2924,
11. Regan, P.H. et al., Acta Phys. Pol. **B28** (1997) 431.
12. D'Auria, J.M. et al., Phys. Lett. **B66**, 233 (1977)
13. Audi, G. and Wapstra, A.H., Nucl. Phys. **A595** (1995) 409.
14. Poves, A. and Martinez-Pinedo, G., Phys. Lett.**B430** (1998) 203.
15. Hardy, J.C. et al., Nucl. Phys. **A509**, 429 (1990)
16. Towner, I.S., Nucl. Phys. **A540**, 478 (1992)
17. Wilkinson, D.H., Nucl. Inst. Meth. Phys. Res. **A335**, 172 (1993)

Study of the β^+ decay of the nucleus ^{75}Rb

Beatriz Fuentes[1,8], A.J. Aas[2], M.J.G. Borge[3], C. Fernández[1],
B. Fogelberg[4], A. Giannatiempo[6], K. Gulda[7], H. Mach[4],
A. Nannini[6], B. Rubio[5], P. Sona[6], O. Tengblad[3,8]
and the ISOLDE Collaboration[8]

1) Dept. of Particle Physics, University of Santiago de Compostela, E-15706 Santiago, Spain.
2) Dept. of Chemistry, University of Oslo, P.O. Box 1033, Blindern, N-0315 Oslo, Norway
3) Instituto de Estructura de la Materia, CSIC, Serrano 113 bis, E-28006 Madrid, Spain.
4) Dept. of Neutron Research, Univ. of Uppsala, S-61182 Nyköping, Sweden.
5) Instituto de F. Corpuscular, CSIC-Univ. Valencia, E-46100 Burjassot, Spain.
6) Dipartimento di Fisica, Universita di Firenze, Italy.
7) Department of Physics, University of Warsaw, PL - 00 681, Poland
8) EP Division, CERN, CH-1211 Geneva 23, Switzerland.

Nuclei in the region A$\approx$70-80 in the proximity of N = Z show many interesting features. Both strong prolate and oblate deformations are possible and the same nucleus may exhibit different shapes at low excitation energies. One of the challenging cases is the proton rich nucleus ^{75}Kr situated in the so called shape-coexistence region. The closest isotopes ^{74,76}Kr have their low spin level structure dominated by shape coexistence and mixing between prolate and oblate states. This is mainly due to the gaps between deformed single states (Z, N = 36 favouring oblate and Z, N = 38 strongly pushing to prolate deformation).

Recent in-beam studies [1] have extended the collective bands of ^{75}Kr to spin 45/2 and found evidence for prolate-oblate coexistence. The spin of the ground state has been established by collinear laser spectroscopy [2] to be 5/2 and the positive parity was deduced from β-decay data.

In order to investigate complementary aspects brought up by β-decay studies, an experiment dedicated to study ^{75}Kr by the β^+ decay of ^{75}Rb was done at the mass separator ISOLDE (CERN). The experimental setup consisted of two independent but simultaneously operated measurement stations interconnected by a tape transport system. The system at the beam deposition point was designed for fast timing measurements by the $\beta\gamma\gamma$(t) method [3] and included four detectors: a β detector, two HPGe detectors and a BaF$_2$ crystal. They allowed $\gamma\gamma$ coincidence measurements and determination of half-lives of excited states in the ns and ps ranges. The second station was specially designed for internal conversion coefficient measurements.

Direct gamma spectra, β-gated gamma and internal conversion electron spectra were measured. The analysis of the $\gamma\gamma$ coincidence data did allow to identify the low spin states in ^{75}Kr up to 3254 keV in the excitation energy (Q$_{EC}$ = 7020(17) keV).

CP495, Experimental Nuclear Physics in Europe, edited by B. Rubio et al.
© 1999 American Institute of Physics 1-56396-907-6/99/$15.00

The results of the present work are displayed in figure 1. Only γ-lines from high energy states ($E^* > 1200$ keV) are shown due to lack of space. This work agrees with previous β-decay studies [4] enriching the level scheme by 25 new levels and 65 new transitions. Our study, characterized by a higher sensitivity than the previous one, does not confirm the presence of the 848.2 and 1026.4 keV transitions, previously proposed to de-excite the 1026.4 keV level(despite poor energy matching).

Considering the observe gamma intensity as complete, we have deduced the logft values. Our results are consistent with the previous observation [4] of strong feeding to the negative parity states at 178.95 and 358.05 keV. Equally strong feeding is observed for levels around 1.85 MeV, 2.4 MeV and 2.5 MeV, respectively, therefore indicating their negative parity character.

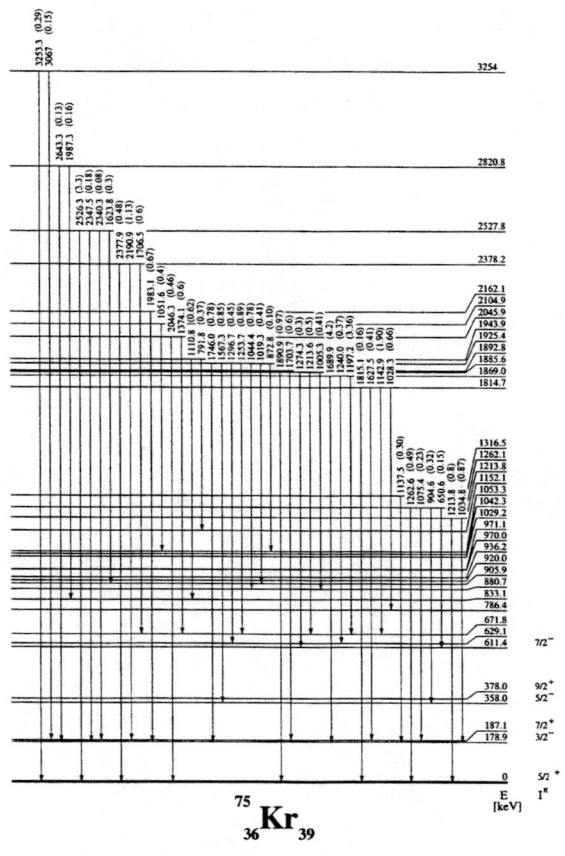

FIGURE 1. Proposed partial decay scheme

In summary, we present a highly enriched level scheme for ^{75}Kr obtained from β-decay studies. Our results are consistent with those obtained for low-spin states in recent in-beam measurements [1].

REFERENCES

1. Skoda, S. et al., *Nucl. Phys.* **A 633** (1998) 565.
2. Keim, M. et al., *Nucl. Phys.* **A 586** (1995) 219.
3. Mach, H., Gill, R.L. and Moszyński M., *Nucl. Instr. Meth.* **A280** (1989) 49-72
4. Kern, B.D. et al., *Phys. Rev.* **C 28** (1983) 2168.

A deformed approach to β^+ decay of proton rich nuclei

P. Sarriguren, E. Moya de Guerra and A. Escuderos

Instituto de Estructura de la Materia, CSIC,
Serrano 123, 28006 Madrid, Spain

Abstract. QRPA calculations are performed in a deformed Hartree-Fock+BCS single-particle basis to study ground state properties and beta decay rates. Consistency of the method is achieved by deriving the mean field and residual interaction from the same effective Skyrme force. We apply this scheme to various isotope chains approaching the $N = Z$ line in the $A \simeq 70$ mass region, where deformation is known to play an important role. We get good agreement with experimental half-lives and predict the behavior of the energy distributions of the Gamow Teller strength.

INTRODUCTION

Reliable theoretical estimates of $\beta-$decay strength distributions are an essential input in nuclear astrophysics to understand better nucleosynthesis and to model the late phases of the stellar life [1]. A consistent and systematic approach to study $\beta-$decay is needed to extrapolate with some guarantees to regions where none or very few experimental data are available. Deformation is also an important ingredient to describe those regions of the nuclear chart characterized by the existence of stable deformed shapes. Following these guidelines we have studied ground state and $\beta-$decay properties of various proton rich chains of even-even isotopes (Ge, Se, Kr and Sr) approaching $N = Z$ in the mass region $A \simeq 70$, where competing nuclear shapes are expected. This is an interesting region where the Gamow Teller strength have been studied experimental [2–4] and theoretically [5–7].

We use a deformed Hartree-Fock approximation with density-dependent Skyrme forces and include pairing correlations between like nucleons in BCS approximation. A separable residual spin-isospin interaction consistent with the mean field is added and treated in QRPA. The residual force is obtained from the exact particle-hole residual interaction corresponding to the Skyrme force after averaging over the nuclear volume. It should be noted that no free parameters enter in the calculation, since both the mean field and the residual interaction are obtained from the Skyrme interaction. Details of this method can be found in Ref. [6,7].

CP495, *Experimental Nuclear Physics in Europe*, edited by B. Rubio et al.
© 1999 American Institute of Physics 1-56396-907-6/99/$15.00

RESULTS

By performing a constrained Hartree-Fock calculation, we investigate first shape isomerism in the mass region above mentioned and find clear signatures of shape coexistence in most of the isotopes studied [7]. We also find [7] good agreement with available experimental data on ground state properties (charge radii, quadrupole moments, Q_{EC} values, etc.).

We calculate the energy distributions of the Gamow-Teller strength, the half-lives, and the strength contained in the experimental Q_{EC} window, studying their dependence on the residual interaction, pairing and deformation. We perform calculations for all the shapes for which we obtain HF minima in the constrained HF calculation.

We find that, compared to the uncorrelated bare two-quasiparticle distribution, RPA shifts the Gamow-Teller strength to higher excitation energies and reduces the total strength increasing the half-lives. Its effects should therefore not be neglected. Pairing correlations are also important because they allow the existence of transitions forbidden otherwise. Finally, the effect of deformation is twofold. One effect of deformation is that, compared to the spherical case, the strength distributions become much more fragmented. Other effect is that, depending on the nuclear shape, the sequence of states in a Nilsson-like diagram can be very different, giving rise to easily distinguishable profiles of the Gamow-Teller strength distributions. This fact can obviously be exploited to find signatures of the nuclear shape. Actually we have identified the most interesting cases to explore experimentally in a search for these deformation effects on the Gamow-Teller strength distributions. It is also important to notice that the results do not depend much on which effective Skyrme interaction is used provided that the minima of the HF energies occur at similar deformations.

We can see in Fig. 1 the energy distributions of the Gamow Teller strength for all the isotopes considered in this work. The results correspond to the Skyrme force SG2 [8] in RPA and for the various shapes that minimize the HF energy. We can see from Fig. 1 that the profiles of the strength distributions corresponding to different shapes are very similar in the Ge and Se isotopes and therefore these isotopes are not good candidates to look for deformation effects in their GT strength distributions. On the other hand we can see that in most Kr and Sr isotopes the differences are very pronounced and this feature can be used to infer the shape of the parent nucleus from their β-decay properties. Obviously the information from β-decay is limited to the Q_{EC} windows, but this would be enough in most cases to distinguish between one shape or another.

Another quantity of interest that can be used in β-decay to differenciate the shapes is the total Gamow Teller strength contained in the Q_{EC} window. It turns out [7] that there are cases where one finds a strong correlation between the nuclear shape and the summed GT strength in the experimentally accessible window.

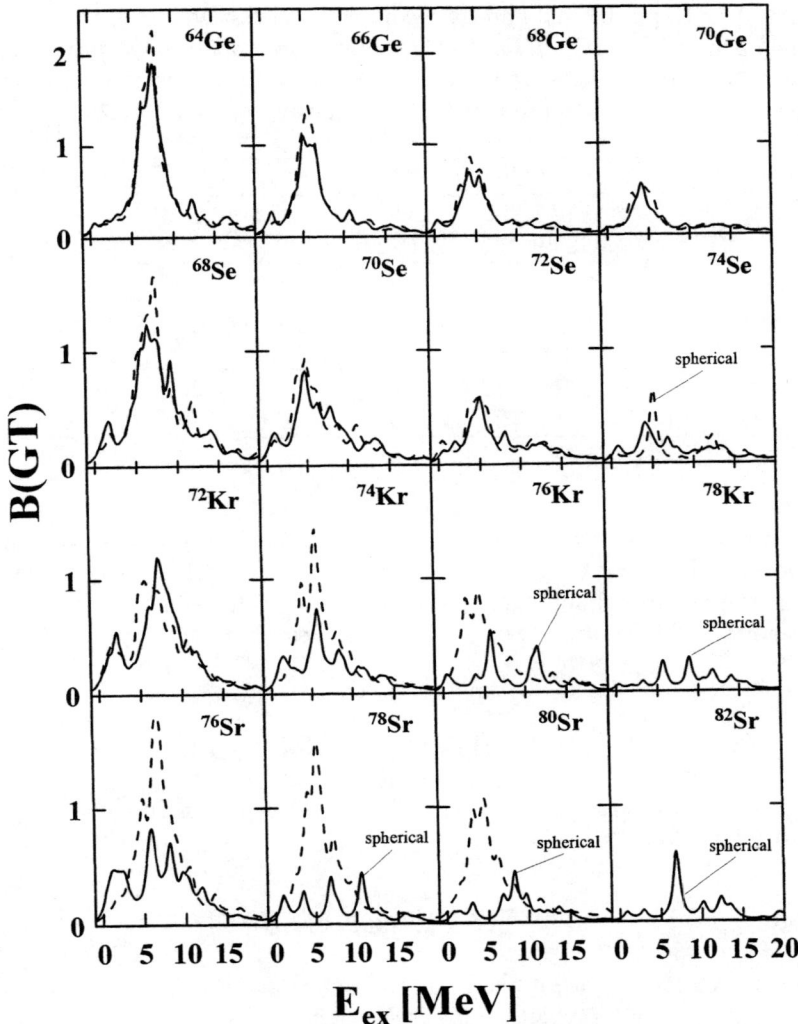

FIGURE 1. Gamow Teller strength distributions $[g_A^2/4\pi]$ as a function of the excitation energy of the daughter nucleus. The results are for the force SG2 in RPA. Solid lines correspond to oblate shapes and dashed lines to prolate shapes unless otherwise specified.

TABLE 1. Experimental half-lives compared to the theoretical values obtained with the Skyrme force SG2 in RPA. The results correspond to the prolate solutions. Also shown within brackets are the results obtained for the oblate solutions, except for ^{76}Kr and ^{78,80}Sr, which are for spherical shapes.

	^{64}Ge	^{66}Ge	^{68}Se	^{70}Se
exp.	63.7 s	2.26 h	35.5 s	41.1 m
theo.	259 (176)	4.1 (3.6)	128 (187)	44 (90)

	^{72}Kr	^{74}Kr	^{76}Kr	^{76}Sr	^{78}Sr	^{80}Sr
exp.	17.2 s	11.5 m	14.8 h	8.9 s	2.7 m	1.8 h
theo.	13 (15)	14 (10)	3.1 (6.4)	35 (11)	5.1 (3.3)	1.5 (6.8)

We can see in Table 1 the half-lives of the unstable isotopes studied in this work. The theoretical values calculated in QRPA with the force SG2 contain the standard quenching factor (0.77) and agree reasonably well with the experimental ones with the exception of ^{64}Ge and ^{68}Se, where we overestimate the half-lives by a factor 3-5.

ACKNOWLEDGMENTS

This work was supported by DGICYT (Spain) under contract number PB95/0123.

REFERENCES

1. NuPECC report on *Nuclear and Particle Astrophysics*, F.K. Thielemann et al. (1997).
2. G. de Angelis et al., *Proposal to the Isolde Committe*, CERN/ISC 98-20 (1998).
3. Ph. Dessagne et al., contribution to this Conference.
4. J.L. Tain, contribution to this Conference.
5. I. Hamamoto and X.Z. Zhang, *Z. Phys.* **A353**, 145 (1995); F. Frisk, I. Hamamoto, and X.Z. Zhang, *Phys. Rev.* **C52**, 2468 (1995).
6. P. Sarriguren, E. Moya de Guerra, A. Escuderos and A.C. Carrizo, *Nucl. Phys.* **A635**, 55 (1998).
7. P. Sarriguren, E. Moya de Guerra and A. Escuderos, *Nucl. Phys.* **A**, (1999) in press.
8. N. Van Giai and H. Sagawa, *Phys. Lett.* **B106**, 379 (1981).

Nuclear Explorations Beyond the Proton Drip-Line

Philip J. Woods

Department of Physics and Astronomy, Edinburgh University, EH9 3JZ UK

Abstract. This paper reviews the recent developments in the study of proton-radioactivity. A large and wide ranging data base now exists to study this phenomenon. Theoretical models of proton emission from spherical nuclei can predict the detailed trends of proton decay spectroscopic factors very well. The first examples of proton decay from highly deformed nuclei have been discovered and are well reproduced by a theoretical approach based on Nilsson states. Such measurements provide an insight into the fragmentation of single particle strength in deformed nuclei. In the case of ^{131}Eu proton decay fine structure has been identified for the first time, thereby providing direct information on the degree of deformation. The technique of Recoil Decay Tagging and its particular application to the study of the structure of deformed proton radioactive nuclei is discussed.

INTRODUCTION

The proton drip-line represents the dividing line between isotopes that are either bound or unbound to the emission of a proton from their ground-states. For light elements, nuclei lying beyond the proton drip-line only exist in the form of resonances, ^{39}Sc being the heaviest system to be studied to date [1]. The Coulomb barrier experienced by an unbound proton increases with element number, Z .Progressing beyond Z=50 it becomes more probable than not that odd Z nuclei have at least one proton-radioactive isotope. This is due to a combination of the large height of the Coulomb barrier and the decrease in the rate of change of proton decay Q-value with neutron number $\triangle Q_p/\triangle N$ which varies with an approximate inverse dependence on the mass number [2]. For proton radioactivity to be observed experimentally it must have a significant decay branch which means that in practise the drip-line must be crossed by several isotopes. For example, ^{171}Ir is the heaviest proton unbound Ir isotope but proton emission is first observed for the isotope ^{167}Ir [3]. A continuous chain of odd-Z proton emitting elements has been identified from Z=67-83 [4], these results along with other examples now constitute a large and wide ranging data base of proton transitions from which to explore the phenomenon of proton radioactivity.

CP495, *Experimental Nuclear Physics in Europe*, edited by B. Rubio et al.
© 1999 American Institute of Physics 1-56396-907-6/99/$15.00

SPHERICAL PROTON EMITTERS

In its simplest form the proton decay transition probability can be calculated using a semi-classical WKB approach [5]. A theoretical review of proton emission from spherical nuclei by Aberg et al. [6] has shown that such calculations agree well with more exact DWBA and two-potential treatments when the same proton potential parameter set is used. The choice of a realistic potential introduces an uncertainty ~ 2 into the calculation [7]. Proton decay rates are extremely sensitive to the orbital angular momentum of the unbound proton and shell model assignments can be confidently made despite the uncertainty in the choice of potential.

A very attractive aspect of proton decay is that the proton can be considered to be preformed inside the nucleus thereby avoiding some of the uncertainty associated with alpha-decay transition rate calculations. Nonetheless, in order to calculate proton decay rates correctly a spectroscopic factor must be introduced . Davids et al. [3] used a simple low seniority shell model calculation to predict proton radioactivity spectroscopic factors for spherical nuclei with $64<Z<82$. In general the spectroscopic factors are reproduced remarkably well although the experimental values for the $d_{3/2}$ transitions are consistently lower than the model predictions possibly indicating the influence of significant mixing of other configurations. Subsequent calculations of spectroscopic factors using a more realistic BCS approach [6] are also able to reproduce the trends of spectroscopic factors well although the discrepancies for $d_{3/2}$ transitions remain

DEFORMED PROTON EMITTERS

It has been known for some time that the decay rates of the proton emitters ^{109}I and ^{113}Cs cannot be reproduced using calculations of the type described above [8]. One possible explanation for this behaviour was the onset of modest prolate deformations in the region of the proton drip-line above Z=50. Bugrov and Kadmensky [9] developed a model for proton emission from deformed nuclei using Nilsson wavefunctions with quadrupole deformation treated as a free parameter. They were able to reproduce the anomalous half-lives of ^{109}I and ^{113}Cs using relatively modest deformations $\beta \sim 0.1$ consistent with values expected for this transitional region. The macroscopic-microscopic mass model of Moller et al. [10] predicted the onset of much higher prolate deformations ($\beta \sim 0.3$) immediately below Z=69 along the region of the proton drip-line. Experiments using the Fragment Mass Analyzer (FMA) [11] at Argonne have identified ground-state proton radioactivity from ^{141}Ho (Z=67) lying just inside this region, and ^{131}Eu (Z= 63) [12] lying at the heart of the region, with predicted quadrupole deformation parameters of 0.29 and 0.33 [10], respectively. Spherical proton decay calculations signally fail to reproduce the half-lives. Davids et al. [12] demonstrated that the decay rates could be well reproduced using the calculational approach of Bugrov and Kadmensky obtaining Nilsson configurations and quadrupole deformations consistent with the

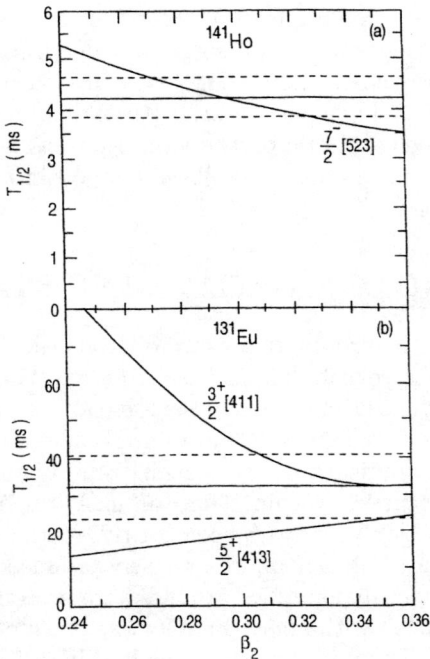

Fig 1. Deformed DWBA calculations for Nilsson states of ^{141}Ho and ^{131}Eu compared to measured proton decay half-lives.

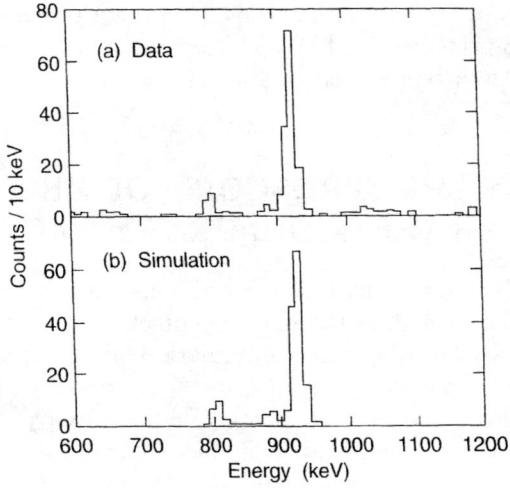

Fig 2. Proton decay energy spectrum showing fine structure in ^{131}Eu compared with a Monte Carlo simulation.

predictions of Moller et al. [10] (see Figure 1). This successful extension of the theory to highly deformed nuclei represents a significant deepening of our understanding of the proton decay phenomenon. These transitions provide a unique insight into the fragmentation of single particle strength in deformed nuclei. In the case of [141]Ho a weakly produced isomeric proton transition has been identified in experiments at Oak Ridge which can also be well understood in terms of a highly deformed Nilsson configuration [13].

DISCOVERY OF PROTON DECAY FINE STRUCTURE

Following the identification of proton radioactivity from the highly deformed nucleus [131]Eu it was decided to revisit this nucleus in an experiment at Argonne in order to search for the previously unobserved phenomenon of proton decay fine structure on the basis that the first excited 2^+ level in the daughter nucleus [130]Sm should be low enough for a significant decay branch [14]. Figure 2a shows the energy spectrum for decays occurring within 100ms of an A = 131 ion implanting into the same quasi-pixel of a Double-sided Silicon Strip Detector situated behind the focal plane of the Argonne FMA. The more intensely produced peak at higher energy corresponds to the previously identified ground-state proton transition from [131]Eu. The second peak produced with approximately one tenth of the intensity is at an energy $\sim$120 keV lower. The peak has the same half-life within errors as the previously identified transition and is assigned to the proton decay fine structure of [131]Eu. Using the Grodzin's formula [15,16] this implies a value of $\beta \sim 0.34$ for the daughter nucleus [130]Sm in excellent agreement with the value of 0.33 predicted by Moller et al. [10]. This also provides a consistency check on the high deformation necessary to reproduce the partial half-life of the main ground-state proton transition from [131]Eu using the deformed DWBA calculational approach [14]. The proton branching ratio is well reproduced by the calculation for a $3/2^+[411]$ configuration (see Figure 3).

GAMMA-RAY SPECTROSCOPY OF PROTON EMITTERS USING RECOIL DECAY TAGGING

Although theories of spherical and deformed proton emitters are now being tested over a wide range of nuclei, including the new phenomenon of proton decay fine structure, it is desirable to have independent information on the structure of these nuclei. In particular high resolution in-beam gamma-ray studies can provide insights into the nuclear deformation. The technique of Recoil Decay Tagging (RDT) [17,18] is an ideal tool that was developed with this particular goal in mind and is now used extensively in the study of neutron-deficient and heavy nuclei exhibiting charged particle decay modes. The RDT technique was successfully applied to ground and isomeric proton emission from [147]Tm in an experiment on the Argonne FMA [19]. The gamma-ray band built on the ground-state is consistent

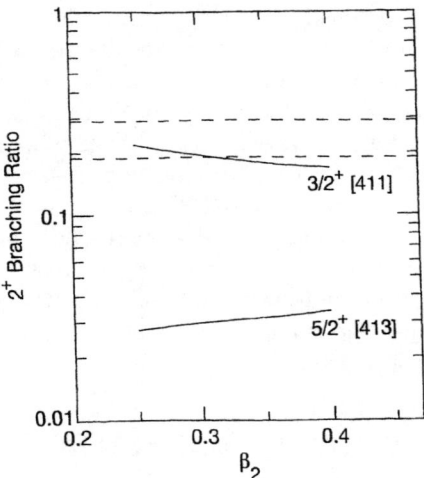

Fig 3. A comparison of the proton branching ratio with deformed DWBA calculations for possible Nilsson configurations in ^{131}Eu.

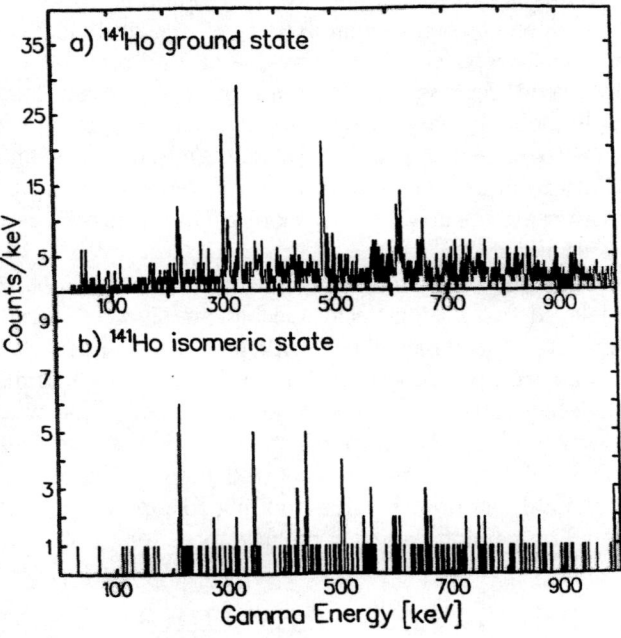

Fig 4. Gamma ray spectra obtained using the RDT technique. Rotational bands are observed populating the ground and isomeric proton decaying states in ^{141}Ho.

with $\beta=0.13$. Interestingly, Kadmensky and Bugrov have applied their model for deformed proton emission to this case [20] and the results do not disagree significantly from spherical calculations, unlike the effect of such a deformation on proton decay rates in the region above $Z = 50$. It appears that Tm and Ho proton emitters lie right at the interface of the region of rapid shape change to high prolate deformations. Clearly it is desirable to identify in-beam gamma-rays from the highly deformed proton emitters. In a very recent RDT experiment using the Argonne FMA coupled to Gammasphere, gamma-rays were successfully identified from the ground and isomeric states in ^{141}Ho, the latter having a cross-section~50nb [21]. These data are shown in Figure 4 and clearly demonstrate the existence of rotational bands built on the ground and isomeric states providing further evidence for the rapid increase in deformation below $Z = 69$.

TWO PROTON EMISSION

Two proton radioactivity has yet to be observed, it is expected to occur in highly neutron - deficient even Z nuclei when the proton pairing energy suppresses single proton emission. The most promising candidate isotope discovered to date is ^{45}Fe identified using high energy heavy ion fragmentation on the FRS at GSI by Blank et al. [22], with the yet to be discovered isotope ^{48}Ni being another possible candidate if it indeed exists. As with one proton radioactivity these relatively low Z nuclei will have decay rates very sensitive to the decay Q-value and an alternative approach would be too look in heavier regions using fusion-evaporation reactions, although present techniques will probably have to improve in sensitivity by approximately two orders of magnitude to make such searches feasible. One should also not rule out two proton radioactivity from an isomeric state in a nucleus not lying beyond the proton drip-line, this after all was how the first example of one proton radioactivity was serendipitously discovered [23].

The two proton decay mechanism can be studied from resonant states. Pioneering studies of the beta-delayed two proton decay mechanism showed a predominance of sequential emission [24]. More recent experiments using radioactive beams to populate two proton-unbound resonances [25,26] have also yielded similar results. In order to suppress the sequential process an experiment at MSU has used inelastic scattering of a radioactive ^{17}Ne beam to populate an excited state that is bound to one proton emission. The initial results reported here at this conference [27] indicate the existence of a decay branch corresponding to simultaneous two proton emission whose large width cannot presently be accounted for by theory.

CONCLUDING REMARKS

The recent years have produced an explosion of information on the phenomenon of proton radioactivity. Theoretical models are able to reproduce in detail the systematic variation of proton decay spectroscopic factors for a wide range of spherical nuclei thereby sensitively testing the nuclear shell model at the extreme edge of stability. The first examples of proton emission from highly deformed nuclei have been discovered. These transitions have provided a direct insight into the fragmentation of single particle strength within highly deformed nuclei. Decay rates from these nuclei are found to agree well with theoretical calculations assuming Nilsson states. Proton decay fine structure is reported here for the first time providing independent confirmation of the high deformations involved. In-beam studies of the gamma-rays using the RDT technique are providing complementary nuclear structure information on proton-radioactive nuclei that will assist in constraining theoretical calculations of proton decay rates. Furthermore, such studies will provide new insights into the behaviour of proton unbound nuclei at high spin and excitation energy. In summary there has been a great advance in our detailed understanding of the phenomenon of proton radioactivity. This has derived from the large increase in known transitions and the varied nuclear landscape in which this process is now found.

ACKNOWLEDGEMENTS

I would like to thank all my colleagues involved in experiments reviewed here. In particular I would like to acknowledge Cary Davids, Darek Seweryniak and Bill Walters (thanks for the 2 pages Bill).

REFERENCES

1. M.F. Mohar et al., *Phys. Rev.* **C38**, 747 (1988).
2. P.J. Woods, Proceedings of the 4th International School of Heavy Ion Physics, Erice, 315 (1997).
3. C. N. Davids, P. J. Woods, J. C. Batchelder, C. R. Bingham, D. J. Blumenthal, L. T. Brown, B. C. Busse, L. F. Conticchio, T. Davinson, S. J. Freeman, D. J. Henderson, R. J. Irvine, R. D. Page, H. T. Pentilla, D. Seweryniak, K. S. Toth, W. B. Walters and B. E. Zimmerman, *Phys. Rev.* **C55**, 2255 (1997).
4. P.J. Woods and C.N. Davids, *Ann. Rev. Nucl. Part. Sci.* **47**, 541 (1997).
5. S. Hofmann, W. Reisdorf, G. Munzenberg, F.P. Hessberger, J.R.H. Schneider, P. Armbruster, *Z. Phys.* **A305**, 111 (1982).
6. S. Aberg, P.B. Semmes, W. Nazarewicz, *Phys. Rev.* **C56**, 1762 (1997).
7. P.J. Sellin, P.J. Woods, T. Davinson, N.J. Davies, A.N. James, K. Livingston, R.D. Page, A.C. Shotter, *Phys. Rev.* **C47**, 1933 (1993).
8. A. Gillitzer, T. Faestermann, K. Hartel, P. Kienle and E. Nolte, *Z. Phys.* **A326**, 107 (1987).
9. V.P. Bugrov and S.G. Kadmensky, *Sov. J. Nucl. Phys.* **49**, 967 (1989).
10. P. Moller, J.R. Nix, K.-L. Kratz, *At. Data Nucl. Data Tables* **66**, 131 (1997)
11. C. N. Davids et al., *Nucl. Instrum. Methods* **B70**, 358 (1992).

12. C.N. Davids, P.J. Woods, D. Seweryniak, A.A. Sonzogni, J.C. Batchelder, C.R. Bingham, T. Davinson, D.J. Henderson, R.J. Irvine, G.L. Poli, J.Uusitalo, W.B. Walters, *Phys. Rev. Lett.* **80**, 1849 (1998).
13. J.C. Batchelder et al., Proceedings of this Conference.
14. C.N. Davids. P.J. Woods, D. Seweryniak, A.A. Sonzogni, M. Carpenter, J. Ressler, J. Schwarz, J. Uusitalo and W.B. Walters, paper in preparation.
15. L. Grodzins, Phys. Lett. 2, 88 (1962).
16. F. Stephens et al., Phys. Rev. Lett. 29, 438 (1972).
17. E.S. Paul, P. J. Woods, T. Davinson, R. D. Page, P. J. Sellin, C. W. Beausang, R. M. Clark, R. A. Cunningham, S. A. Forbes, D. B. Fossan, A. Gizon, J. Gizon, K. Hauschild, I. M. Hibbert, A. N. James, D. R. LaFosse, I. Lazarus, H. Schnare, J. Simpson, R. Wadsorth and M. D. Waring, *Phys. Rev.* **C51**, 78 (1995).
18. R. S. Simon, K-H. Schmidt, F. P. Hessberger, S. Hlavae, M. Honusek, G. Munzenberg, H. G. Clerc, U. Gollerthan, W. Schwab, *Z. Phys.* **A325**, 197 (1986).
19. D. Seweryniak, C. N.Davids, W. B. Walters, P. J. Woods, I. Ahmad, H. Amro, D. J. Blumental, L.T. Brown, M. P. Carpenter, T. Davinson, S.M. Fischer, D.J. Henderson, R.V.F. Janssens, T. L. Khoo, I. Hibbert, R.J. Irvine, R. J. Irvine, C.J. Lister, J. A. Mckenzie, D. Nisius, C. Parry, and R. Wadsworth *Phys. Rev.* **C55**, R2137 (1997).
20. S.G. Kadmensky and V.P. Bugrov, *Phys. At. Nucl.* **59**, 399 (1996).
21. D. Seweryniak et al., paper in preparation.
22. B. Blank et al., *Phys. Rev. Lett.* **77**, 2893 (1996).
23. K.P. Jackson et al., *Phys. Lett.* **B33**, 281 (1970).
24. R. Jahn et al., *Phys. Rev.* **C31**, 1576 (1985).
25. R. A. Kryger, A. Azhari, M. Hellstrom, J. H. Kelley, T. Kubo, R. P. Pfaff, E. Ramakrishnan, B. M. Sherril, M. Thoenesson, S. Yokoyama, R. J. Charity, J. Dempsey, A. Kirov, N. Robertson, D. G. Sarantites, L. G. Sobotka, J. A. Winger, *Phys. Rev. Lett.* **74**, 861 (1995).
26. C. R. Bain, P. J. Woods, R. Coszach, T. Davinson, P. Decrock, M. Gaelens, W. Galster, M. Huyse, R. J. Irvine, P. Leleux, M. Loiselet, C. Michotte, R. Neal, A. Ninane, G. Ryckewaert, A. C. Shotter, G. Vancraeynest, J. Vervier and J. Wauters, *Phys. Lett.* **B373**, 35 (1996).
27. M. Chromik, P.G. Thirolf, M Thoenesson, M. Fauerbach, T. Glasmacher, R. Ibbotson, R.A. Kryger, H. Scheit, P.J. Woods, Proceedings of the 2nd International Conference on Exotic Nuclei and Atomic Masses, Shanty Creek 286 (1998).

β-Strength Measurements in Nuclei

J.L. Tain, D. Cano-Ott, B. Rubio

Instituto de Física Corpuscular, Centro Mixto C.S.I.C.-Univ. Valencia, 46071 Valencia, Spain

Abstract. A knowledge of the β-strength distribution in nuclei is both of fundamental character and great importance in related fields. Despite intrinsic difficulties, the measurement of this quantity from the subsequent electro-magnetic de-excitation is in most cases the only possible method. In this case and for nuclei far from stability, total absorption γ-ray spectroscopy appears to be the only reliable technique.

RELEVANCE AND MEASUREMENT OF β-STRENGTH DISTRIBUTIONS

β-decay is the process that governs the transmutation of most of the known nuclear species. Knowledge of the probability of occurrence of the basic process mediated by the weak interaction is clearly of fundamental importance. A knowledge of the distribution of this probability over the final nuclear states gives information on the complex nuclear wave functions involved. At the same time, this knowledge is the key to understanding other phenomena, both in nuclear physics and in other fields such as particle physics, astrophysics and reactor technology. These issues were stressed in review articles by Klapdor [1] in the mid 80s. The persistence of the validity of such arguments and the continuously renewed interest in measurements aimed at determining these quantities can be shown in two recent examples. The ICARUS liquid argon detector [2] is designed to study the so called solar neutrino problem [3] by counting the flux of high energy neutrinos coming from the Sun as well as being sensitive to neutrino oscillations. The calibration of such a detector requires a knowledge of the β-decay probabilities of ^{40}Ti into the excited states of ^{40}Sc (the mirror process of the reaction $\nu_e + ^{40}Ar \rightarrow e^- + ^{40}K$) which could only be measured recently with the necessary accuracy (see [4] and references therein). The second example comes from the field of reactor technology. Despite improvements down the years there remains a sizeable discrepancy between the measurements of the decay heat (energy released by fission products after reactor shutdown) due to γ-ray emission and the summation calculations based on data libraries combining theoretical and experimental quantities. In a recent careful analysis [5] it has been pointed out that most of the discrepancy is probably due to inaccurate measurements of the β-decay probability distribution of a few isotopes. Understanding the

CP495, *Experimental Nuclear Physics in Europe*, edited by B. Rubio et al.
© 1999 American Institute of Physics 1-56396-907-6/99/$15.00

discrepancy would have an impact on safety regulations and reactor modelling.

The quantity which is usually determined is the β-strength S_β as a function of the final level excitation energy E_x. It is proportional to the average over an energy interval ΔE_x of the β-decay reduced transition probabilities $B_{i \to f}$ to the levels in that interval [6]:

$$S_\beta(E_x) = \frac{1}{D} \frac{1}{\Delta E_x} \sum_f B_{i \to f} \tag{1}$$

Here D is a constant related to the vector coupling constant g_V ($D = 2 \ln 2 \pi^3 \hbar^7 / m_e^5 c^4 g_V^2$) and $B_{i \to f}$ is given in units of $g_V^2/4\pi$.

The β-strength can be experimentally determined from the β intensity or feeding probability (per unit energy interval) I_β:

$$S_\beta(E_x) = \frac{I_\beta(E_x)}{f(Q_\beta - E_x)T_{1/2}} \tag{2}$$

where f stands for the statistical rate Fermi function which depends on the energy $Q_\beta - E_x$ available to the decay, and $T_{1/2}$ is the β-decay half-life.

The methods of measuring S_β can be classified as direct or indirect. The latter do not measure the β-decay directly but make use of other processes, governed by transition operators which are in close relation to that governing β-decay. The most popular methods by far have been the study of charge exchange reactions with light ions: (p, n), $(^3\text{He},t)$ and their inverse reactions. They have the advantage over the direct methods of not being restricted by an energy window (Q_β), and they allow the study of transitions with orbital angular momentum transfer greater than $\Delta l = 0$. On the other hand they are limited to giving information close to the β-stability line. The technique itself suffers from systematic uncertainties coming from the reaction mechanism and the background subtraction [7].

The direct methods are based on the detection of the particles (mostly p or n) or electro-magnetic radiation emitted from the states populated in the β-decay process. The former happens for nuclei far enough from stability and for states high enough in excitation energy. When particle emission is possible, its detection is facilitated by the low multiplicity (few states of the final nucleus are directly populated). In contrast, many different γ-ray cascades can develop from the highly excited states. This, together with the limited efficiency of germanium γ-ray detectors, leads to an increasing underestimation of the feeding probability with excitation energy. This is the well known *Pandemonium* problem [8] which affects much of the measured data, and is quite unfortunate since most of the information on β-strength distribution can only be gained in this way.

There is an alternative to germanium high resolution spectroscopy to obtain information from electro-magnetic de-excitation. It is the so called total absorption γ-ray spectroscopy technique first applied to β-decay studies in the early 70s [6]. The principle of the measurement is to use a large 4π scintillation detector. For an ideal spectrometer of efficiency 1, every cascade will be registered as a count at the cascade (or level) energy and the recorded spectrum will be the distribution sought. For real spectrometers the response of the apparatus has to be taken into account and the complication introduced in the analysis has cast some doubts on the reliability of the results and hindered wider dissemination of the technique.

We have carried out a thorough investigation of several aspects of the technique, part of which has already been published [9,10]. In this contribution we report on the reliability of the analysis.

THE TOTAL ABSORPTION γ-RAY SPECTROSCOPY TECHNIQUE

In order to illustrate our results we present the analysis of the decay of the ^{150}Ho 2^- isomer [11], measured with the Total Absorption Spectrometer (TAS) installed at the GSI On-line Mass-separator [12]. This spectrometer has been used in a series of experiments to investigate the problem of the missing Gamow-Teller strength in the rare earth region as well as on nuclei in the neighbourhood of ^{100}Sn.

For the purpose of the analysis the relation between the measured spectrum and level feeding distribution can be represented by

$$d_i = \sum_{j=0}^{j_{max}} R_{ij} f_j , \quad i = 1, i_{\max} \quad \text{or} \quad \mathbf{d} = \mathbf{R} \cdot \mathbf{f} \tag{3}$$

where d_i represents the content of channel spectrum i, f_j represents the β-decay feeding to the level labelled j, and the column labelled j of the matrix $\mathbf{R}$, which will be designated by $\mathbf{R}_j$, represents the response of the spectrometer to the decay into level j, which can be constructed by recursive convolution:

$$\mathbf{R}_j = \sum_{k=0}^{j-1} b_{jk} \mathbf{g}_{jk} \otimes \mathbf{R}_k \tag{4}$$

b_{jk} represents the branching ratio for the transition from level j to level k and $\mathbf{g}_{jk}$ the response to the emitted γ-ray.

We have shown in Ref. [9] that an accurate γ-ray response can be obtained from Monte Carlo simulations. The problem of the analysis is thus reduced to obtaining the branching ratios b_{jk} in order to build $\mathbf{R}$ and then to perform the inversion of the relation represented by Eq. (3).

We concentrate first on the second problem. The inversion of Eq. (3) belongs to the class of *ill-conditioned linear inverse problems* common to many fields of science. They basically share the numerical indeterminacy of the solution, which manifests itself in wild oscillations of the result or simply in computer overflow. In recent years a great deal of understanding has been accumulated on the basic approach to the solution of these problems as well as on practical aspects of the numerical algorithms. In practice a great variety of methods has evolved, coming from fields as different as reconstruction of astronomical images, analysis of seismographic data, positron emission tomography or X-ray crystallography. It is not surprising that they have apparently little in common. So the first question to be answered is, whether different methods would lead to different results or more precisely to quantify the systematic uncertainty coming from the inversion method.

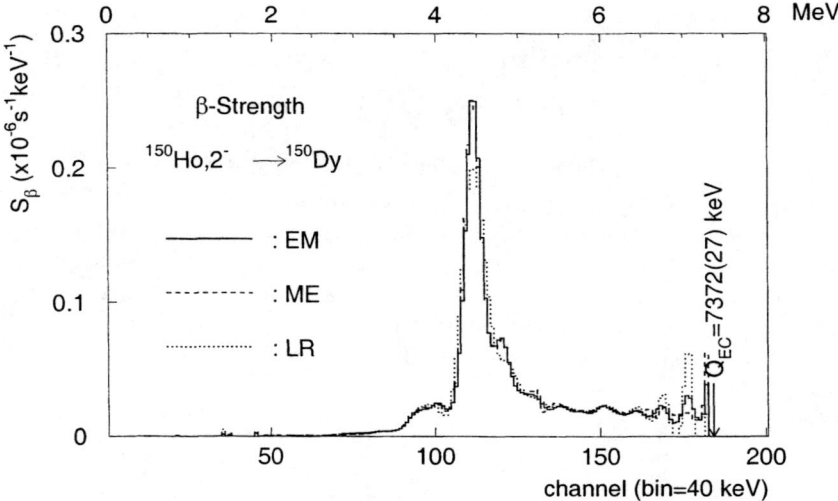

FIGURE 1. The ^{150}Ho (2^-) β-strength distributions obtained from the analysis of the same TAS spectra using the Expectation-Maximization method (solid line), the Maximum Entropy method (dashed line), and the Linear Regularization method (dotted line).

In order to investigate the issue we have implemented three widely employed methods based on quite different principles and leading to quite different algorithms: a) The *Linear Regularization* (LR) method [13], which searches for a solution that reproduces the data in the least squares sense, under the condition that the solution has to have a smooth character (in the case of the β-strength function it derives from the average character of the quantity). This leads to a set of linear equations which must be solved. b) The *Maximum Entropy* (ME) method [14], which follows from the principle that the less biased solution is the one maximizing the information theoretic entropy of Shannon, subject to the condition that the data are reproduced. The maximization is performed in this case using a simple iterative algorithm. c) The *Expectation-Maximization* (EM) method [15] which is a general iterative method for maximum likelihood estimation of parameters from incomplete data.

In Figure 1 we compare the results of the analysis using the three methods when applied to the same set of data with the same response matrix. Their similarity is remarkable taking into consideration the very different nature of the algorithms. In particular the EM and ME methods are very close except at the high energy end. The larger difference of the LR result is a consequence of over smoothing the solution in an attempt to damp the strong oscillations at the higher energies. The oscillations are consequence of the very low statistics in this part of the spectrum. It should be noticed that the LR method can produce (unphysical) negative value results while the other methods do not, and should therefore be preferred.

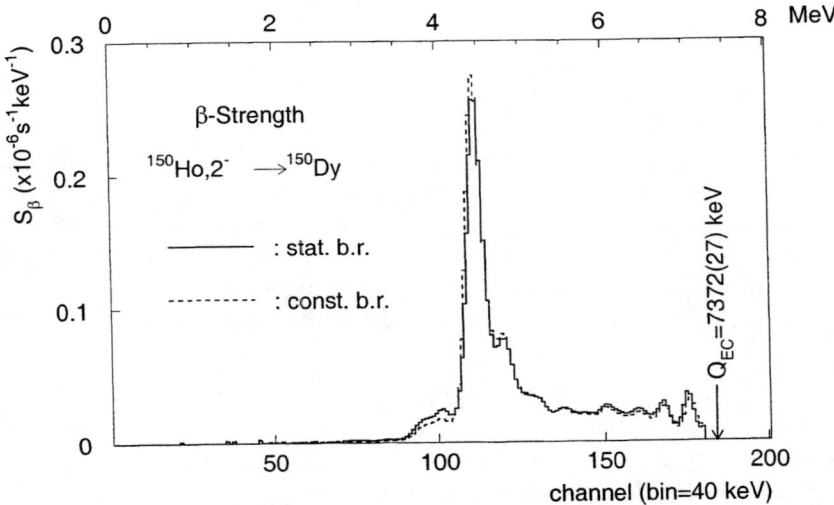

FIGURE 2. The ^{150}Ho (2^-) β-strength distributions obtained from the analysis of the same TAS spectra using two different assumptions about the branching ratios (see text for explanation).

We turn now to the question of the appropriate value of the branching ratios needed to construct the response matrix. In general the information on the b_{jk} is only available (and reliable) for the lowest states in excitation energy since it comes from high resolution germanium detector spectroscopy. For decays with large Q_β windows the only possibility would be to make assumptions about the branching ratios of the high energy levels. This may seem to affect seriously the reliability of the result. On the other hand one should remember that in the limit of an ideal TAS, the dependence of the response function on the detailed branching ratios disappears. In order to quantify this dependence we have analyzed the same set of data with different assumptions for the branching ratios. We compare in Figure 2 the results obtained for two quite different assumptions on the b_{jk}. The same data set as in Figure 1 was analyzed in this case with the EM method. The branching ratios up to $E_x = 2.741$ MeV were taken from the level scheme obtained in a high resolution experiment performed with 6 closely arranged Euroball CLUSTER detectors [11]. The dashed line distribution of Figure 2 was obtained with the rather unrealistic assumption that *all* levels above 2.8 MeV decay with equal probability and exclusively to the levels below. The solid line distribution was obtained using a statistical model for the electro-magnetic de-excitation of the levels above 2.8 MeV (which could be regarded as the best possible assumption). The parameters for the model were taken from the literature [16]. In the model, level densities were obtained from the Fermi gas back shifted formula, and γ-ray strength functions are related to the E1, E2 and M1 giant resonance parameters. As can be observed the difference between the results based on very different assumptions is rather small. This proves that the spectrometer performs as a good TAS and that reliable results can indeed be obtained.

SUMMARY

The study of β-strength distributions in nuclei remains a topic of extraordinary interest in its own right and also regarding its impact on other fields. Accurate measurements are strongly needed. In many cases the only possibility of obtaining the information is from the γ-ray de-excitation of the states of the daughter nucleus. For nuclei far from the β-stability the technique of γ-ray total absorption spectroscopy is then the only one capable of providing reliable results.

REFERENCES

1. Klapdor H.V., *Prog. Part. Nucl. Phys.* **10**, 131 (1983); *ibid* **17**, 419 (1986).
2. *http://www.aquila.infn.it:80/icarus/main.html*
3. Zuber K., *Phys. Rep.* **305**, 295 (1998).
4. Liu W. *et al.*, *Phys. Rev.* **C 58**, 2677 (1998).
5. Yoshida T. *et al.*, *J. Nucl. Sci. Technol.* **36**, 135 (1999).
6. Duke C.L. *et al.*, *Nucl. Phys.* **A 151**, 609 (1970).
7. Negele J. W., and Vogt E., *Adv. Nucl. Phys.* **24**, 1 (1998).
8. Hardy J. C. *et al.*, *Phys. Lett.* **B 71**, 307 (1977).
9. Cano-Ott D. *et al.*, *Nucl. Instrum. and Meth.* **A 430**, 333 (1999).
10. Cano-Ott D. *et al.*, *Nucl. Instrum. and Meth.* **A 430**, 488 (1999).
11. Agramunt J. *et al.*, *Proceedings SGR97*, World Scientific, Singapore, 150 (1998).
12. Karny M. *et al.*, *Nucl. Instrum. and Meth.* **B 126**, 411 (1997).
13. Press W. H. *et al.*, *Numerical Recipes*, Cambridge University Press (1992).
14. Collins D. M., *Nature* **298**, 49 (1982).
15. Shepp L. A., and Vardi Y., *IEEE Trans. Med. Imaging* **1**, 113 (1982).
16. Kopecky J., and Uhl M., *Phys. Rev.* **C 41**, 1941 (1990).

β-decay half-lives of new neutron-rich isotopes of elements from Pm to Tb

S. Ichikawa[1], M. Asai[1], K. Tsukada[1], A. Osa[2], M. Sakama[3],
Y. Kojima[5], M. Shibata[4], I. Nishinaka[1], Y. Nagame[1], Y. Oura[3],
and K. Kawade[4]

[1] *Japan Atomic Energy Research Institute, Tokai-mura, Ibaraki 319-1195, Japan*
[2] *Japan Atomic Energy Research Institute, Takasaki 370-1292, Japan*
[3] *Department of Chemistry, Tokyo Metropolitan University, Hachioji, Tokyo 192-0364, Japan*
[4] *Department of Energy Engineering and Science, Nagoya University, Nagoya 464-8603, Japan*
[5] *Applied Nuclear Physics, Faculty of Engineering, Hiroshima University, Hiroshima 739-8527, Japan*

Abstract. Eight new neutron-rich lanthanide isotopes produced in the proton-induced fission of ^{238}U have been identified using the JAERI on-line isotope separator (JAERI-ISOL) coupled to a gas-jet transport system. For six of these, each half-life was determind: ^{159}Pm $(2 \pm 1 \text{ s})$, ^{161}Sm $(4.8 \pm 0.8 \text{ s})$, ^{165}Gd $(10.3 \pm 1.6 \text{ s})$, ^{166}Tb $(21 \pm 6 \text{ s})$, ^{167}Tb $(19.4 \pm 2.7 \text{ s})$ and ^{168}Tb $(8.2 \pm 1.3 \text{ s})$. The observed half-lives were compared with theoretical calculations. The recent calculation by the gross theory with the new one-particle strength function shows quite good agreement with the experimental half-lives.

INTRODUCTION

Information on nuclear-decay properties such as half-lives, nuclear masses and neutron binding energies of neutron-rich isotopes is particularly important in astrophysical calcultions involving the rapid neutron capture process as well as in the field of nuclear structure. Therefore, considerable progress has been achieved in the study of neutron-rich medium to heavy mass nuclei [1–4] using various production, separation and detection methods. However, no information currently exists on the decay properties of neutron-rich lanthanide isotopes with $A = 160 - 170$ and $Z = 61 - 65$. This comes from the fact that these isotopes which lie off the line of stability cannot be easily produced in the neutron-induced fission of ^{235}U. In proton-induced fission of actinides, the fission yields of the neutron-rich lanthanide isotopes with $A = 160 - 170$ and $Z = 61 - 65$ are significantly enhanced with respect to those in the neutron-induced fission of ^{235}U. To study decay properties of these unknown isotopes in the proton-induced fission of ^{238}U, we developed the gas-jet coupled on-line isotope separator (JAERI–ISOL) system and eight new isotopes were successfully identified.

CP495, *Experimental Nuclear Physics in Europe*, edited by B. Rubio et al.

EXPERIMENT

The experments were performed at the JAERI (Japan Atomic Energy Research Institute) tandem accelerator facility. A stack of eight ^{238}U targets (4 mg/cm^2 thick each) set in a multiple-target chamber was bombarded with 20 MeV proton beams with the intensity of about 1 μA. Fission products emitted from the targets were transported into the ISOL with the gas-jet transport system. The transported nuclides were ionized as the monoxide ions (^{A}M^{16}O$^+$) in the thermal ion source and then mass-separated [5,6]. The mass-separated products were collected on an aluminized Mylar tape and transported to the detection port at prescribed time intervals. The detection port was equipped with two plastic scintillators for β-ray measurements, a planar and a coaxial HPGe detectors for X/γ–ray measurements. β-γ and γ-γ coincidence data were taken with these four detectors and recorded event by event together with time information used in a half-life analysis.

RESULTS AND DISCUSSION

The observed K x-ray following the β^-–decay of the products in the mass-separated fraction provided direct identification of isotopes. New eight isotopes observed, together with the values of their half-lives and detected X and γ rays, are summarized in Table 1 [7–10]. The γ rays observed through the β^-–decay of ^{166,167,168}Tb (see Table 1) are assigned to be $2^+ \rightarrow 0^+$, $4^+ \rightarrow 2^+$ transitions in ^{166,168}Dy and $3/2^- \rightarrow 1/2^-$, $5/2^- \rightarrow 1/2^-$ transitions in ^{167}Dy , respectively. Figure 1 shows the new limits of the nuclear chart around the neutron-rich lanthanides.

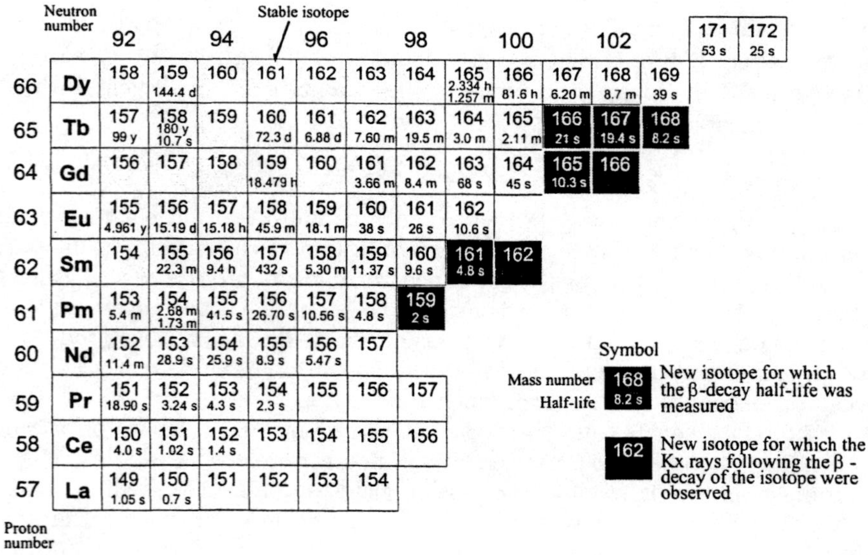

FIGURE 1. New neutron-rich lanthanide isotopes discovered at the JAERI-ISOL facility.

Table 1: Measured half-lives and detected X and γ rays for the new lathanide isotopes.

Isotope	Half-life	detected X and γ rays		
^{159}Pm	2 ± 1	Sm Kx ray		
^{161}Sm	4.8 ± 0.8	Eu Kx ray	263.7 keV	
^{162}Sm		Eu Kx ray		
^{165}Gd	10.3 ± 1.6	Tb Kx ray	50.4 keV	
^{166}Gd		Tb Kx ray		
^{166}Tb	21 ± 6	Dy Kx ray	76.58 keV	177.13 keV
^{167}Tb	19.4 ± 2.7	Dy Kx ray	57.2 keV	69.7 KeV
^{168}Tb	8.2 ± 1.3	Dy Kx ray	74.96 keV	173.37 keV

Table 2: Observed and predicted half-lives for six new isotopes.

Isotope	Experiments	Calculations			
		Gross theory[11]	pn-QRPA[12]		
	$T_{1/2}$ (s)	$T_{1/2}$, Q_β[13] (s), (MeV)	$T_{1/2}$, Q_β[14] (s), (MeV)	$T_{1/2}$, Q_β[15] (s), (MeV)	$T_{1/2}$, Q_β[16] (s), (MeV)
^{159}Pm	2 ± 1	3.08, 5.52	2.93, 5.12	2.54, 4.89	2.80, 5.29
^{161}Sm	4.8 ± 0.8	6.72, 4.80	10.7, 4.71	13.0, 4.51	12.6, 4.98
^{165}Gd	10.3 ± 1.6	16.0, 4.19	20.6, 3.92	27.4, 3.77	18.4, 4.14
^{166}Tb	21 ± 6	33.6, 4.89	83.7, 4.80	166, 4.54	82.8, 4.81
^{167}Tb	19.4 ± 2.7	18.2, 4.10	67.3, 3.54	130, 3.41	63.0, 3.86
^{168}Tb	8.2 ± 1.3	7.25, 5.97	37.1, 5.74	68.4, 5.37	28.6, 5.78

In Table 2, the measured half-life values are compared with the theoretical predictions calculated by Tachibana et al.[11] using the gross theory with the selected Q_β values and by Staudt et al.[12] with the proton-neutron quasiparticle random-phase approximation (pn-QRPA). As shown in Table 2, the calculated half-lives of the pn-QRPA for ^{159}Pm, ^{161}Sm and ^{165}Gd are in agreement with the experimental ones within a factor of about 2-3, while those for ^{166,167,168}Tb are 3-8 times longer than the experimental ones. On the other hand, the calculated half-lives of the gross theory are in good agreement with the experimental ones.

The calculated half-life value depends on the input data of the Q_β value as well as a number of factors associated with the β-strength function that can affect the accuracy of these calculations. According to Tachibana et al., the modification of the β-strength function in the gross theory made the calculated half-life shorter than the provious one; if the same Q_β value is used as the input parameter, the gross theory estimates the half-life about 0.5 times as short as that of the previous one. The overestimation by the pn–QRPA is partly due to the input Q_β values. The half-lives of ^{166,167,168}Tb calculated by the pn–QRPA (Groote [15]) are 7-8 times longer than the experimental ones. If the evaluated Q_β values (Audi and Wapstra [13]) are used in the calculation, it is estimated that the calculated half-lives for ^{166,167,168}Tb are in agreement within a factor of 2–3.

As one of the nuclear structure information, the first 2_1^+ and 4_1^+ states of the daughter nuclides ^{166}Dy and ^{168}Dy were derived through the β^-–decay studies of ^{166}Tb and ^{168}Tb[10]. The level energies of the first 2_1^+ and 4_1^+ states in ^{166}Dy and ^{168}Dy are 76.58, 253.71 keV and 74.96, 248.33 keV, respectively. The level energy of the first 2_1^+ state in ^{168}Dy is lower than that in ^{166}Dy. It indicates that the deformation of ^{168}Dy is larger than that of ^{166}Dy. The enery ratio of $E(4_1^+)/E(2_1^+)$ in both the nuclei approaches 3.33 for a deformed symmetric rotor.

In the study of proton-induced fission of ^{248}Cm, the results indicate that the yields of the fission products with $A \geq 145$ were largely enhanced in comparison with those observed in proton-induced fission of light and medium actinides [17]. The present investigation can be extended to more neutron-rich lanthanide isotopes in the $A \approx 170$ region produced in the proton-induced fission of ^{248}Cm.

ACKNOWLEDGMENTS

We would like to thank the crew of the JAERI tandem accelerator for supplying intense and stable beams for the experiments. This work was carried out along a line of the JAERI-University Collaboration Research Project.

REFERENCES

1. M. Bernas *et al.*, Nucl. Phys. **A616**, 325c(1997).
2. M. Huhta *et al.*, Phys. Lett. B **405**, 230(1997).
3. M. Becker *et al.*, Nucl. Phys. **A522**, 557(1991).
4. G. Gautherin *et al.*, Eur. Phys. J. A **1**, 391(1998).
5. S. Ichikawa *et al.*, Nucl. Instrum. Metods A **274**,59(1989).
6. S. Ichikawa *et al.*, Nucl. Instrum. Methods A **374**, 330(1996).
7. M. Asai *et al.*, J. Phys. Soc. Jpn. **65**, 1135(1996).
8. S. Ichikawa *et al.*, Phys. Rev. C **58**, 1329(1998).
9. S. Ichikawa *et al.*,Proc. of ENAM98, AIP Conf. Proc. No. 455, p.540.
10. M. Asai *et al.*, Phys. Rev. C **59**, 3060(1999).
11. T. Tachibana and M. Yamada, Proc. of ENAM95, *Arles, 1995* (Editions, Frontiéres, Gif-sur-Yvette, 1995), p.763.
12. A. Staudt *et al.*, At. Data Nucl. Data Tables 44, 79(1990).
13. G. Audi and A. H. Wapstra, Nucl. Phys. **A595**, 409(1995).
14. E. R. Hilf *et al.*, CERN Report No. 76-13, 1976, p. 142.
15. H. v. Groote *et al.*, At. Data Nucl. Data Tables **17**, 418(1976).
16. P. Möller and J. R. Nix, At. Data Nucl. Data Tables **26**, 165(1981).
17. Z. Qin *et al.*, Radiochem. Acta (in press).

Total Absorption Spectroscopy of ^{150}Ho 2^- and ^{150}Ho 9^+ decays

D. Cano-Ott[a], J. Agramunt[a], L. Batist[d], R. Collatz[c], A. Gadea[a],
M. Gierlik[b], M. Górska[c], H. Grawe[c], M. Hellström[c], Z. Hu[c],
Z. Janas[b], M. Karny[b], R. Kirchner[c], P. Kleinheinz[a], F. Moroz[d],
A. Płochocki[b], E. Roeckl[c], B. Rubio[a], K. Rykaczewski[b], M. Shibata[c],
J. Szerypo[b], J.L. Tain[a] and V. Wittmann[d]

[a] Instituto de Física Corpuscular, Dr. Moliner 50, E-46100 Burjassot-Valencia, Spain
[b] Institute of Experimental Physics, Warsaw University, PL-00-681 Warsaw, Poland
[c] Gesellschaft für Schwerionenforschung, D-64220 Darmstadt, Germany
[d] St. Petersburg Nuclear Physics Institute, 188-350 Gatchina, Russia

Abstract. The problem of the Gamow–Teller (GT) quenching consists in the overestimation of the predicted Strength with respect to the experimental one. A retardation of the allowed GT transitions has been explained in terms of core polarisation, pairing correlations ant higher order phenomena [1], but its understanding requires reliable experimental results.

THE TOTAL ABSORPTION RESULTS

It is known that the GT–Strengths obtained from high resolution spectroscopy measurements of complex decays suffer from an experimental hindrance due to the limited efficiency of the detectors: the GT–Strength to states at high excitation energy is not observed and instead is incorrectly assigned to lower lying levels. An alternative is the use of a detector of efficiency ≈ 1, called Total Absorption Spectrometer [2]. The Total Absorption Spectroscopy is a very powerful technique, but the extremely complex inverse problem associated to the analysis of the data has not contributed to its popularity. A detailed discussion on the analysis procedures and the reliability of the results can be found in [3].

The β-decays of ^{150}Ho 2^- and ^{150}Ho 9^+ were measured with the Total Absorption Spectrometer installed at the GSI On-line Mass-separator [2]. These two cases present a very clean and simple interpretation in terms of the Extreme Single Particle Shell Model (ESPSM), since there is only one allowed GT transition $\pi h_{11/2} \rightarrow \nu h_{9/2}$ involved. The low spin isomer, with a $(\pi h_{11/2})^2_{0^+} (\pi d_{3/2} \nu f_{7/2})_{2^-}$ valence particle configuration around the ^{146}Gd core, decays only to the 1^-, 2^-

CP495, *Experimental Nuclear Physics in Europe*, edited by B. Rubio et al.
© 1999 American Institute of Physics 1-56396-907-6/99/$15.00

and 3^- 4-quasi particle (qp) states in ^{150}Dy. On the other hand, the high spin isomer is characterised by the $(\pi h_{11/2})^2_{0^+}$ $(\pi h_{11/2} \nu f_{7/2})_{9^+}$ configuration and can decay to the 8^+ 2-qp state and to the 8^+, 9^+ and 10^+ 4-qp states. The shape of the Strength was measured in both cases. Figure 1 shows the comparison between the theoretical B(GT) (lines) taken from [1], the two values obtained in this work (squares) and previous values from high resolution experiments. The B(GT)'s found experimentally in this work are ^{150}Ho 2^- : B_{exp}(GT) = 0.58(4) and ^{150}Ho 9^+ : B_{exp}(GT) = 0.79(9), presenting hindrance factors of $h_{^{150}\text{Ho} 2-}$ = 6.3(3) and $h_{^{150}\text{Ho} 9+}$ = 6.9(9) with respect to the ESPSM B(GT)'s.

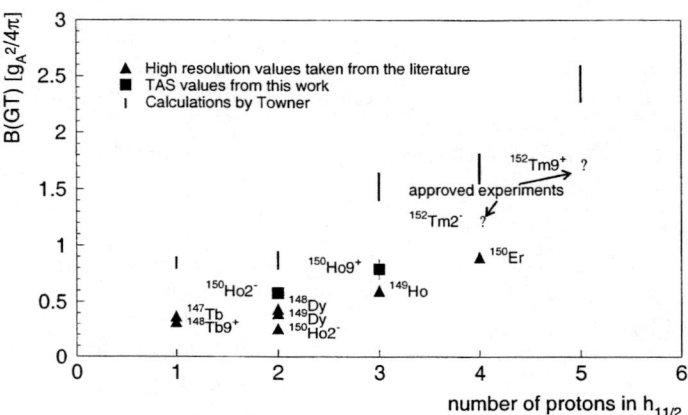

FIGURE 1. Comparison of B(GT) as a function of the number of protons in $h_{11/2}$: theoretical values including hindrance factors (solid lines) taken from [1], TAS results of this work (squares) and values coming from high resolution experiments (triangles).

The experimental technique guarantees that all the GT–Strength inside the Q_{EC} window was detected, and the three independent analysis methods [3] used, provide the correctness of the results. Thus, the experimental hindrance has been eliminated, and the existing discrepancies between the theoretical B(GT) values and the ones from this work, indicate a higher quenching than it is theoretically expected. Further approved experiments on ^{152}Tm 2^- and ^{152}Ho 9^+, signed by question marks in Figure 1, will bring new information on this topic and hopefully motivate further theoretical research.

REFERENCES

1. I.S. Towner, *Nucl. Phys.* **A444**, 402 (1985)
2. M. Karny it et al., *Nucl. Instr. and Meth. in Phys. Res.* **B 126**, 411 (1997)
3. J.L. Taín *et al.*, *Proceedings to this conference.*

Short-Lived Proton Radioactivity Studies at HRIBF

J. C. Batchelder[1], C. R. Bingham[2,3], T. N. Ginter[4], C. J. Gross[1,2],
R. Grzywacz[3], M. Karny[5], Z. Janas[5], F. Mas[2], J. W. McConnell[2],
A. Piechaczek[6], K. Rykaczewski[2], P. Semmes[7], K. S. Toth[2], and
E. F. Zganjar[6]

1 Oak Ridge Institute for Science and Education, Oak Ridge, TN 37831 USA
2 Physics Division, Oak Ridge National Laboratory, Oak Ridge, 37831 TN USA
3 University of Tennessee, Knoxville TN 37996 USA
4 Vanderbilt University, Nashville TN 37235 USA
5 Warsaw University, Warsaw, Hoza 69 Poland
6 Louisiana State University, Baton Rouge, LA 70803 USA
7. Tennessee Technological University, Cookeville, TN 38505

Abstract: An accurate determination of the experimental spectroscopic factor of proton emitting nuclei precisely defines the main component of the proton wave function for the unbound state. However, this has proven difficult for nuclei with $Z \leq 71$ due to the unknown beta-branching ratios involved. One way to solve this problem is to study proton-emitters with half-lives far too short for beta-emission to compete. Recent work at the Holifield Radioactive Ion Beam Facility has produced information on ^{141m}Ho, ^{145}Tm, ^{150m}Lu and ^{151m}Lu, all of which have half-lives in the μs region. A comparison between calculated and experimental spectroscopic factors for these nuclei is given.

INTRODUCTION

The study of the decay of proton-emitting isotopes allows one to study nuclear structure effects in nuclei that are inaccessible during in-beam experiments. The emitted proton tunnels through the Coulomb and centrifugal barriers, and the decay probability depends strongly on the energy of the proton and its angular momentum ℓ. Because of this, the ℓ value of the emitted proton can often be determined through the use of a simple spherical WKB calculation of the expected rate of the tunneling process. However, this calculation does not take into account the details of the nuclear structure effects. Therefore, the difference between the calculated and experimental proton half-lives is due to nuclear structure effects (the overlap between the wave functions of the parent and daughter states) and is known as the spectroscopic factor. The ratio of the calculated half-life to the measured value is defined as the experimental spectroscopic factor: $S_p^{exp} = (T_{1/2}^{calc} / T_{1/2}^{exp})$. The theoretical spectroscopic factor (S_p^{th})

CP495, *Experimental Nuclear Physics in Europe*, edited by B. Rubio et al.
© 1999 American Institute of Physics 1-56396-907-6/99/$15.00

TABLE 1. A comparison of half-lives calculated from different approaches to the WKB approximation.

Nucleus	Q_p (MeV)	orbital	WKB1	WKB2	TPA
$^{145}_{69}$Tm	1.753(11)	$0h_{11/2}$	1.8(3) µs	$1.6^{+.3}_{-.2}$ µs	$1.5^{+.3}_{-.2}$ µs
$^{150}_{71}$Lu	1.283(4)	$0h_{11/2}$	39(4) ms	33(3) ms	30(3) ms
	1.317(15)	$1d_{3/2}$	$6.3^{+2.6}_{-1.8}$ µs	$7.9^{+3.2}_{-2.2}$ µs	$7.9^{+3.3}_{-2.3}$ µs
$^{151}_{71}$Lu	1.255(3)	$0h_{11/2}$	74(6) ms	60(5) ms	58(5) ms
	1.332(10)	$1d_{3/2}$	$4.3^{+1.1}_{-0.9}$ µs	$5.5^{+1.4}_{-1.1}$ µs	$5.5^{+1.4}_{-1.1}$ µs

can be calculated via an independent quasi-particle BCS approximation [1]. The theoretical spectroscopic factor is given by: $S_p^{th} = u_j^2$, where u_j^2 is the probability that the orbital ($n\ell j$) is empty in the daughter. In the region of $64 < Z < 82$, the orbitals $0h_{11/2}$, $1d_{3/2}$, and $2s_{1/2}$ are expected to lie close together at the ground state. Of these orbitals, only the $h_{11/2}$ is expected to have rather small admixtures to the wave function, as opposed to the $d_{3/2}$ which can have a component of the $s_{1/2}$ state coupled to the 2^+ core. Therefore to rigorously test the predictive power of the BCS theory, precise S_p^{exp} values for $h_{11/2}$ nuclei are essential.

A further consideration when comparing experimental and calculated $T_{1/2}$ values is the method of calculation of the $T_{1/2}$ value from which S_p^{exp} is deduced. A simple semiclassical model [2] calculation describes the decay rate as the product of a barrier penetration probability P (calculated using the WKB approach and the optical model potential of Bechetti and Greenlees [3]) and a frequency factor ν. The frequency factor can be estimated by replacing the real potential with a square well and ignoring angular momentum effects (s-wave emission). This calculation, denoted WKB1 in this paper, has been used for many years (see Ref. [4]).

Recently, the semiclassical calculation of $T_{1/2}^{calc}$ has been reinvestigated [5] and compared with more realistic calculations with the DWBA and the two potential approach (TPA) [6]. In this work, the frequency factor ν in the semiclassical model was calculated using the same optical model potential inside the barrier, rather than using a square well. The results of this method, using the normalization condition given by Eq. (25) of Ref. [5], are denoted WKB2. A comparison of WKB1, WKB2 and TPA calculations for recent results at HRIBF is shown in Table 1.

A comparison of S_p^{th} values with S_p^{exp} from $h_{11/2}$ nuclei is rather inconclusive for nuclei with $Z < 72$ (see figure 1) because the known proton emitters in this region have either large error bars for their branching ratios (as is the case for ^{147}Tm: 15(5)% [7]), or the β-decay branch is unknown (^{146}Tm [8] and ^{150,151}Lu [9]) and is estimated from the gross β-decay theory of Takahashi *et al* [10]. If the estimate is in error by a factor of 2 or more, its use would result in a corresponding factor of 2 uncertainty in the

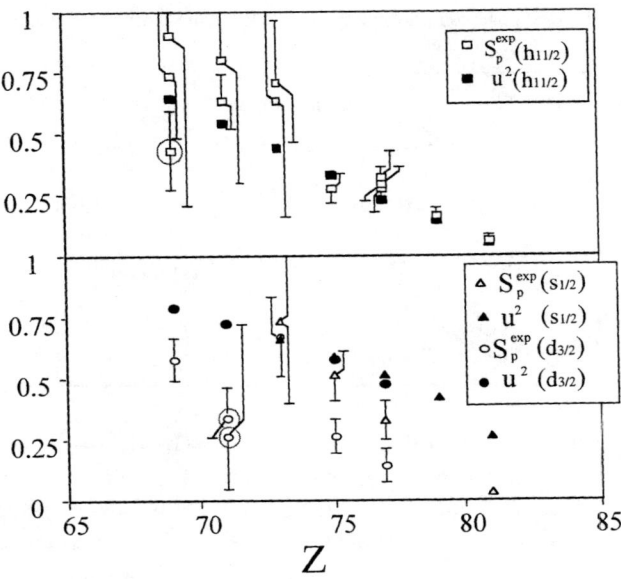

FIGURE 1. Comparison of S_p^{exp} and u^2 for known proton emitters with $69 \le Z \le 81$. Part (a) compares the values for protons emitted form the $\pi s_{1/2}$ and $\pi d_{3/2}$ orbitals, while (b) contains the values for the $\pi h_{11/2}$ orbitals. The circled values (^{145}Tm, ^{151m}Lu, ^{150m}Lu) are from this work. The values for u_j^2 are taken from reference 5.

spectroscopic factor. These nuclei lie to the neutron-deficient side of the N=82 closed shell, so that the major competition with proton emission is β rather than α decay. Because β-decay branching ratios are difficult to measure accurately, large uncertainties exist in the corresponding partial proton half-lives.

Recent experiments at the Holifield Radioactive Ion Beam Facility have yielded the discovery of several new proton emitters (^{141m}Ho [11], ^{145}Tm [12], ^{150m}Lu[13] and ^{151m}Lu[14]) with half-lives of a few µs. The above approach has been shown [5] to reproduce the half-lives fairly well for spherical nuclei, while poorly reproducing these rates for nuclei with a high degree of deformation[15] (^{141m}Ho), such as that between Z = 50 and 67. As such, this paper will only consider those nuclei believed to be spherical. Table 2 lists the proton half-lives of these new states with values calculated using the TPA approximation for the expected orbitals of $h_{11/2}$, $d_{3/2}$, and $s_{1/2}$. From Table 1, it can clearly be seen that the $0h_{11/2}$ ($\Delta\ell = 5$) TPA value matches most closely to the experimental half-life of 3.5(10) µs for ^{145}Tm. In both ^{151m}Lu, and ^{150m}Lu nuclei, the $\pi d_{3/2}$ ($\Delta\ell = 2$) TPA transition fits best with the experimental values.

Comparing the measured half-life of 3.5(10) µs of ^{145}Tm with the calculated TPA value of 1.5(3) µs results in a spectroscopic factor of 0.43(15) for this $0h_{11/2}$ emitter.

TABLE 2. Comparison of the observed proton half-lives with values calculated using the TPA approximation.

species	E_p (MeV)	Exp $T_{1/2}$	calculated $T_{1/2}$ (TPA)			S_p^{exp}
			$0h_{11/2}$	$1d_{3/2}$	$2s_{1/2}$	
^{145}Tm	1.728(10)	**3.5(10)** μs	**1.5(2)** μs	$0.96^{+0.16}_{-0.14}$ ns	0.12(2) ns	0.43(15)
^{150m}Lu	1.295(15)	30^{+95}_{-15} μs	15^{+6}_{-5} ms	$7.9^{+3.3}_{-2.3}$ μs	$0.98^{+0.41}_{-0.28}$ μs	$0.27^{+0.46}_{-0.22}$
^{151m}Lu	1.310(15)	**16(1)** μs	$9.2^{+2.4}_{-1.9}$ ms	$5.5^{+1.4}_{-1.0}$ μs	$0.69^{+0.18}_{-0.13}$ μs	$0.34^{+0.12}_{-0.08}$

This value is slightly lower than that of 0.64 for u^2 [5]. However, the large error bars on the half-life preclude any positive statement on the deviation from the theoretical value. (Also, it should be noted that this nucleus is predicted [16] to be deformed with $\beta_2 = 0.249$). Overall, the data on the $h_{11/2}$ proton emitters are consistent with a spherical picture. The S_p^{exp} values for protons emitted from the $\pi s_{1/2}$ orbitals also agree with the calculated u^2 values.

The experimental values for the $\pi d_{3/2}$ spectroscopic factors, however, are consistently lower than the spherical shell model values. (Note that the data point for ^{156}Ta decay represents the upper limit for S_p^{exp}). This may indicate that the $3/2^+$ states have a significant amount of mixing with other configurations such as $\pi s_{1/2}$ coupled to the first 2^+ state in ^{150}Yb in the case of ^{151}Lu.

REFERENCES

1. R. A. Sorenson and E. D. Lin, Phys. Rev. **142**, 729 (1966).
2. H. A. Bethe, Rev. Mod. Phys. **9**, 62 (1937).
3. F. D. Becchetti, Jr. and G. W. Greenlees, Phys. Rev. **182**, 1190 (1969).
4. S. Hofmann, in *Nuclear Decay Models,* edited by D. N. Poenaru, Inst. of Phys. Publishing, Bristol and Philadelphia, 1996, pp. 143-203.
5. S. Åberg, P. B. Semmes, and W. Nazarewicz, Phys. Rev. C **56**, 1762, (1997).
6. S. A. Gurvitz And G. Kalbermann, Phys Rev. Lett. 59, 262 (1987).
7. K. S. Toth, *et al.*, Phys. Rev. C. **47** 1804 (1993).
8. K. Livingston, *et al.*, Phys. Lett. **312B** 46, (1993).
9. P. J. Sellin, *et al.*, Phys Rev C **47**, 193 (1993).
10. K. Takahashi, *et al.*, At. Data Nucl. Data Tables **12**, 101 (1973).
11. K. Rykaczewski, *et al.*, Phys. Rev. C **60**, R2970 (1999).
12. J. C. Batchelder *et al.*, Phys Rev C **57**, R1042 (1998).
13. T. N. Ginter *et al*, submitted to Phys. Rev. C.
14. C. R. Bingham, *et al.*, Phys. Rev. C **59**, R2984 (1999).
15. C. N. Davids, *et al.*, Phys. Rev. Lett, **80**, 1849 (1998).
16. P. Möller, J. R. Nix, W. D. Myers and j. Swiatecki, At. Data Nucl. Data Tables **59**, 185 (1995).

New Spectroscopy of Heavy Neutron Rich Nuclei : Isomeric Studies Following Relativistic Fragmentation

M. Pfützner[1], P.H. Regan[2], Zs. Podolyák[2], J. Gerl[3], M. Hellström[4], M. Caamaño[2], P. Mayet[3], M. Mineva[4], M. Sawicka[1] and Ch. Schlegel[3] for the GSI Isomer Collaboration

[1] *Institute of Experimental Physics, Warsaw University, PL-00-861 Warsaw, Poland*
[2] *Dept. of Physics, University of Surrey, Guildfort, GU2 5XH, UK*
[3] *GSI, Planckstrasse 1, D-64291 Darmstadt, Germany*
[4] *Div. of Cosmic and Subatomic Physics, Lund University, SE-22100, Sweden*

Abstract. Gamma spectroscopy methods have been used to search for microsecond isomers among fragmentation products of 1 GeV/nucleon ^{208}Pb beam. Decays of several known K-isomers in the rare earth region of $A \sim 180$ were observed, including the $K = 35/2$ isomer in ^{179}W. Several new isomeric decays in the very neutron rich systems, such as ^{190}W, have been identified. In addition, in the course of this work, a number of neutron rich rare earth isotopes have also been synthesised and identified for the first time.

INTRODUCTION

Projectile fragmentation of relativistic heavy ions has proven to be a highly successful method of producing nuclei very far from the line of stability. Indeed, previously inaccessible areas have been opened for spectroscopic studies in both very neutron rich and deficient regions of the nuclear chart. Moreover, it was found that in fragmentation reactions, both at intermediate and relativistic energies, near-yrast isomeric states are populated abundantly [1,2]. It follows that the combination of a projectile fragment separator, allowing the separation and in-flight event-by-event identification of selected exotic ions, together with gamma detector arrays mounted at the final focus of the spectrometer may be a powerful technique for spectroscopy studies of isomers at the limits of known nuclei.

The GSI Darmstadt SIS/FRS facility currently offers the unique possibility to study relativistic projectile-like fragments of *heavy* primary beams, such as ^{208}Pb or ^{238}U. In the first experiment devoted to isomeric spectroscopy of heavy nuclei following relativistic fragmentation, a beam of ^{238}U at 1000 MeV/nucleon was used

CP495, *Experimental Nuclear Physics in Europe*, edited by B. Rubio et al.
© 1999 American Institute of Physics 1-56396-907-6/99/$15.00

to produce neutron-rich nuclei in the lead region [2]. Among the results, were the observation of 7 new neutron-rich isotopes and the identification of 9 known and 4 new isomeric decays with lifetimes in the microsecond range.

Recently, an experiment has been carried out at GSI in which the fragmentation of the relativistic ^{208}Pb beam was used to produce and study neutron-rich rare earth nuclei around $A \sim 180$. The main goal was to search for K-isomers predicted to occur abundantly in this region and which are too neutron rich to be populated by fusion-evaporation or deep inelastic reactions. In this contribution, we present selected preliminary results from this experiment.

EXPERIMENTAL DETAILS

The nuclei of interest were produced by the fragmentation of the ^{208}Pb beam at 1000 MeV/nucleon impinging on a 1.6 g/cm^2 beryllium target. Projectile fragments were separated by the Fragment Separator (FRS), operated in the standard achromatic mode, and identified in-flight, ion by ion by time-of-flight, energy-loss and magnetic rigidity measurements. After slowing down in a variable aluminum degrader at the final focus of the FRS, identified ions were implanted in an aluminum plate surrounded by four *Segmented Clover* type germanium detectors. Signals from the gamma detectors were recorded only during a 80μs gate opened by an incoming heavy ion, allowing a clean correlation of the gamma radiation with an ion. A time delay of the gamma events with respect to the implantation was measured by means of time-to-digital converters (8 μs range) as well as by the time-to-amplitude modules (80 μs range). In a single FRS setting about 16 different ions (4 elements, 4 isotopes each) were transmitted to the final focus. The region of $N \geq 102$ and $Z \leq 78$ was covered with six FRS settings. For calibration and future studies on the proton drip line using this method, one additional setting was centered on the very neutron deficient nucleus, $^{139}_{65}$Tb.

RESULTS

In order to verify the heavy ion identification procedure and also to collect data on the population probability of various K-isomers, a region of Hf-W-Os isotopes close to stability line was scanned in three FRS settings. In this region, several high-spin K-isomeric states are well known from previous studies using fusion-evaporation and deep inelastic reactions. Indeed, we observed decays from about 20 of these known isomers. Since their decay schemes are known, the isomeric ratios can be deduced, giving unprecedented information for testing models of angular momentum distribution following fragmentation reactions. Here, we mention only the observation of the decay of the $I = 35/2\hbar$ isomer in ^{179}W [3], populated with a probability of about 2% (see figure 1). This isomer corresponds to the highest discrete spin yet identified in a fragmentation product.

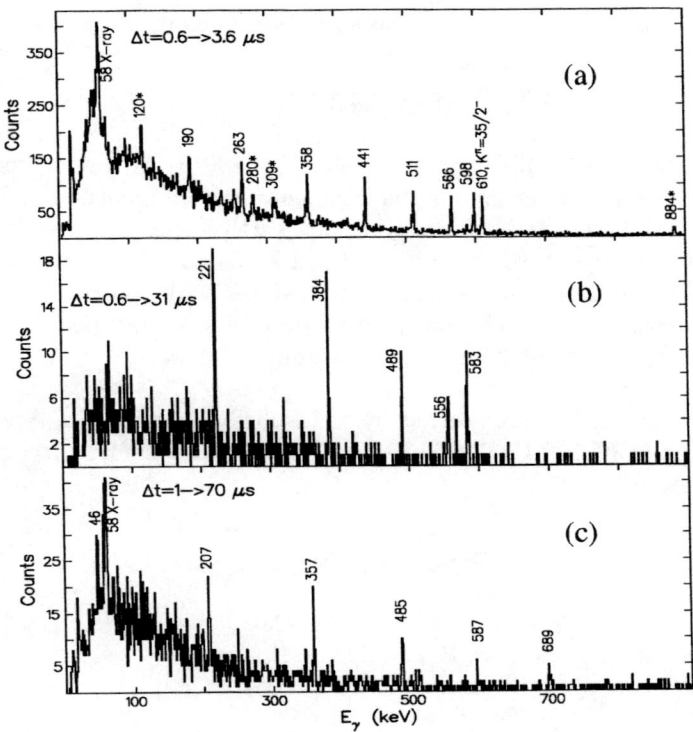

FIGURE 1. Gamma-ray spectra following the isomeric decays in (a) ^{179}W, (b) ^{138}Gd and (c) ^{190}W from three settings of the FRS following the fragmentation of a ^{208}Pb beam. The time ranges over which each of these spectra were taken after the ion implantation are also labelled.

In the course of three other FRS settings, covering the boundary of known neutron-rich nuclei, about ten new isotopes were identified and decays of a few new isomers were observed. As a illustrating example, we give the case of ^{190}W$_{116}$. Until this work, ^{190}W was the most neutron rich isotope of this element which had been identified and no excited states were known for this nucleus. Figure 1 shows the gamma spectrum recorded in coincidence with ^{190}W ions giving evidence for the decay of, most probably, a $K^{\pi}=10^{-}$ isomer directly into the ground-state band.

In the final setting, to test the production of *proton* drip line nuclei by ^{208}Pb fragmentation, the FRS was tuned to the very neutron deficient Z=65 nucleus, ^{139}Tb. This nucleus was successfully identified for the first time. Additionally, the previously identified $K^{\pi} = 8^{-}$ isomer in ^{138}Gd [4] was observed, (see fig. 1).

This result suggests that the relativistic fragmentation technique may be a promising alternative method to access this region which is currently of great interest due to reports of deformed ground-state proton emitters [5].

SUMMARY

Isomeric states with halflives in the range from 100 ns to several hundred microseconds were searched for among the fragmentation products of 1 GeV/nucleon ^{208}Pb beam. The stable and neutron rich rare earth nuclei with between A~180 and 200, as well as the proton drip-line around Z~65 have been studied. A number of new isotopes and isomeric decays from states with spins up to $K^\pi = \frac{35}{2}\hbar$ have been observed. This work clearly opens up promising perspectives for future nuclear structure studies very far from the stability line in heavy nuclei.

This work was partially supported by the Polish Committee of Scientific Research under grant KBN 2 P03B 036 15, EPSRC(UK) and the EU Access to Major Facilities program.

REFERENCES

1. Grzywacz, R. et al., *Phys. Lett.* **B 355**, 439 (1995).
2. Pfützner, M. et al., *Phys. Lett.* **B 444**, 32 (1998).
3. Walker, P.M. et al., *Nucl. Phys.* **A568**, 397 (1994).
4. Bruce, A.M. et al., *Phys. Rev.* **C 55**, 620 (1997).
5. Rykaczewski, K. et al., *Phys. Rev.* **C 60** *in press* (1999).

Large deformation change in iridium isotopes from laser spectroscopy

D. Verney*, L. Cabaret†, J. Crawford‡, H.T. Duong†, J. Genevey‖,
G. Hubert¶, F. Ibrahim‖1, M. Krieg¶, F. Le Blanc*, J.K.P. Lee‡,
G. Le Scornet§, D. Lunney§, J. Obert*, J. Oms*, J. Pinard†,
J.C. Putaux*, B. Roussière*, J. Sauvage*, V. Sebastian¶
and the ISOLDE collaboration**

*Institut de Physique Nucléaire, IN2P3-CNRS, 91406 Orsay Cedex, France
†Laboratoire Aimé Cotton, 91405 Orsay Cedex, France
‡Physics Department, Mc Gill University, H3A2T8 Montréal, Canada
‖Institut des Sciences Nucléaires, IN2P3-CNRS, 38026 Grenoble Cedex, France
¶Institut für Physik der Universität Mainz, 55099 Mainz, Germany
§ CSNSM, IN2P3-CNRS, 91405 Orsay Cedex, France
**CERN, 1211 Genève 23, Switzerland

Abstract. Laser spectroscopy measurements have been performed on neutron-deficient iridium isotopes. The hyperfine structure and isotope shift of the optical Ir I transition $5d^7 6s^2 \; ^4F_{9/2} \to 5d^7 6s6p \; ^6F_{11/2}$ have been studied for the $^{182-189}$Ir, ^{186}Ir^m and ^{191,193}Ir isotopes. The nuclear magnetic and quadrupole moments were obtained from the hyperfine splitting measurements and the changes of the mean square charge radii from the isotope shift measurements. A large deformation change between ^{187}Ir and ^{186}Ir and between ^{186}Ir^m and ^{186}Ir^g has been observed.

Introduction

It is well known that the determination of the nuclear moments and nuclear radii is an important source of information about nuclear shape. Laser spectroscopy allows the determination of magnetic and quadrupole moments from the measurement of the hyperfine structure of atomic lines. The mean square charge radius variation between isotopes is measured by the shift of those lines along isotopic series. Thus, this technique, that is a powerful tool to study deformation instabilities, has already been used to study the mercury transitional region [1-5]. In the present work, the laser spectroscopy is applied to Ir isotopes and the results are presented.

---

1) Permanent address : Institut de Physique Nucléaire, 91406 Orsay Cedex, France

CP495, *Experimental Nuclear Physics in Europe*, edited by B. Rubio et al.
© 1999 American Institute of Physics 1-56396-907-6/99/$15.00

Experimental procedure

The COMPLIS experimental setup (figure 1) installed at the ISOLDE facility at CERN is based on the Resonant Ionisation Spectroscopy technique [6]. The iridium

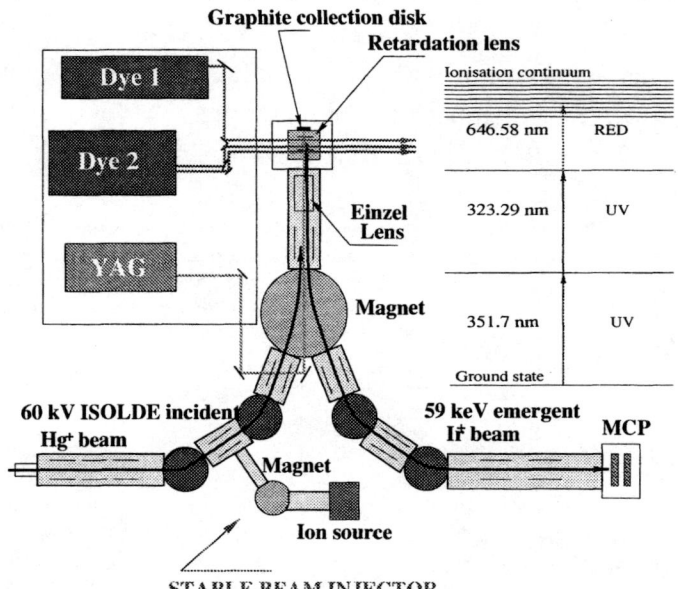

FIGURE 1. COMPLIS experimental setup.

isotopes were obtained as β-decay descendants of the mercury isotopes delivered by ISOLDE as radioactive ion beams after extraction and mass separation. The Hg ions entering the COMPLIS beam line under 60 kV are guided through a series of electrostatic quadrupoles and deflectors towards the ionisation chamber, where they are focused and slowed down to 1 kV in order to be deposited in the first atomic layers of a graphite substrate. After waiting the β-decay delay necessary to obtain an optimized amount of Ir isotopes, they are desorbed as atoms by heating the graphite surface with a pulsed Nd:YAG laser beam. The iridium atoms are then ionized in three steps at the crossing point of three laser beams. The hyperfine spectroscopic information are provided by the first excitation step at 351.7 nm $(5d^7 6s^2\ {}^4F_{9/2} \rightarrow 5d^7 6s6p\ {}^6F_{11/2})$: allowed atomic transitions are induced between the hyperfine levels for the resonant frequency values when scanning the frequency of the light emission of a single-mode pulsed dye laser [7]. Transitions to a second excitation step and then to the ionization continuum (see figure 1) are induced respectively by the frequency-doubled radiation and the radiation of a second pulsed dye laser (commercial type Lambdaphysik). The ions are then accelerated under 59 kV toward the emergent beam line of COMPLIS. They are finally detected using microchannel plates and mass-identified by their time of flight.

Results and interpretation

Table 1 shows the magnetic dipole moments μ_I and the spectroscopic quadrupole moments Q_S determined from A and B hyperfine constants extracted from the experimental spectra. They are consistent with the values available in literature, furthermore the signs for the magnetic moments of $^{182-189}$Ir and ^{186}Ir^m could be assigned for the first time. A very first step towards an interpretation of the experimental results is done by comparing the experimental magnetic moments to thoses calculated using an axial rotor+quasiparticle model. In this approach, the quasiparticle states of the even-even Os core were calculated self-consistently using the constraint Hartree-Fock method with the BCS treatment for the pairing [8]. The total hamiltonian including the Coriolis interaction term is diagonalized to get the core-coupled quasi-particle states [9]. Magnetic moments are calculated with

TABLE 1. Magnetic and quadrupole moments values for Ir isotopes.

		COMPLIS		LITERATURE [11]	
A	I	$\mu_I(\mu_N)$	Q_S(b)	$\mu_I(\mu_N)$	Q_S(b)
193	3/2+	+0.1658(76)	+0.750(69)	+0.1637(6)	+0.751(9)
191	3/2+			+0.1507(6)	+0.816(9)
189	3/2+	+0.145(7)	+0.87(7)	0.13(4)	+0.878(10) [12]
188	1−	+0.31(1)	+0.47(5)	0.302(10)	+0.484(6) [12]
187	3/2+	+0.171(10)	+0.88(10)		+0.941(11) [12]
186^g	5+	+3.69(15)	−2.6(9)	3.88(5)	−2.548(31) [12]
186^m	2−	−0.634(29)	+1.5(1)	0.638(8) [13]	+1.456(17) [12]
185	5/2−	+2.55(7)	−1.8(6)	2.605(13)	−2.06(14)
184	5−	+0.690(32)	+2.6(4)	0.696(5)	+2.1(3)
183	5/2−	+2.36(8)	−1.9(6)		
182	3+	+2.64(19)	−1.7(5)	2.28(8)	

TABLE 2. Experimental magnetic moments and rotor+quasiparticle model calculation results for odd neutron-deficient iridium isotopes.

Ir	spin	$\mu_{exp}(\mu_N)$	β	state K[Nn$_z\Lambda$]	$\mu_{calc}(0.6g_{s,free})$	$\mu_{calc}(g_{s,free})$
193	3/2+	+0.1637(6)	0.110	3/2 [402]	+0.51	−0.12
			0.213	3/2 [402]	+0.49	−0.14
191	3/2+	+0.1507(6)	0.112	3/2 [402]	+0.51	−0.11
			0.217	3/2 [402]	+0.48	−0.16
189	3/2+	+0.144(7)	0.115	3/2 [402]	+0.51	−0.10
			0.210	3/2 [402]	+0.48	−0.15
187	3/2+	+0.17(1)	0.217	3/2 [402]	+0.48	−0.16
186^m	2−	−0.63(3)		π3/2[402] $\otimes\nu$7/2[503] [15]	−0.71 [14]	−0.49 [14]
186^g	5+	+3.69(15)		π1/2[541] $\otimes\nu$1/2[?] [15]		
185	5/2−	+2.55(7)	0.251	1/2 [541]	+1.98	+1.31
183	5/2−	+2.36(8)	0.267	1/2 [541]	+1.89	+1.26

those latter states using g_R=Z/A values [10]. As shown in table 2, the satisfactory agreement obtained confirms the labels for the proton state for odd iridium isotopes. From the $\pi \otimes \nu$ configurations proposed for ^{186}Ir^g and ^{186}Ir^m in references

[14,15], we can underline a change in the label of the proton state which is identified as $\pi 3/2\ 3/2$ [402] from ^{193}Ir to ^{186}Ir^m and $\pi 5/2\ 1/2$ [541] from ^{186}Ir^g to ^{182}Ir. As shown in figure 2, the sudden breaking up between ^{187}Ir and ^{186}Ir^g in the regular decrease of the nuclear radii, as well as the important radius difference between ^{186}Ir^g and ^{186}Ir^m (already suggested in [12]) reflect this change in the proton state.

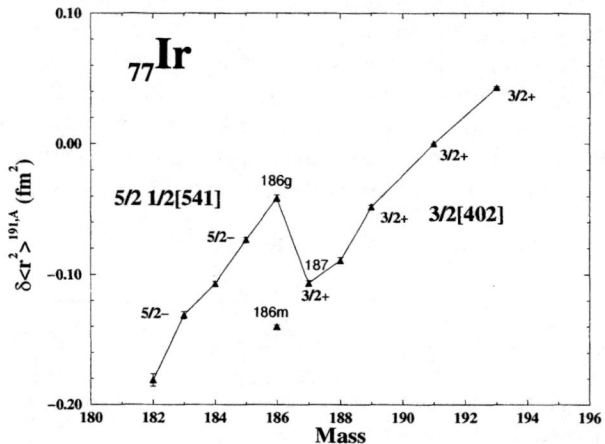

FIGURE 2. Preliminary curve of the mean square charge radius variations of neutron-deficient iridium isotopes from isotope shift measurements. The F atomic factor was approximated using the $\lambda^{191-193}$ given by G. Sawatzky and R. Winkler Z. Phys. **D14**, 9 (1989).

REFERENCES

1. Ulm G. *et al.*, *Z. Phys.* **A325**, 247 (1986).
2. Wallmeroth K. *et al.*, *Nucl. Phys.* **A493**, 224 (1989).
3. Le Blanc F. *et al.*, *Phys. Rev. Let.* **79**, 2213 (1997).
4. Hilberath T. *et al.*, *Z. Phys.* **A342**, 1 (1992).
5. Le Blanc F. et al., to be published in Phys. Rev. C.
6. Hurst G. S. *et al.*, *Rev. Mod. Phys.* **51**, 767 (1979).
7. Pinard J. and Liberman S., *Opt. Commun.* **20**, 344 (1977).
8. Flocard H. *et al.*, *Nucl. Phys.* **A203**, 433 (1973).
9. Meyer M. *et al.*, *Nucl. Phys.* **A316**, 93 (1979).
10. Libert J. *et al.*, *Phys. Rev.* **C25**, 586 (1982).
11. *Table of Isotopes*, eighth edition, Eds R. B. Firestone and V. S. Shirley, New York: John Wiley and sons inc., 1996.
12. Seewald G. *et al.*, *Phys. Rev. Let.* **77**, 5016 (1996).
13. Eder R. *et al.*, *Hyp. Int.* **59**, 83 (1990).
14. Ekström C. *et al.*, *Phys. Scr.* **14**, 199 (1976).
15. Ben Braham A. *et al.*, *Nucl. Phys.* **A533**, 113 (1991).

Experimental identification of intruder bandheads in odd-mass $^{187-193}$Pb

A.N. Andreyev[1], J.F.C. Cocks[2], O. Dorvaux[2], K. Eskola[3],
P. Greenlees[2], P. Jones[2], R. Julin[2], S. Juutinen[2], K. Helariutta[2],
M. Huyse[1], H. Kettunen[2], P. Kuusiniemi[2], M. Leino[2], M. Muikku[2],
W.H. Trzaska[2], K. Van de Vel[1], P. Van Duppen[1], R. Wyss[4]

[1] *Instituut voor Kern- en Stralingsfysica, University of Leuven, Celestijnenlaan 200 D, B-3001 Leuven, Belgium*
[2] *Department of Physics, University of Jyväskylä, FIN-40351 Jyväskylä, Finland*
[3] *Department of Physics, University of Helsinki, Helsinki, Finland*
[4] *Department of Physics, Royal Institute of Technology, 104 05 Stockholm, Sweden*

Abstract. Fine-structure α-decays of the odd mass $^{191-197}$Po identifying proton based intruder states in the daughter lead nuclei have been observed, leading to a systematics of intruder states in odd mass lead isotopes from ^{197}Pb down to ^{187}Pb. The interpretation of these states involves the coupling of the $i_{13/2}$ or $p_{3/2}$ odd neutron to the oblate deformed even lead core.

Within the last decade the neutron-deficient even lead isotopes became the subject of extensive experimental and theoretical studies, see [1–4] for a review. The low-lying 0^+ π(2p-2h) intruder states, associated with weakly deformed oblate shapes have been found to coexist with the spherical 0^+ π(0p-0h) states from ^{202}Pb down to ^{190}Pb and, recently, in ^{188}Pb [2–4]. In contrast, the information on the low-lying deformed bandheads in odd-mass lead nuclei is scarce: they are known only in ^{195,197}Pb [5,6]. In our recent study [7] evidence was given on the identification of oblate $1i_{13/2} \otimes \pi$(2p-2h) intruder state in ^{187m}Pb. In this contribution, we report on the observation through the fine structure in the polonium α decay, of intruder states in $^{189-193}$Pb, completing the systematics of the intruder bandheads in the odd-mass nuclei from ^{197}Pb down to the midshell ^{187}Pb nucleus.

The experiment has been performed at the RITU gas-filled recoil separator [8] at the University of Jyväskylä (JYFL). The ^{193}Po nuclei were studied in the ^{166}Er(^{32}S,5n)^{193}Po reaction, while ^{195}Po - in ^{169}Tm(^{32}S,p5n)^{195}Po and ^{160}Dy(^{40}Ar,5n)^{195}Po reactions. ^{197}Po was produced in the latter reaction on the admixtures of the heavier dysprosium isotopes in the ^{160}Dy target. Pulsed beams (2 ms beam ON/8 ms beam OFF) were delivered by the K=130 cyclotron. The

CP495, *Experimental Nuclear Physics in Europe*, edited by B. Rubio et al.
© 1999 American Institute of Physics 1-56396-907-6/99/$15.00

evaporation residues after separation were implanted into a position-sensitive silicon strip detector. Four germanium detectors were installed behind the strip detector for α-γ and α-X ray coincidence measurements. The strip detector was surrounded by six silicon detectors (called further electron detectors) used to detect conversion electrons in coincidence with the α-particles. In the data analysis, α-decays in the strip detector were searched for in a prompt coincidence with either the low-energy signals in the electron detectors or with γ events in the Ge detectors.

An example of an α-spectrum, collected for the ^{166}Er(^{32}S,5n)^{193}Po reaction, is shown in Fig.1a. Figure 1b shows the same spectrum, but under a condition of a prompt ($\Delta T_{\alpha-e^-}$ <30 ns) coincidence with a low-energy signal ($E_{electron} \leq 600$ keV) from the electron detectors. The line at $E_\alpha = 6200(20)$ keV represents a known fine structure ($I_\alpha = 0.22\%$) α-decay of ^{194}Po towards an excited ($E^*(0_2^+)=658$ keV) 0^+ intruder state in ^{190}Pb [9]. The line at $E_\alpha = 6390(20)$ keV was assigned to the fine structure α-decay of ^{193m}Po towards an excited level at $E^*=637(1)$ keV in the daughter ^{189}Pb nucleus. This assignment is supported by the observation of 6378(15) keV-637(1) keV α-γ coincident pairs in the prompt ($\Delta T_{\alpha-\gamma}$ <30 ns) α-γ matrix. The Q-value for the sum of the 6378(15) keV and 637(1) keV $\alpha - \gamma$ pairs is close to the Q_α-value of the g.s.$\rightarrow$g.s. α-decay of ^{193m}Po.

Similarly, fine structure α-decay of ^{193g}Po was identified by observing eight 6420(20) keV-548(1) keV α-γ coincident pairs with the sum Q-value, close to the Q_α-value for the g.s.$\rightarrow$g.s. α-decay of ^{193g}Po. In the α-e$^-$ coincidence spectrum (Fig.1b) a corresponding weak α-transition at $E_\alpha = 6420(20)$ keV could be tentatively observed as well.

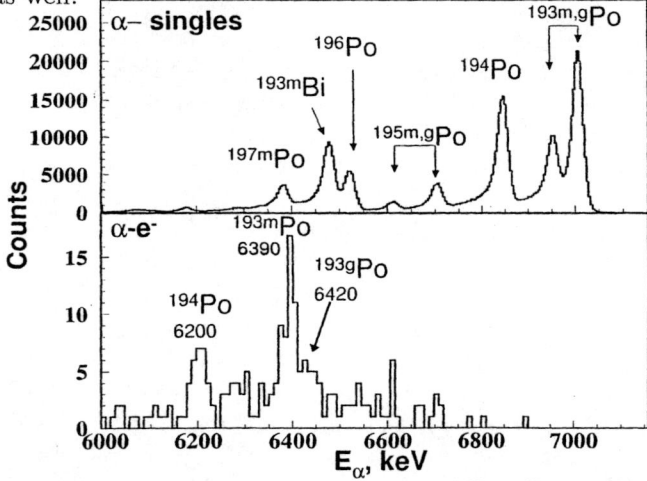

FIGURE 1. a) Total α-spectrum, collected in the ^{166}Er(^{32}S,5n)^{193}Po reaction between beam pulses. b) the same as a), but under a condition of a prompt ($\Delta T_{\alpha-e^-} \leq 30$ ns) coincidence with low energy electrons ($E_{e^-} \leq 600$ keV). Some α-lines are labelled with their energy in keV.

By comparing the numbers of α-γ and α-e$^-$ events for these α decays corrected for the γ and electron efficiency, total conversion coefficients of $\alpha_{tot}=1.12(37)$ and

α_{tot}=0.34(16) for the 637 keV, respectively, 548 keV γ transitions can be deduced. These values are clearly in excess to the theoretical values [10] for possible competing multipolarities, indicating a strong E0 component in both of these transitions. On the basis of this data branching ratios for 6378 keV and 6420 keV α decays were derived and the corresponding hindrance factor values (HF) were calculated, see Table. The spin and parity assignments for ^{193m,g}Po and ^{189m,g}Pb have been adopted from the well-established systematics of heavier odd mass lead and polonium isotopes in this region. The assignement is also supported by the low HF values [7].

TABLE. Alpha-decay energies, intensities, HF, spin and parity, energies of the excited 13/2$^+$ and 3/2$^-$ states and total conversion coefficients identified in this work. The hindrance factors were calculated as described in [7].

	^{A}Po			$^{A-4}$Pb		
	E_α (keV)	I_α%	HF	I^π	E^*(keV)	α_{tot}
^{191g}Po→^{187g}Pb	7334	93	2.3	3/2$^-$	0	
	6960(15)	7(2)	1.9(5)	3/2$^-$	375	
^{191m}Po→^{187m}Pb	7378	54	24	13/2$^+$	0	
	6888(15)	46(9)	0.64(15)	13/2$^+$	494	$\geq$0.3
^{193g}Po→^{189g}Pb	6949	99.3	2.5	3/2$^-$	0	
	6420(20)	0.7(3)	3.7(1.6)	3/2$^-$	548	0.34(16)
^{193m}Po→^{189m}Pb	7003	99.2	2.1	13/2$^+$	0	
	6378(15)	0.8(3)	1.2(4)	13/2$^+$	637	1.12(37)
^{195g}Po→^{191g}Pb	6606	99.83	2.43	3/2$^-$	0	
	6030(20)	0.17(5)	6.5(1.9)	3/2$^-$	597	
^{195m}Po→^{191m}Pb	6699	99.8	1.96	13/2$^+$	0	
	6050(20)	0.20(12)	2.8(1.5)	13/2$^+$	670	0.81(27)
^{197m}Po→^{193m}Pb	6385	99.3	2	13/2$^+$	0	
	5622(20)	0.05(3)	1.7(1)	13/2$^+$	757	

Similarly, fine-structure α-decays in ^{195m,g}Po and in ^{197m}Po (see Table) were found on the basis of the sum energy balance by analysing prompt α-γ matrixes collected for corresponding reactions. For ^{195m}Po α-e$^-$ events were found as well and a conversion coefficient of α_{tot}=0.81(27) for 670 keV γ-transition was deduced, proving the presence of the strong E0 component in this decay. Due to lower statistics and a rather high background we were not able to extract the numbers of α-e$^-$ coincidences for ^{195g,197m}Po.

The systematics of low-lying excited states in the lead nuclei with 184$\leq$A$\leq$200 is shown in Fig.2. For A$\geq$190 the spherical 17/2$^+$, 21/2$^+$ and 25/2$^+$ states in the odd mass isotopes closely follow the 2$^+$, 4$^+$ and 6$^+$ spherical states in the neighboring nuclei, respectively. This is a well-established pattern, resulting from the weak coupling of an odd neutron to the *spherical* states in the even core [11].

The parabolic behavior of the oblate 0$^+$ π(2p-2h) intruder bandheads in the even lead nuclei with A$\geq$188 is a well-established phenomenon [1,2]. In our work for the first time the systematics of the corresponding intruder states in the odd mass lead nuclei is presented. It is clearly seen that the excitation energy of the 13/2+ states in $^{187m-197m}$Pb (our data and [5,6]) follows closely the excitation energy of the 0+ intruder states in the neighboring even cores and deviates strongly from the behaviour of the 2+ and 17/2+ spherical states. The same trend is observed for the 3/2− states in $^{187g-191g}$Pb. Moreover, similar to the 0+ intruder states in $^{188-194}$Pb the 13/2+ and 3/2− states become the first excited states in the $^{187-195}$Pb isomers.

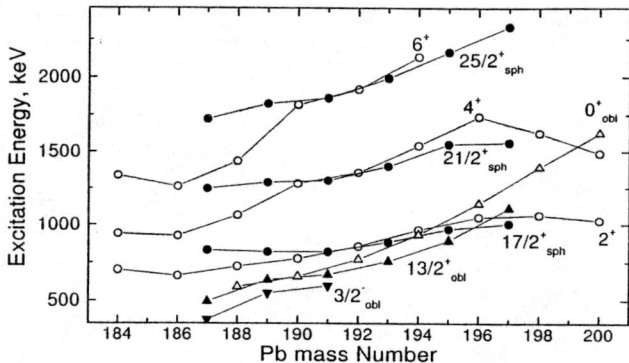

FIGURE 2. The systematics of the yrast 2+, 4+ and 6+ states (o) in the even mass lead nuclei and of the spherical 17/2+, 21/2+ and 25/2+ states (●) in the odd mass isotopes. The data on the oblate-deformed 0+ (△), 13/2+ (▲) and 3/2− (▼) proton-based indtruder states are shown as well.

In conclusion, $\nu i13/2^+ \otimes \pi(2p\text{-}2h)$ and $\nu p3/2^- \otimes \pi(2p\text{-}2h)$ proton based intruder bandheads have been observed in $^{187-193}$Pb nuclei, completing the systematics of intruder states in odd mass lead nuclei from ^{197}Pb down to a midshell nucleus ^{187}Pb.

This work was supported by the Academy of Finland, the Access to Large Scale Facility program under the Training and Mobility of Researchers program of the European Union, by the FWO-Vlaanderen and by the IUAP program, Belgium. M.H. is Research Director of the FWO-Vlaanderen.

REFERENCES

1. Wood, J. L. *et al.*, *Phys. Rep.* **215**, 101 (1992).
2. Bijnens, N.*et al.*, *Z. Phys.* **A356**, 3 (1996).
3. Allatt, R.G. *et al.*, *Phys. Lett.* **B437**, 29 (1998).
4. Andreyev, A.N. *et al.*, *J. Phys.* **G25**, 835 (1999).
5. Griffin, J.G. *et al.*, *Nucl. Phys.* **A530**, 401 (1991).
6. Vanhorenbeeck, J. *et al.*, *Nucl. Phys.* **A531**, 63 (1991).
7. Andreyev, A.N. *et al.*, *Phys. Rev. Lett.* **82**, 1819 (1999).
8. Leino, M. *et al.*, *Nucl. Instr. & Meth.* **B99**, 653 (1995).
9. Wauters, J. *et al.*, *Phys. Rev. Lett.* **72**, 1329 (1994).
10. Rosel, F. *et al.*, *Atomic Data and Nuclear Data Tables*, **vol.21**, 91 (1978).
11. Fotiades N. *et al.*, *Phys. Rev.* **C57**, 1624 (1998).

Fine Structure in the Alpha Decay of ^{191g,m}Bi and ^{193g,m}Bi

H. Kettunen[1], J. F. C. Cocks[1], A. N. Andreyev[2], O. Dorvaux[1], K. Eskola[3],
P. T. Greenlees[1], K. Helariutta[1], M. Huyse[2], P. Jones[1], R. Julin[1], S. Juutinen[1],
P. Kuusiniemi[1], M. Leino[1], W. H. Trzaska[1], J. Uusitalo[1], K. Van de Vel[2], and
P. Van Duppen[2]

[1] *Dept. of Physics, University of Jyväskylä, FIN-40351, Jyväskylä, Finland*
[2] *University of Leuven, Celestijnenlaan 200 D, B-3001, Leuven, Belgium*
[3] *Dept. of Physics, University of Helsinki, FIN-00014, Helsinki, Finland*

Abstract. *Two new hindered α transitions are presented for ^{191}Bi and ^{193}Bi. The α-γ concidence method was used to search for these weak transitions. The new alpha particle energies for ^{191}Bi are 6581(9) keV and 6343(5) keV and for ^{193}Bi the energies are 6160(10) keV and 5860(15) keV.*

Some of the most famous examples of shape coexistence occur in very neutron deficient even-even Pt, Hg, and Pb isotopes [1]. Odd-Z, even-N nuclei near the Z=82 closure also provide excellent test cases for this phenomenon. In odd-mass Au (Z=79) and Tl (Z=81) isotopes low-lying 9/2$^-$ states have been identified in α-decay measurements. These intruder states result from proton excitations across the Z=82 shell gap to $\pi h_{9/2}$ orbitals and are associated with an oblate-deformed minimum which coexists with the near-spherical 1/2$^+$($\pi s_{1/2}$) ground state. In Bi (Z=83) isotopes, the ground states are $\pi h_{9/2}$ particle states and the intruder states are $\pi s_{1/2}$ hole states. Alpha-decay studies of odd-mass Bi isotopes show that the decay rates are sensitive to the proton particle-hole character of the initial and final states [2], [3], [4]. That is, the 9/2$^-$ ground states decay in an unhindered manner to the 9/2$^-$ Tl intruder states and the intruder states decay in a similar way to the 1/2$^+$ Tl ground states. Decays from 9/2$^-$ states to 1/2$^+$ states, and vice versa, are hindered.

In the present work, fine structure in the α-decay of the 9/2$^-$ and 1/2$^+$ states in ^{191,193}Bi has been studied. The nuclei were produced using fusion evaporation reactions: ^{160}Dy(^{36}Ar,p4n)^{191}Bi, ^{166}Er(^{32}S,p4n)^{193}Bi, and ^{160}Dy(^{40}Ar,p6n)^{193}Bi. The fusion evaporation products were separated from the primary beam and fission products using the RITU gas-filled recoil separator [5]. The recoiling nuclei were implanted in a position-sensitive Si detector placed at the focal plane. A single TESSA-type (25 % relative efficiency) Ge detector was placed adjacent to the Si detector in order to detect γ-rays in prompt coincident with α particles. In order to search for new α transitions in ^{191}Bi and ^{193}Bi, α-γ matrices were constructed. This information leads to the construction of the decay schemes shown in figure 1. The γ-ray transitions of energy 300 keV and 319 keV are assigned as the known decays from the 3/2$^+$ states to 1/2$^+$ states in ^{191}Bi and ^{193}Bi, respectively [2]. The hindrance factors for the new α transitions (found in this work), which are feeding these 3/2$^+$ states, are shown in figure 1 (illustrated by stars), along with those for

CP495, *Experimental Nuclear Physics in Europe*, edited by B. Rubio et al.
© 1999 American Institute of Physics 1-56396-907-6/99/$15.00

other transitions. The systematic behaviour of hindrance factors can be used to extract nuclear structure information. The transitions represented in fig 1. can be grouped into two sets: unhindered (small hindrance factors) and hindered (large hindrance factors). The $9/2^-$ to $9/2^-$ and $1/2^+$ to $1/2^+$ transitions are unhindered. The $9/2^-$ to $3/2^+$, $1/2^+$ to $3/2^+$ and $9/2^-$ to $1/2^+$ transitions, on the other hand, are hindered. This can be understood in terms of proton particle-hole configurations. Decay of the $\pi(1p)9/2^-$ Bi ground state to the $\pi(2p-1h)\,1/2^+$ Tl intruder state proceeds via the removal of 2 protons from the Z=82 core. It is a Pb-like decay and is thus unhindered. Similarly, decay of the $\pi(2p-1h)\,1/2^+$ intruder state to the $\pi(1h)\,1/2^+$ Tl ground state can be thought of as a Po-like decay. Large hindrance factors are measured for the $9/2^-$ to $1/2^+$ transitions. The strong hindrance can be understood since these decays are transitions between $\pi(1p)$ and $\pi(1h)$ states, and thus involve the breakup of a pair of protons. The $9/2^-$ to $3/2^+$ transitions are also strongly hindered. This is not surprising since the breakup of a proton pair is again involved, but here the odd proton in the daughter nucleus occupies the $d_{3/2}$ orbital rather than the $s_{1/2}$ orbital.

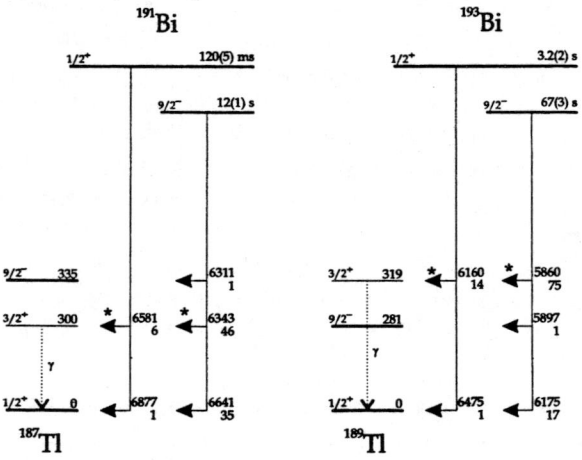

Figure 1. *Decay schemes for ^{191}Bi and ^{193}Bi.*

*This work was supported by the Academy of Finland and the Access to Large Scale Facility program under the Training and Mobility of Researchers (TMR) program of the European Union (EU).

References
[1] M. Muikku et al., Phys. Rev. C **58**, R3033 (1998) and references therein.
[2] E. Coenen et al., Phys. Rev. Lett. **54**, 1783 (1985)
[3] J. Wauters et al., Phys. Rev. C **55**, 1192 (1997)
[4] J. C. Batchelder et al., Eur. Phys. J. A **5**, 49 (1999)
[5] M. Leino et al., Nucl. Instrum. Methods B Phys. res. **99**, 653 (1995).

B(E1) rates in ^{229}Ra and traces of octupole correlations at the upper border of the A=225 island

L.M. Fraile[a], A.J. Aas[b], M.J.G. Borge[a], B. Fogelberg[c],
L.M. García-Raffi[d], I.S. Grant[e], K. Gulda[f], E. Hagebø[b],
W. Kurcewicz[f], G. Løvhøiden[g], H. Mach[c], T. Martínez[d], B. Rubio[d],
J.L. Taín[d], O. Tengblad[a,h], T.F. Thorsteinsen[i] and the ISOLDE
Collaboration.

[a] *Instituto de Estructura de la Materia, CSIC, Serrano 113 bis, E-28006 Madrid, Spain*
[b] *Dept. of Chemistry, Univ. of Oslo, P.O.Box 1033, Blindern, N-0315 Oslo, Norway*
[c] *Dept. of Neutron Research, Univ. of Uppsala, S-61182 Nyköping, Sweden*
[d] *Instituto de Física Corpuscular, CSIC–Univ. Valencia, E-46100 Burjassot, Spain*
[e] *The Schuster Laboratory, Univ. of Manchester, Manchester M13 9PL, UK*
[f] *Dept. of Physics, Univ. of Warsaw, Pl-00 681 Warsaw, Poland*
[g] *Dept. of Physics, Univ. of Oslo, P.O.Box 1048, Blindern, N-0316 Oslo, Norway*
[h] *EP Division, CERN, CH-1211 Geneva 23, Switzerland*
[i] *Dept. of Physics, Univ. of Bergen, N-5000 Bergen, Norway*

Abstract. The introduction of the fast timing $\beta\gamma\gamma$(t) method at ISOLDE (CERN) has opened the heavy actinide region to lifetime measurements down to tenths of picoseconds, providing a direct experimental means of measuring the E1 rates. Recently, this method has been successfully applied to ^{229}Ra, located at the border of the A=225 island of octupole deformation. In the light of the new results, structural similarities are established between ^{229}Ra and its N=141 isotone ^{231}Th, investigated with the same technique. The comparison shows that some collective aspects of both nuclei are related, but that the strength of the E1 transitions presents interesting differences. The collective B(E1) rates are also discussed in comparison with neighbouring nuclei in the upper border of the island of octupole deformation at A=225.

INTRODUCTION

One of the most characteristic fingerprints of the presence of octupole correlations in odd A nuclei is the appearance of negative and positive parity bands, close in energy and interconnected by enhanced E1 transitions. The magnitude of the E1 rates can be found up to two orders of magnitude higher than those in nuclei without octupole correlations. The fast timing $\beta\gamma\gamma$(t) method [1] constitutes a

CP495, *Experimental Nuclear Physics in Europe*, edited by B. Rubio et al.

key tool in the study of this kind of nuclei, as it provides a direct experimental technique for the measurement of E1 rates. The $\beta\gamma\gamma(t)$ method has been introduced [2] at the PSB ISOLDE separator for the study of the heavy actinide region and successfully applied to ^{229}Ra [3], at the upper border of the A=225 island of octupole deformation.

The excited structure of ^{229}Ra was populated by beta decay from ^{229}Fr, produced at the ISOLDE (CERN) facility. The experimental setup consisted of two independent stations. The first one allowed the use of the fast timing $\beta\gamma\gamma(t)$ method and the collection of $\gamma\gamma$ coincidences. At the second station the internal conversion coefficients were measured by means of a mini-orange spectrometer combined with a HPGe telescope (see Ref. [3] for details). The new experimental data enrich the previously known level scheme of ^{229}Ra by more than twenty new states, among which the low-lying ones provide important information on the band structure of the nucleus. Three parity pairs of rotational bands with $K^\pi = 5/2^\pm$, $1/2^\pm$ and $3/2^\pm$ have been found. The half-lives of five excited states have been measured via the $\beta\gamma\gamma(t)$ method. The experimental values for the transition rates de-exciting such states are obtained from the level lifetimes and the branching ratios.

E1 TRANSITION RATES

The B(E1) values obtained for ^{229}Ra (see Table 1) reveal the enhancement of E1 transitions between the partners of a given parity pair of rotational bands. This characteristic seems to be correlated to the number of nucleons away from the "octupole deformed" A=222 nuclei. The B(E1) rate for the 137.5 keV ΔK=0 transition between the lowest-lying 5/2 bands of opposite parity has been established to be $1.90(12) \times 10^{-4} e^2 fm^2$. This is a moderately fast value which can be taken as the reference for the expected E1 rates in this nucleus.

TABLE 1. Experimental E1 rates for transitions connecting parity doublets in ^{229}Ra

Initial state (keV)	$T_{1/2}^{(i)}$	Final state (keV)	$b_i b_f$[a]	γ-ray E_γ (keV)	$B(\mathbf{E1}; i \to f)$ (e^2fm^2)
137.5	0.66(4) ns	41.3	Aa	96.2	$5.7(6) \times 10^{-5}$
		0.0	Aa	137.5	$1.90(12) \times 10^{-4}$
479.0	≥ 30 ps	213.0	Cb	266.0	$\geq 5.19(18) \times 10^{-5}$
		168.8	Cc	310.1	$\geq 2.6(3) \times 10^{-4}$
		142.7	Cc	336.2	$\geq 1.39(5) \times 10^{-4}$

[a] Band assignment for the *initial* and *final* bands: A: $\frac{5}{2}^-[752]$, C: $\frac{1}{2}^-[501]$, a: $\frac{5}{2}^+[633]$, b: $\frac{3}{2}^+[631]$, c: $\frac{1}{2}^+[631]$.

The equivalent E1 transition rate between the parity doublet band-heads in the closest odd isotope ^{227}Ra [4] is ten times faster, $B(E1)_{90} = 1.41(8) \times 10^{-3} e^2 fm^2$, whereas in the isotone ^{231}Th [5] is much slower, $B(E1)_{186} = 4.6(3) \times 10^{-5} e^2 fm^2$. In the case of the even isotope ^{228}Ra [6] the E1 transitions between the $K^\pi=0^\pm$ bands have $B(E1) \geq 1.2 \times 10^{-4} e^2 fm^2$. These values can be compared with those of a Ra isotope well inside the octupole island as ^{223}Ra with B(E1)$\sim$ $2.5 \times 10^{-3}\, e^2\, fm^2$, and with those of nuclei without octupole correlations, where B(E1)$\sim 10^{-5} - 10^{-6}\, e^2\, fm^2$. This comparison reveals the enhancement of the E1 transition rates, and shows that there is no abrupt disappearance of the octupole correlations in heavy Ra. The B(E1) values obtained for the ΔK=0 E1 transitions between the 479.0 keV 1/2$^-$ bandhead in ^{229}Ra and the 142.7 and 168.8 keV states in the 1/2$^+$ band (see Table 1) are of the same order of magnitude.

A second important feature is the expected retardation of about one order of magnitude of the B(E1) rates for the $\Delta K \geq 1$ E1 transitions in comparison to the ΔK=0 transitions, arising from the K-number selection rules [7]. This aspect can be observed by the fact that the rate for the 266.0 keV E1 transition from the 1/2[501] state at 479.0 keV is a factor of 5 lower than the rate for the 310.1 keV E1 transition feeding the 3/2 state at 168.8 keV, member of the 1/2[651] doublet band.

The dipole moment D_0, although not strictly applicable to octupole transitional nuclei, constitutes a suitable parameter for phenomenological comparison of the B(E1) values for different transitions and nuclei in this region. It can be obtained from the expression [8]

$$B(E1; I_i \rightarrow I_f) = \frac{3}{4\pi}|\langle I_i K_i 10|I_f K_f\rangle D_0 + (-1)^{I_i+1/2}\langle I_i - K_i 11|I_f K_f\rangle D_1|^2,$$

which, using the B(E1) limits given in Table 1, yields a limit of $|D_0| \geq 0.038$ e·fm for the 310.1 and 336.2 keV E1 transitions between the K=1/2$^-$ and the K=1/2$^+$ bands in ^{229}Ra.

This value is lower than the one obtained for the closest odd isotope ^{227}Ra [4] ($|D_0| \geq 0.11$ e·fm), but higher than the one obtained for the isotone ^{231}Th [5] ($|D_0| = 0.010(1)$ e·fm), providing an indication that the octupole correlations are intermediate between both nuclei.

For the 137.5 keV transition connecting the K=5/2$^-$ and K=5/2$^+$ bands in ^{229}Ra a value of $|D_0|$= 0.0334(11) e·fm is obtained, which is very close to the limit for the K=1/2$^\pm$ bands. There again, a lower value of $|D_0|$= 0.017(1) e·fm is obtained for the K=5/2$^\pm$ bands in ^{231}Th.

In order to summarize the behaviour of the $|D_0|$ values, Fig. 1 illustrates the ones obtained for the ground state parity partner bands in Ra and Th isotopes. It clearly shows how the $|D_0|$ value for Ra isotopes smoothly diminishes when the number of nucleons increases. In turn, the $|D_0|$ for Th isotopes seems to have reached an end point already in ^{231}Th.

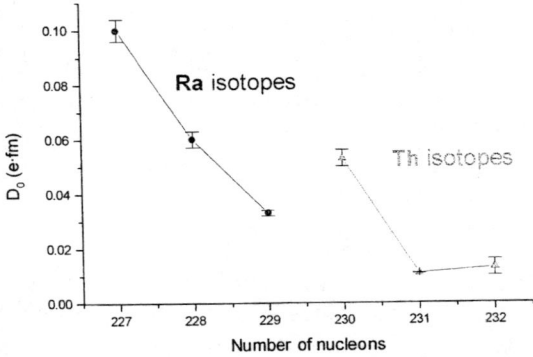

Another revealing feature of the presence of octupole correlations is the energy gap existing between the ground state parity doublet bandhead, in the case of odd-A nuclei, or between the positive parity ground state and the first negative parity excited state, in the case of even nuclei. As shown in Fig. 2 the trend followed by even Ra isotopes is very similar to that of Th isotopes with the same number of neutrons. Odd Ra isotopes seem to have the same tendency as the odd Th isotopes. In order to confirm it data from ^{231}Ra are needed. The analysis of the beta decay of ^{231}Fr to ^{231}Ra is presently being carried out (preliminary results are presented elsewhere in the proceedings).

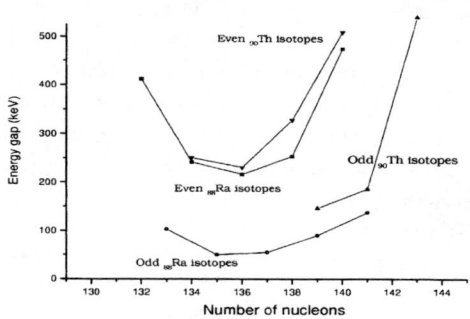

The previous discussion gives a coherent picture of the region, not only because of the smooth disappearance of the octupole correlations, but also because of the similarities between the low-lying quadrupole structures of the isotones. In fact, the quadrupole features can be shown by intercomparing the E1 transitions between a given state of one band and members of its parity doublet. In the case of the K=5/2$^{\pm}$ bands in ^{229}Ra the ratio B(E1)$_{5/2\rightarrow7/2}$ / B(E1)$_{5/2\rightarrow5/2}$ is 0.30, whereas the value obtained from the angular momentum selection rules is 0.40. The value obtained for the same ratio in ^{231}Th is 0.41. For the K=1/2$^{\pm}$ bands in ^{229}Ra the ratio B(E1)$_{1/2\rightarrow3/2}$ / B(E1)$_{1/2\rightarrow1/2}$ is 1.9(2), in agreement with the value of 2.0 obtained from the angular momentum selection rules. The same transitions in the isotope ^{227}Ra give a ratio of precisely 2.0.

CONCLUSIONS

The study of the E1 transition rates in ^{229}Ra and neighbouring nuclei provides a way of understanding the role of the quadrupole and octupole collectivities in the heavy actinide region. The study shows that the effect of octupole correlations in ^{229}Ra is more intense than in ^{231}Th, but not so pronounced as in ^{227}Ra. The comparison with the closest nuclei in the region reveals a smooth disappearance of octupole correlation effects.

REFERENCES

1. Mach, H. et al., *Nucl. Instr. Meth.* **A280** (1989) 49.
2. Mach, H. et al., in *Nuclear Shapes and Nuclear Structure at Low Excitation Energies* (Editions Frontières, Gif-sur-Yvette, 1994) 391.
3. Fraile, L.M. et al., *Nucl. Phys.* **A** (1999), in press.
4. Aas, A.J. et al., *Nucl. Phys.* **A611** (1996) 281.
5. Aas, A.J. et al., *Nucl. Phys.* **A** (1999), in press.
6. Gulda, K. et al., *Nucl. Phys.* **A636** (1998) 28 .
7. Löbner, K.E.G., *Phys. Lett.* **B26** (1968) 369.
8. Bohr, A., Mottelson, B., *Nuclear Structure, II*, (W.A. Benjamin, London, 1975) 114.

Nuclear structure of ^{231}Ra

R. Boutami[a], L.M. Fraile[a], M.J.G. Borge[a], A.J. Aas[b],

B. Fogelberg[c], L.M. García-Raffi[d], I.S. Grant[e], K. Gulda[f],

E. Hagebø[b], W. Kurcewicz[f], M.J. López-Jiménez[a], G. Løvhøiden[g],

H. Mach[c], T. Martínez[d], B. Rubio[d], J.L. Taín[d], A.G. Teijeiro[a],

O. Tengblad[a,h], T.F. Thorsteinsen[i] and the ISOLDE Collaboration.

[a] Instituto de Estructura de la Materia, CSIC, Serrano 113 bis, E-28006 Madrid, Spain
[b] Dept. of Chemistry, Univ. of Oslo, P.O.Box 1033, Blindern, N-0315 Oslo, Norway
[c] Dept. of Neutron Research, Univ. of Uppsala, S-61182 Nyköping, Sweden
[d] Instituto de Física Corpuscular, CSIC–Univ. Valencia, E-46100 Burjassot, Spain
[e] The Schuster Laboratory, Univ. of Manchester, Manchester M13 9PL, UK
[f] Dept. of Physics, Univ. of Warsaw, Pl-00 681 Warsaw, Poland
[g] Dept. of Physics, Univ. of Oslo, P.O.Box 1048, Blindern, N-0316 Oslo, Norway
[h] EP Division, CERN, CH-1211 Geneva 23, Switzerland
[i] Dept. of Physics, Univ. of Bergen, N-5000 Bergen, Norway

Abstract. The study of the upper border of the octupole deformation region near A=225, where the octupole deformation vanishes in the presence of a well developed quadrupole field, is of great relevance in order to understand the interplay of octupole and quadrupole collectivities. Within the IS322 collaboration at CERN we carry out a systematic investigation of the heavy Fr – Th nuclei that presently includes ^{227}Fr, ^{227,228,229}Ra, ^{229}Ac and ^{229,231}Th. The heaviest Ra isotope we have studied so far and in which the fast timing $\beta\gamma\gamma$(t) method has been applied is ^{231}Ra.

In the last two decades experimental signatures of strong octupole collectivity have been found in the region around A=225, in whose upper border the nucleus ^{231}Ra is located. Experimental studies of this octupole transitional region are essential in order to understand the interplay of quadrupole and octupole collectivities and to reveal the exact mechanism by which the octupole deformation disappears in the presence of a well developed quadrupole field. The theoretical expectations for ^{231}Ra point towards a nucleus already outside the octupole region, with strongly reduced octupole correlations in comparison to lighter radium isotopes.

Previous knowledge on the excited states of ^{231}Ra consisted of the assignment of the three strongest gamma lines obtained in beta decay studies [1]. In the present work, the levels in ^{231}Ra have been populated in the β^- decay of ^{231}Fr. Our study significantly extends the spectroscopic information and includes the first level lifetime measurements in ^{231}Ra using the fast timing $\beta\gamma\gamma$(t) method [2].

The experimental setup consisted of two independent, but simultaneously

CP495, *Experimental Nuclear Physics in Europe*, edited by B. Rubio et al.
© 1999 American Institute of Physics 1-56396-907-6/99/$15.00

operated, measurement stations interconnected by a tape transport system. The first station, *fast timing station*, was situated at the point of beam deposition and was specifically designed for the fast timing measurements with the $\beta\gamma\gamma(t)$ method providing at the same time the necessary $\gamma\gamma$-coincidence information to build the level scheme. The second station, *conversion electron station*, situated one meter away from the first one, was especially designed for internal conversion coefficient measurements. It consisted of an electron spectrometer including a Si(Li) detector and a MINI-ORANGE filter for the detection of conversion electrons. A gamma telescope consisting of a planar HPGe and a coaxial 20 % HPGe allowed for detection of both gamma and X-rays. A β detector was included as well.

Measurements at the *conversion electron* and *fast timing stations* were conducted simultaneously with the tape cycle times of 26 s (short cycle) and 200 s (long cycle). In this way, a radioactive sample was collected for 26 s or 200 s at the first station and thereafter transported in a very short time (<0.5 s) to the second station, where it was measured before a new sample arrived. Singles spectra for the planar, coaxial and electron detectors, as well as β-gated γ singles spectra for the coaxial HPGe were collected. By dividing the acquisition time into 8 consecutive subgroups of 3.25 s (short cycle) or 25 s (long cycle) each, γ-multispectra from the planar detector and from the β-gated coaxial detector were also recorded allowing for time evolution studies.

The data analysis from the *conversion electron station* is underway. The γ transitions in ^{231}Ra have been identified by comparing to the time evolution of ^{231}Ra X-rays and by comparison between short and long cycles. Table 1 summarizes the results. The half-life of the parent nucleus ^{231}Fr has been determined to be $T_{1/2}=17.7(6)$ s, from the ten most intense γ lines and X-rays. This value is in good agreement and more precise than the previously reported in Ref [1].

TABLE 1. Identified gamma transitions in ^{231}Ra

E (keV)	Intensity	E (keV)	Intensity	E (keV)	Intensity	E (keV)	Intensity
49.3(1)	7 ± 1	276.1(2)	7 ± 1	454.1(2)	793±25	706.5(4)	30± 4
66.2(1)	52 ± 2	285.7(2)	70 ± 3	492.7(2)	22 ± 2	730.0(4)	30± 5
82.5(1)	14 ± 1	309.8(2)	59 ± 2	525.3(2)	85 ± 4	746.9(4)	25± 2
83.3(1)	25 ± 1	314.4(5)	5 ± 2	539.2(2)	59 ± 6	834.7(4)	16± 2
86.2(2)	41 ± 3	331.1(2)	51 ± 2	551.3(3)	22 ± 3	1583.1(5)	62± 2
95.5(1)	151± 6	346.5(2)	17 ± 1	572.7(4)	19 ± 2	1605.5(4)	35± 2
96.8(3)	4 ± 1	363.4(2)	51 ± 2	617.2(3)	58 ± 3	1642.5(5)	43± 2
123.1(2)	8 ± 1	415.7(2)	35 ± 2	674.5(4)	50 ± 3	1682.8(6)	31± 2
239.4(2)	18 ± 1	432.6(2)	1000±33	684.6(5)	44 ± 6	1686.1(5)	39± 2
273.7(2)	12 ± 1	445.1(2)	49 ± 2	694.4(5)	18 ± 3	1876.4(6)	22± 1

REFERENCES

1. Hill, P. et al., *Z. Phys.* **A320** (1985) 531.
2. Mach, H. et al., *Nucl. Instr. Meth.* **A280** (1989) 49.

Identification of Excited States in ^{226}U: Evidence for Octupole Deformation

P.T. Greenlees[1,2], N. Amzal[1], A. Andreyev[3], P.A. Butler[1],
K.J. Cann[1], J.F.C. Cocks[1,2], O. Dorvaux[2], T. Enqvist[2,7], P. Fallon[4],
B. Gall[6], M. Guttormsen[5], D. Hawcroft[1], K. Helariutta[2],
F.P. Hessberger[7], F. Hoellinger[6], G.D. Jones[1], P. Jones[2], R. Julin[2],
S. Juutinen[2], H. Kankaanpää[2], H. Kettunen[2], P. Kuusiniemi[2],
M. Leino[2], S. Messelt[5], M. Muikku[2], S. Ødegård[5], R.D. Page[1],
A. Savelius[2], A. Schiller[5], S. Siem[5], W.H. Trzaska[2], T. Tveter[5],
J. Uusitalo[2]

[1] *Oliver Lodge Laboratory, University of Liverpool, Liverpool L69 7ZE, UK*
[2] *Department of Physics, University of Jyväskylä, FIN-40351 Jyväskylä, Finland*
[3] *Joint Institute for Nuclear Research, SU-141980 Dubna, Russia*
[4] *Lawrence Berkeley National Laboratory, Berkeley, California 94720, USA*
[5] *Department of Physics, University of Oslo, N-0316 Oslo, Norway*
[6] *Centre de Recherches Nucléaires, F-67037 Strasbourg Cedex, France*
[7] *Gesellschaft für Schwerionenforschung, D-64220 Darmstadt, Germany*

Abstract. The level scheme of ^{226}U has been deduced from the results of two experiments carried out at the University of Jyväskylä, Finland. Both α- and γ-ray-spectroscopic techniques have been employed. The interleaved states of positive- and negative-parity indicate the octupole nature of this nucleus, and the behaviour of the difference in aligned angular momentum between the positive- and negative-parity bands as a function of rotational frequency is consistent with that expected for a rotating reflection-asymmetric shape.

INTRODUCTION

A long-standing prediction by Nazarewicz and co-workers is that actinide nuclei with $Z = 88 - 92$ and $N \simeq 134$ should possess deep minima in the potential energy surface for non-zero β_3, with gains in potential energy over that for a reflection-symmetric shape of approximately 0.2 to 0.5 MeV [1]. Nuclei in this region are difficult to study using standard γ-ray spectroscopic techniques, due to strong fission competition and a lack of suitable beam and target combinations. Recent studies, employing multi-nucleon transfer reactions, have shown that only five nuclei in this

CP495, *Experimental Nuclear Physics in Europe*, edited by B. Rubio et al.
© 1999 American Institute of Physics 1-56396-907-6/99/$15.00

region (^{222,224,226}Ra and ^{224,226}Th) exhibit rotational alignment properties expected for octupole deformation [2]. The calculations of Nazarewicz et al. predict that ^{224,226}U should possess gains in potential energy similar to those for the isotopes of Ra and Th listed above. A study of ^{226}U is possible using the ^{208}Pb(^{22}Ne,4n)^{226}U reaction, for which the maximum production cross-section is approximately 6μb. In order to extract the γ-rays of interest from the high fission background, the technique of recoil-decay tagging (RDT) [3,4] was utilised.

RECOIL-DECAY TAGGING MEASUREMENT

The RDT study used the ^{208}Pb(^{22}Ne,4n)^{226}U reaction at a bombarding energy of 112 MeV, and employed the RITU gas-filled recoil separator [5] together with the JUROSPHERE array of Ge detectors. The γ-ray data obtained allowed the level scheme of ^{226}U (shown in figure 1) to be deduced for the first time. It can be seen that there are two interleaved bands of opposite parity, connected by strong electric dipole transitions. Using measured $E1/E2$ γ-ray branching ratios and rotational model formulae, a value of the ratio of the intrinsic dipole to quadrupole moment, $|D_0/Q_0|$, of $7.9(5)\times10^{-4}$ fm^{-1} was extracted. Due to strong internal conversion and the low efficiency of the JUROSPHERE device at low energies, it was not possible to confidently assign the energy of the 2^+ to 0^+ transition, though a candidate with an energy of 80.5 keV was observed. Further details of this experiment can be found in ref. [6].

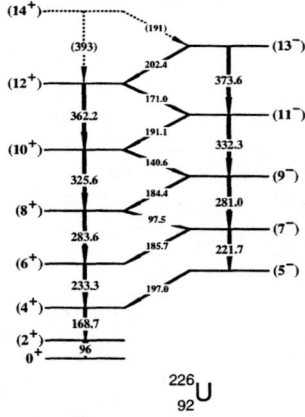

FIGURE 1. Level Scheme of ^{226}U deduced from the present work.

ALPHA DECAY OF ^{230}PU

A further experiment was carried out in order to obtain an improved measurement of the 2^+ to 0^+ transition energy in ^{226}U, through observation of fine structure in the α decay of ^{230}Pu. The experiment again employed the RITU gas-filled recoil separator, and the ^{208}Pb(^{26}Mg,4n)^{230}Pu reaction. A silicon-strip detector at the focal plane of the RITU device was used to observe the decay of fusion-evaporation products implanted after separation from primary beam and fission products. The α-α correlation technique is then used to extract events corresponding the α decay of ^{230}Pu. The spectrum of events correlated to the α decay of either ^{226}U or ^{222}Th within a search time of 900 ms is shown in figure 2.

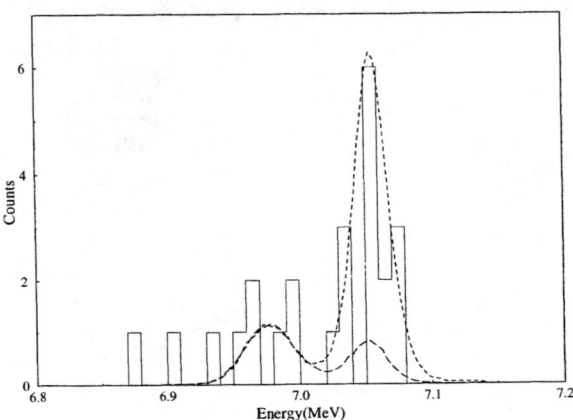

FIGURE 2. Spectrum of events correlated to the α decay of ^{226}U or ^{222}Th, within a maximum time interval of 900 ms. The dotted and dashed lines are output from a Monte Carlo simulation.

It can be seen that there are two groups of events, the lower energy group corresponding to the α decay of ^{230}Pu into the 2^+ state of ^{226}U. Since the decay of ^{230}Pu occurs within the detector, any internal conversion electrons emitted following population of the 2^+ state will sum their energy with that of the α particle, shifting the centroid of the fine structure peak to higher energy. To deduce the energy of the 2^+ to 0^+ transition this effect must be taken into account. This is done through use of a Monte Carlo simulation.

The simulation calculates the energy of an event through knowledge of: the ground-state to ground-state α-particle energy, the excitation energy of the excited state, the detector resolution, the relevant internal conversion coefficients and electron binding energies, and the implantation depth. In this case, the transition energy is not known, and is varied. The transition energy which best reproduces the experimental centroid separation of the two peaks is then taken as the excitation energy of the 2^+ state. The results of this procedure yield an energy of 96(25) keV for the 2^+ to 0^+ transition in ^{226}U. The output from the Monte Carlo simulation is shown as the dotted line in figure 2. Also in figure 2 (dashed line) is output from the simulation which shows those events which originated from population of the 2^+ state. A large fraction of the events are shifted below the ground-state to ground-state α-decay peak, thus the measured intensity ratios must be corrected in order to obtain the α-decay branching ratios. Further analysis and discussion of this experiment can be found in ref. [7].

ROTATIONAL ALIGNMENT PROPERTIES

A plot of the difference in aligned angular momentum, $\Delta i_x = i_x^- - i_x^+$, between the negative- and positive-parity bands as a function of rotational frequency for several $N = 134$ nuclei is shown in figure 3. The value of Δi_x is expected to be

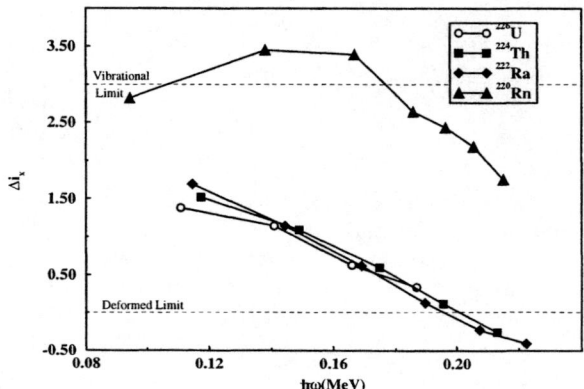

FIGURE 3. Plot of the difference in aligned angular momentum, Δi_x, between the negative- and positive-parity states for several $N = 134$ isotones as a function of rotational frequency.

approximately three for an octupole-vibrational nucleus, and zero for an octupole-deformed nucleus. It can be seen that the behaviour of Δi_x for ^{226}U closely follows that of ^{222}Ra and ^{224}Th. Such behaviour is consistent with that expected for a rotating reflection-asymmetric shape. In the case of ^{220}Rn, the value of Δi_x is approximately three over the full frequency range. This behaviour is due to the rapid alignment of an octupole phonon with the rotation axis. See ref. [2] for a more detailed discussion of these rotational alignment properties.

REFERENCES

1. Nazarewicz, W. *et al.*, *Nucl. Phys.* **A429**, 269 (1984).
2. Cocks, J.F.C. *et al.*, *Nucl. Phys.* **A645**, 61 (1999).
3. Simon, R.S. *et al.*, *Z. Phys. A* **325**, 197 (1986).
4. Paul, E.S. *et al.*, *Phys. Rev.* **C51**, 78 (1995).
5. Leino, M. *et al.*, *Nucl. Instrum. Methods Phys. Res.* **B99**, 653 (1995).
6. Greenlees, P.T. *et al.*, *J. Phys. G* **24**, L63 (1998).
7. Greenlees, P.T. *et al.*, *Accepted for publication in Eur. Phys. J. A.*

Studies of superheavy elements – status and prospects

Sigurd Hofmann

Gesellschaft für Schwerionenforschung (GSI),
Planckstrasse 1, D-64220 Darmstadt, Germany

Abstract. The recent progress in experimental work for the exploration of superheavy elements resulted in the identification of the elements 110 to 112 at GSI Darmstadt and gave evidence for the synthesis of element 114 at JINR Dubna and element 118 at LBNL Berkeley. The data are compared with theoretical predictions. An outlook is given on planned continuation of the experimental work using further improved set-ups.

DECAY PROPERTIES OF SHE

Already in early calculations of the stability of superheavy elements (SHEs) using the Strutinski method, a region of minimum shell-correction energy was obtained for nuclei near Z=108 and N=162 (see Fig. 1 in [1]). This minimum resulted for deformed nuclei at deformation parameters $\epsilon_2 \approx 0.21$ and relatively large values $\epsilon_4 \approx 0.09$. At the time the region of spherical SHEs near Z=114 and N=184 was in the center of interest. Therefore no attention was paid to the region of deformed heavy elements. However, it was the relative stability of the nuclei there which led to the synthesis of the elements 107 to 109 in the years 1981 to 1984 [2] and of the elements 110 to 112 in the years 1994 to 1996 [3].

The new elements were identified by α-decay chains which allowed for a generic parent-daughter correlation. Because the decay chains could be followed down into the region of nuclei of known binding energy, it was possible to determine also the binding energy of the new isotopes. From these values the shell-correction energy could be extracted by subtraction of the macroscopic binding energy calculated from a liquid-drop model. The result was convincing experimental evidence that the stabilizing negative shell-correction energy decreases up to element 109, a result, which created new interest of theoreticians in the region of deformed heavy elements. During the years a number of theoretical work was published which based on the macroscopic-microscopic approach [4, 5, 6]. Recently, also attempts were made to understand the shell structure properties of SHEs on the grounds of self-consistent theories [7, 8].

CP495, *Experimental Nuclear Physics in Europe,* edited by B. Rubio et al.
© 1999 American Institute of Physics 1-56396-907-6/99/$15.00

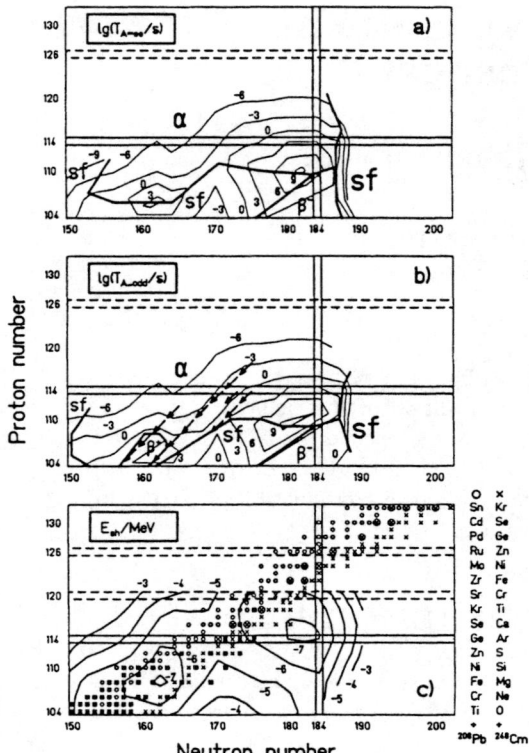

FIGURE 1. Dominating partial α, β or fission half-life for even-even nuclei (a) and odd nuclei (b). The bold lines separate regions of dominant α decay, β decay and spontaneous fission. In (b) the measured decay chains of the isotopes [277]112 [9], [287]114 [10], [289]114 [10], and [293]118 [11] are also drawn. Diagram (c) shows the ground-state shell-correction energy and compound nuclei, which can be reached in reactions with targets of [208]Pb or [248]Cm and stable projectile isotopes. The nuclei presently known are marked by filled squares.

There are several factors which influence the calculations, thus making them uncertain especially in the region of SHEs. The actual location of the three low spin proton subshells $2f_{5/2}$, $3p_{3/2}$ and $3p_{1/2}$, which are filled between Z=114 and 126, may result in a washing out of the shell effect for SHEs. Qualitatively, we may expect a wide and less deep minimum of the negative shell-correction energy in the case that the low spin proton levels are equally distributed in energy between Z=114 and 126. Then, also the fission barriers will be flat and narrow, their height and width is mainly determined by the ground-state shell-correction energy. As a result, the fission half-lives will be relatively short. On the other hand, if only one, but wide energy gap will exist beyond one of the proton numbers 114, 120 or 126, then the shell-correction energy will be pronounced for that element. In combination with the neutron shell effect at N=184 a sharp and deep minimum will be formed, similar to that of the doubly magic [208]Pb, resulting in a high fission barrier and relatively long fission half-life. Also the α half-lives will be stronger modulated by great shell effects resulting in relatively long α half-lives below and short half-lives above the magic number.

Fig. 1 shows the predictions of the macroscopic-microscopic calculations [5, 6] for the half-lives and shell-correction energy of nuclei in the region of SHEs. The most recent decay chains measured at Darmstadt [9], Dubna [10], and Berkeley [11] are also plotted.

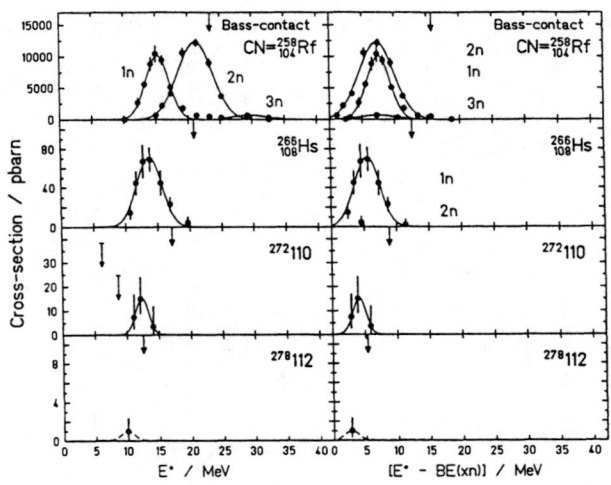

FIGURE 2. Measured even-element excitation functions

SYNTHESIS OF SHE

A summary of recently measured even-element excitation functions is shown in Fig. 2. On the left part, the cross-sections are plotted as a function of the dissipated energy E*, calculated from the center-of-mass beam-energies in the middle of the target thickness and the Q-values. The Q-values were determined from the experimental masses of projectile and target [12] and the mass predictions for the compound nucleus [4]. The arrows mark the energy which is necessary to reach the contact at the mean radii according to the fusion model by Bass [13]. On the right part, the neutron binding energies according to Myers and Swiatecki [4] are subtracted. The resulting free reaction energy is a sum of the kinetic energy of the emitted neutrons and the energy of emitted γ rays. The continuous curves are gaussian fits through the data points, the dashed curve (Z=112) is extrapolated.

In all cases where excitation functions are known, the cross-section maxima on the right in Fig. 2 are approximately centered between zero and an energy resulting from the contact configuration according to the Bass model [13]. This empirical result seems to present a sound means for the determination of the position of the cross-section maxima in cold fusion reactions.

The cross-section trend of the 1n evaporation channel is plotted in Fig. 3. Extrapolation of the curve into the region of heavier elements exhibits a cross-section of about 1 fb for the synthesis of element 116. Higher cross-sections may be expected, if the trend of the data, namely increasing cross-section with increasing isospin, will continue. This effect was proved in the case of element 110 production by use of

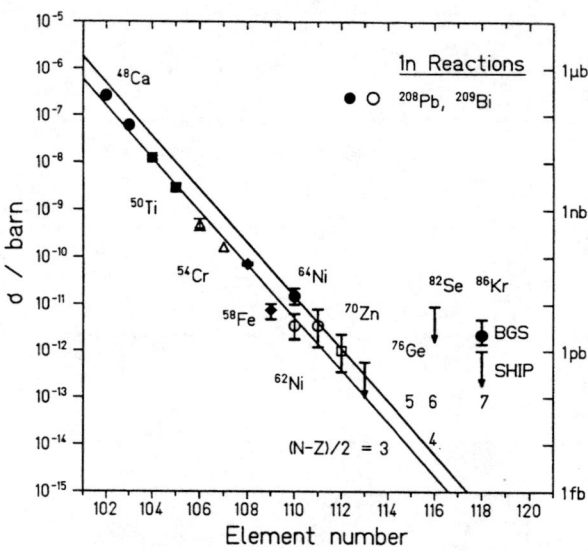

FIGURE 3. Plot of cross-section data for cold fusion reactions

the projectiles ^{62}Ni and ^{64}Ni. The systematics was not confirmed in the case of the synthesis of the elements 112 and 113 using the $T_z=5$ projectile ^{70}Zn.

However, the recent results obtained at Berkeley seem to establish a positive trend of the cross-sections. Based on a prediction by Smolanczuk [14], who calculated a value of 670 pb for the reaction ^{86}Kr + ^{208}Pb → ^{293}118 + 1n, this reaction was investigated at the new Berkeley gas-filled separator BGS. Three decay chains were measured and assigned to the decay of ^{293}118 [11]. A cross-section of $(2.2^{+2.6}_{-0.8})$ pb was obtained.

In a confirmation experiment at SHIP this value could not be repeated yet. During an irradiation time of 24 days a beam dose of 2.9×10^{18} projectiles was reached, using the same reaction and a beam energy resulting at the same excitation energy as in the Berkeley experiment. No event chain similar to the Berkeley data was measured which results in an upper limit of 1.0 pb (1 event would be 0.6 pb, preliminary anaylsis). However, due to statistical uncertainties, the negative SHIP result is still in agreement with a cross-section at the lower end of the range given by the Berkeley data (see Fig. 3).

A comparison of excitation energies at the barrier for cold and hot fusion reactions over a wide range of SHEs is shown in Fig. 4. A remarkable transition is observed from a region of high excitation energies (>40 MeV) for reactions with ^{238}U target resulting in elements up to Z≈114 into a region of low excitation energies, down to 6 MeV for element 126. This reflects a change from hot fusion to cold fusion with regard to the excitation energy. It is however likely that the excitation energy is not the most important quality parameter for distinguishing the two types of reaction. Another, possibly more important difference could be that ^{208}Pb as a target

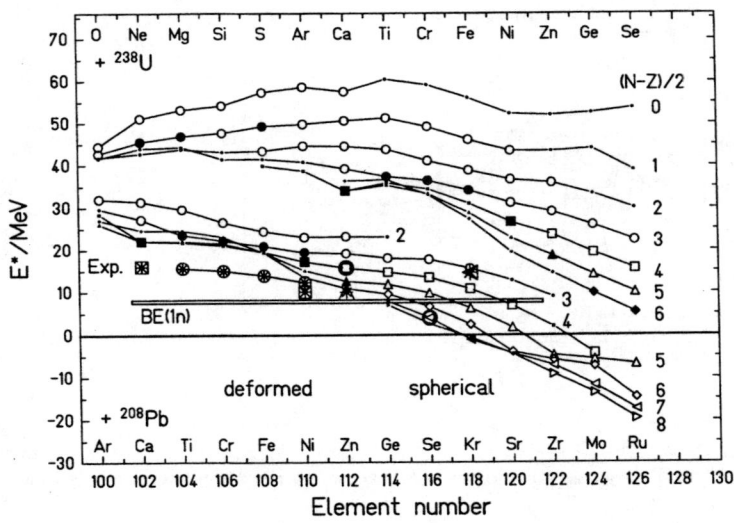

FIGURE 4. Diagrams of excitation energies at the interaction distance according to the fusion model of Bass [13].

is a spherical closed shell nucleus with empty shells above the closure, whereas ^{238}U is well deformed and midshell.

A pronounced structure shows the curve of $T_z=4$ projectile nuclei with a local minimum excitation energy for production of element 112 with a ^{48}Ca beam. A systematic investigation of hot fusion reactions with ^{48}Ca was started recently at the U400 at Dubna. In irradiation of ^{244}Pu and ^{242}Pu targets, 1 and 2 decay chains were measured which were assigned to the isotopes ^{289}114 and ^{287}114, respectively [10].

Following the rule of maximum cross-sections of cold fusion reactions worked out by means of Fig. 2, the curves in Fig. 4 allow for the extrapolation of the trend beyond element 112. There is, e.g., only a very narrow window of excitation energy left for the production of element 114 with a ^{76}Ge beam and 1n emission.

At the barrier, the kinetic energy in the center of mass system is converted in potential energy, and the reaction partners come to rest in a central collision in a touching configuration. Fig. 5 shows the relations in an energy-distance diagram for the reaction ^{64}Ni + ^{208}Pb. The initially existing kinetic energy of 236.2 MeV, at which the cross-section maximum was measured, is exhausted by the Coulomb potential at a distance of 14 fm between the reaction partners. At that distance only nucleons at the outer surface are just in contact. Nevertheless, at that energy this heavy system has highest probability to fuse and to survive.

Calculations of the fusion barrier by Möller et al. [15] show evidence for the

141

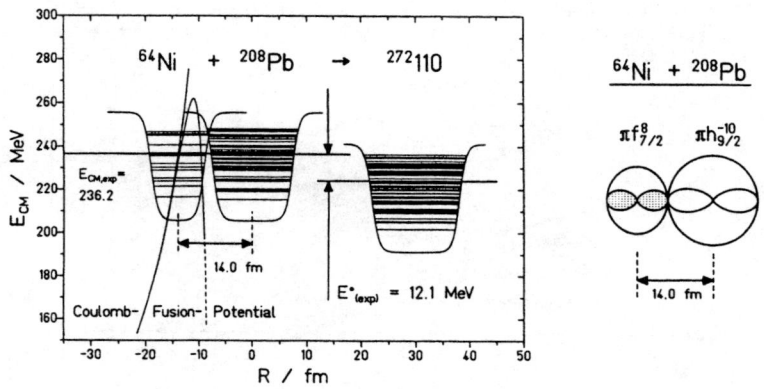

FIGURE 5. Energy against distance diagram for the synthesis of element 110.

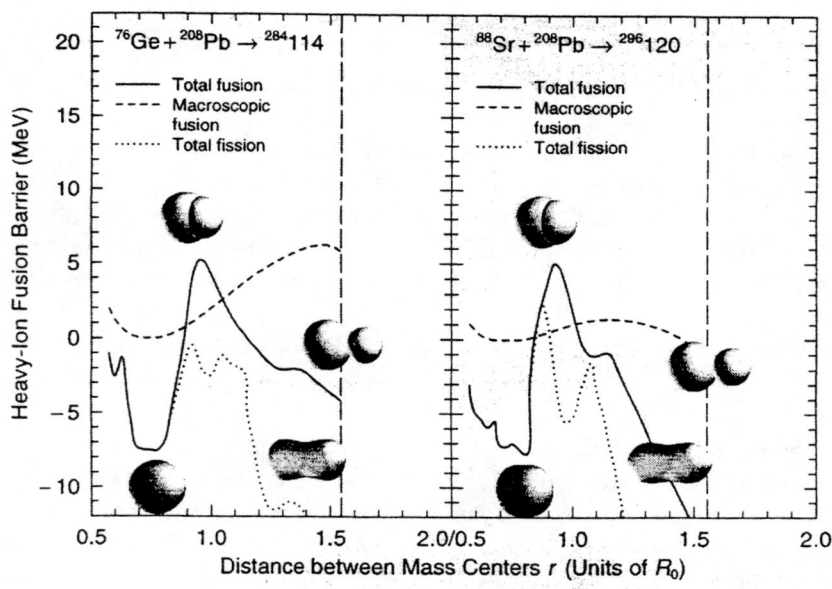

FIGURE 6. Calculated barriers for the synthesis of element 114 and 120 via cold fusion [15].

appearance of an inner barrier in cold fusion reactions when the compound nucleus is beyond element 112. The inner barrier hinders fusion at energies which are just high enough to reach the contact point. The height of the barrier for the fusion of the elements from 114 to 120 is about 13 MeV above the ground-state of the compound nucleus (see Fig. 6). The existence of the inner barrier may explain the negative result of our search for the 0n channel in the case of element 116 [3] and also the positive Berkeley result for the synthesis of element 118 by 1n emission at an excitation energy of 13.3 MeV [11].

OUTLOOK

The continuation of experiments into the region of SHEs at decreasing cross-sections will be provided by further accelerator developments. At increased beam currents, values of few 100 pμA may become available, the cross-section level to perform experiments can be shifted down into the region of 1 fb in principal. But these high currents, again, demand new target and separator developments. An efficient target cooling and further suppression of background is needed. Supplementary equipment as proposed by the VEGA and SHIPtrap projects [16, 17] will open the region of SHEs also for gamma and high resolution mass spectroscopy.

The experimental program will be significantly influenced by the recent results on the synthesis of the elements 112 and 114 by hot fusion at Dubna and the synthesis of element 118 by cold fusion at Berkeley. Despite the exciting new results, many questions of more general character are still open and are waiting to be answered by future improved experiments. A short list is given in the following: 1. Proof of the shell effect at Z=114, 120 or 126. 2. Ground-state to ground-state α decay of even-even nuclei for more accurate evaluation of nuclear binding energies. 3. Search for α transitions of even-even nuclei into rotational levels for determination of the degree of deformation, especially in the region of nuclei near N=162. 4. Fission branchings of even-even nuclei for comparison with theoretically predicted fission half-lives. 5. Extension of the cross-section systematics by measurement of complete excitation functions. 6. Comparison of cross-sections of various combinations of odd and even reaction partners as probably the best approach to understand the reaction mechanism on a microscopic level. 7. Search for radiative capture processes (0n channel).

We have good reasons to believe that the investigation of SHEs is not a closed field of research, but that during the coming years more data will be measured in order to better understand the stability of the heaviest elements and the processes which lead to fusion.

ACKNOWLEDGEMENTS

The experiments at SHIP were performed together with P. Armbruster, H.G. Burkhard, H. Folger, F.P. Heßberger, B. Kindler, B. Lommel, G. Münzenberg, V. Ninov, S. Reshytko, H.J. Schött, C. Stodel (GSI Darmstadt), A.N. Andreyev, A.Yu. Lavrentev, A.G. Popeko, A.V. Yeremin (FLNR-JINR Dubna), R. Janik, S. Šaro (University Bratislava) and M. Leino, University Jyväskylä. The most important prerequisite to realize the experimental program was a stable and high current beam of accurate energy from the UNILAC. It is a great pleasure to thank for their efforts the people from the ion-source group, the accelerator, the target laboratory and the department of experimental electronics and data acquisition. Concerning the theory of the synthesis and stability of SHEs, I am particularly grateful for fruitful discussions with N.V. Antonenko, S. Ćwiok, W. Greiner, P. Möller, W. Nazarewicz, W. Nörenberg, W. von Oertzen, W. Reisdorf, R. Smolanczuk and A. Sobiczewski.

REFERENCES

[1] Nilsson, S.G., et al., *Nucl. Phys. A* 131, 1 (1969).

[2] Münzenberg, G., *Rep. Prog. Phys* 51, 57 (1988).

[3] Hofmann, S., *Rep. Prog. Phys.* 61, 639 (1998).

[4] Myers, W.D., and Swiatecki, W.J., *Nucl. Phys. A* 601, 141 (1996).

[5] Möller, P., et al., *Atomic Data and Nucl. Data Tables* 59, 185 (1995).

[6] Smolanczuk, R., and Sobiczewski, A., Proc. XV. Nucl. Phys. Conf., St. Petersburg, 313 (1995).

[7] Ćwiok, S., et al., *Nucl. Phys. A* 611, 211 (1996).

[8] Rutz, K., et al., *Phys. Rev. C* 56, 238 (1997).

[9] Hofmann, S., et al., *Z. Phys. A* 354, 229 (1996).

[10] Oganessian, Yu.Ts., et al., *Nature* 400, 242 (1999).

[11] Ninov, V., et al., *Phys. Rev. Lett.* (1999), submitted.

[12] Audi, G., and Wapstra, A.H., *Nucl. Phys. A* 565, 1 (1993).

[13] Bass, R., *Nucl. Phys. A* 231, 45 (1974).

[14] Smolanczuk, R., *Phys. Rev. C* 59, 2634 (1999).

[15] Möller, P., et al., Proc. of the Int. Conf. on Exotic Nuclei and Atomic Masses, ENAM 98, Bellaire, Michigan, USA 698 (1998).

[16] Gerl, J., et al., GSI proposal (1998), unpublished.

[17] Äystö, J., et al., (1998), see http://www.gsi.de/~shiptrap.

Nuclear Structure Investigations of Neutron Deficient Nuclei in the Region Z = 103 to 105

F.P. Heßberger[1], S. Hofmann[1], D. Ackermann[1,2], P. Armbruster[1],
G. Münzenberg[1], Ch. Stodel[1], A.Yu. Lavrentev[3], A.G. Popeko[3],
A.V. Yeremin[3], S. Saro[4], M. Leino[5]

[1]GSI, Darmstadt, Germany, [2]Johannes Gutenberg-Universität Mainz, Germany [3]JINR, Dubna, Russia,
[4]Comenius University, Bratislava, Slovakia, [5]University Jyväskylä, Jyväskylä, Finland

Abstract. The isotopes ^{257,255}Rf, ^{257,256}Db, ^{253,252}Lr have been produced in bombardments of ^{207,208}Pb and ^{209}Bi target nuclei with ^{50}Ti and identified by their α-decay. New or improved decay data could be obtained. Analysis of the fine structure of the α-decay pattern of ^{257}Rf allowed the construction of a first tentative level scheme for the daughter nucleus ^{253}No and also the identification of a low lying high spin isomeric state, while from α-γ- coincidence measurements for ^{255}Rf a first tentative level scheme of the daughter nucleus ^{251}No was derived.
For ^{257}Db we found that two nuclear levels decay by α-emission and populate also different levels in the daughter nuclues ^{253}Lr. The levels are produced by the reaction process.
In bombardments of ^{209}Bi with ^{50}Ti at E^{*}_{CN} = 26.4 MeV and 30.8 MeV the previously unknown isotopes ^{256}Db and ^{252}Lr were identified.

INTRODUCTION

The search for superheavy elements (SHE) predicted close to the double magic nucleus ^{298}114 is still one of the most fascinating motivations for extensive experimental work in the field of nuclear physics. Much effort to produce and to detect SHE in the laboratory as well as to search for them in nature has been brought up during the past thirty years. Although some possibly promising results have been reported recently (1,2) the unambiguous identification of the first superheavy nucleus is yet to be done.

Rececent theoretical calculations (3,4) predict two areas of enhanced stability in the transactinide region: One around Z = 114 and N = 184, representing the 'classical' SHE region and in addition a second one centered around Z = 108 and N = 162 with almost equal maximum shell correction energies but β_2 – values of up to ≈0.24, representing 'deformed' nuclear shells.

Numerous nuclei in the region around Z = 108 and N = 162 have been identified during the past years. Although for most of them only basic decay properties are known presently, for some cases a more detailed study was possible allowing for a deeper insight in

CP495, *Experimental Nuclear Physics in Europe*, edited by B. Rubio et al.
© 1999 American Institute of Physics 1-56396-907-6/99/$15.00

their nuclear structure. These results can be compared with theoretical predictions, which certainly will have a feedback to further theoretical investigations.

The experiments were performed at the velocity filter SHIP at GSI using a ^{50}Ti beam and ^{207,208}Pb, ^{209}Bi targets. The separated evaporation residues (ER) from complete fusion were implanted into a position sensitive 16 - strip silicon detector where their α - decay or spontaneous fission (sf) was registered. In one case (^{255}Rf) also γ - rays in coincidence to α - events were measured. Details on the experimental set-up are to be found in (5).

ISOTOPE ^{257}Rf

The isotope ^{257}Rf was studied in detail(6). Twelve α-lines were attributed to this isotope, produced by ^{208}Pb(^{50}Ti,n)^{257}Rf (fig.1a). When produced by α-decay of ^{265}Hs via ^{208}Pb(^{58}Fe,1n)^{265}Hs -- α --> ^{261}Sg -- α --> ^{257}Rf (fig. 1b), however, the α-spectrum had a completely different shape. In the energy region of the two most intense lines at

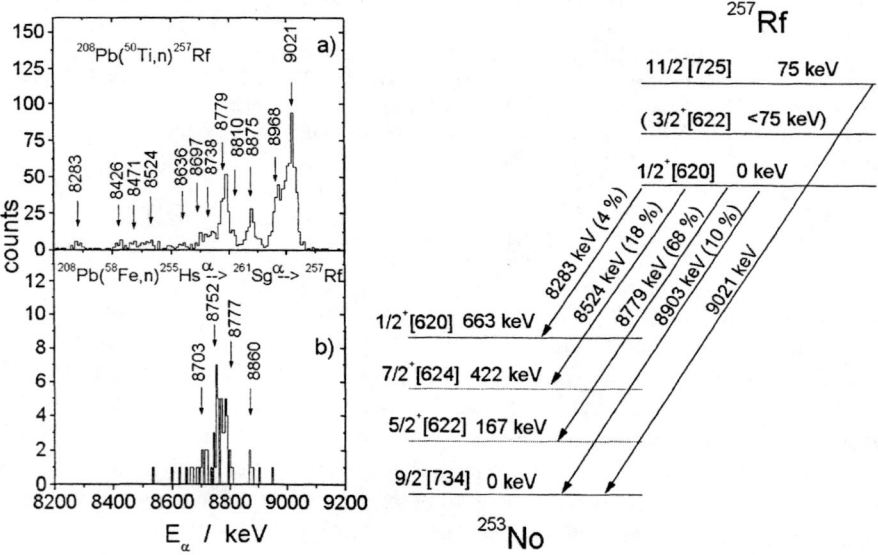

FIGURE 1. α-decay spectrum of ^{257}Rf   **FIGURE 2.** Tentative level schemes for ^{253}No and ^{257}Rf

E_α = 8968, 9021 keV only one decay was observed. It thus seemed evident, that these decays essentially origin from a nuclear level, which is not populated by the α-decay of ^{261}Sg. The one count at E ≈ 8950 keV, however, may indicate also a weak decay from the groundstate (gs) having an energy similar to the E_α = 8968 keV decay from the isomeric state. Based on published level assignments and recent calculations by Cwiok et al.(7) the tentative decay scheme shown in fig. 2 is proposed. Improved data to confirm it are necessary, however. It differs slighty from that proposed in (6) by placing E_α = 8950 keV as the gs to gs transition due to the one correlated decay discussed above. The favoured transition connecting the gs of ^{257}Rf ($1/2^+$) to the analogous state in ^{253}No is strongly suppressed by an unfavourable Q_α - value. The isomeric state is attributed to an $11/2^-$ - state for which γ-decay to the gs is strongly suppressed due to the high spin difference. Comparison of experimental partial half-life and a calculated value results in a hinfrance factor HF$\approx$40, which suggests the $9/2^-$ gs of ^{253}No as the daughter level, since this value would be abnormally low for a transition including both, a change in the parity and a change in the principal quantum number.

ISOTOPE ^{255}Rf

For ^{255}Rf, produced by ^{207}Pb(^{50}Ti,2n)^{255}Rf, besides α - decay also α -γ - coincidences were measured: γ - decays with $E_{\gamma,mean}$ = 203 keV were coincident to α-decays of 8722 keV and those of 142 keV coincident to α- decays of 8773 keV (8). The sums $E_\alpha + E_\gamma$ = 8925 keV and $E_\alpha + E_\gamma$ = 8915 keV are equal within 10 keV, and also close to the mean energy E_α = (8924 $\pm$ 10) keV of α- decays being the most energetic ones correlated to ^{251}No. So it seems evident that both γ-transitions lead to the gs of ^{251}No. Due to the lowest HF value of HF $\approx$ 4 the decay E_α = 8722 keV is attributed to the favoured transition, leading to the analogous Nilssen state. Since the gs of ^{255}Rf is predicted as $9/2^-$[734] (7) the analogous state in ^{251}No must be located above the $5/2^+$[622] level opposite to the theoretical prediction. This again suggests to assign the E_α = 8773 keV line to the transition $9/2^-$[734] $\rightarrow$ $5/2^+$[622], which results in an abnormally low value of HF $\approx$ 13 for this transition. One thus may assume that the $5/2^-$[622] state is populated strongly by internal conversion of the $9/2^-$[734] - state and that the E_α = 8773 keV line is strongly influenced by energy summing of α - particles and conversion electrons, which also could explain the lower sum energy $E_\alpha + E_\gamma$ = 8915 keV.

ISOTOPES ^{257}Db, ^{253}Lr

Improved decay data for ^{257}Db and ^{253}Lr were obtained in a recent experiment aimed to measure an excitation function for ^{50}Ti + ^{209}Bi $\rightarrow$ $^{259-x}$Db + xn. While the previously reported α-energies of ^{257}Db and ^{253}Lr (9) could be reproduced within the error bars, as a new result it was realized that ^{257}Db[1] - α-decays of E_α = 9163 keV ($T_{1/2}$ = 0.76 s) were only followed by ^{253}Lr[1] - α-decays of E_α = 8723 keV ($T_{1/2}$ = 1.49 s) while ^{257}Db[2] - α- decays of E_α = 9074, 8967 keV ($T_{1/2}$ = 1.50 s) were only followed by ^{253}Lr[2] - α- decays of E_α = 8794 keV ($T_{1/2}$ = 0.57 s). This means that two levels in ^{257}Db which also have different half-lives are populated by the reaction process. They decay by α - emission into different levels in the daughter nucleus ^{253}Lr having also different half -lives. No statistically significant difference in the decay energies of the granddaughter ^{249}Md α -particles following either decays of ^{257}Db[1], ^{253}Lr[1] or ^{257}Db[2] ^{253}Lr[2] was found so far. On the basis of calculated level schemes at low excitation energies (7) a possible candidate for an isomeric state is the 1/2$^-$[521] level. The observed decay sequences presently can only be understood if the gs of ^{249}Md is assumed as 7/2$^-$[514] opposite to 1/2$^-$[521] as predicted in (7). Yet, to draw definite conclusions thorough α - γ - and α - conversion electron measurements are necessary.

ISOTOPES ^{256}Db, ^{252}Lr

The new isotopes ^{256}Db and ^{252}Lr were produced by ^{209}Bi(^{50}Ti,3n)^{256}Db -- α -> ^{252}Lr at $E^* = 26.4$, 30.8 MeV. They were identified by α - α - correlations to their decay products ^{248}Md, ^{244}Es, ^{248}Fm (via EC of ^{248}Md), ^{244}Cf. α - energies and half-lives are: ^{256}Db: $E_\alpha = 9017\pm20$ keV (≈0.64), 8891 ± 20 keV (≈0.12), 9075 ± 20 keV (≈0.12), 9120 ± 20 keV (≈0.12), $T_{1/2} = 1.5$ +0.5/-0.3 s. ^{252}Lr: $E_\alpha = 9018\pm20$ keV (0.75), 8974 ± 20 keV (0.25), $T_{1/2} = 0.36$ +0.11/-0.07 s. Also nine sf events were observed at these excitation energies and attributed to the 3n deexcitation channel. Since the odd- odd nucleus ^{256}Db is not expected to have a notable (>0.001) sf branch, the sf events are assigned to ^{256}Rf, produced by ^{256}Db -- EC -> ^{256}Rf -sf ->. We obtain a value of $b_{EC} = 0.35\pm 0.12$ for ^{256}Db.

References

1. Oganessian, Yu.Ts, Yeremin, A.V., Gulbekian, G.G., Bogomolov, S.L., Buklanov, G.V., Chelnokov, M.L., Chepigin,V.I., Gikal, B.N., Gorshkov, V.A., Gulbekian, G.G., Itkis, M.G., Kabachenko ,A.P., Lavrentev, A.Yu., Malyshev, O.N., Rohac, J., Sagaidak, R.N., Hofmann, S., Münzenberg, G., Saro S., Giardina G., Morita K., *Nature* **400**, 242-245 (1999)
2. Ninov, V., Gregorich, K.E., Loveland, W., Ghiorso, A., Hoffman, D.C., Lee, D.M., Nitsche, H., Swiatecki, W.J., Kirbach, U.W., Laue, C.A., Adams, J.L., Patin, J.B., Dhaughnessy, D.A., Strellis, D.A. Wilk, P.A., submitted to *Phys. Rev. Lett.*
3. Smolanczuk, R., Sobiczewski, A. "Shell Effects in the properties of Heavy and Superheavy Nuclei," in *Proceedings of the EPS Conference Low Energy Nuclear Dynamics*, St.Petersburg (Russia), April 18-22, 1995, World Scientific, New Jersey, London, Hong Kong 1995, pp. 313-320 and private communication
4. Möller, P., Nix, J.R., Myers, W.D., Swiatecki, W.J., *Atomic Data and Nuclear Data Tables* **59**, 185 - 381 (1995)
5. Hofmann, S., Ninov, V., Heßberger, F.P., Armbruster, P., Folger, H., Münzenberg, G., Schött, H.-J., Popeko, A.G., Yeremin, A.V., Andreyev, A.N., Saro, S., Janik, R., Leino, M.Z. *Phys. A* **350**, 277-280 (1995)
6. Heßberger, F.P., Hofmann, S., Ninov, V., Armbruster, P., Folger, H., Münzenberg, G., Schött, H.-J., Popeko, A.G., Yeremin, A.V., Andreyev, A.N., Saro, S., *Z. Phys. A* **359,** 415-425 (1997)
7. Cwiok, S., Hofmann, S., Nazarewicz, W., *Nucl. Phys. A* **575,** 356-394 (1994)
8. Heßberger, F.P., "Experiments on the Synthesis of New Superheavy Element", in *Proceedings of the 4th International Conference on Dynamical Aspects of Nuclear Fission,* Casta Papiernicka (Slovakia), October 19-23, 1998, *acta physica slovaka* **49**, 43-52 (1999)
9. Heßberger, F.P., Münzenberg, G., Hofmann, S., Agarwal, Y.K., Poppensieker, K., Reisdorf, W., Schmidt, K.-H., Schneider, J.R.H., Schneider, W.F.W., Schött, H.J., Armbruster, P., Thuma, B., Sahm, C.-C., Vermeulen, D., *Z. Phys. A* **322**, 557-566 (1985)

Identification of Heavy and Superheavy Nuclides Using Chemical Separator Systems

Andreas Türler*

*Paul Scherrer Institute, CH-5232 Villigen PSI, Switzerland
for a Paul Scherrer Institute (PSI) - Bern University - Gesellschaft für Schwerionenforschung (GSI) -
Forschungszentrum Rossendorf (FZR) - Mainz University - Flerov Laboratory of Nuclear Reactions
(FLNR) - Lawrence Berkeley National Laboratory (LBNL) collaboration

Abstract. With the recent synthesis of superheavy nuclides produced in the reactions $^{48}Ca+^{238}U$ and $^{48}Ca+^{242,244}Pu$, much longer-lived nuclei than the previously known neutron-deficient isotopes of the heaviest elements have been identified. Half-lives of several hours and up to several years have been predicted for the longest-lived isotopes of these elements. Thus, the sensitivity of radiochemical separation techniques may present a viable alternative to physical separator systems for the discovery of some of the predicted longer-lived heavy and superheavy nuclides. The advantages of chemical separator systems in comparison to kinematic separators lie in the possibility of using thick targets, high beam intensities spread over larger target areas and in providing access to nuclides emitted under large angles and low velocities. Thus, chemical separator systems are ideally suited to study also transfer and (HI, αxn) reaction products. In the following, a study of (HI, αxn) reactions will be presented and prospects to chemically identify heavy and superheavy elements discussed.

INTRODUCTION

The identification of transactinide nuclides ($Z \geq 104$) using chemical separator systems is interesting from the standpoint of nuclear physics and, of course, also from the standpoint of chemistry. Due to the strong influence of relativistic effects on the electronic structure of the heaviest elements, their place in the Periodic Table is not *a priori* determined only by their nuclear charge. In terms of nuclear physics, a chemical separation can be regarded as a somewhat slow, but very efficient Z separator. Since a chemical experiment is not affected by the reaction kinematics, relatively fast chemical separations ($t_{sep.} \geq 10$ min) have successfully been used to study simultaneously a multitude of transfer reaction products induced by heavy ions on e.g. actinide targets (1). Another domain, where chemistry is very efficient, are (HI, αxn) reactions and (HI, xn) reactions, where the recoil nucleus is produced with a relatively low kinetic energy and emitted under large angles relative to the beam axes. Current chemical separators, such as the **On-Line Gas** chromatography **Apparatus** (OLGA) are able to continuously isolate nuclides with half-lives as short as 5 s, which are produced with cross sections as low as about 25 pb (2). With the discovery of the deformed shells at $Z=108$ and $N=162$ (3), chemical investigations could be extended to Sg ($Z=106$) (4). The half-lives of the nuclides ^{265,266}Sg were determined for the first time in these ex-

CP495, *Experimental Nuclear Physics in Europe*, edited by B. Rubio et al.
© 1999 American Institute of Physics 1-56396-907-6/99/$15.00

periments (2). With the recent synthesis of superheavy nuclei with Z=114, 112, 110, and 108 in the reactions ^{48}Ca+^{244}Pu and ^{48}Ca+^{242}Pu (5) much longer lived nuclei than the previously known neutron-deficient isotopes of the heaviest elements have been identified, suggesting that the island of stability near Z=114 (or Z=120-126) and N=184 has been reached for the first time. Half-lives (or life times) of several seconds to several minutes have been reported for isotopes of elements 108 through 114 (5). However, since the observed α-decay chains lie entirely in an unknown region of the nuclear chart, no final assignment of Z and A of the produced superheavy nuclei could be made. Here, the sensitivity of radiochemical methods may present a viable alternative in conclusively identifying the Z of one of the members of the observed decay chains.

CHEMICAL STUDIES OF (HI, αxn) REACTIONS

In (HI, αxn) reactions very neutron-rich heavy elements can be synthesized, which otherwise are not accessible without using neutron-rich radioactive beams. Therefore, studies of (HI, αxn) reactions have been initiated by our group at PSI. The excitation functions of the reactions ^{248}Cm(^{18}O, α3n)^{259}No and ^{244}Pu(^{22}Ne, α3n)^{259}No were investigated and compared with the 5n deexcitation channels ^{248}Cm(^{18}O, 5n)^{261}Rf and ^{244}Pu(^{22}Ne, 5n)^{261}Rf, respectively. The irradiations were carried out at the PSI Injector I cyclotron. Reaction products were collected directly behind the target in thin Au foils and after the end of the bombardment chemically processed in order to isolate 59-min ^{259}No. The 78-s ^{261}Rf was transported with an aerosol gas-jet to the PSI tape detection system and identified by searching for time correlated α-α decay chains of the ^{261}Rf-^{257}No pair. The measured excitation functions are shown in Fig. 1 (6). In contrast to (HI, 5n) reactions, which are described quite accurately (mostly within a factor of 2) by the HIVAP (7) model, production cross sections of (HI, α3n) reactions are underestimated by about two orders of magnitude. Thus the measured cross sections of the (HI, α3n) reactions were compared with predictions of the "break-up fusion" model (8,9), which yielded much better agreement with the experimental data for the reaction ^{248}Cm(^{18}O, α3n)^{259}No. With the ^{22}Ne projectile ($Q_α$=9.7 MeV) the α3n cross section was underestimated by at least one order of magnitude. Therefore, it was assumed that due to the high $Q_α$, Coulomb excitation was not sufficient for the break-up of the projectile and that the emission of the α-particle occurred only after the fusion barrier has already been penetrated. In this case, the emission of an α-particle efficiently contributes to the cooling of the compound nucleus. However, with this assumption the α3n cross section was considerably overestimated. A satisfactory agreement with experimental data was obtained, when only in about 10% of the collisions the break-up occurred after the fusion barrier had already been penetrated (6). In order to test this concept the reactions ^{248}Cm(^{20}Ne, α3n)^{261}Rf and ^{248}Cm(^{22}Ne, α3n)^{263}Rf were investigated. Only an upper limit of ≤0.77 nb could be established for the cross section of the ^{248}Cm(^{20}Ne, α3n)^{261}Rf reaction at 117 MeV beam energy (6). In the reaction ^{248}Cm(^{22}Ne, α3n) the new nuclide ^{263}Rf is produced, which is expected to decay primarily by α-particle emission with a half-life of about

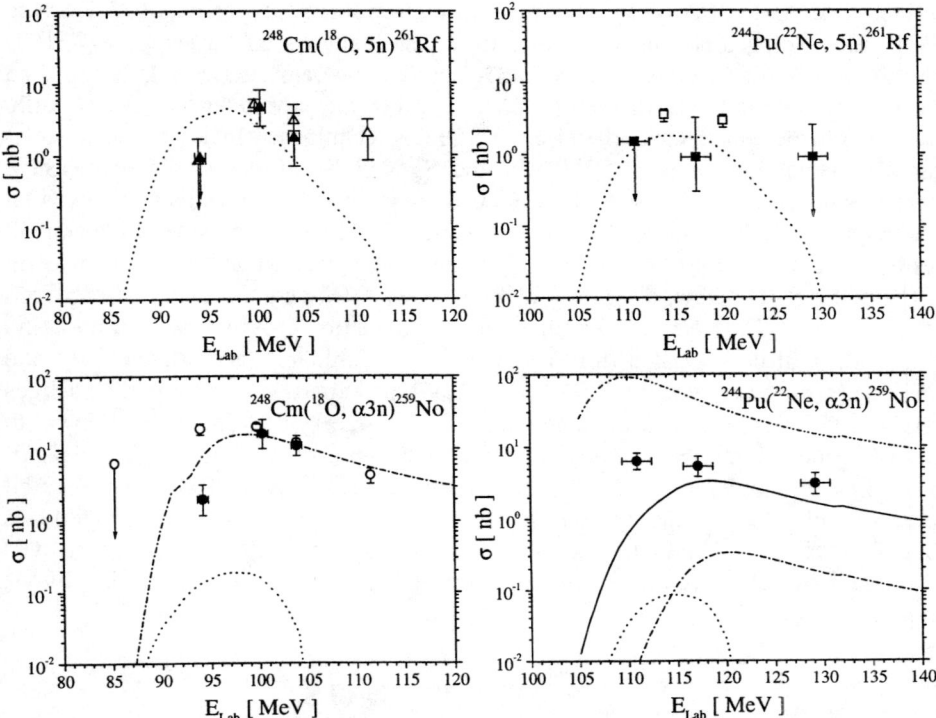

FIGURE 1. Measured excitation functions of the reactions $^{248}Cm(^{18}O, 5n)^{261}Rf$, $^{244}Pu(^{22}Ne, 5n)^{261}Rf$, $^{248}Cm(^{18}O, \alpha3n)^{259}No$, and $^{244}Pu(^{22}Ne, \alpha3n)^{259}No$. Open symbols are data from the literature: Δ, O Ref.(11) + 5MeV, $\square$ Ref.(12). The lines show model predictions: HIVAP (dotted lines), "break-up fusion" with projectile break-up before barrier penetration (dash dotted lines), "break-up fusion" assuming projectile break-up after fusion barrier penetration (dash dot dot line) and logarithmic mean of the two assumptions (solid line).

30 min (10). After chemical separation, which lasted about 30 min, a total of two α decays with the expected α-decay energies (7.79 and 7.87 MeV) and four spontaneous fission (SF) events were recorded (6), which were attributed to SF of ^{259}Md the EC-decay daughter of ^{259}No, which is the α-decay daughter of ^{263}Rf. Further experiments with an improved separation from ^{214}Po (E_α =7.69 MeV) will have to be conducted in order to unambiguously establish the discovery of ^{263}Rf. Future investigations should focus also on other projectiles with particularly high Q_α such as ^{26}Mg (Q_α=10.6 MeV) or ^{48}Ca (Q_α=14.4 MeV) in order to establish the usefulness of (HI, αxn) reactions to synthesize neutron-rich superheavy elements with sizable production cross sections.

CHEMISTRY WITH HEAVY AND SUPERHEAVY ELEMENTS

Due to the very low production cross sections of heavy and superheavy elements, long experiments have to be envisaged. Therefore, also chemical separator systems must be designed to run fully automatically with as few experimenters as possible. Due to their expected place in the Periodic Table, most of the transactinide elements (Rf–118) are ideally suited for gas chemical investigations. The elements Rf through Bh (Z=107) form highly volatile halide and/or oxyhalide species, whereas Hs (Z=108) is expected to form very volatile HsO_4. Elements 112 through 118 are expected to be quite volatile already in the elemental state. Therefore, the OLGA technique, which was already successful in isolating isotopes of Rf, Db, and Sg will be ideally suited to continue chemical studies of the heaviest elements. Currently, chemical investigations focus on three transactinides, namely Bh, Hs, and element 112. At PSI, an experiment lasting 32 days to study the chemical properties of Bh has been approved and will take place in the fall of 1999. The new, neutron-rich Bh isotopes ^{267}Bh and ^{266}Bh will be produced in the reaction ^{249}Bk(^{22}Ne; 4,5n) with expected production cross sections of about 50 pb (13). Bh will be separated from less volatile contaminants (actinides, Pb, Bi, and Po) in the form of very volatile BhO_3Cl. At FLNR, an experiment to study the nuclides ^{270}Hs and ^{269}Hs produced in the reactions ^{238}U(^{36}S; 4,5n) or ^{248}Cm(^{26}Mg; 4,5n) with estimated production cross sections of about 3 to 10 pb (13) is scheduled for spring of 2000. Hs will rapidly be isolated in the form of very volatile HsO_4. The overall efficiency of the chemistry set-up to detect a correlated decay sequence ^{269}Hs $\xrightarrow{\alpha}$ ^{265}Sg $\xrightarrow{\alpha}$ ^{261}Rf could be as high as 30%. Experiments to isolate short-lived Hg isotopes (which serve as a model for element 112) have been conducted at FLNR and look very promising (14).

ACKNOWLEDGMENTS

This work was supported in part by the Swiss National Science Foundation.

REFERENCES

1. von Oertzen, W., "Cold multinucleon transfer and synthesis of new elements" in *Heavy Elements and Related New Phenomena,* eds. Gupta, R.K. and Greiner, W., Singapore: World Scientific, 1998.
2. Türler, A. et al., *Phys. Rev.* **C57**, 1648-1655 (1998).
3. Lazarev, Y.A. et al., *Phys. Rev. Lett.* **73**, 624-627 (1994).
4. Türler, A. et al., *Angew. Chem. Int. Ed.* **38**, 2212-2213 (1999).
5. Oganessian, Yu.Ts., *Nature* **400**, 242-245 (1999).
6. Dressler, R., *Synthese neutronenreicher Isotope schwerster Elemente in xn- und α-xn-Reaktionen,* PhD thesis, University of Bern, 1999.
7. Reisdorf, W., Schädel, M., *Z. Phys.* **A343,** 46 (1992).
8. Kermann, A.K., McVoy, K., *Ann. Phys. (NL)* **122**, 197 (1979).
9. Udagawa, T., Tamura, T., *Phys. Rev. Lett.* **45**, 1311 (1980).
10. Cwiok, S., Hofmann, S., Nazarewicz, W., *Nucl. Phys.* **A573**, 356-394 (1994).
11. Silva, R.J., *Nucl. Phys.* **A216**, 97-108 (1973).
12. Utyonkov, V.K. et al., "Discovery of Enhanced Nuclear Stability near the Shell Closures N=162 and Z=108" in *Proceedings of the VI International School Seminar on Heavy Ion Physics,* eds. Oganessian, Yu.Ts. and Kalpakchieva, R., Singapore: World Scientific, 1998, pp. 400-408.
13. Sagaidak, R., private communication 1999.
14. Yakushev, A.B., private communication, 1999.

Low energy fission of ^{256}No, ^{270}Sg, ^{271}Hs and ^{286}112 nuclei formed in reactions with ^{22}Ne and ^{48}Ca ions

M. G. Itkis[1], N. A. Kondratiev[1], E. M. Kozulin[1], L. Krupa[1], Yu. Ts. Oganessian[1], I. V. Pokrovsky[1], E. V. Prokhorova[1] and A. Ya. Rusanov[2]

[1] *Flerov Laboratory of Nuclear Reactions, Joint Institute for Nuclear Research, 141980, Dubna, Moscow region, Russia*
[2] *Institute of Nuclear Physics of the National Nuclear Center of Kazakhstan, 480082 Alma-Ata, Kazakhstan.*

One of the most interesting aspects of modern nuclear physics is synthesis of heavy and superheavy elements and study of their properties. In this connection experimental investigation of the fusion-fission process of compound nuclei with $Z > 100$ at low excitation energies is of special importance. With this purpose at the U-400 accelerator of FLNR the fission cross sections, mass and energy distributions (MED) of fission fragments were measured in the reactions ^{48}Ca+^{208}Pb→^{256}No, ^{48}Ca +^{238}U →^{286} 112, ^{22}Ne+^{248}Cm→^{270}Sg and ^{22}Ne+^{249}Cf→^{271}Hs in the compound nucleus excitation energy range from 12 to 60 MeV.

The interest in the reactions ^{22}Ne + ^{248}Cm and ^{22}Ne + ^{249}Cf is connected first of all with the fact that the number of neutrons in the compound nuclei ^{270}Sg and ^{271}Hs is 164 and 163 respectively. In this case one can expect that at low excitation energies the fission fragment mass distribution demonstrates properties of bimodal fission determined by structural peculiarities of the deformation potential energy surface, much as it takes place in spontaneous fission of ^{258}Fm and ^{260}Md ($Z \geq 50, N \approx 160$). The interest in reactions with ^{48}Ca ions is explained by their importance for the presently realized at the FLRN program of the superheavy ion synthesis in the region of nuclei with $Z = 114 - 116$ and $N = 182 - 184$, for which a considerable increase in their stability to spontaneous fission and α - decay has been predicted. That's why it is very important to measure the compound nucleus cross section for reactions of the ^{48}Ca + U, Pb - type, on the basis of which one is able to predict the fission cross section and survivability of superheavy nuclei.

We investigated the fusion-fission reactions ^{22}Ne + ^{248}Cm and ^{22}Ne + ^{249}Cf at the projectile energies equal to 102 and 127 MeV. These energies result in the initial excitation energy of the compound nucleus of ^{270}Sg $E^* = 28$ and 50 MeV. As for the fission of ^{271}Hs, only the preliminary results will be presented.

CP495, *Experimental Nuclear Physics in Europe*, edited by B. Rubio et al.
© 1999 American Institute of Physics 1-56396-907-6/99/$15.00

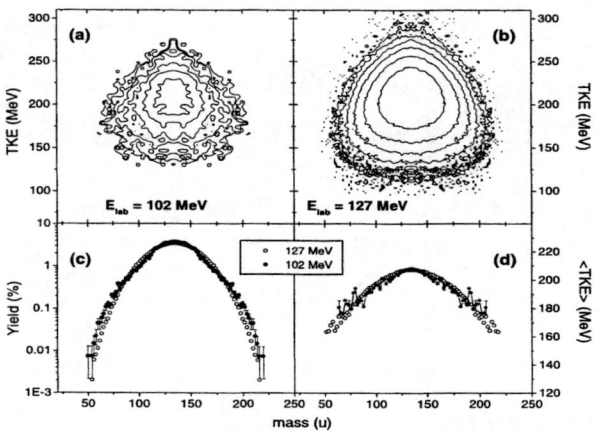

FIGURE 1. Mass-energy distribution of the ^{270}Sg fission fragments for the ^{22}Ne progectiles of 102 MeV and 127 MeV energies

Figure 1 presents the main results of the measuring of the MED of the fragments of the compound nucleus of ^{270}Sg for two values of the energies of ^{22}Ne projectiles. For energy of 127 MeV, $9.2 \cdot 10^5$ fission events were registered; for energy of 102 MeV, $1.2 \cdot 10^4$. Shown in Figs. 1a and 1b are the two-dimensional TKE-mass matrices of the fragments. For high energy (Fig. 1b), we see a triangular distribution with rounded edge slopes typical for rather strongly heated nuclei. Its properties are close to those predicted by the liquid-drop model. For the neon energy of 102 MeV (Fig.1a) $E^* = 28$ MeV, the two-dimensional matrix is different in shape from

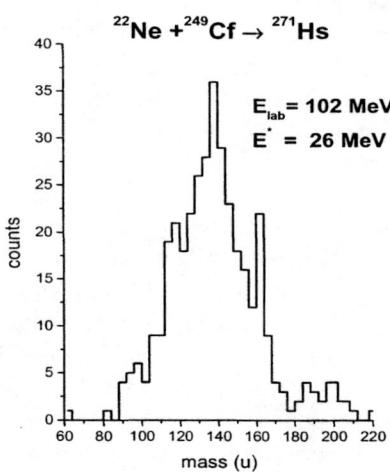

FIGURE 2. The ^{271}Hs fission fragment mass distribution for the ^{22}Ne energy of 102 MeV.

154

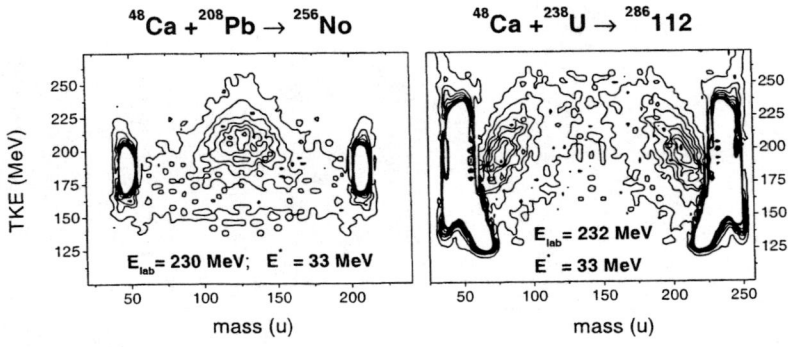

FIGURE 3. Two-dimensional plots TKE vs fragment mass for the reactions ^{48}Ca+^{208}Pb→^{256}No and ^{48}Ca+^{238}U→^{286}112

the above-described one. It has a sharp top and is definitely elongated towards the higher TKE. Fig. 1c presents a comparison of the mass distribution of the fragments for both energies. It is clearly seen that for the lower energy unlike for the high energy, the mass distribution is definitely nongaussian in shape and can be considered as manifestation of different fission modes. One mode, rather narrow, is associated with masses near $A_{CN}/2$, the other mode, rather wide, with asymmetric masses, the latter being so wide as for the mass yields at its edge slopes to be higher than those for high excitation energy.

Figure 2 presents the data on the mass distributions of the fragments of ^{271}Hs at E_{lab} = 102 MeV. Only 365 pair fission events were collected in the experiment, but a sharp peak is clearly visible at M $\approx$ A/2, because, as for neutrons, this nucleus is practically similar to ^{270}Sg. So, we probably also observe manifestation of the super short mode.

Let us consider the investigation results of the reactions ^{48}Ca + ^{208}Pb and ^{48}Ca+^{238}U leading to formation of the nuclei ^{256}No and ^{286}112 at different excitation energies. Figure 3 presents symmetrized plots of TKE-M for both reactions at several values of the ion energy. One can see that in the case of ^{256}No, Y(M,TKE) at any excitation energy is of distinctive triangle shape, which is typical for the classic symmetric fission of a compound nucleus. In the case of the ^{286}112 decay, the Y(M,TKE) matrices have a more complicated structure, determined by domination of quasifission reactions with a characteristic grouping of fragments around the fission fragment mass 208 and complementary to it masses, and thus extraction of events of the compound nucleus fission becomes quite a complicated task.

GAMMA RAY SPECTROSCOPY

γ-Ray Spectroscopy of N = Z Nuclei

C. J. Lister

Physics Division, Argonne National Laboratory, Argonne, IL 60439

Abstract. The use of γ-ray spectroscopy to probe the properties of marginally bound nuclear states has evolved from being a curiosity a decade ago to being the mainstream use for these devices. The key to this success has been the development of ultra-sensitive channel selection techniques which allow the parentage of each emitted γ-ray to be established. With these techniques, and the enhanced efficiency of the arrays themselves, the level of sensitivity for nuclear spectroscopy has increased by several orders of magnitude, in some special cases reaching the 10's nanobarns level, 1000 times more sensitive than was possible a decade ago. In this paper I will discuss some recent developments in light nuclear spectroscopy, on nuclei with N = Z, below mass 100. These examples have been chosen to compliment other presentations at this conference which have covered similar experiments in heavier nuclei.

INTRODUCTION

At the turn of the millennium it is natural to review what we have achieved culturally in general and more specifically what we have achieved in our 100 years of nuclear physics. One common theme is that of exploration and the broadening and deepening of our knowledge which comes with such travels. Geographically, we have explored every corner of our planet during this millennium. On the nuclear landscape we have come to know our local environment and our journeys have started to take us away from the valley of stability, but we still have far to travel. It is appropriate that we meet in Seville, a beautiful and historic center for global exploration, to review our journeys across this new nuclear territory and our plans for the future.

I am going to present some results from our exploration of marginally bound nuclear states which lie right at the periphery of the nuclear landscape. I am going to discuss making γ-ray measurements of these states to probe their wavefunctions. I will present measurements made with Gammasphere, the U.S. national γ-ray facility, as this is what I am most familiar with. With Gammasphere, a great deal of progress has been made especially along the proton dripline and in heavy nuclei which I do not have time to discuss. Instead, I have chosen to report on the exploration of light nuclei in order to balance the other talks on heavier systems at this meeting. Consequently, from the start, I must make it clear that this is the report of a few journeys, not a review of state of our knowledge of the whole landscape. However, this paper taken with the other reports from this meeting give a fair view of where we stand, and where we are going.

THE MOTIVATION

Before I start discussing specific problems, it is worth revisiting the motivation for these travels. One of the key intellectual challenges of our time is to better understand the evolution of our universe to the state it has reached today. To have any hope of

CP495, *Experimental Nuclear Physics in Europe*, edited by B. Rubio et al.
1999 American Institute of Physics 1-56396-907-6

understanding this evolution requires factual input from a wide variety of scientific disciplines. In the nuclear sector, we are about to add vital information in two domains: at the highest energies when quarks become deconfined, through the study of heavy-ion collisions at relativistic energies, and at low energies where radioactive beams are making astrophysically important reactions and nuclei accessible for study. Information on the high-energy domain will improve our understanding of the very early universe, shortly after the Big Bang, and information on the low energy domain will improve our understanding of the subsequent synthesis of the elements which exist around us now. As most of the synthesis mechanisms involve nuclei very far from stability, especially neutron-rich nuclei, that is where our exploration must go. Both these contributions are vital for making progress in building the "big picture" of the evolution of our world.

Much attention has been dedicated to the theoretical understanding of the structure, binding energies and decays of dripline nuclei. The results of these investigations, though often not in complete agreement, indicate that nuclei far from stability, especially neutron-rich nuclei, may be quite different from those along the valley of stability. These investigations, and the diversity of their conclusions, enrich the need to make and investigate these nuclei, as it has become clear that our understanding of nuclear structure is still to a significant extent "experimentally driven", and extrapolations from what we know about near-stable nuclei will not give a realistic view of dripline nuclei. Finally, reaching out to the most exotic nuclei presents a challenge to the dedicated experimentalists. A great deal of innovative investigation has led us far from the valley of stability, but many challenges lie ahead, especially the task of successfully harnessing the potential of radioactive beams.

THE EXPERIMENTS

I am going to start with the $N = Z$ nucleus ^{24}Mg. This nucleus does not lie far from stability, indeed it is stable and we must go to high excitation, above 10 MeV, to reach marginally bound states which mainly decay through particle emission. ^{24}Mg is a very important nucleus, however, as it has been a benchmark for testing new theoretical models. Its importance lies in the fact that it lies in the mid-sd shell and exhibits considerable collectivity, and its state wavefunctions can be calculated in many ways including traditional shell models, deformed Nilsson-Strutinsky calculations, HFB calculations, Monte-Carlo shell models, relativistic meanfield methods, cluster models, etc. Indeed, there are more than 2600 papers dedicated to ^{24}Mg in the data bases, more than half of them theoretical. Comparison of these calculations reveals deep insight both into the models, and into nuclear structure in general. One of the intriguing aspects of ^{24}Mg concerns the location and collectivity of the fully aligned sd-shell $J = 10$ and 12 natural parity states which are expected to lie at about 20 MeV in excitation. These levels may be "band-terminating" states of low collectivity, as suggested by spherical shell models, but their collectivity may be regained through fp-shell admixtures, or more complicated cluster amplitudes. The experimental situation is not clear, as the states are more than 10 MeV particle unbound, so proton and alpha decays dominate the decays, and the radiative branches are expected to be below 10^{-3}. However, it is the radiative branches which can most easily quantify the collectivity and reveal the details of the state wavefunctions, although the competition between γ-decay and the various particle decay channels is also very important (1).

Spin $J = 12$ does not sound very high to most γ-ray spectroscopists, but it is extremely difficult to reach in such a light nucleus. Using $A^{5/3}$ scaling of the moment of inertia, $J = 12$ in ^{24}Mg corresponds to a similar rotational frequency to a $J = 200$ state in a mass

A = 200 nucleus. However, reactions which populate the states in the excitation-energy and spin range of interest are known, for example the $^{12}C(^{16}O,\alpha)^{24}Mg$ reaction at about 62 MeV. Here, the entry state is correct but many experimental problems remain: ^{24}Mg constitutes less than 10% of the fusion cross-section, the states of interest are only populated with cross-sections of a few millibarns, the decay is mainly by particles, so the effective radiative cross-sections are sub-micorbarn and worse, the decay γ-rays of interest are of high energy, > 6 MeV, so are difficult to detect. Not surprisingly, despite many years of intensive effort (2,3), and great experimental innovation, the γ-decay of the J = 10 and 12 states have never been seen.

In principle, one may think that the use of large arrays like Eurogam or Gammasphere should make such measurements possible. However, the high backgrounds, high γ-ray energies, huge Doppler shifts and low multiplicities all make "standalone" γ-γ or γ-γ-γ experiments seem untenable. What is needed is an alternative approach which is triggered by surviving ^{24}Mg nuclei which do not break-up. With this selection, the experiment becomes one of seeking the few individual γ-rays linking the high spin states to the well-known J = 8 states below.

We are working on an alternative technique which may provide the required sensitivity. In this experiment we use the Argonne Fragment Mass Analyzer (FMA) (4), a zero-degree mass separator, to not only select the ^{24}Mg ions but to select individual states through time-of-flight spectroscopy. Identification of the ^{24}Mg ions is straightforward, using an A/Q = 24/10 FMA setting and measuring the recoil nuclei in a 3-electrode ion-chamber. With the mass and de/dx selection, and careful time-random subtraction, spectra with only ^{24}Mg γ-rays can be constructed. The two-body reaction we use is particularly appropriate for state selection as the kinematics are very tightly constrained with the ^{24}Mg ions all recoiling into a 1 degree cone and the alpha-particle emitted upstream, so each state populated in the magnesium ion has its own unique time of flight through the 8.6 m spectrometer. On average, the flight time is about 550 ns, but has a spread of about 50 ns corresponding to an excitation-energy range of 10 MeV. As the accelerator beam can be bunched to less than 500 ps FWHM, and the channel-plate FMA focal plane has sub-ns timing for these ions, state selection of about 200 keV FWHM can be reached. In practice, to increase the yield of gamma rays, we used thick targets (120 $\mu g/cm^2$) which compromised this resolution by an order of magnitude, and used the complementary gamma-ray data to more accurately locate the states of interest.

Preliminary experiments have been encouraging. I cannot show you the J = 10 states yet, but we have candidates and are approaching the necessary level of sensitivity. In a pre-Gammasphere study we successfully tested the FMA time-of-flight state selection technique by correlating recoil ^{24}Mg ions with their corresponding alpha-particles detected in an annular detector mounted at 180 degrees. In our first Gammasphere experiment we found two candidates for two J = 10 states which both appear to decay to more than one of the known low-lying J = 8 states. On the first day of this conference we completed a high-statistics experiment which we hope will confirm these candidates, but we will have to wait for confirmation at a future meeting when our analysis is complete. Our initial estimates indicate that the candidate gamma-ray branches are at the 10^{-4} level, corresponding to radiative cross-sections of a few microbarns. One thing which is absolutely clear from these experiments is that there are NO states in ^{24}Mg above 17 MeV which have γ-ray branches bigger than 0.1%, a confirmation of a conclusion of the Oxford/Manchester group (1-3). This is because the states are so far above the particle decay thresholds that the confining Coulomb and centrifugal barriers are only a few femtometers thick so even the most favored wavefunctions cannot survive tunneling long enough to gamma decay.

N = Z Nuclei Above ^{56}Ni

The competition between particle and γ-ray decays makes a natural bridge to N = Z nuclei far from stability, my second topic. Between ^{56}Ni and ^{100}Sn the N = Z line approaches the proton dripline, so the excitation energy at which proton emission successfully competes with gamma decays falls to much lower energies. The competition is very interesting, as the physics of particle decay and γ-decay are quite different, with the particle decay depending on single-particle aspects of the wavefunction and the tunneling probability through the Coulomb and centrifugal barriers, while the electromagnetic decay is enhanced by collective parts of nuclear wavefunction. A first example (5) of this competition has been found recently using Gammasphere and the microball (6) particle detector to study the decay of a high-spin deformed band in ^{58}Cu to both low-lying levels in ^{58}Cu and to levels in ^{57}Ni. This case is particularly favored as the Q-value for the particle decay is high due to the unusually large binding energy of the daughter nucleus. However, it has many interesting physics aspects: the γ-decays are hindered by electromagnetic isospin selection rules, and the particle decays hindered by a parent-daughter shape change. Thus, it is not surprising that the decaying state may be long-lived on the ns-scale. A follow up experiment has recently been carried out in which an additional wall of silicon detectors was added downstream to detect the mono-energetic decay protons with higher resolution and more precisely measure the particle gamma-competition. The results should be most enlightening, as this type of competition should be a ubiquitous feature of dripline nuclei, applying to both the low-lying yrast levels and to the quasi-bound continuum above. A related study in the Os region (7) has made a measurement to compare the "entry-region" (the energy-spin domain below which γ-ray emission dominates) in nuclei which are beyond the proton dripline (signaled by groundstate proton radioactivity) and in neighboring nuclei which are more bound. These data are still undergoing analysis, but indicate a surprising similarity between nuclei with particle-bound and unbound groundstates.

Another long-standing topic of interest has been the evolution of nuclear shapes between the spherical shell closures at N = Z = 28 and N = Z = 50. Considerable recent progress has been made. In the lighter nuclei, around N = Z = 30 a new region of very deformed high-spin bands has been found (8) in nuclei which are near-spherical in their groundstates. These superdeformed shapes are very interesting and seem to be a natural link between the 4p-4h deformed bands which are well known in light nuclei like ^{16}O and ^{40}Ca, and the traditional superdeformed bands known in heavier regions. It is also interesting to compare these very collective bands to less-deformed states recently discovered in ^{56}Ni (9) to ascertain the critical requirements for superdeformation to be sustained. The "decay-out" mechanism is also interesting and seems to be quite different to that found in heavier nuclei which should give broader insight into the general question of the issues involved in electromagnetic-shape changing processes.

Moving to heavier N = Z nuclei, the production cross-sections fall rapidly with the presently available stable beams, and the experiments rapidly become more difficult. During the last decade, since these nuclei were first studied by the recoil-gamma coincidence method (10,11), rather little progress has been made, despite considerable effort (12,13). The exception is ^{72}Kr (14) where the groundstate band has been extended to clearly reveal shape coexistence effects. However, in the weeks proceeding this meeting, we have performed experiments on N = Z nuclei ^{68}Se (15), ^{80}Zr (16), ^{84}Mo and possibly ^{88}Ru (17). Only ^{68}Se has been studied in any detail so I will concentrate on that, but I expect considerable progress will be made on the others when data analysis is done.

^{68}Se is almost unique as it is predicted to be one of the few cases in the periodic table predicted to have a large ($\beta_2 \sim -0.3$) oblate groundstate caused by a deformed shell-gap at N = Z = 34. This oblate shape is calculated to be sufficiently well bound that it is expected to be yrast with rotation for several MeV, when it is crossed by a prolate configuration, and the configurations are expected to co-exist with rather little mixing due to a high oblate-prolate barrier. We find strong evidence for rotational bands based on both of these configurations. The candidate groundstate oblate band has all the expected characteristics; exceptionally low moment of inertia, no apparent backbending, and quite small mixing with its co-existing prolate partner. The prolate band shows a much higher moment of inertia and a backbending. Thus, although it is difficult to find definitive proof for an oblate groundstate shape, the evidence is quite convincing and will be published this fall.

In ^{80}Zr we have collected a large data set (16), in a similar study to the original Daresbury measurement (11) and all the known states could be seen "online". However, the enhanced efficiency of Gammasphere will allow γ-γ coincidence measurements, so progress through the backbending region around J = 8 should be achieved. In a further study, attempting to extend N = Z "inbeam" information to ^{88}Ru we used the ^{32}S(^{58}Ni,2n) reaction using a MoS$_2$ target on a rotating target wheel. From the yields of ^{88}Mo and ^{88}Tc, we think there is a reasonable chance of finding the first excited state in ^{88}Ru. A surprising side-product of the experiment was good population of ^{84}Mo through the ^{32}S(^{58}Ni,α2n) reaction. The known first excited state in ^{84}Mo was clearly seen "online", so again the power of Gammasphere should allow considerable progress.

So I must conclude. This meeting has come at an interesting time for γ-ray spectroscopy, in the USA, as during the last year since Gammasphere arrived at ANL we have been extensively studying marginally bound states in many nuclei from ^{24}Mg to ^{254}No (18) in experiments triggered by the FMA. In some experiments we have been making gamma-ray spectroscopic studies on states produced with cross-sections of 10's nb. For states produced at the few microbarn level quite detailed correlation measurements are now possible. These experiments are 2-3 orders of magnitude beyond what was possible ten years ago. One new application of Gammasphere has been to use both the BGO and HpGe detectors to perform high efficiency calorimetry. This has proven especially useful for studying fission barriers in very heavy nuclei (18). We are still learning how best to trigger the experiments especially to minimize the deadtime and maximize to data we collect. As you can tell from this report, much of the data is not yet analyzed, so we hope many new results will come in the new millennium.

Looking further in the future, the prospects for nuclear structure seem brighter than for many years. The generalization of our understanding of nuclear structure to encompass all nuclei has become the driving theme for our field and has relinked our research with many other fields of physics. The prospects for accelerated radioactive beams have turned from a dream to a reality, both in fragmentation and ISOL-type facilities, and the prospects for even more powerful facilities are good. More sophisticated instruments continue to improve the sensitivity of our studies. In all, the first few decades of the next millennium should be a very exciting time in nuclear for nuclear physics.

This research was supported by the U.S. Department of Energy, Nuclear Physics Division, under Contract W-31-109-Eng-38 and many other DOE and NSF grants.

REFERENCES

1. Vermeer, W. J. et al., *Nucl. Phys.* **A485**, 380 (1988).
2. Smith, A. E. et al., *Phys. Lett.* **B176**, 292 (1986).
3. Catford, W., *Nucl. Instrum. Methods* **A247**, 367 (1986).
4. Davids, C. N. et al., *Nucl. Instrum. Methods* **B70**, 358 (1992).
5. Rudolph, D. et al., *Phys. Rev. Lett.* **80**, 3018 (1998).
6. Sarantites, D. G. et al., *Nucl. Instrum Methods* **A381**, 481 (1996).
7. Carpenter, M. P. et al., Private Communication (1999).
8. Svensson, C. E. et al., *Phys. Rev. Letts.* **82**, 3400 (1999).
9. Rudolph, D. et al., *Phys. Rev. Letts.* **82**, 3763 (1999).
10. Varley, B. J. et al., *Phys. Lett.* **B194**, 463 (1987).
11. Lister, C. J. et al., *Phys. Rev. C* **42**, 1191 (1990).
12. Skoda, S. et al., *Phys. Rev. C* **58**, R5 (1998).
13. Bucurescu, D. et al., *Phys. Rev. C* **56**, 2497 (1997).
14. de Angelis, G. et al., *Phys. Lett.* **B45**, 217 (1997).
15. Fischer, S. J. (DePaul University) et al., Private Communication (1999).
16. Bernstein, L. (LLNL) et al., Private Communication (1999).
17. Vincent, S. (University of Notre Dame) et al., Private Communication (1999).
18. Reiter, P. et al., *Phys. Rev. Lett.* **82**, 509 (1999).

New Results from Euroball

Santo Lunardi

Dipartimento di Fisica dell'Universita' and INFN, Sezione di Padova, Padova, Italy

Abstract. The study of nuclear structure at the limits of angular momentum and of the neutrons to protons ratio requires detector systems of unprecedented sensitivity in order to detect the very weak signals characteristic of such regimes. The Euroball γ–ray spectrometer has been built to address the physics of high-spin states and of exotic nuclei lying far from stability. Euroball came into operation in spring 1997 and has been used for more than one year at the Legnaro National Laboratories, Italy. The first results obtained with this powerful instrument are presented in this talk.

INTRODUCTION

The progress of nuclear structure through in beam γ–ray spectroscopy studies has been enormous in the last 10-15 years. This is mainly due to the great advances in detector technology which allowed the construction of γ–ray spectrometers of increasing efficiency and sensitivity up to the latest generation, such as Euroball in Europe and Gammasphere in the USA. These instruments reach a total photopeak efficiency of $\approx 9\%$ at 1.3 MeV and allow to study nuclear phenomena at the level of 10^{-5} of the production cross-section in a heavy-ion reaction.

A unique feature of Euroball with respect to all other γ–detector arrays is the presence of 15 CLUSTER detectors, each one composed of seven large volume encapsulated Ge crystals tightly packed in a commom cryostat. The CLUSTER detectors, which cover a solid angle of 1π in the backwards direction, give almost half of the efficiency of the array and are especially designed to detect high-energy γ–rays. Euroball has been conceived in such a way that various ancillary detectors, such as a Silicon ball, a neutron-wall, a plunger and other kinds of heavy-ion detectors can be easily added, allowing to increase the sensitivity of the spectrometer especially for very weak reaction channels on both sides of the valley of stability.

The array has been operational at the Legnaro National Laboratories, Italy, for more than one year using the heavy-ion beams provided by the Tandem XTU + ALPI complex. Euroball has than been moved to the Vivitron accelerator at IReS Strasbourg where its performance has been further improved adding an inner ball of BGO detectors.

The main physics themes that have been addressed during the Legnaro period concern the structure of the atomic nucleus in extreme conditions (e.g. very high

CP495, *Experimental Nuclear Physics in Europe*, edited by B. Rubio et al.

angular momenta, exotic nuclear shapes and the limits of nuclear stability) which allow to learn more about the underlying nuclear force and to improve the models developed up to now to describe the nucleus. In this talk some of the results obtained so far (the analysis of many experiments is still going on) on the above topics will be presented.

HIGH SPIN STATES IN NORMAL DEFORMED NUCLEI

Large multi-detector γ-ray spectrometers like Euroball have been originally designed with the main goal to push the study of nuclei towards the limits of angular momentum. The use of high-fold coincidence data allows to extract from the enormous background produced in a nuclear reaction the weak sequences of γ-ray transitions connecting states at high spins. Among these are the superdeformed (SD) rotational bands where the spectroscopic study has now reached, in some cases, the quality we were used to have, with previous detector systems, at low spin in normal deformed nuclei. The main physics themes that are addressed, in the "normal deformed" regime, when spinning the atomic nucleus to high angular momentum (up to spin $60\hbar$), are the quenching or disappearance of nuclear pairing correlations and the so called "termination" of rotational bands, i.e. the disappearance of quantum many-body collectivity.

The transitional nuclei with neutron number around N=90 are one of the best battleground where such phenomena have been first observed. At Euroball high-spin rotational bands in ^{161,162}Er have been excited using the reaction ^{130}Te(^{36}S,xn) [1]. Bands in ^{162}Er and ^{161}Er have been identified up to 56 $\hbar$ and 123/2 $\hbar$, respectively. In two of the bands of ^{161}Er tentative evidence is found for unpaired band-crossings. Such crossings, which are not correlated in rotational frequency and cannot be explained through pairing correlations, constitute the best indication of the decline of static pairing correlations at high spins.

The process of band termination at high spins has been addressed also in the A=130 mass region. Here, in the nucleus ^{127}La the first evidence has been found of a "smooth band termination" in valence space. The collective rotational band based on a specific configuration, with no holes in the Z=50 core, gradually looses its collectivity and could be observed up to four units less than the maximum angular momentum that can be reached within the valence space [2]. These results are well described theoretically in the framework of cranked Nilsson-Strutinski calculations.

EXOTIC NUCLEAR SHAPES

At high angular momentum nuclei can assume shapes very different from spherical or deformed with $\beta \leq 0.3$, being often called "exotic". In this classification we include prolate superdeformed (ellipsoids with major to minor axis ratio 2:1), hyperdeformed (3:1 axis ratio) and triaxial superdeformed shapes. Hyperdeformed shapes are predicted to exist in ^{152}Dy together with the spherical, prolate normal- and super- deformed shapes. The nucleus ^{152}Dy is in fact one of the best examples of the coexistence of very different shapes at similar excitation energy and spin, since all the three shapes mentioned above are observed with their characteristic nuclear spectra. This is also the only nucleus where the presence at very high spin of the hyperdeformed shape has been inferred from the observation of ridges in γ–γ matrices [3,4] a few years ago. The identification of a discrete rotational band built on such a shape was anyway not achieved and had to be postponed to future experiments with bigger arrays. In an experiment performed at Euroball using the same reaction as in ref. [3,4], namely ^{120}Sn+^{37}Cl at 187 MeV, the 30 keV ridges have been confirmed, but it was not possible to associate them to a discrete rotational band based on a hyperdeformed shape. The population of the third minimum in the potential energy surface at high spin appears to be still below the detection limits of the present big arrays and awaits further developments in detector technology.

Superdeformed shapes with pronounced triaxiality have been recently discovered [5] in nuclei with N≈94 and Z≈71. Theory predicts such exotic shapes in a rather large number of nuclei of this region. In a search for exotic structures in the odd-odd nucleus ^{164}Lu, performed in one of the first Euroball experiments, eight new, presumably triaxial, SD bands were found [6]. The structure of the two lowest bands for which spin and parity have been determined corresponds most likely to shapes with $\gamma \approx$ 20°. In the same experiment also the ^{163}Lu nucleus, where a triaxial SD band was known, has been populated. The band has now been firmly linked to normal-deformed states [8]; furthermore a new band with similar moment of inertia has been found. For large γ deformation theory predicts the so-called wobbling mode which gives rise to bands with higher moment of inertia at higher excitation energies. The new band discovered could be the first excited band associated to this mode. A characteristic decay pattern from the "wobbling" band to the ground state band built on the triaxial SD minimum should be observed. The search for the wobbling motion in nuclei will be pursued in an experiment already approved at Euroball in Strasbourg [7].

DECAY-OUT OF SUPERDEFORMED NUCLEAR SHAPES

The understanding of the decay-out mechanism of SD bands towards normal-deformed states is still an open problem in nuclear structure physics. Progress in this field has been achieved in the A=130 and A=190 region through the discovery, in some cases, of the transitions connecting the SD minimum with the ND one. In this way it has been possible to fix the excitation energy and spin of the SD states allowing for a more precise comparison with theory.

Several experiments have been performed at Euroball addressing the question of the decay-out of SD bands in the three regions with A≈130, A≈150 and A≈190. The attempts in the A=150 region have brought to confirm the linking transition already known in ^{143}Eu and to propose some candidates both in ^{143}Eu and ^{144}Gd. A firm assignment of excitation energy and spins for SD bands in the A=150 region is however still not possible.

In the A=130 region, SD bands have been connected to ND states in a series of Nd nuclei (with the exception of ^{136}Nd) in previous experiments performed at GASP and Gammasphere. It is important to know the excitation energy of the ^{136}Nd SD band because from the odd-even mass differences of the series of Nd nuclei from ^{132}Nd to ^{137}Nd it is possible to estimate the pairing strength in the SD configuration, as it was done in the past for the neutron (proton) pairing gap $\Delta_n(\Delta_p)$ in normal-deformed matter.

The SD bands of the even-even Nd nuclei, which are based on two-quasiparticle excitations, have an intensity of about 1% of the total population of the nucleus, as in the mass A=150 region, and therefore the identification of discrete linking transitions becomes as difficult as in the other heavier mass regions. The higher efficiency and better selectivity of the Euroball array has allowed to clearly identify two discrete decay paths from the lowest states of the SD band in ^{136}Nd [9]. The resulting decay-out pattern is similar to the one of the yrast SD band of ^{134}Nd [10]. The lowest state of the ^{136}Nd SD band is thus located at 7.03 MeV of excitation energy and has spin and parity $17^{(-)}$.

Using a first-order Taylor expansion and the three-point formula, the experimental neutron pairing gap in the normal-deformed ground state and in the SD configuration for the ND nuclei has been extracted [9]. The pairing gap is reduced by approximately a factor of 2 for the SD states with respect to the normal-deformed ground state. These findings, while confirming the predictions of various theoretical approaches about the quenching of pairing in the second well of nuclei, prove that a sizeble pairing gap is still present at SD shapes in A≈130 nuclei.

NUCLEI FAR FROM β–STABILITY

The study of exotic nuclei on both sides of the valley of stability is nowadays one of the major themes of research pursued both by taking advantage of the first radioactive beams becoming available and by the use of stable heavy-ions beams. In this last case, the exotic nuclei of interest are produced with very low cross sections and, in order to overcome this difficulty, the use of selective devices is compulsory. Euroball is a powerful instrument to select extremely weak γ-ray sequences from the background produced in a nuclear reaction as it can exploit high fold events. It can however be more useful in the study of very exotic nuclei if coupled with selective ancillary devices such as light charged particle detectors, neutron detectors and fission fragments detectors. Almost half of the experiments performed with Euroball at Legnaro made use of various ancillary detectors having the goal to explore both the neutron-rich and proton-rich nuclei.

The spectroscopy of nuclei far from β-stability is providing new data which reveal unexpected changes in the shell structure; for example some shell gaps for light nuclei (N=20 and N=28) are vanishing, new regions of deformation are appearing and the known nuclear models have to be continuosly improved in order to explain the wealth of new phenomena discovered in experiment.

Neutron-rich Nuclei

Experimentally, neutron-rich nuclei can be reached by different methods such as multinucleon transfer reactions, heavy-ion induced fission and spontaneous fission. By the first method, through the use of the ^{48}Ca+^{48}Ca reaction at 140 MeV it has been possible to populate very neutron-rich nuclei close to ^{48}Ca. The fusion-evaporation reaction channels are still dominant, with multinucleon transfer processes accounting for less than 0.1% of the total reaction cross section. Despite this very low production yield, investigations of excites states in some nuclei around ^{48}Ca were possible due to the high resolving power of the Euroball array. The interest in studying these nuclei lies in the fact that calculations suggest that the major N=28 shell gap, which plays a definite role in the structure of nuclei with Z$\geq$20, disappears at Z=16 and therefore nuclear deformation sets in. Such prediction has been confirmed by measuring the energies and the B(E2) values for the lowest 2^+ states in neutron-rich radioactive nuclei of S and Ar via intermediate-energy Coulomb excitation [11]. First evidence has been provided of moderate deformation near N=28 for S nuclei, whereas the effects of the N=28 shell closure persist in the Z=18 Ar nuclei. In the Euroball experiment excited states in ^{44,46}Ar have been populated through multinucleon transfer processes: the 2^+ energy has been determined very precisely as compared with the Coulomb excitation experiment [12]. The identification of excited states above 2^+ in ^{44}Ar will provide also more stringent tests for the large shell model calculations.

The discrete spectroscopy of secondary fission fragments has been pursued at Euroball using the ^{12}C+ ^{238}U reaction. The goal was to further understand the nuclear structure of neutron-rich nuclei in a region not accessible by the spontaneous fission process and to search for isomers in a wide range of proton and neutron number. As selective device a fission-fragment detector composed of photovoltaic cells (called SAPhIR) has been used. Among the results obtained in this study one can mention the new data on excited states in the series of Cd isotopes. In ^{118}Cd the vibrational states belonging to the 3-phonon quintuplet have been populated and their spin and parity determined [13]. The spectroscopy of neutron-rich Pd nuclei (states observed up to spin $\approx$14 $\hbar$) has been extended up to mass 119 with no evidence of the prolate-oblate phase transition predicted by theory.

Proton-rich Nuclei

By coupling to Euroball a neutron wall and a silicon ball many nuclei on the N=Z line can be reached up to mass A≈90. Nuclei close to the proton drip-line with N=Z represent the best case where the neutron-proton pairing interaction is enhanced and therefore could manifest itself in the low-excitation properties. In nuclei with N=Z it is also possible to study the violation of isospin symmetry induced by the Coulomb interaction.

Experiments have been performed starting from $f_{7/2}$ nuclei up to ^{80}Zr, with the main goal of studying the T=0 n-p pairing. Some of the results, see for example the ^{64}Ge nucleus [14], are presented as contributions to this Conference. For the others the analysis is in progress.

CONCLUSIONS

In the first year of operation Euroball has demonstrated its power, in terms of efficiency and selectivity, by extending the spectroscopy of nuclei towards high spins, exotic deformations and far from stability. The first results obtained are already promising, but much more will surely be provided by the analysis of the Legnaro experiments and of those being now performed in Strasbourg, where the new experimental programme, with the enhanced Euroball configuration, has been started with success in June 1999.

The results presented come from the work of all the colleagues who have used Euroball during the Legnaro phase. I would like to thank them for providing the material for this talk.

REFERENCES

1. Simpson, J.,Talk given at the Conference "The Nucleus, New Physics for the New Millenium", Cape Town, South Africa, January 1999.
2. Parry, C.M., et al., preprint 1999.
3. Galindo-Uribarri, A. et al., Phys. Rev. Lett. **71**, 231 (1993).
4. Viesti, G., et al., Phys. Rev. **C51**, 2385 (1995).
5. Schmitz, W., et al., Nucl. Phys. **A539**, 112 (1992).
6. Tormanen, S., et al., Phys. Lett. **B454**, 8 (1999).
7. Hagemann, G., et al., Euroball proposal 1999.
8. Domscheit, J., et al., Legnaro Annual Report 1998, p. 39.
9. Perries, S., et al., submitted to Phys. Rev. C.
10. Petrache, C.M., et al., Phys. Rev. Lett. **77**, 239 (1996).
11. Scheit, H., et al., Phys. Rev. Lett. **77**, 3967 (1996).
12. Fornal, B., et al., Annual Report, Institute of Nuclear Physics, Kracow 1998, p.47.
13. Durell, J., et al., Legnaro Annual Report 1998, p. 47.
14. Gadea, A., et al., Contribution to this Conference.

Shell structure of neutron-rich light nuclei: New Vista

Faisal Azaiez

Institut de Physique Nucléaire, IN2P3-CNRS, 91406 Orsay Cedex France

Abstract. The structure of neutron-rich light nuclei can be investigated by means of in-beam gamma spectroscopy using both stable and radioactive beams. Two of such experiments have been recently performed at GANIL. i) Secondary beams of neutron-rich Ge, Zn and Ni isotopes around N=40 were produced from the fragmentation of primary ^{70}Zn and ^{86}Kr beams, and analyzed by means of the LISE3 spectrometer. For the transitions between the 0^+ ground state and the first 2^+ state, B(E2) values have been extracted from Coulomb excitation cross-section measurement. Some of the obtained results related to the effectiveness of the N=40 spherical sub-shell closure are discussed. ii) In-beam γ-spectroscopy of neutron-rich nuclei around ^{32}Mg produced by fragmentation of ^{36}S have been recently performed. Gamma decay of relatively high-lying excited states have been measured in a large number of exotic light nuclei. New results obtained in a number of neutron-rich nuclei around N=20 are presented.

INTRODUCTION

One of the most challenging goals of experiments with radioactive beams is to determine how the structure of atomic nuclei changes near the drip lines. Recently, elastic and inelastic secondary reactions, induced by radioactive beams on stable targets, has been used in order to extract nuclear structure information. Among them, inelastic scattering in inverse kinematics and Coulomb excitation are known to provide valuable nuclear structure information such as the energy and the collectivity of the first excited states of nuclei. Inelastic proton scattering to the first excited 2^+ state of the doubly magic ^{56}Ni nucleus is a nice example of such experiment with radioactive secondary beams [1]. On the other hand, Coulomb excitation experiments of unstable beams such as ^{11}Be [2,3], ^{32}Mg [4] and neutron rich Ar and S isotopes [5,6] has been recently demonstrated. Coulomb excitation of secondary radioactive beams as well as a novel method based on in-beam γ-spectroscopy of exotic nuclei produced by projectile fragmentation have been recently used at GANIL. The description of the two experiments together with the obtained results on the spectroscopy of neutron rich nuclei around N=40 and N=20 will be presented.

CP495, *Experimental Nuclear Physics in Europe*, edited by B. Rubio et al.

NEUTRON-RICH NUCLEI AROUND N=40

The ^{68}Ni$_{40}$ nucleus is known to exhibit a high 2^+ energy of 2.03 MeV [7] in contrast to its neighboring isotopes ^{66}Ni [7] and ^{70}Ni [8] which have lower 2^+ energies of 1425 keV and 1259 keV, respectively. The sudden increase of the 2^+ energy at N=40 suggests that ^{68}Ni is spherical and can be considered as a good core to modelize more neutron-rich Ni isotopes up to ^{78}Ni [9]. The unknown reduced transition probability B(E2: 0^+ to 2^+) of ^{68}Ni should then be measured in order to confirm this shell effect. In this respect, it would be interesting to compare this B(E2) value to that of the doubly magic ^{56}Ni, which was found to be 600 (120) e^{2}fm^4 [1]. For heavier N=40 isotones, proton excitations in the fp-shell increase the collectivity to B(E2)=1600(140) e^{2}fm^4 for ^{70}Zn [10] and 2130(60) e^{2}fm^4 for ^{72}Ge [10]. Recently, an experimental program has started at GANIL in order to address the open questions concerning the spherical gap at N=40. In the following, we will present preliminary results from a Coulomb excitation experiment, performed at GANIL, with neutron-rich nuclei in the vicinity of ^{68}Ni.

The Coulomb excitation of secondary beams has been induced by the Coulomb field of a thick target placed in the center of a large gamma-array detector. The nuclei ^{76}Ge, ^{70,72}Zn and ^{68}Ni were produced at an energy of about 50MeV/u by the fragmentation of a ^{86}Kr beam at 65 MeV/A. They were identified event-by-event in two large-area (25cm^2) silicon detectors mounted at a distance of 50 cm from the secondary lead target. Two clover Ge-detectors were placed around the implantation detector in order to measure the γ-rays originating from the decay of isomers produced in the fragmentation of the primary beam and transmitted by the LISE3 spectrometer. The scattered fragments were detected up to an angle of 3 degrees in the laboratory frame. At these small deflection angles, the Coulomb inelastic contribution strongly dominates the total cross section. The mean production rate of the fragments was of about 100 particles/second for ^{76}Ge, ^{72}Zn and 20 particles/second for ^{68}Ni and ^{70}Zn. The lead target of 220 mg/cm^2 thickness was surrounded by the γ-array of 70 BaF$_2$ detectors of Chateau de Cristal, mounted in the 4π geometry. The γ-rays of interest, subsequent to the coulomb excitation, are emitted in flight with a velocity of v/c=0.3. Consequently, the Doppler effect induces a broadening of the γ-lines of the ordre of 15% at 1 MeV.

The ^{76}Ge nucleus is a well-known case of deformed nucleus, exhibiting a low 2^+ state (563 keV) and a strong excitation probability, B(E2) = 2680 e^{2}fm^4. It has been used as a calibration measurement in order to determine B(E2) values of ^{70,72}Zn and ^{68}Ni. Even if the spectrometer was not optimized for ^{70}Zn, an excitation probability B(E2) = 1500(400) e^{2}fm^4 has been determined for the 2^+ state at 885 keV. The agreement between the present value and the one measured at low energy [10], where the coulomb excitation is the dominant fraction of the 2^+ excitation, is a clear indication of the validity of other extracted B(E2) values from our data. The B(E2) values of the Zn isotopic chain indicate that the collectivity is increasing at N=40, the maximum of collectivity occurring at N=42 [11]. This shows that protons are quickly washing out the N=40 effect due to the large number of 2qp-excitations in the fp shell above Z=28. For the coulomb excitation of ^{68}Ni, very few counts are found around the expected energy E(2$^+$) = 2.03 MeV. Therefore, only an upper limit of B(E2) = 800 e^{2}fm^4 could be derived [11].

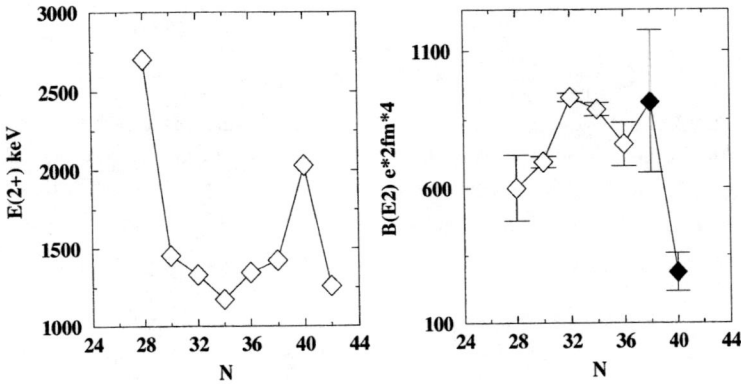

FIGURE 1. Systematics of the 2$^+$ energies and B(E2) values in Ni isotopes. New results from our experiments are indicated by filled symbols.

More recently, this experiment has been repeated using more favorable conditions: (i) A neutron-rich primary beam of ^{70}Zn, closer to the nuclei of interest, has been used at 60 Mev/u and the production rate of ^{68}Ni has been increased by almost three orders of magnitude. (ii) Instead of the BaF$_2$-array, a higher resolution γ-array consisting of four segmented clover Ge detectors has been used as four faces of a cube around the lead target. The immediat consequense was a much higher resolving power for the γ-ray detection. This is nicely demonstrated in the spectra of figure 2, where gamma-lines corresponding to the 2$^+$ to 0$^+$ transitions in ^{66}Ni and ^{68}Ni are clearly visible. Despite a relatively smaller efficiency (5.5% at 1.3 MeV) of the four segmented clover detecors (placed at 5.5 cm from the lead target), the increase of the secondary beam intensities allowed an unambiguous determination of the B(E2: 0$^+$ to 2$^+$) value in ^{66}Ni and ^{68}Ni. The obtained values [11] have been added to the systematics in figure 1. This indicates for the first time the effectiveness of the N=40 spherical shell effect in the Ni isotopes.

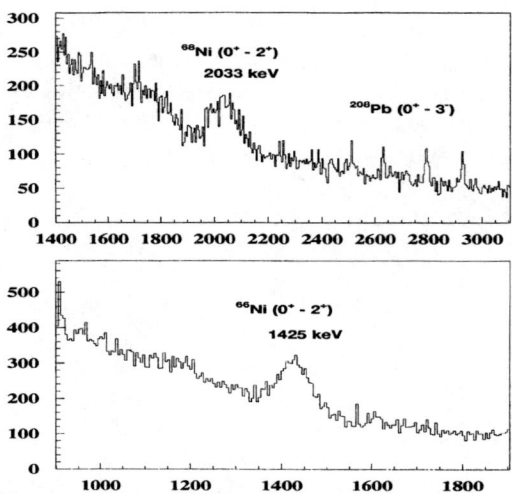

FIGURE 2. Coulomb excitation spectra of ^{66}Ni (lower spectrum) and ^{68}Ni (upper spectrum), obtained after Doppler correction using the segmentation of the Clover detectors. The four sharp lines in the ^{68}Ni represent the coulomb excitation to the 3^- state of the ^{208}Pb target (the four lines correspond to the four different angles of the segments).

NEUTRON RICH NUCLEI AROUND N=20

From the study of the structure of light neutron-rich nuclei, it has been recently suggested that some major shell-gaps are weakened when large isospin values are encountered. The typical cases of ^{32}Mg and ^{44}S, where a large quadrupole collectivity has been found [4–6] have brought some evidence for such a shell-gap weakening at large neutron excess. Though, information on the excitation energies of the first 2^+ states and on the B(E2) values of the 0^+ to 2^+ transitions is not sufficient to fully understand the structure of these nuclei. For instance the measurement of the E(4^+)/E(2^+) ratio should shed some light on the origin of the large quadrupole collectivety observed. In order to bring more spectroscopic information on ^{32}Mg and neighboring nuclei, a novel experimental method has been used. This method is based on the production of very neutron-rich nuclei in relatively higher excited states, through the projectile fragmentation process, and on the detection of their in-beam γ-decay. Such experiment aiming for the measurement of the E(4^+)/E(2^+) ratio in $^{30-32}$Mg, $^{26-28}$Ne and $^{20-22}$O has been recently performed at GANIL. A ^{36}S beam, at 77MeV/u was used on a 2.77 mg/cm^2 Be target. The target was located at the entrance of the SPEG spectrometer which was used to analyze the different fragments produced in the reaction. Many neutron rich exotic nuclei have been produced and identified. The produced nuclei are identified in a time of flight versus energy-loss plot. It is worth pointing out that most of the produced nuclei are TERRA INCOGNITA for nuclear spectroscopy and thus γ-spectroscopy of these nuclei (such as $^{22-23}$O, $^{27-28}$Ne , ^{33}Mg) is completely unknown. Gamma-spectroscopy for all the produced exotic nuclei is obtained by performing coincidences between the analyzed fragments and γ-rays emitted in flight during their decay to the ground state. For that purpose a highly efficient (25 % at 1.33MeV) γ-array consisting of 74 BaF$_2$ crystals (the same used for the Coulomb excitation experiment) was used around the target covering symmetrically the upper and lower hemispheres (roughly 80 % of the solid angle around the target is covered). This array is supposed to provide γ-fragment as well as γ-γ-fragment coincidences. The latter is needed to build-up a level scheme for each fragment. In addition to the BaF$_2$ array, four 70 % high resolution Ge detectors were used at the most backward angles (in between the two hemispheres) in order to help identifying some more complex BaF$_2$ spectra (see spectra below). In the following, part of the results obtained in even-even nuclei are presented. After gating on the proper fragment and on the true γ-fragment coincidences (subtracting the random coincidences contribution), Doppler corrected γ-spectra are presented and commented.

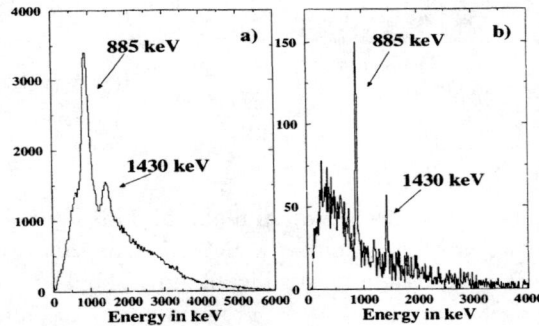

FIGURE 3. BaF_2 (a) and Ge (b) spectra of the in-beam γ-decay of ^{32}Mg.

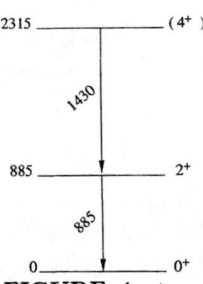

FIGURE 4. tentative level scheme of ^{32}Mg. Energies are given in keV.

i) In figure 3, the obtained BaF$_2$ and Ge spectra of ^{32}Mg are shown. Like for many other produced fragment, the gamma spectra (Ge and BaF$_2$ spectra) of ^{32}Mg exhibit more than one line. For all these fragments, γ-angular distribution and γ-γ coincidence between BaF$_2$ detectors has to be analyzed in order to deduce a level scheme. This type of analysis has been done for ^{32}Mg and reveals that the two lines: the 885 Kev (the well known 2^+ to 0^+ transition in ^{32}Mg [12]) and the 1430 Kev, already observed in the β-decay of ^{32}Na [13] were found to be in coincidence. The analysis of the gamma-ray angular distributions and angular correlations were found, within the error bars, to be independant of the nature (multipolarity) of known gamma-rays from the data. This is suggesting that the orientation of the fragment angular momenta is not larger then 20 %. Though, the analysis of the gamma-ray intensities, in known cases such as ^{16}C and ^{20}O, shows that the fragmentation reactions in our experiment favor the population of the Yrast states. For that reason we are suggesting that the 1430 Kev gamma-ray corresponds to the 4^+ to 2^+ transition. It is worth pointing out that Monte-Carlo Shell Model calculations [14], made prior to our experiment, reproduce very well the excitation energy of $2_1{}^+$ states observed in ^{26,28}Ne and ^{32}Mg. Furthermore these calculations, in agreement with our spin assignment, report a 4^+ state at 1.4 MeV above the 2^+ state in this nucleus. The level scheme we are proposing for ^{32}Mg (see figure 4) is not typical of a rigid axially symetric rotor, as one may assume from the 2^+ excitation energy and B(E2) values in ^{32}Mg [4].

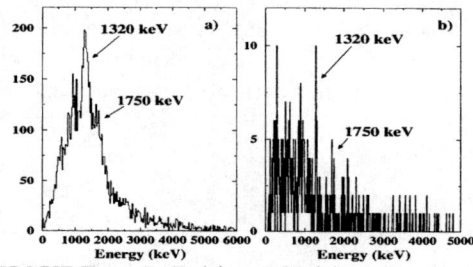
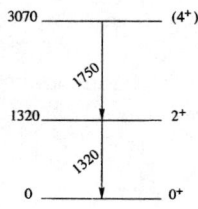

FIGURE 5. BaF_2 (a) and Ge (b) spectra of the in-beam γ-decay of ^{28}Ne.

FIGURE 6. Tentative level scheme of ^{28}Ne.

ii) The BaF_2 and Ge spectra of ^{28}Ne are shown in figure 5. This figure illustrates the importance of a high efficiency for gamma-ray detection in this experiment. While the Ge spectrum does not show any significant line, the BaF_2 spectrum exhibits a quite convincing structure at 1.3 MeV. From intensity arguments, this γ-line is very likely to represent the 2^+ to 0^+ transition in ^{28}Ne which shows for the first time that, approaching N=20, the 2^+ energies in the Ne isotopes decrease dramatically. One can see in figure 9 that the 2^+ energy drops from around 2 MeV in ^{24}Ne and ^{26}Ne to 1.3 Mev in ^{28}Ne (it is worth pointing out that the 2^+ excitation energy of ^{26}Ne has been already measured in a β-decay experiment at GANIL [15]). This behavior is presumably a sign of shell structure change for neutron rich Ne isotopes similar to the one observed long time ago in the Mg isotopes [12].

iii) The obtained γ-spectra of ^{22}O from both the BaF_2 and Ge detectors is presented in figure 10. From intensity argument, the γ-line observed for the first time at 3.1 MeV represents the 2^+ to 0^+ transition which extends the systematic of the 2^+ transition energies of Oxygen isotopes up to N= 14. One can see in figure 8, that Oxygen isotopes exhibit an increase of the 2^+ energy at N=14 just like Ne

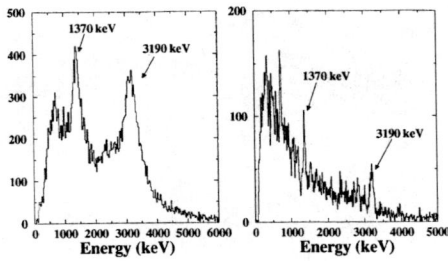 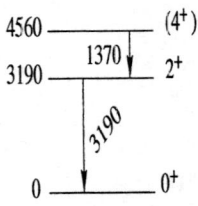

FIGURE 7. BaF_2 and Ge spectra of the γ-decay of ^{22}O.

FIGURE 8. Tentative level scheme of ^{22}O.

and Mg do. This is suggesting that a similar shell effect weakening is developing at N=20 for oxygen isotopes . This should be confirmed by the measurement of the 2^+ excitation energy in ^{24}O which in turn will help understanding why ^{26}O and ^{28}O isotopes seem to be unbound [17,18].

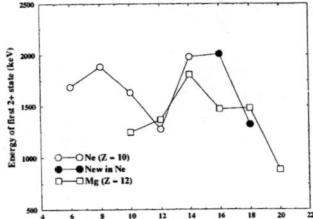 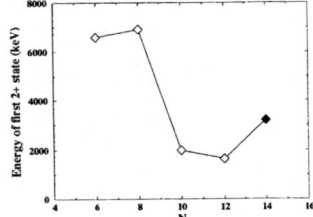

FIGURE 9. Systematics of the first 2^+ energy in Mg (open squares) and Ne (open circles) isotopic chains. The filled circles represent the energies measured in this experiment.

FIGURE 10. Systematics of the first 2^+ energy in O. The filled diamond represent the energy measured in this experiment.

Beside the importance of the obtained preliminary results, this experiment shows that the in-beam-γ-spectroscopy from fragmentation reactions is very promising for exploring nuclear structure far from stability. It also highlights the needs for a dedicated gamma detection system from the point of view of efficiency, resolution and Doppler Broadening reduction. These three features are the basic requirements for which EXOGAM [19] was built and thus make it the ideal gamma detection system for such experiment.

REFERENCES

1. G. Kraus *et al.*, *Phys. Rev. Lett.* **73** , 1773 (1994)
2. R. Anne *et al.*, *Z. Phys.* **A352**, 397 (1995)
3. T. Nakamura *et al.*, *Phys. lett.* **B394**, 11 (1997)
4. T. Motobayashi *et al.*, *Phys. Lett.* **B346**, 9 (1995)
5. H. Scheit *et al.*, *Phys. Rev. Lett.* **77**, 3967 (1996)
6. T. Glasmacher *et al.*, *Phys. Lett.* **B395**, 163 (1997)
7. R. Broda *et al.*, *Phys. Rev. Lett.* **74**, 868 (1995)
8. R. Grzywacz *et al.*, *Phys. Rev. Lett.* **81**, 766 (1998)
9. H. Grawe *et al.*, *Prog. in Part. and Nucl. Phys.* **38**, 15 (1997)
10. P. H. Stelson and F. K. McGowan, *Nucl. Phys.* **A32**, 652 (1962)
11. S. Leenhardt, *thesis work I.P.N Orsay*
12. D. Guillemaud-Mueller *et al.*, *Phys.Rev.* **C41**, 937 (1990)
13. G. Klotz *et al.*, *Phys. Rev.* **C47**, 2502 (1993)
14. T. Otsuka *et al.*, *Nuclear Structure '98, Gatlinburg,tennessee, USA (1998)*
15. A. T. Reed *et al. private communication*
16. D. Guillemaud-Mueller *et al.*, *Nucl. Phys.* **A426**, 37 (1984)
17. M. Fauerbach *et al.*, *Phys. Rev.* **C53**, 647 (1996)
18. O. Tarasov *et al*, *Phys. Lett.* **B409**, 64 (1997)
19. F. Azaiez and W. Korten., *Nucl. Phys. News*, **Vol.7**, No.4 (1997)

The Production and Study
of Neutron-rich Nuclei

J. L. Durell

Department of Physics and Astronomy
The University of Manchester, Manchester M13 9PL, UK

Abstract.
Over the last few years there has been a great increase in our knowledge of neutron-rich nuclei resulting from the use of large γ-ray arrays to investigate prompt γ rays from fission and deep inelastic reactions. In this paper we shall discuss various aspects of the physics underlying the fission and deep-inelastic reaction mechanisms that are relevant to the spectroscopic investigation of neutron-rich nuclei.

INTRODUCTION

Fusion-evaporation reactions have led to extensive investigations of the nuclear structure of nuclei on the proton-rich side of the stability line over a wide range of atomic number. Neutron-rich nuclei have been less easy to study. However, the advent of large γ-ray arrays gave rise to new initiatives in the investigation of neutron-rich nuclei produced in the fission process. This was possible because the arrays have sufficient sensitivity and resolving power to disentangle the very many prompt γ rays from the broad range of fragments formed by fission. Both spontaneous and induced fission have been used to produce a wide range of nuclei in the A = 80 to 160 mass region.

Further advances in the study of neutron-rich nuclei are being made by the use of deep-inelastic processes in heavy-ion interactions. The rapid N:Z equilibration that occurs in such reactions can be utilised by having heavy targets as a "source" of neutrons that can be absorbed by the beam nucleus. Deep-inelastic reactions open up the possibility of widening the range of nuclei that can be investigated.

To study the new and different physics that theoretical calculations are predicting for neutron-rich nuclei it is necessary to study them over as wide a range as possible. A wide range of different kinds of states also needs to be studied. Experimenters therefore need to understand the reaction mechanisms that determine the nuclei and states that are populated in fission and deep-inelastic reactions. The aim of this presentation is to discuss some of the features of these mechanisms that are relevant to the spectroscopic investigation of neutron-rich nuclei.

CP495, *Experimental Nuclear Physics in Europe*, edited by B. Rubio et al.
© 1999 American Institute of Physics 1-56396-907-6/99/$15.00

It is expected that radioactive beam facilities will provide a considerable leap forward in the investigation of neutron-rich nuclei. Some comments will be made concerning the use of radioactive beams in the study of nuclei with an excess of neutrons.

FISSION

Spontaneous fission provides, in general, the most neutron-rich fragments. The low-energy character of this process means that the mass distribution is asymmetric. The heavy-mass yield distribution stays fairly constant as a function of the mass of the fissioning system [1], with the light-mass peak varying. This gives the experimenter some choice in the fragments produced, at least for the light-mass fragments. The nuclei ^{252}Cf and ^{248}Cm have been the most exploited systems for the study of fission fragment spectroscopy and reviews of the progress that has been made can be found in Ref. [2,3].

There is one limitation of mass-asymmetric fission: whereas the study of γ rays following spontaneous fission has allowed an increase in our knowledge of nuclei *above* the doubly magic ^{132}Sn [4,5], it has not been possible to learn anything about the levels of equivalent nuclei with Z = 48,49. This is because these nuclei lie in the minimum of the yield distribution. One way to overcome this problem is to use heavy-ion induced fusion-fission as the production mechanism.

Heavy-ion fusion reactions lead to compound nuclei at high excitation energy. If the compound nucleus is sufficiently heavy, fission competes with neutron evaporation. Fission from such a "hot" nucleus will be mass symmetric, leading to a peak in the fission yield at an atomic number half that of the compound nucleus. There is unfortunately a price to pay, in terms of the wish to produce fission fragments as neutron rich as possible: pre-fission neutron emission will reduce the neutron richness of the fissioning system and hence that of the fission fragments produced. The systematics of pre-fission neutron emission have been well-studied [6] and experimenters are able to predict which isotopes of a given atomic number they are going to populate. We can illustrate this by considering the fusion of ^{12}C with ^{238}U, which forms the compound nucleus ^{250}Cf. The systematics tell us that there will be six prefission neutrons, and that each primary fragment will evaporate two neutrons on average. This means that the most probable secondary fission product will be ^{120}In. This prediction is consistent with experimental observation. In the spontaneous fission of ^{252}Cf we should expect the most probable In product to have A = 125. In a heavy-ion induced fission process, there is therefore a significant reduction in the neutron-richness of the observed fission fragments. It should be noted that there would be insufficient yield of ^{125}In from spontaneous fission to perform any prompt γ spectroscopy.

The spectroscopic data from prompt γ-ray experiments allow a much more detailed investigation of fission yields. The fact that γ rays from a given product are in prompt coincidence with γ rays from the complementary products, means that

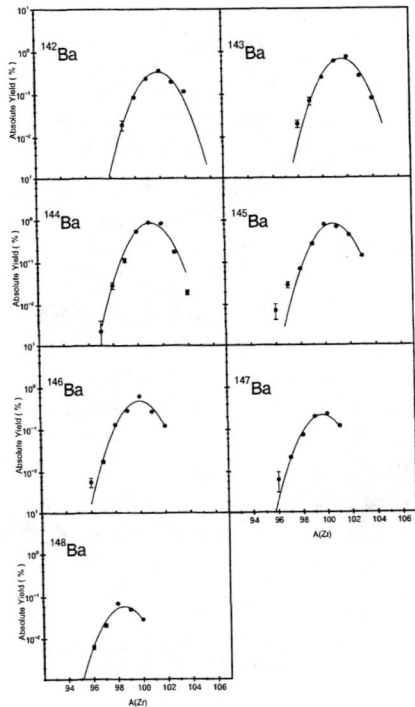

FIGURE 1. Yields of Ba isotopes from the spontaneous fission of ^{248}Cm.

we are able to look at the distribution of the yield of a particular isotope over the range of associated fission fragments. This is illustrated in Figure 1 which shows the yields of Ba isotopes from the spontaneous fission of ^{248}Cm as a function of the mass number of the complementary Zr isotopes. The gaussian curves superimposed on the experimental data have been calculated from parameters given in Ref. [1]. It appears that the yield distributions from "cold" fissioning systems have, to a good approximation, a universal shape.

Experimentalists are thus able to predict what range of nuclei that they are likely to see in any fission process. The next question is to ask what γ rays and levels are likely to be observed? Figure 2 shows a partial level scheme deduced for ^{118}Cd from data taken following the ^{12}C + ^{238}U reaction. It can be seen from this example that yrast states are the mostly strongly populated; that excited, near-yrast bands are less intense, but clearly observable; and that decays from levels with angular momentum approaching $20\hbar$ can be seen.

This simple, yet typical, example of ^{118}Cd shows us that yrast states can be seen to high enough spin to observe the first quasi-particle alignments. Also excited bands with reasonably high angular momentum will be more easily seen: in deformed nuclei γ bands will be populated in preference to β bands; $K = \Omega_1 + \Omega_2$ 2-quasi-particle bands will be populated more strongly than $K = \Omega_1 - \Omega_2$ bands.

The construction of a complete decay scheme with known angular momenta of the discrete states permits the calculation of the average angular momentum of the fission product from the γ-ray intensities. It is found that fission products from heavy-ion induced fission have an average angular momentum about $2\hbar$ greater than products from spontaneous fission. The average spins of primary fragments produced in heavy-ion fusion reactions vary as $A^{\frac{5}{6}}$, which is what would be expected if statistical processes dominate. In the case of spontaneous fission the average spins appear to vary with mass in approximately the same way. In spontaneous fission the relatively cold collective modes at scission should determine the angular momentum sharing between the primary fragments.

Experimental studies of prompt γ rays from fission have not only extended our knowledge of neutron-rich nuclei, but have also provided information about the fission process. The experiments are beginning to reach new levels of sophistication: angular correlations and linear polarisations of prompt γ rays from fission are measured [7]; lifetimes of yrast states have been determined from Doppler lineshapes

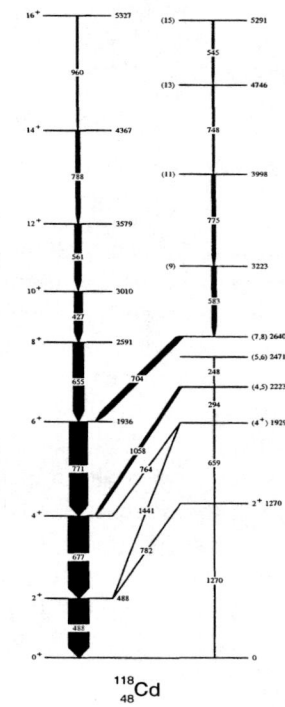

FIGURE 2. Levels of ^{118}Cd. The width of the arrows indicate the γ-ray intensities.

[8]; more recently the first measurements of g-factors have been made by aligning recoiling fission fragments with magnetic fields [9]. There will be continuing advances that will lead to deeper investigations of the nuclei that can be formed in spontaneous fission and by fusion-fission reactions. However, without some big step in sensitivity it appears that we may have reached the limit to how neutron-rich we can go if we are restricted to long-lived fission sources and fission induced by stable beams. Radioactive beam facilities offer the opportunity of using fission induced by neutron-rich beams. The nature of the yield curves shown above (similar yield curves are obtained for heavy-ion induced fission) imply that the addition of just a few neutrons to the fissioning system can give a large increase in the yield of more neutron-rich fission fragments.

DEEP-INELASTIC REACTIONS

Deep-inelastic reactions (DIR) involving heavy beams at energies just above the barrier are characterised by three features which are of importance in the present context:
1) the entrance channel kinetic energy in excess of the Coulomb potential energy at close contact is dissipated into excitation energy of the reaction products;
2) there is rapid N:Z equilibration to the n:p ratio of the combined system;
3) for non-zero impact parameters, "tangential friction" acts to transfer angular momentum from relative motion to intrinsic rotation of the reaction products.

These imply that, when an intermediate mass nucleus interacts with a heavy target (which is intrinsically neutron-rich) we can produce excited, rotating, neutron-rich products with atomic numbers close to that of the beam nucleus. The experimenter would like to produce nuclei at high angular momentum. Experimental analyses [10] of heavy-ion reaction data show that the tangential friction is of the "rolling" type and that $\frac{2}{7}$ of the relative angular momentum in each partial wave contributing to the DIR is transferred to intrinsic rotation. The newly gained angular momentum is expected to be shared between the two reaction products in the ratio of their radii. The magnitude of the partial waves contributing to DIR increases with bombarding energy, therefore to obtain high angular momentum experiments should be done at high beam energies.

The excitation energy of the reaction products also increases with bombarding energy. Therefore a consequence of maximising angular momentum would be an increase in the number of neutrons evaporated from the reaction products, reducing their neutron-richness. This problem is still an open question since the degree of thermal equilibration between the exit channel products has not yet been fully established. Some compromise is required between the wish to study new nuclei to high spin, and how far these new nuclei will be from the line of stability. Examples of the two approaches can be found in Ref. [11,12].

The use of neutron-rich radioactive beams offers the opportunity of extending the experimenter's ability to produce new nuclei further from stability. However, the gains may not be as large as might first be thought. If a heavy target is used, then its N:Z ratio dominates because of its larger contribution to the di-nuclear system. If a beam of ^{68}Ni is used with a ^{208}Pb target instead of ^{64}Ni, then the equilibration line only moves by 0.9 neutrons. The situation will be better with a heavier beam, and therefore the use of secondary beams of fission fragments provides a more exciting prospect for future spectroscopy with DIR.

Most experiments are being done using thick targets so that γ rays are emitted from stopped nuclei. This technique has some limitations as far as the observation of unknown nuclei are concerned. Future work, particularly with radioactive beams, is likely to involve the detection and identification of reaction products with gas detectors used sometimes with a spectrometer. Such a scenario is foreseen in the plans for SPIRAL, where the EXOGAM array will be designed to be run in conjunction with VAMOS. The need to use thin targets will reduce data rates, but will ensure the correct association of γ rays with new nuclei.

SUMMARY

Much recent progress has been made in our knowledge of the structure of intermediate mass neutron-rich nuclei. This progress has come about by using fission and deep-inelastic reactions as production tools, and the power of large γ-ray arrays to determine spectroscopic properties of the nuclear states. New ideas are being applied to these approaches, using fission sources or stable beams, which will increase our understanding. The availability of radioactive beams should lead to new impetus in taking advantage of fission and deep-inelastic reactions.

REFERENCES

1. Vandenbosch, R., and Huizenga, J.R., *Nuclear Fission*, New York: Academic Press, 1973.
2. Ahmad I., and Phillips W.R., *Rep. Prog. in Phys.* **58**, 1415 (1995).
3. Hamilton J.H. et al., *Prog. Nucl. Part. Phys.* **35**, 635 (1995).
4. Zhang C.T. et al., *Phys. Rev. Lett.* **77**, 3743 (1996)
5. Zhang C.T. et al., *Z. Phys.* **A358**, 9 (1997)
6. Hinde D.J. et al., *Nucl. Phys.* **A452**, 550 (1986)
7. Jones M.A. et al., *Nucl. Phys.* **A605**, 133 (1996)
8. Smith A.G. et al., *Phys. Rev. Lett.* **77**, 1711 (1996)
9. Smith A.G. et al., *Phys. Lett.* **B453**, 206 (1999)
10. Bock R. et al., *Nucl. Phys.* **A388**, 334 (1982)
11. Broda R. et al., *Phys. Lett.* **B251**, 245 (1990)
12. Ishii T. et al., Contribution to this Conference

Projected Hartree-Fock + BCS spectra of Mg isotopes

Angel Valor*, Paul-Henri Heenen* and Paul Bonche†

*Physique Nucléaire Théorique, C.P. 229, Université Libre de Bruxelles,
1050 Brussels, Belgium
† SphT, C.E.A. Saclay, 91191 Gif sur Yvette Cedex, France

Abstract. We have developed a method to project simultaneously on angular momentum and particle number Hartree-Fock+BCS nuclear wave functions. The method generalizes previous developments in which mean field wave functions were mixed as a function of a collective variable by the Generator Coordinate Method (GCM). Calculations have been performed for several Mg isotopes, where spectra and transition probabilities have been determined. The axial quadrupole moment has been used as a generator coordinate. We show here test results obtained for ^{24}Mg.

Mean field wave functions break several symmetries of the nuclear Hamiltonian. The breaking of rotational invariance precludes the determination in a clean way of transition probabilities. The breaking of the particle number symmetry, due to pairing correlations, smoothes the variation of properties of nuclei as a function of the neutron and proton numbers. To overcome these problems, we have implemented a simultaneous projection on good angular momentum, neutron and proton numbers of mean field wave functions. These wave functions are generated by Hartree-Fock+BCS+Lipkin-Nogami (HF+BCS+LN) calculations with a constraint on a collective variable. We use a Skyrme interaction in the mean field channel and a density dependent, zero range interaction in the pairing channel. Afterwards, projected diagonal and non diagonal overlap and Hamiltonian matrix elements are calculated, and a configuration mixing calculation using a discretized form of the generator coordinate method is performed on the collective variable. For first applications, we have imposed time reversal invariance and axial symmetry on the mean field wave functions. The axial quadrupole moment q_0 has been chosen as the collective coordinate.

A computer code, called PROMESSE (PROjected Matrix Elements of mean field Self-consistent Solutions with Effective interactions) has been developed to calculate these matrix elements. It is based on the technique of calculation on a 3-dimensional cartesian mesh that has already been used in several studies of exotic nuclear states. It permits to calculate spectra and transition probabilities.

CP495, *Experimental Nuclear Physics in Europe*, edited by B. Rubio et al.
© 1999 American Institute of Physics 1-56396-907-6/99/$15.00

As a first test, we have applied the method to ^{24}Mg. We use the Sly4 Skyrme interaction and a surface delta pairing interaction of a intensity $G = 1000 MeV fm^3$. On fig. 1.a, we show the intrinsic and the projected energy curves, as a function of q_0. The spectra generated by a configuration mixing is also indicated. The configuration mixing does not modify significantly the energies. The restoration of symmetries, on the other hand, shifts the mean field minimum to larger deformations.

On the right part of figure 1, are compared to the experimental data the spectra and the B(E2) transition probabilities obtained within the yrast band, by taking the projected states corresponding to the minima of the energy curves shown on fig 1.a and between the states obtained in the GCM calculation. The overall agreement with the data is good. The mixing of configurations has the effect to decrease the 2^+ to 0^+ B(E2) value, an effect that we have found quite general.

Our plans for the future are to relax the symmetry restrictions imposed on the mean field wave functions. The code PROMESSE has been constructed in such a way that these generalizations should be rather easy.

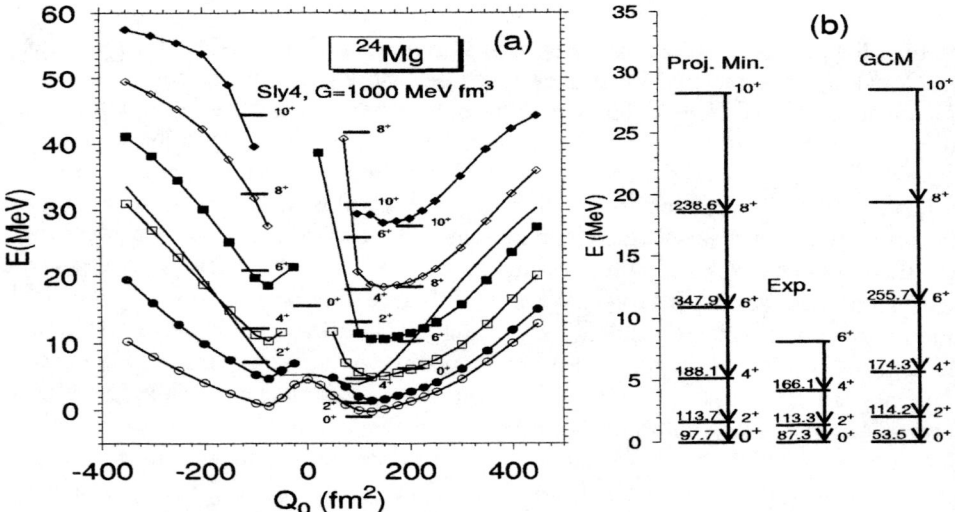

FIGURE 1. Projected results for the nucleus ^{24}Mg. (a) Energies as a function of the axial quadrupole moment. Solid line: HF+BCS+LN results; the curves displaying circles, filled circles, boxes, filled boxes, diamonds and filled diamonds correspond, respectively, to projected energies in the $0^+,2^+,4^+,6^+,8^+,10^+$ spin representations. The horizontal bars are the projected energies of the collective intrinsic states. They are located at the axial quadrupole moment in which the collective wave functions reach their global maximum. (b) excitation energies and transition probabilities between: intrinsic states minimizing each projected energy curve (left), yrast collective states (right), experimental levels (center)

High-Spin States in Odd-Odd $N=Z$ ^{46}V

C. D. O'Leary[*,†], M. A. Bentley[†], D. E. Appelbe[*1], R. A. Bark[‡2],
D. M. Cullen[*], S. Ertürk[*3], A. Maj[‡,||], J. A. Sheikh[¶4] and
D. D. Warner[§]

[*] Oliver Lodge Laboratory, University of Liverpool, Liverpool L69 7ZE, United Kingdom
[†] School of Sciences, Staffordshire University, Stoke-on-Trent, ST4 2DE, United Kingdom
[‡] Tandem Accelerator Laboratory, Niels Bohr Institute, Risø, DK-4000 Roskilde, Denmark
[||] Niewodniczanski Institute of Nuclear Physics, 31-342 Krakow, Poland
[¶] Physics Department, University of Surrey, Guildford GU2 5XH, United Kingdom
[§] CCLRC Daresbury Laboratory, Warrington WA4 4AD, United Kingdom

Abstract. High-spin states up to the $f_{\frac{7}{2}}$-shell band termination at $J^\pi=15^+$ have been observed for the first time in the odd-odd $N=Z=23$ nucleus ^{46}V. The new level scheme has two separate structures corresponding to spherical and prolate shapes. A rotational band has very simliar energies to the yrast sequence in ^{46}Ti and is therefore assumed to be a $T=1$ configuration.

Odd-odd $N=Z$ fpg-shell nuclei are an important region for the study of pairing correlations between both like and unlike nucleons. The relative closeness of the neutron and proton Fermi surfaces in these nuclei promotes the occurrence of np pairing to such a degree that it can compete favourably with nn and pp pairing modes.

High-spin states in one such nucleus, ^{46}V, were investigated using the PEX detector apparatus at the Niels Bohr Institute Tandem Accelerator Laboratory. A ^{28}Si beam at a laboratory energy of 87 MeV incident upon a 500 μg/cm^2 ^{24}Mg target produced ^{46}V via the ^{24}Mg(^{28}Si, $p\alpha n$)^{46}V reaction. Greater detail on the experiment and analysis can be found in [1].

The new level scheme derived from this experiment is shown in Fig. 1. The $f_{\frac{7}{2}}$ shell band terminating state at $J^\pi=15^+$ has been established for the first time. As in previous work [2] the ground state spin and isospin of ^{46}V are assigned to be

---

1) Present Address: Department of Physics and Astronomy, McMaster University, Hamilton, Canada L8S 4M1
2) Present Address: Department of Nuclear Physics, Research School of Physical Sciences and Engineering, Australian National University, Canberra, ACT 0200, Australia
3) Present Address: Nigde Universitesi, Fen-Edebiyat Fakültesi, Fizik Bölümü, Nigde, Turkey
4) Present Address: Tata Institute of Fundamental Research, Colaba, Bombay, 400 005, India

$J^\pi=0^+$ and $T=1$, in common with all other fpg-shell odd-odd $N=Z$ nuclei except ^{58}Cu.

An interesting feature of this new level scheme are the two apparently separate structures. The yrast structure **A** demonstrates non-collective behaviour. In contrast structures **C** and, in particular, **B** display collective characteristics, with rotational-like level spacings. As ^{46}V is an ^{40}Ca core plus 3 protons and 3 neutrons, this behaviour might be expected and has been observed in other nuclei near the centre of the shell. Thus it is assumed that ^{46}V may have a prolate deformation in common with other mid $f_{\frac{7}{2}}$ shell nuclei.

The band labelled **C** in Fig. 1 has energies very close (915, 2052 and 3363 keV) to those of the yrast band in the isobaric analogue nucleus ^{46}Ti (at 889, 2010 and 3299 keV). This structure is therefore assigned as a $T=1$ configuration with the other states in Fig. 1 probably corresponding to $T=0$ structures.

REFERENCES

1. C. D. O'Leary *et al.*, Phys. Rev. Lett. **79** (1997) 4349.
2. A. R. Poletti, E. K. Warburton and J. W. Olness, Phys. Rev. C **23** (1981) 1550.

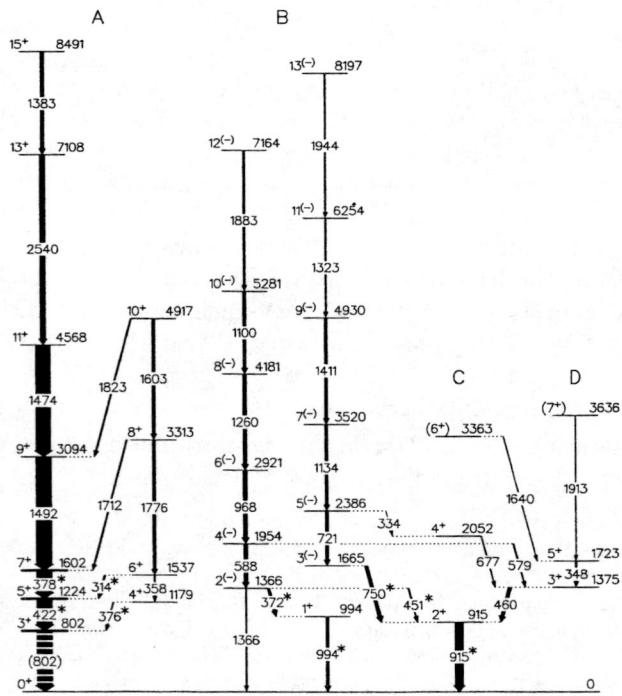

FIGURE 1. Level scheme for ^{46}V deduced from this work.

Variety of collective motions in the middle of the $1f_{7/2}$ shell

F. Brandolini[1], S.M. Lenzi[1], N.H. Medina[3], D.R. Napoli[2],
R.V.Ribas[3], C.A. Ur[1], D. Bazzacco[1], J.A. Cameron[4], G. de Angelis[2],
M. De Poli[2], A. Gadea[2], S. Lunardi[1], C. Rossi Alvarez[1],
H. Somacal[2], C. Svensson[4], A. Poves[5], J. Sanchez-Solano[5]

[1] *Dipartimento di Fisica dell'Universita di Padova and INFN, Padova, Italy*
[2] *Laboratori Nazionali di Legnaro - INFN, Legnaro, Italy*
[3] *Instituto de Fisica, Universidade de Sao Paulo, Sao Paulo, Brasil*
[4] *Department of Physics and Astronomy, Mc Master University, Hamilton, Canada*
[5] *Departamento de Física Teorica, Universidad Autónoma, Cantoblanco, Madrid, Spain*

Abstract.
A variety of collective phenomena is proven to occur at the middle of the $1f_{7/2}$ shell, which can be microscopically described by Shell Model calculations. For this purpose a detailed spectroscopy as well as precise DSAM lifetime measurements were performed. The cases of ^{48,49,50}Cr are illustrated.

After a long quiescence, in the last few years a spectacular progress has been made by nuclear spectroscopy in the middle of $1f_{7/2}$ shell, both theoretically and experimentally. Large scale shell model (SM) calculations in the full pf configuration space have been very successful in explaining the building up of deformation for natural parity states of low spin [1–3]. As the configuration space is not too large, the middle of $1f_{7/2}$ shell turns out to be a unique workbench where collective properties, such as axial and triaxial deformation, band crossing, band termination, particle-rotor coupling etc, can be microscopical described. Recently, large scale SM calculations for $1f_{7/2}$ nuclei have been extended to encompass unnatural parity bands, which are described as due to the presence of a hole in the $1d_{3/2}$ orbital [4]. From the experimental side, the new generation γ-detector arrays and improved DSAM analysing procedures [5] provide the good quality information required for a comparison with modern SM calculations. We will consider the case of ^{48}Cr [6–8], ^{50}Cr [8–10] and ^{49}Cr [10,11].

The reaction ^{28}Si on ^{28}Si at 115 MeV bombarding energy was performed at the LNL Tandem accelerator. The target consisted of 0.8 mg/cm^2 of ^{28}Si on 15 mg/cm^2 of Au and concidence γ-rays were collected with the spectrometer GASP. For some aspects in ^{49}Cr the reaction ^{46}Ti$(\alpha,n)^{49}$Cr at a bombarding energy of 12 MeV has also been performed at the CN van de Graaff accelerator at LNL.

CP495, *Experimental Nuclear Physics in Europe*, edited by B. Rubio et al.
© 1999 American Institute of Physics 1-56396-907-6/99/$15.00

48Cr: 48Cr is the elective starting point for a general discussion, as it is exactly in the middle of the shell and it should have the largest deformation. The agreement of predicted level energies of ground state (gs) band with experimental ones is excellent up to band termination in the $1f_{7/2}$ at $I^{\pi}=16^+$ [6,7], as shown in Fig. 1.

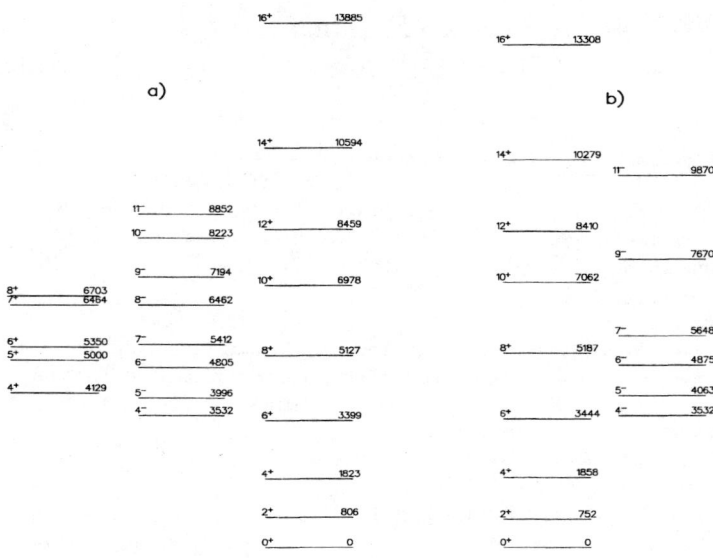

FIGURE 1. Calculated levels in 48Cr a), as compared with experimental ones b).

The level spacing does not increase linearly with energy and drops down drastically at $I^{\pi}=12^+$. This is interpreted as due to a dominance of seniority 4 states, indicating that rotational collectivity is contrasted by pairing correlations. B(E2) values, obtained from DSAM lifetimes measurements, were well reproduced theoretically [8] and lead to a deformation parameter $\beta \simeq 0.28$ at low spin. It is also of great interest in 48Cr to understand the nature of the observed sideband. Owing to the large deformation at rather low spin, it is meaningful to use the Nilsson scheme. In the $f_{7/2}$ shell we thus expect a K=4+ band from the excitation of one nucleon (with equal probability proton or neutron), from the [321]3/2− orbital to [312]5/2− one. SM model calculations however predicts such a band to be about one MeV higher than experiment, as displayed in Fig. 1. The band levels are expected to decay remarkably towards the gs band via E2 transitions, owing to band mixing, but this is not observed. The alternative possibility of a K=4− band, obtained by exciting a nucleon from the $2d_{3/2}$ orbital [202]3/2+ fully agrees with experiment. A band K=3− of similar origin was known in 46Ti. The E1 forbiddeness for Z=N explains the isomerism of the 4− bandhead ($\tau = 4.7$ ns). In Fig. 1 the calculated energy of the 4− state was adjusted to the experimental one. Signature splitting is theoretically predicted both for K=4$^+$ and K=4$^-$ bands.

^{50}Cr and ^{49}Cr. An unnatural parity band of similar origin was also observed in ^{49}Cr and ^{50}Cr, as shown in Fig. 2. In ^{50}Cr a band K= 4$^-$ was seen, which is proposed to have a proton particle-hole nature. In ^{49}Cr the proton particle-hole pair couples with the unpaired neutron to the maximum K giving rise to a band K=13/2$^+$. One has also to note the presence, at lower energy, of a band K=3/2$^+$ and that the 13/2$^+$ state acts as a yrast trap for the lower positive parity band. The two bands can be regarded as the coupling of the nucleon hole to the K$^\pi$ = 0$^+$, T=1 and K$^\pi$ = 5$^+$, T=0 bands in ^{50}Mn [12]. For all mentioned bands, the K$^\pi$=5$^+$ one in ^{50}Mn included, $\beta \simeq 0.26$ was inferred from DSAM measurements [8,11].

The observation of the bandheads of the predicted normal parity bands with K=4$^+$ in ^{50}Cr and K=13/2$^-$ in ^{49}Cr, respectively, was favored in the heavy ion reaction by their population from the corresponding levels with opposite parity. This takes place via K-favored E1 transitions, while the more energetic E1 transition to the gs band are K-hindered. This was not the case in ^{48}Cr.

The yrast band in ^{48}Cr and ^{50}Cr are shown in Fig. 2 up to the 1f$_{7/2}$ band termination. An interesting question in ^{50}Cr is to establish the nature of the yrast 10$^+$ state, which decays with a small B(E2). This is interpreted as due to the simultaneous alignment of both proton and neutron pairs to K=4$^+$ and K=6$^+$, respectively. The yrast 10$^+$ state is thus described as having K=10$^+$. The involved proton and neutron states are identified with the 4$^+$ level at 3324 keV and the (6$^+$) one at 3824 keV, respectively. The yrast band in ^{49}Cr shows signature splitting at

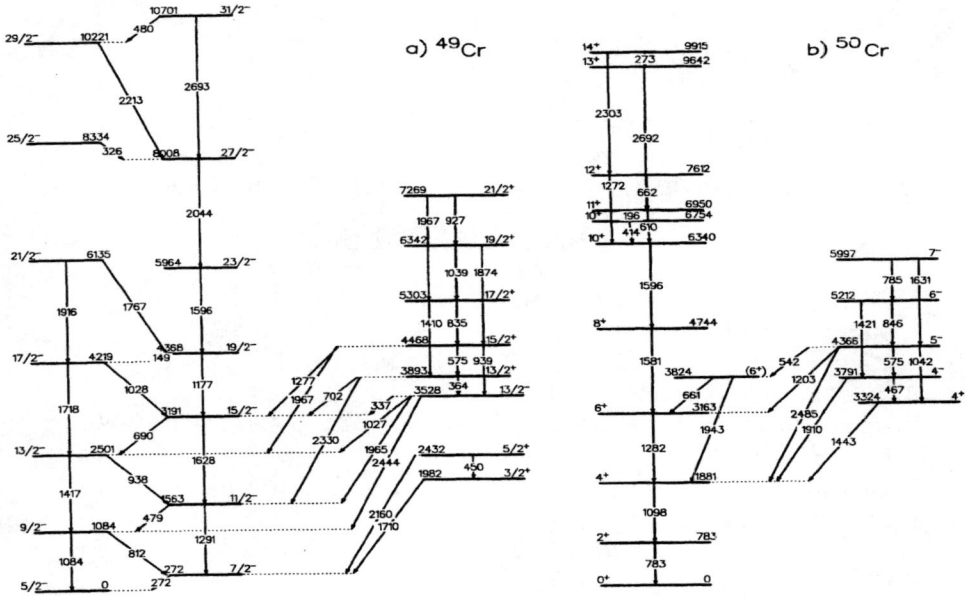

FIGURE 2. Simplified level scheme in ^{49}Cr a) and in ^{50}Cr b), respectively.

low spin, pointing to triaxiality. Furthermore the 19/2⁻ level is quite depressed, showing the dominance of seniority 3 configurations, in analogy to ^{48}Cr.

There is not space for a detailed comparison with SM calculations. We note that this has been reported for ^{50}Cr [8], where the predicted level energies agree within 100-200 keV. Energy level predictions for the gs band in ^{49}Cr [3], as well as for the sideband, do have a similar precision. B(E2) and B(M1) values are correctly predicted in ^{50}Cr [8]. A comparison with SM predictions of the B(E2) and B(M1) values in ^{49}Cr for stretched transitions from both negative and positive parity levels, is shown in Fig. 3, as obtained from our DSAM analysis [11]. The agreement is good. The yrare 13/2⁻ state decays via retarded M1 and E2 transitions, as expected, owing to K-hindrance (The corresponding points are displayed slightly on the right with respect to those for the yrast state).

As a final comment, we stress the paradox that spherical large scale SM predictions in the $1f_{7/2}$ shell result to be better for deformed nuclei in the middle of the shell than for more spherical nuclei near to its borders.

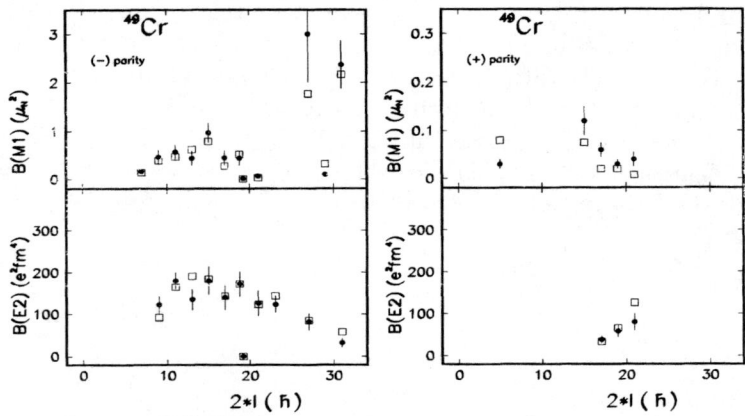

FIGURE 3. B(E2) and B(M1) values for negative parity levels a) and positive ones b) in ^{49}Cr.

REFERENCES

1. Caurier E., et al., *Phys. Rev. C* **50**, 225 (1994).
2. Martinez-Pinedo G., et al., *Phys. Rev. C* **54**, R2150 (1996).
3. Martinez-Pinedo G., et al., *Phys. Rev. C* **55**, 187 (1997).
4. Poves A., and Sanchez-Solano J., *Phys. Rev. C* **58**, 179 (1998).
5. Brandolini F., and Ribas R.V., *Nucl. Instr. and Meth., A* **417**, 150 (1998).
6. Lenzi S.M., et al., *Z. Phys. A* **354**, 117 (1996).
7. Cameron J.A., et al., *Phys. Lett. B* **387**, 266 (1996).
8. Brandolini F., et al., *Nucl. Phys. A* **642**, 387 (1998).
9. Lenzi S.M., et al., *Phys. Rev. C* **56**, 1313 (1997).
10. Cameron J.A., et al., *Phys. Rev. C* **58**, 808 (1998).
11. Brandolini F., et al., *contribution at INPC98*, Paris 1998 and to be published.
12. Svensson C.E., et al., *Phys. Rev. C* **58**, R2621 (1998).

Absence of blocking effect in the $N = Z$ odd-odd nucleus ^{62}Ga: a new signature of the n-p pairing?

N. Belcari[a], G. de Angelis[a], J. Eberth[b], S. Skoda[b], R. Wyss[c],

E. Farnea[d,1], A. Gadea[d,2], A. Algora[a], M. De Poli[a], E. Fioretto[a],

T. Martínez[a,3], D.R. Napoli[a], G. Prete[a], P. Spolaore[a], D. Bazzacco[e],

F. Brandolini[e], A. Buscemi[e], R. Isocrate[e], S. Lenzi[e], S. Lunardi[e],

R. Menegazzo[e], C.M Petrache[e], P. Pavan[e], C. Rossi Alvarez[e],

Zs. Podolyák[f], P.G. Bizzeti[g], A.M. Bizzeti-Sona[g].

[a] INFN Laboratori Nazionali di Legnaro, 35020 Legnaro, Italy
[b] IKP, Universität zu Köln, 50937 Köln, Germany
[c] Royal Institute of Technology, Stockholm, Sweden
[d] Instituto de Física Corpuscular, 46100 Valencia, Spain
[e] Dipartimento di Fisica and INFN Sezione di Padova, 35131 Padova, Italy
[f] University of Surrey, GU2 5XH Guildford, United Kingdom
[g] Dipartimento di Fisica and INFN Sezione di Firenze, 50125 Firenze, Italy

Abstract. The high-spin structure of the $N \simeq Z$ nuclei ^{62,64,65}Ga have been investigated using the EUROBALL array coupled to the ISIS Si-ball. The collective bands of such nuclei are then compared showing an unexpected behaviour.

A strong experimental signature of the neutron-proton pairing correlations in the $T = 0$ channel can be provided by the absence of the blocking effect in the alignment processes for odd-odd $N = Z$ nuclei. Such alignments are blocked for usual pairs of identical particles due to the Pauli principle, but are permitted for pairs of unlike particles.

The high-spin structure of the $N \simeq Z$ nuclei ^{62,64,65}Ga has been investigated in an experiment at the Laboratori Nazionali di Legnaro using the EUROBALL spectrometer coupled to the ISIS Si-ball [1]. The gallium isotopes have been populated respectively via the ^{40}Ca(^{32}S;$2\alpha pn$, $\alpha 3pn$, $\alpha 3p$) reaction at 155 MeV. A stack of two 0.5 mg/cm^2 self-supporting calcium targets was used.

---

[1] Fellowship supported by the EC under contract n°ERBFMBICT983126
[2] Fellowship supported by the EC under contract n°ERBFMBICT983127
[3] Under EC contract n°ERBFMGECT980110

CP495, *Experimental Nuclear Physics in Europe*, edited by B. Rubio et al.
© 1999 American Institute of Physics 1-56396-907-6/99/$15.00

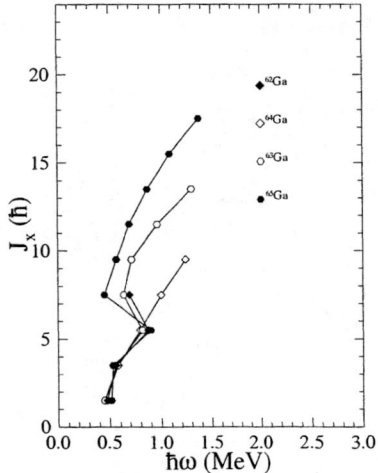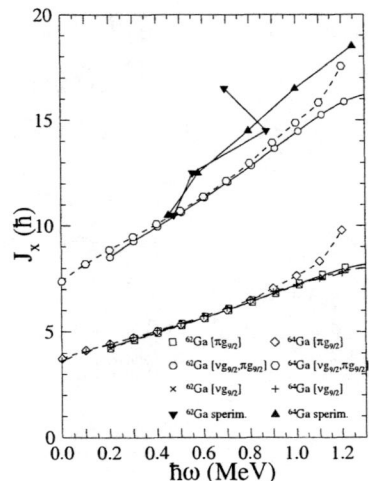

FIGURE 1. Left: the spin component on rotational axis J_x (aligned momentum) versus rotational frequency $\hbar\omega$ extracted from experimental data for ^{62,63,64,65}Ga; right: experimental J_x values of ^{62,64}Ga compared to HFB+Cranking calculation (only $T = 1$ pairing component) for the $[\nu 1g_{9/2} \otimes \pi 1g_{9/2}]$ band.

The level schemes of ^{64,65}Ga have been extended up to an excitation energy of, respectively, 9.7 MeV and 18.0 MeV and to $J^\pi = 17^+\hbar$ and $41/2^+\hbar$. The data confirm the ^{62}Ga level scheme previously studied by Skoda et al. [2]

Figure shows a comparison of the aligned angular momentum J_x versus rotational frequency for the collective band, built promoting particles in the shape-driving $g_{9/2}$ orbitals, for the ^{62,64,65}Ga and the ^{63}Ga [3] isotopes. It can be noticed that the first $g_{9/2}$ alignment is observed at $\hbar\omega \approx 0.7$ MeV in the odd-A ^{63,65}Ga nuclei and it is absent in the odd-odd ^{64}Ga due to the blocking effect. Evidence for an unexpected alignment at $\hbar\omega \approx 0.7$ MeV is observed in the odd-odd N=Z nucleus ^{62}Ga. This alignment can be explained in terms of a promotion of a $T = 0$ correlated n-p pair in the same signature $g_{9/2}$ orbitals. HFB+Cranking calculations including only the isovector component of the pairing do not predict any aligment at this rotational frequency. The calculations also predict a non-collective structure becoming yrast above $\hbar\omega = 1.2$ MeV. Therefore it cannot be excluded that the lowering of such structure can cause this effect.

REFERENCES

1. E. Farnea et al., Nucl. Inst. and Meth. **A400**, 87 (1997)
2. S. Skoda et al., to be published.
3. M. Weiszflog, private communication

Spectroscopy at N=Z with EUROBALL III

A.Gadea[a,b] [1], E.Farnea[a] [2], G.De Angelis[b], N.Belcari[b], T.Martinez[b] [3], B.Rubio[a],
J.L. Tain[a], D.R. Napoli[b], M. De Poli[b], P. Spolaore[b], G. Prete[b], E. Fioretto[b], D.
Bazzacco[c], F. Brandolini[c], S.M. Lenzi[c], S. Lunardi[c], P. Pavan[c], C. Rossi Alvarez[c],
P.G. Bizzeti[d], A.M. Bizzeti-Sona[d], J. Eberth[e], T. Steinhardt[e], O. Thelen[e],
R. Wyss[f], C. Fahlander[g], D. Rudolph[g], A. Atac[h], A. Axelsson[h], J. Nyberg[h],
J. Persson[h], M. Weiszflog[h], W. Gelletly[i], P. Regan[i], Zs. Podolyák[i], J. Garcés Narro[i]
and the EUROBALL III and NEUTRON-WALL collaborations.

[a] *IFIC, Burjassot, 46100 Valencia, Spain*
[b] *LNL, Legnaro (PD) 35020, Italy*
[c] *INFN, Sezione di Padova and Universita' di Padova, Italy*
[d] *Dipartimento di Fisica dell'Università and INFN, Sezione di Firenze, Italy*
[e] *Institut für Kernphysik, Univerität zu Köln, Germany*
[f] *Royal Institute of Technology, Stockholm, Sweden*
[g] *Division of Cosmic and Subatomic Physics, University of Lund, Sweden*
[h] *The Svedberg Laboratory and Department of Radiation Sciences, Uppsala University, Sweden*
[i] *University of Surrey, Surrey GU2 5XH, United Kingdom*

Abstract. A complete study of the nuclear structure by means of γ spectroscopy requires, in addition to the high resolution γ measurement and accurate DCO's or angular distributions, the information concerning the Electric or Magnetic character of the transition.
This information for transitions in nuclei far from stability valley is now reachable in the new generation of Ge-arrays based in composite detectors. EUROBALL III is a good example with the high polarization sensitivity of the 90° ring of Clovers. The Polarization correlations PCO's measured in coincidence with the Cluster detectors permits to investigate transitions in weakly populated nuclei. In this contribution we present results on medium mass N=Z nuclei measured with EUROBALL III coupled with light particle ancillary detectors.

The study of N=Z nuclei between the double magic A=56 and A=100 can provide important information concerning nuclear structure. Exploring this region it is possible to find, between the two spherical extremes, a large variety of nuclear shapes like largely deformed bands in A≈60 [1–3] or the strong variation in shape

---

1) Fellowship supported by the EC under contract n^o ERBFMBICT983127
2) Fellowship supported by the EC under contract n^o ERBFMBICT983126
3) Supported by the EC under contract n^o ERBFMGECT980110

CP495, *Experimental Nuclear Physics in Europe*, edited by B. Rubio et al.

with small changes in nucleon number in the A≈70 region [4,5].

The isospin impurities resulting from the coulomb interactions, increase with Z and is maximum for N=Z nuclei, the medium mass and heavy N=Z are excellent cases to study the isospin admixture and its relation with the nuclear deformation [6]. A possible way to learn about isospin impurities in the N=Z nuclei is to investigate the presence of E1 transitions, forbidden in these nuclei between states with T=0.

Experiments

In the framework of EUROBALL III it was possible for the first time last year to perform experiments using ancillary detectors. In our contribution to this conference we have presented results from experiments performed at LNL Legnaro on nuclei at N=Z and in the vicinity, but due to space limitations, in the proceedings we will include only one nucleus, the ^{64}Ge produced by ^{32}S beam at 125 MeV on ^{40}Ca target with Au backing. The experiment was performed with EUROBALL, the ISIS Si-ball [7] and the Neutron-Wall [8]. The high selectivity of this detector combination has been used to study proton rich reaction products with N=Z.

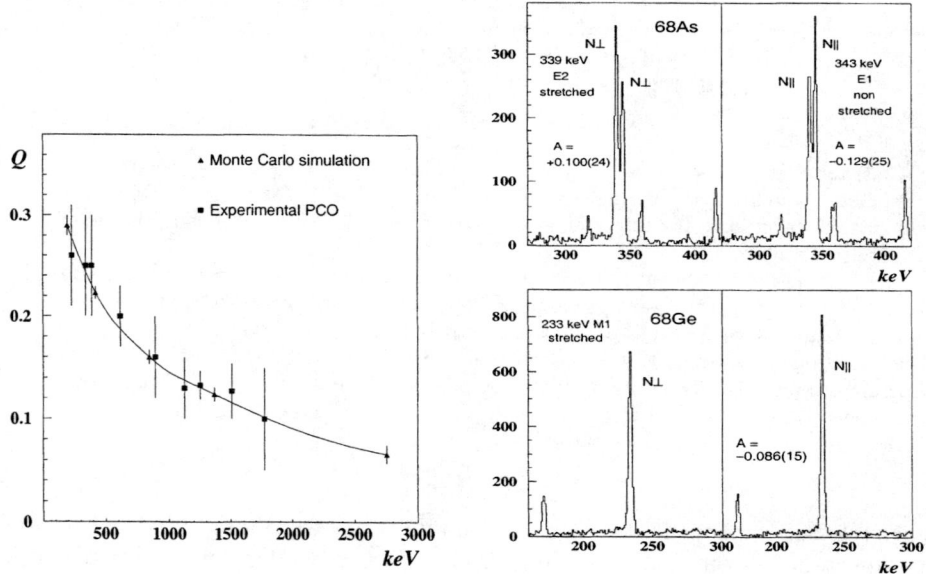

FIGURE 1. EUROBALL III Polarization sensitivity "Q". The experimental sensitivity is determined from PCO's in the ^{32}S (125 MeV) + ^{40}Ca reaction, using lines from ^{68}As, ^{68}Ge, ^{66}Ge and ^{63}Ga. The asymmetry is defined as $A = (N_\perp - N_\parallel)/(N_\perp + N_\parallel)$.

In addition to the conventional particle-γ-γ and γ-γ-γ coincidences and DCO's we have analyzed the polarization of the γ transitions emitted by the oriented nuclei, using the Clover [9] detectors as γ-polarimeters. EUROBALL-Clover coincidences (PCO's) gated with the particle information provided by the ancillary detectors have been used to select the weakly populated channels. In order to calculate the theoretical PCO's the pertinent software has been developed based on the formalism used by R.M.Steffen and K.Alder [10]. A similar work has been done in parallel by Ch.Droste and coworkers [11]. The comparison between the extracted polarization sensitivity "Q" and the Monte Carlo simulation performed by L.M.Garcia-Raffi et al. [12] is shown in Fig.1, together with an example of experimental data.

The ^{64}Ge nucleus

The N=Z=32 nucleus ^{64}Ge has been previously studied in in-beam experiments by P.J.Ennis et al. [13] and we have now extended the level scheme to higher spins [14]. As it is already pointed in [13] one important feature of this nucleus is the high branching of the (5^-) to 4_1^+ (1665 keV) and 4_2^+ (1048 keV) E1 transitions [14], in principle forbidden in N=Z nuclei [15] and competing with the (5^-) to (3^-) E2 (see Fig.2). The light even-even Ge isotopes present the same de-excitation pattern with two E1 transitions going into the 4_1^+ and 4_2^+ competing with the E2 intraband transition.

The B(E2) measured in the neighbour even-even nuclei, ranges between few W.u. and 0.5 W.u. for ^{66}Ge. These low values indicate low collectivity and different structure of the states involved in the transition, the E1 is therefore favoured even at N=Z where it should be quenched and makes difficult any discussion based on the relative branching between the E2 and E1 transitions de-exciting the (5^-), without half life information.

TABLE 1. Preliminary intensities, DCO ratios and polarization values for the transitions in ^{64}Ge.

E_γ	I_γ	DCO	POL	POL$_{th}$	E_γ	I_γ	DCO	POL	POL$_{th}$
528	780(25)	1.02(5)	+0.8(4)	+0.50	1127	570(50)	0.88(6)	+0.7(4)	+0.50
677	102(6)				1151	750(18)	1.20(12)	+0.6(5)	+0.63
747	89(6)	1.3(4)	+1.2(6)	+0.53	1234	330(30)	0.66(5)		
848	220(25)	0.93(7)			1580	< 50			
873	340(30)	1.07(11)			1665	567(18)	0.64(5)	-0.7(6)	-
902	1000		0.11(6)	+0.63	1820	300(30)	0.97(8)		
1048	130(9)	0.53(17)	-0.8(7)	-	2068	96(6)			
1090	105(8)				2206	250(30)	0.54(7)		

The polarization analysis for the higher intensity transitions in ^{64}Ge shows a clear electric character for all the transition except the two de-exciting the (5^-) state (see Fig.2 and Table 1). For this transitions the data suggest a dominant magnetic character which can be a large M2 mixing ($\delta \approx -6$)in the E1 transition established in ^{64}Ge by systematics and DCO's. Nevertheless, due to the large error bars we can not exclude a pure E1 transition. An excerpt of the ^{64}Ge level scheme compared with a reduced space shell model calculation with excellent agreement for the positive parity low lying states is also shown in Fig. 2.

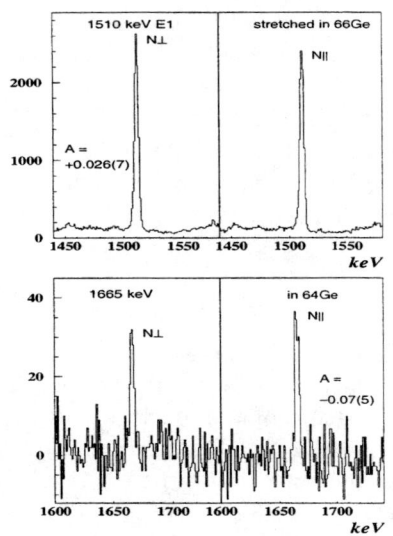
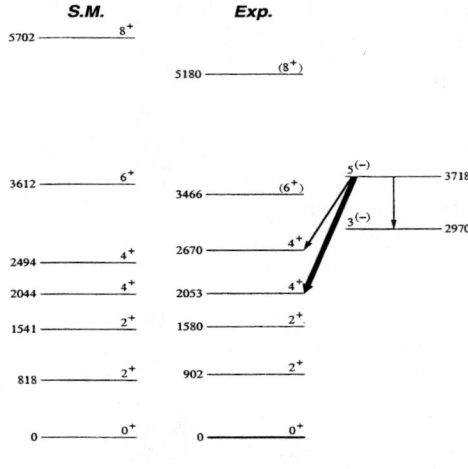

FIGURE 2. Comparison of the polarization experimental data for the (5^-) to 4^+ transitions in ^{66}Ge and ^{64}Ge (left). The low lying positive parity states observed in ^{64}Ge together with the 3^- and 5^- states and a shell model calculation in the $f_{5/2}$, $p_{3/2}$, $p_{1/2}$ configuration space done with OXBASH [16] and with the matrix elements for the Ni and Zn region given in ref. [17]

REFERENCES

1. D.Rudolph et al. *Phys. Rev. Lett.* **80**, 3018 (1998).
2. C.E.Svensson et al. *Phys. Rev. Lett.* **79**, 1233 (1997).
3. C.E.Svensson et al. *Phys. Rev. Lett.* **82**, 3400 (1999).
4. R.B.Piercey et al. *Phys. Rev. Lett.* **47**, 1514 (1981), *Phys. Rev. C* **25**, 1914 (1982)
5. W.Nazarewicz et al. *Nucl. Phys. A* **435**, 397 (1985).
6. J.Dobaczewski and I.Hamamoto *Phys. Lett. B* **345**, 181 (1995).
7. E.Farnea et al. *Nucl. Inst. and Meth. A* **400**, 87 (1997).
8. Ö.Skeppstedt et al. *Nucl. Inst. and Meth. A* **421**, 531 (1999).
9. P.M.Jones et al. *Nucl. Inst. and Meth. A* **362**, 556 (1995).
10. R.M.Steffen and K.Adler in: The electromagnetic interaction in nuclear spectroscopy. ed. W.D.Hamilton (North-Holland, Amsterdam, 1975) ch. 12.
11. Ch.Droste et al. *Nucl. Inst. and Meth. A* **378**, 518 (1996).
12. L.M.Garcia-Raffi et al. *Nucl. Inst. and Meth. A* **391**, 461 (1997).
13. P.J.Ennis et al. *Nucl. Phys. A* **535**, 461 (1997).
14. E.Farnea et al. The following contribution to this proceedings.
15. E.K.Warburton and J.Weneser in: *Isospin in nuclear physics.* ed. D.Wilkinson (North-Holland, Amsterdam, 1969) ch. 5.
16. B.A.Brown at al. *MSU-NSCL report 524.*
17. J.F.A Van Hienen et al. *Nucl. Phys. A* **269**, 159 (1976).

High spin studies of ^{64}Ge with EUROBALL

E. Farnea[a,1], A. Gadea[a,2], G. de Angelis[b], J. Eberth[c], T. Steinhardt[c],
O. Thelen[c], S. Skoda[c], A. Algora[b], D. Bazzacco[d], N. Belcari[b],
P.G. Bizzeti[e], A.M. Bizzeti-Sona[e], F. Brandolini[d], A. Buscemi[d],
M. De Poli[b], J. Garcés Narro[f], W. Gelletly[f], R. Isocrate[d],
T. Martínez[b,3], D.R. Napoli[b], P. Pavan[d], Zs. Podolyák[f],
C. Rossi Alvarez[d], B. Rubio[a], P. Spolaore[b], J.L. Tain[a], R. Wyss[g]

[a] Instituto de Física Corpuscular, 46100 Valencia, Spain
[b] INFN Laboratori Nazionali di Legnaro, 35020 Legnaro, Italy
[c] IKP, Universität zu Köln, 50937 Köln, Germany
[d] Dipartimento di Fisica and INFN Sezione di Padova, 35131 Padova, Italy
[e] Dipartimento di Fisica and INFN Sezione di Firenze, 50125 Firenze, Italy
[f] University of Surrey, GU2 5XH Guildford, United Kingdom
[g] Royal Institute of Technology, Stockholm, Sweden

Abstract. The high-spin states of the $N = Z$ nucleus ^{64}Ge have been investigated using the EUROBALL array coupled to the ISIS Si-ball and to a highly efficient n-Wall. Spins and parities were assigned on the basis of a DCO and of a polarization analysis.

The $N = Z$ nucleus ^{64}Ge was studied using the EUROBALL γ-ray spectrometer coupled with the ISIS Si-ball [1] and with a highly-efficient array of neutron detectors [2] to obtain the required channel selectivity. The reaction was ^{32}S (125 MeV) + ^{40}Ca, using a 1 mg/cm^2 target evaporated on a 12 mg/cm^2 gold backing. The quality of the data was such that some ambiguities in the known decay scheme [3,4] were solved, as shown in figure 1. It was considered of great importance to get a clear experimental assignment of the multipolarity of the 1665 keV transition, which was reported [3,4] to have an E1 character. The experimental confirmation of its character and a measurement of the lifetime of the level at 3718 keV excitation energy would help addressing the question of isospin mixing [5].

Spins were assigned on the basis of a DCO ratio analysis. An initial alignment

---

1) Fellowship supported by the EC under contract n°ERBFMBICT983126
2) Fellowship supported by the EC under contract n°ERBFMBICT983127
3) Under EC contract n°ERBFMGECT980110

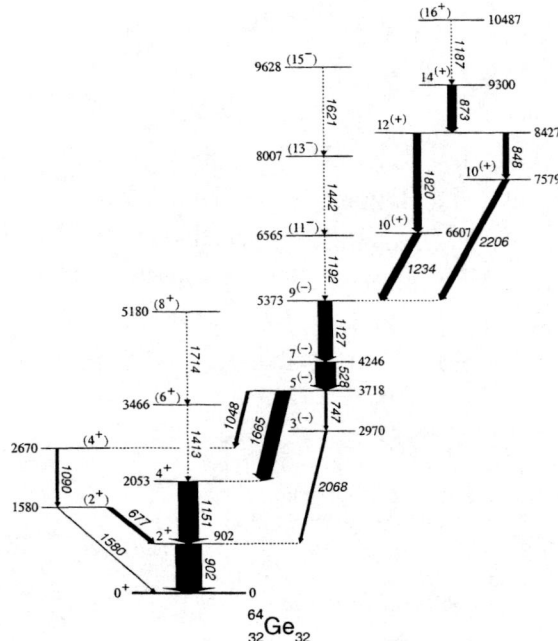

FIGURE 1. Decay scheme for the ^{64}Ge nucleus, deduced from the GASP and the EUROBALL data. The widths of the arrows are proportional to the intensities measured in the EUROBALL experiment, dashed lines were seen in the GASP data but not in the present data. The ground-state band is based on a $(f_{5/2}p_{1/2}p_{3/2})^{4\pi+4\nu}$ configuration. The negative-parity band involves the occupation of the $g_{9/2}$ orbital. The positive-parity band above 5 MeV excitation energy requires the promotion of two or four quasiparticles into the same $g_{9/2}$ orbital.

$\sigma/J = 0.30(5)$ was estimated from the experimental DCO ratios for known lines in ^{63}Ga. The results for the 1665 keV transition are 0.64(5) gating from above and 0.68(4) gating from below, which are not compatible with a $\Delta I = 0$ spin difference and are in better agreement with a $\Delta I = 1$ difference. Either small ($\arctan\delta \approx 0$) or large ($|\arctan\delta| \approx 90$) mixing ratios are compatible with the data.

Details on the polarization analysis are given in the contribution by A. Gadea et al. [6]. Although the statistical errors were large, the experimental data suggest a magnetic character for the 1665 keV transition. This, together with the systematics of the neighbouring even germanium isotopes, slightly favours a stretched E1 transition with a large M2 mixing, leading to $\delta = -5.7^{+1.9}_{-5.7}$. It should be stressed that such an estimate for δ is quite different from the one of Ennis et al. [3], even if our assignment for the level at 3718 keV excitation energy is in agreement with the results of [3,4], $I^\pi = 5^-$. At the moment we are still working on an angular distribution analysis to confirm the present results.

REFERENCES

1. E. Farnea et al., Nucl. Inst. and Meth. **A400**, 87 (1997)
2. Ö. Skeppstedt et al., Nucl. Inst. and Meth. **A421**, 531 (1999)
3. P.J. Ennis et al., Nucl. Phys. **A535**, 392 (1991)
4. G. de Angelis et al., Nucl. Phys. **A630**, 426c (1998)
5. J. Dobaczewski and I. Hamamoto, Phys. Lett. **B345**, 181 (1995)
6. A. Gadea et al., contribution to these proceedings

Shape changes induced by quasiparticle alignment

A. Algora[a,b], G. de Angelis[a], F. Brandolini[c], R. Wyss[d],
C. Fahlander[a], A. Gadea[e], E. Farnea[e], W. Gelletly[f], A.
Aprahamian[g], D. Bazzacco[c], F. Becker[h], P.G. Bizzeti[i], A.
Bizzeti-Sona[i], D. de Acuña[a], M. De Poli[a], J. Eberth[h], D. Foltescu[a],
S.M. Lenzi[c], S. Lunardi[c], T. Martinez[a], D.R. Napoli[a], P. Pavan[c],
C.M. Petrache[c], C. Rossi Alvarez[c], D. Rudolph[j], B. Rubio[e],
W. Satuła[k], S. Skoda[h], P. Spolaore[a],H.G. Thomas[h], and C.A. Ur[a]

[a] *Istituto Nazionale di Fisica Nucleare, Laboratori Nazionali di Legnaro, Legnaro, Italy*
[b] *Institute of Nuclear Research of the Hungarian Academy of Sciences, Debrecen, Hungary*
[c] *Dipartimento di Fisica dell'Università, and Istituto Nazionale di Fisica Nucleare, Sezione di Padova, Padova, Italy*
[d] *KTH-Kärnfysik, Stockholm, Sweden*
[e] *Instituto de Física Corpuscular, Valencia, Spain*
[f] *Department of Physics, University of Surrey, Guildford, UK*
[g] *Department of Physics, University of Notre Dame, Notre Dame, USA*
[h] *Institut für Kernphysik, Universität zu Köln, Germany*
[i] *Dipartimento di Fisica dell'Università and Istituto Nazionale di Fisica Nucleare, Sezione di Firenze, Firenze, Italy*
[j] *Sektion Physik, Ludwig-Maximilians Universität München, Garching, Germany*
[k] *Institute of Theoretical Physics, University of Warsaw, Warsaw, Poland*

Abstract. Mean lifetimes of high-spin states of ^{74}Kr have been determined using the Doppler shift attenuation method (DSAM). The high-spin states were studied using the ^{40}Ca(^{40}Ca,α2p) reaction at a beam energy of 160 MeV with the GASP array. The investigated ground state band and negative parity side band show the presence of three different configurations in terms of transitional quadrupole deformations. The deduced quadrupole deformation changes are well reproduced by cranked Woods-Saxon Strutinsky calculations.

INTRODUCTION

Kr isotopes have been the ground for extensive research in recent years. The low density of single-particle levels in this mass region of the nuclear chart gives rise to pronounced shape effects [1]. Large shell gaps at $Z, N = 34, 36$ for oblate

CP495, *Experimental Nuclear Physics in Europe*, edited by B. Rubio et al.
© 1999 American Institute of Physics 1-56396-907-6/99/$15.00

shapes and $Z, N = 38, 40$ for prolate shapes result in very deformed ground state configurations and prolate-oblate shape coexistence [2].

In the present work we focused our interest in the determination of lifetimes of high spin states of ^{74}Kr using the Doppler Shift Attenuation Method (DSAM) in order to deduce quadrupole shapes of the states. The main goal was to extend the former knowledge of lifetimes [3] to the states above the bandcrossing in the ground state band and to states in other bands. This will allow us to further study the polarization effects of different configurations and to clarify the importance of the shell gaps.

EXPERIMENTAL PROCEDURE AND ANALYSIS

The high spin states of ^{74}Kr were populated using the ^{40}Ca + ^{40}Ca reaction at a bombarding energy of 160 MeV. The beam was delivered by the Tandem XTU accelerator of Legnaro National Laboratory and the recoiling ions produced in the 0.9 mg/cm^2 ^{40}Ca target were stopped in a 10 mg/cm^2 gold–backing. The emitted γ-rays were detected using the GASP array [4].

For the lifetime analysis, the data were sorted into $\gamma\gamma$ matrices each having on one axis the detectors of one of the rings at 34^0, 60^0, 72^0, 90^0, 108^0, 120^0, and 146^0, and on the second axis any other detectors. For the energy and efficiency calibrations ^{152}Eu and ^{56}Co sources were used.

The MonteCarlo code LINESHAPE [5] was used. This code incorporates a description of the nuclear stopping power according to the LSS theory [6] and MINUIT for the error analysis. For the electronic stopping power a corrected Northcliffe-Schilling parametrization [7] was used. In the particular case of our analysis the slowing process of the nucleus was simulated using 10.000 simulations. The contribution of the side feeding to the lifetime of each level was modelled using a cascade of two transitions of adjustable lifetime. The sidefeeding intensities were determined from the spectra generated at 90^0, in this way the only free parameters in the fit procedure were the level lifetimes τ and the sidefeeding lifetimes.

The spectra for the analysis was generated with the procedure Gate on a Transition Below. In the analysis of the ground state band (GSB) the spectra was generated using two gates: lines 558 and 456 keV. In the case of the negative parity sideband (NPSB) transition 1586 keV was used. Special attention was paid to the fact that some transitions of the GSB have similar energies to the transitions of the NPSB. For this reason we began our study extracting the lifetime information from the NPSB and the obtained results were used in the deconvolution process of the GSB. Our results are presented in Fig. 1. The deduced transitional quadrupole moments show a transition from 3.0 eb to 2.1 eb along the GSB. The lifetime values of the NPSB can be fitted with a rather constant value of 2.8 eb.

INTERPRETATION OF THE RESULTS

In the Woods-Saxon Strutinsky approach, the origin of deformation for the light Kr-isotopes is related to the large deformed shell gaps at prolate ($\beta_2 = 0.35 - 0.4$) for N=Z=38 and oblate shapes ($\beta_2 \leq -0.3$) for N=Z=36. Hence one expects shape coexistence and deformation changes to occur in these nuclei. For the case of ^{74}Kr, Total Routhian Surface (TRS) calculations [8] predict prolate-oblate shape coexistence at low spins. The prolate-oblate energy difference is 200 keV whereas one estimates the experimental value to be $\approx$500 keV [9]. Unfortunately, our data set does not allow the determination of any life-times at low spins. Still, our measurement indicates three distinctly different transitional quadrupole moments for three different configurations, indicating strong polarization effects. The ground band beyond the perturbation at $I = 2\hbar$ is highly-deformed with a Q_t-value of 3eb. In the calculations, we obtain a value of 3.45eb, corresponding to a $\beta_2 = 0.375$ and rather small $\beta_4 = 0.005$. Since the absolute value of the transitional quadrupole moment is connected with uncertainties in both experiment and theory, we focus in the following on the relative changes. At $\hbar\omega = 0.6$ MeV, the minimum at prolate shape becomes very soft and at the next calculated frequency, only the prolate minimum at small deformation remains. The pronounced effect before and after the band crossing is nicely seen in the calculations, see Fig. 1.

Different factors contribute to the shape change at prolate shape. The aligning particles origin from the $g_{9/2}$ unique parity sub-shell and their contribution to the total quadrupole moment is expected to become reduced. The small effect of these orbitals is related to the fact, that the $g_{9/2}$ protons and neutrons are close to the Fermi-surface and are essentially of quasi-particle(qp) nature. The K-mixing does not effect the total quadrupole moment very much and even at $\hbar\omega = 1.0$ MeV, the reduction amounts only to $\delta Q_{20} = 0.15$eb.

We need to understand the reduction of the quadrupole moment in a different manner. Besides the direct contribution from the $g_{9/2}$ protons and neutrons, there exists a shape-driving factor related to the angular momentum gain. The rotational alignment of a given high-j orbit has its maximum at spherical shape. This means that the nucleus can gain additional alignment by shrinking the deformation. Thus, one is dealing with a balance between shell correction, that may favour more deformed shapes and angular momentum which is gained more efficiently at less deformed shapes. This balance has been discussed in a slightly different context for the case of the superdeformed Hg nuclei [10]. There it was a subtle effect, resulting in a smooth shrinking after the neutron alignment, whereas here we deal with a rather pronounced jump. The reason is of course the large $dI/d\beta$ in the band crossing region. One should note here, that the dominant part of this effect is due to the contribution of the protons.

For the negative parity band, we do not experience such strong shape effects. Since we deal with a proton 2qp-particle excitation from the [312]3/2 into the [431]3/2 Nilsson orbit there is a slight increase of the intrinsic quadrupole moment of this configuration. Nevertheless, the measured and calculated shape is somewhat smaller, again indicating the gain in rotational energy at reduced deformation. Since further proton alignment is blocked, we deal only with the effect arising from the neutrons.

By careful gating and analysis we were able to extract life time data of a large set of states in ^{74}Kr. The deduced quadrupole moments show strong configuration dependent shape effects. The deformation of the excited 2qp configuration of negative parity appears rather stable with spin, whereas the ground band experiences quite a dramatic shape change after the S-band crossing. The shape changes are explained by the competition of the contribution of single particle and collective degrees of freedom to the rotational energy, favouring smaller deformation for the S-band.

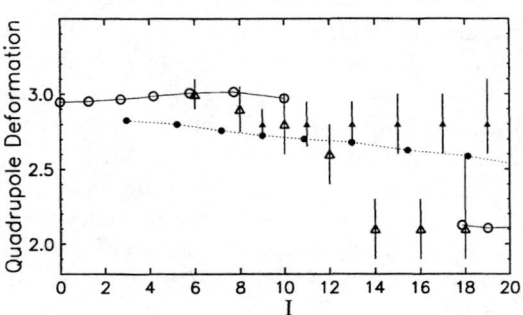

Fig 1. Comparison of the experimentally determined quadrupole shapes with the theoretical calculations. Empty symbols correspond to the ground state band data. Filled symbols label the data for the negative parity sideband. The theoretical points are conected by lines.

REFERENCES

1. W. Nazarewicz *et al.*, Nucl. Phys. **A435** (1985) 397.
2. I. Ragnarsson *et al.*, Phys. Rep **45**, (1978),1., P. Bonche *et al.*, Nucl. Phys. **A443** (1985) 39., P. -H. Heenen *et al.*, Nucl. Phys. **A561** (1993) 367., A. Petrovici *et al.*, Nucl. Phys. **A605** (1996) 290.
3. S. L. Tabor *et al.*, Phys. Rev **C41** (1990) 2658., J. Heese *et al.*, Phys. Rev **C43** (1991) R921.
4. D. Bazzacco, Int. Conf. on Nuclear Structure at High Angular Momentum, Ottawa, 1992, Vol.2, AECL 10613, p.376.
5. J. C. Wells, and N. Johnson, Report No. ORNL-6689 (1991) 44.
6. J. Lindhard *et al.* Mat. Fys. Medd. Dan. Vid. Selsk. **33**, no. 14 (1963).
7. L. C. Northcliffe *et al.*, Nucl. Dat. Tables **A7** (1970) 233., S. H. Sie *et al*, Nucl. Phys **A291** (1977) 443.
8. W. Satuła *et al.* , Nucl. Phys. **A578**(1994)45, W. Satuła and R. Wyss, Phys. Rev. **C50** (1994) 2888, W. Satuła and R. Wyss, Phys. Script. T56 (1995)159
9. C. Chandler *et al.*, Phys. Rev **C56** (1997) R2924.
10. B. Cederwall, *et al.*, Phys. Rev. Lett. **72** (1994) 3150

Magnetic Rotation in the Odd-Odd Nuclei ^{82,84}Rb

H. Schnare[1], R. Schwengner[1], S. Frauendorf[1], F. Dönau[1],
L. Käubler[1], H. Prade[1], A. Jungclaus[2], K. P. Lieb[2], C. Lingk[2],
S. Skoda[3], J. Eberth[3], G. de Angelis[4], A. Gadea[4], E. Farnea[4], D.R.
Napoli[4], C. A. Ur[4], G. Lo Bianco[5]

[1] *Institut für Kern- und Hadronenphysik, FZ Rossendorf, 01314 Dresden, Germany*
[2] *II. Physikalisches Institut, Universität Göttingen, 37073 Göttingen, Germany*
[3] *Institut für Kernphysik, Universität zu Köln, 50937 Köln, Germany*
[4] *INFN, Laboratori Nazionali di Legnaro, 35020 Legnaro, Italy*
[5] *INFN, Sezione di Milano, 20133 Milano, Italy*

Abstract. Rotational bands with strong magnetic dipole transitions have been observed in the doubly-odd nuclei ^{82}Rb and ^{84}Rb. These bands show the characteristic features of magnetic rotation. They are the first evidence of this new kind of nuclear excitation in the $A \approx 80$ region. The results are well reproduced within the framework of the Tilted Axis Cranking model on the basis of four-quasiparticle configurations of the type $\pi(fp)\,\pi(g_{9/2}^2)\,\nu(g_{9/2})$.

INTRODUCTION

The conventional concept of nuclear rotation is based on the existence of a deformed mass distribution of the nucleus. Regular rotational bands are formed by energy levels that depend on the spin I according to $E \propto I(I+1)$ and are connected by electric quadrupole ($E2$) transitions. Recently, a surprising phenomenon has been observed in nearly spherical Pb isotopes around $A = 200$. While the excited states at low spin show irregular multiplet-like structures as expected for nearly spherical nuclei, regular sequences that follow the $I(I+1)$ rule evolve at high spin, indicating a rotational mode. The levels of these sequences are linked by strong magnetic dipole ($M1$) transitions whereas cross-over $E2$ transitions are very weak [1–4]. These observations are well understood in the framework of the Tilted Axis Cranking (TAC) model [5]. Here, the coupling of angular momentum vectors of few high-j nucleons is the basic mechanism for generating the total spin I of the nucleus. Few protons occupy orbitals with long spin vectors above a closed shell (high-j particle-like orbitals) while the neutrons fill up a shell except for few holes

CP495, *Experimental Nuclear Physics in Europe*, edited by B. Rubio et al.
© 1999 American Institute of Physics 1-56396-907-6/99/$15.00

(high-j hole-like orbitals), or vice versa. A perpendicular coupling of their angular momenta is energetically favored because it maximizes the overlap of the spatial density distribution, which is torus-like for particle orbitals and dumbbell-like for hole orbitals. This coupling results in a substantial component of the magnetic dipole moment, which is transverse to the total spin and gives rise to large $M1$ transition probabilities. The total angular momentum is increased by the gradual alignment of the individual particle and hole spins along the axis of the total angular momentum. As a consequence of this alignment the transverse component of the magnetic dipole moment decreases and one expects smoothly decreasing $B(M1)$ transition probabilities with increasing spin. Recent results [4] confirm this trend of the $B(M1)$ values in the Pb region. The TAC model predicts magnetic rotation in several regions of the nuclear chart [6]. These regions are characterized by nuclei of low deformation in the vicinity of shell closures. A mass region, where magnetic rotation is predicted to occur, is located around $A = 80$, where there are few particle-like protons in the high-j $g_{9/2}$ intruder orbital and few $g_{9/2}$ neutron holes in the $N = 50$ shell.

EXPERIMENTAL METHODS AND RESULTS

Excited states in ^{82}Rb and ^{84}Rb were populated using the fusion-evaporation reaction ^{76}Ge(^{11}B,xn) at a beam energy of 50 MeV, delivered by the XTU tandem of the Laboratori Nazionali di Legnaro (Italy). The target consisted of a stack of two thin self-supporting ^{76}Ge foils with a thickness of about 0.2 mg/cm^2 each. Emitted γ-rays were detected with the GASP spectrometer consisting of 40 Compton-suppressed HPGe detectors and an inner ball with 80 BGO elements. Approximately 1.5×10^8 high-fold ($F \geq 3$) events were collected and sorted offline into $E_\gamma - E_\gamma$ matrices as well as an $E_\gamma - E_\gamma - E_\gamma$ cube. Spin and parity assignments are based on $\gamma - \gamma$ directional correlations of oriented states (DCO) and deexcitation modes. The analysis of the data resulted in a considerable extension of the previously known level schemes [7]. In particular, regular sequences of strong magnetic dipole transitions were observed in both nuclei. Partial level schemes of ^{82}Rb and ^{84}Rb, showing the $M1$ bands and their deexcitation to low-lying states are shown in Fig. 1.

INTERPRETATION

A striking feature of the $M1$ bands is the regularity of the level spacings ($I \propto \hbar\omega, \hbar\omega = E_\gamma(\Delta I = 1)$) which can be seen in the upper panels of Fig. 2, indicating the rotational character. Nevertheless, the emission of $M1$ radiation is strongly favored over the emission of $E2$ radiation.

This magnetic character of the rotation is demonstrated by the ratios of transition probabilities $B(M1)/B(E2)$ for each level in the band. These ratios are shown in the lower panels of Fig. 2. They reach values up to about 20 $(\mu_N/eb)^2$, which are comparable with the ratios in other regions of magnetic rotation. The $B(M1)/B(E2)$ ratios decrease smoothly with increasing rotational frequency $\hbar\omega$ in a range of $\hbar\omega \approx 0.4 - 0.7$ MeV, manifesting the shears mechanism. Hence, we have found the main characteristics of magnetic rotation. We have interpreted the $M1$ bands in ^{82}Rb and ^{84}Rb in the framework of the TAC model. In the calcula-

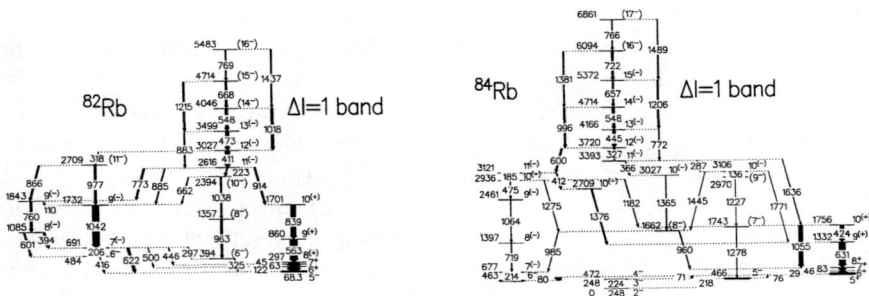

FIGURE 1. Partial level scheme of ^{82}Rb and ^{84}Rb with the width of the γ-transitions proportional to their measured intensities.

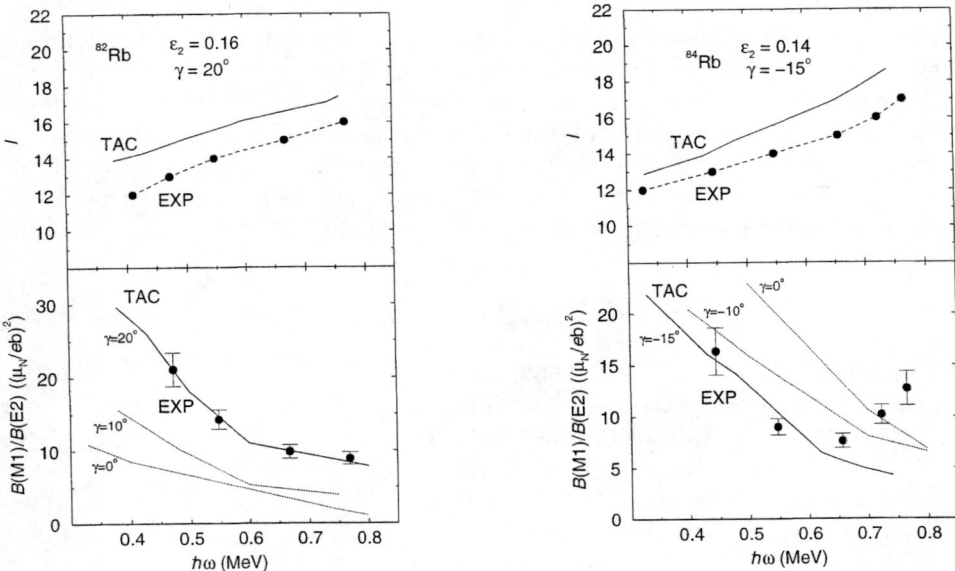

FIGURE 2. Comparison of experimental spins (upper panel) and $B(M1)/B(E2)$ ratios (lower panel) in ^{82}Rb and ^{84}Rb with TAC model predictions as a function of rotational frequency. The lower panel includes TAC calculations for different values of the triaxiality parameter γ.

tions, the lowest-lying four-quasiparticle (4qp) configuration for $Z = 37$ and $N = 45, 47$ turns out to be $\pi(fp)\ \pi(g_{9/2}^2)\ \nu(g_{9/2})$, which has been adopted. This configuration has negative parity which is consistent with our experimental findings. The "shears" mechanism is generated by two protons in the $g_{9/2}$ orbital and one unpaired $g_{9/2}$ neutron. An equilibrium deformation of $\epsilon_2 = 0.16$ and $\epsilon_2 = 0.14$ is obtained in ^{82}Rb and ^{84}Rb, respectively. Both nuclei turn out to be very soft with respect to γ deformation. Therefore, TAC calculations have been carried out for different γ values, which are included in the lower panels of Fig. 2. As can be seen in the figures, the best approximation to the experimental data is obtained by choosing $\gamma = 20°$ and $\gamma = -15°$ for ^{82}Rb and ^{84}Rb, respectively. The calculated $B(M1)/B(E2)$ ratios for the proposed configuration in ^{82}Rb are in excellent agreement with the experimental values. In ^{84}Rb the behavior of the experimental $B(M1)/B(E2)$ ratios is fairly well described in the calculation up to $\hbar\omega = 0.7$ MeV. The upbend in the experimental $B(M1)/B(E2)$ ratios above $\hbar\omega = 0.7$ MeV cannot be described with the chosen 4qp configuration. It may indicate a change of the configuration caused by a break-up of an additional pair of protons or neutrons, which is consistent with the slight upbend of the experimental function $I(\hbar\omega)$ in the upper panel of Fig. 2.

CONCLUSION

We have observed regular magnetic dipole bands in the nuclei ^{82}Rb and ^{84}Rb. These bands are the first examples of magnetic rotation in the $A \approx 80$ region [8]. They show the typical characteristics: (i) the regularity of the bands ($I \propto \hbar\omega$), (ii) the observation of strong $M1$ and weak crossover $E2$ transitions with ratios of the transition probabilities $B(M1)/B(E2)$ of up to about 20 $(\mu_N/eb)^2$, and (iii) a smooth decrease of the $B(M1)/B(E2)$ ratios with increasing frequency, which is evidence for the shears mechanism. The results are well reproduced by TAC calculations on the basis of the lowest-lying 4qp configuration: $\pi(fp)\ \pi(g_{9/2}^2)\ \nu(g_{9/2})$.

REFERENCES

1. Clark, R.M., et al., *Nucl. Phys.* **A562**, 121 (1993).
2. Baldsiefen, G., et al., *Nucl. Phys.* **A574**, 521 (1994).
3. Neffgen, M., et al., *Nucl. Phys.* **A595**, 499 (1995).
4. Clark, R.M., et al., *Phys. Rev. Lett.* **78**, 1868 (1997).
5. Frauendorf, S., *Nucl. Phys.* **A557**, 259c (1993).
6. Frauendorf, S., *Z. Phys.* **A358**, 163 (1997).
7. Döring, J., et al., *Z. Phys.* **A338**, 457 (1991) and Z. Phys. **A339**, 425 (1991).
8. Schnare, H., et al., *Phys. Rev. Lett.* **82**, 4408 (1999).

Investigation of mixed-symmetry states in ^{94}Mo

C. Fransen[1], N. Pietralla[1], P. von Brentano[1], U. Kneissl[2], and H.H. Pitz[2]

[1] Institut für Kernphysik, Universität zu Köln, D-50937 Köln, Germany
[2] Institut für Strahlenphysik, Universität Stuttgart, D-70569 Stuttgart, Germany

Abstract. A powerful combination of a photon scattering experiment and a $\gamma\gamma$-coincidence study following the β-decay of the $J^\pi = (2)^+$ low-spin isomere of ^{94}Tc was used to investigate low spin excitations in ^{94}Mo. We paid special attention to proton-neutron mixed-symmetry (MS) states: The 1^+ scissors mode and the 2^+ MS state are identified from $M1$ transition strengths. A γ transition between MS states was observed for the first time. The $M1$ and $E2$ transition strength are in reasonable agreement with calculations in the O(6) limit of the proton-neutron version of the Interacting Boson Model (IBM–2).

In most low-lying collective states in heavy nuclei protons and neutrons move in phase. However, the proton-neutron version of the Interacting Boson Model (IBM–2) predicts [1] a class of low-lying states, which contain antisymmetric parts with respect to the proton-neutron degree of freedom. These states are called mixed-symmetry (MS) states. It is interesting to investigate experimentally whether the predicted MS states exist in heavy nuclei. Only few examples of MS states are known. One example is the 1^+ MS state, which is called scissors mode due to its geometrical picture in deformed nuclei [2]. The scissors mode was dicovered in 1984 by A. Richter and co-workers in Darmstadt in an electron scattering experiment on the heavy deformed nucleus ^{156}Gd [3]. It was investigated very well during the last 15 years in electron scattering [4] and in systematic photon scattering experiments mostly in the rare earth region [5]. This enabled systematic studies of the $M1$ excitation strength of the scissors mode [6,7] and its excitation energy [8,9] including data for weakly deformed nuclei.

The lowest 2^+ MS state is interpreted as the MS one Q-phonon excitation, which results from an antisymmetric coupling of a proton quadrupole excitation and a neutron quadrupole excitation. It is orthogonal to the 2_1^+ state, which is the proton-neutron symmetric quadrupole excitation. There are few data about this fundamental 2_{ms}^+ state: In some weakly deformed nuclei $J^\pi = 2^+$ MS states were identified from lifetime measurements [10–12]. Other MS states are basically

CP495, *Experimental Nuclear Physics in Europe*, edited by B. Rubio et al.
© 1999 American Institute of Physics 1-56396-907-6/99/$15.00

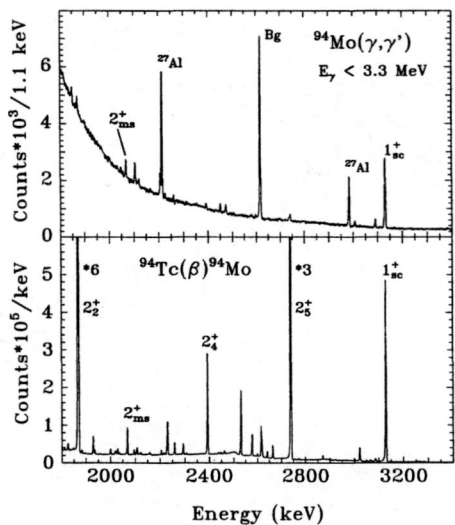

FIGURE 1. γ spectra of ^{94}Mo. The top part shows the photon scattering spectrum off ^{94}Mo in the energy range from 1.8 MeV to 3.3 MeV. At 2067 keV and 3129 keV we observed peaks resulting from the ground state transition of a strongly excited 2^+ and 1^+ state, respectively. Photon scattering cross sections are measured relative to well known cross sections of ^{27}Al [14], which was irradiated simultanously. The bottom part shows the γ spectrum observed off-beam following the β decay of ^{94}Tc to ^{94}Mo. High statistics and low background enable us to determine decay branching ratios and to measure $\gamma\gamma$-coincidences.

unknown.

To investigate mixed-symmetry states in ^{94}Mo we performed [13] a combination of a photon scattering experiment, which was done at the Dynamitron accelerator at the Institut für Strahlenphysik in Stuttgart and a β-delayed $\gamma\gamma$-coincidence measurement at the Tandem accelerator at the Institut für Kernphysik in Cologne. From the photon scattering experiment we determined level energies, spin quantum numbers, integrated scattering cross sections, and lifetimes of excited $J = 1^\pm$ and 2^+ states. From the β-delayed $\gamma\gamma$-coincidence study we got branching ratios and $E2/M1$ multipole mixing ratios for decay transitions from $J^\pi = 1^+$ and 2^+ states in ^{94}Mo. This new combination of γ spectroscopic devices allows a detailed investigation of the $J^\pi = 1^+$ and 2^+ mixed-symmetry states.

Fig. 1 shows the spectra of the photon scattering experiment in the top part and of the β-decay experiment in the bottom part. At 2067 keV we observed a peak stemming from the ground state decay of the 2_3^+ state. From our data [13] we got detailed information about the decay properties of this state: It shows a weakly collective E2 transition to the ground state with a decay transition strength of 1.9(3) W.u. The $E2/M1$ multipole mixing ratio $\delta = 0.15(4)$ gives evidence that the $2_3^+ \rightarrow 2_1^+$ transition has predominantly $M1$ character. The $2_3^+ \rightarrow 2_1^+$ $M1$

transition matrix element amounts to $1.5(1)\mu_N$. As shown in Fig. 3 the 2_3^+ state is the only 2^+ state which decays via an enhanced $M1$ transition to the 2_1^+ state. The enhanced $2_3^+ \rightarrow 2_1^+$ $M1$ transition and the weakly collective $2_3^+ \rightarrow 0_1^+$ transition agree with the MS interpretation of this state.

At 3129 keV a peak stemming from the decay of a 1^+ state was detected. A further 1^+ state was detected at 3512 keV. The total $M1$ excitation strength from the ground state to the 1^+ states amounts to $\Sigma B(M1) \uparrow = 0.61(7)\mu_N^2$. These data fit well into our previously determined systematics of the 1^+ scissors mode: We expect the scissors mode in ^{94}Mo at an excitation energy of 3.2–3.5 MeV with a total excitation strength of $B(M1) \uparrow = 0.55\mu_N^2$ from empirical formulas [6,8,9] extracted from data of the 1^+ scissors mode in the rare earth region. This good agreement is a strong argument that the 1_1^+ state at 3129 keV is in fact the main fragment of the scissors mode in ^{94}Mo.

In the spectrum gated with the $2_3^+ \rightarrow 2_1^+$ transition a peak resulting from the decay of the main fragment of the two Q-phonon scissors mode at 3129 keV to the one Q-phonon 2_{ms}^+ state was detected as shown in Fig. 2. This is an interesting point because it was the first time a transition between MS states was observed: From the Q-phonon structure [13] this transition is to be expected to be a collective $E2$ transition, which should be as strong as the decay of the 2_1^+ state to the ground state because it results from the annihilation of the symmetric Q-phonon of the 1_{sc}^+ two phonon state. Unfortunately it was not possible to determine the multipolarity of this transition because of low statistics. But if we assume a dominant $E2$ character we get a transition strength of $B(E2, 2_{ms}^+ \rightarrow 1_{sc}^+) = 372(42)e^2\text{fm}^4$, which would be equal to the collective $2_1^+ \rightarrow 0_1^+$ strength of $B(E2, 2_1^+ \rightarrow 0_1^+) = 406(8)e^2\text{fm}^4$.

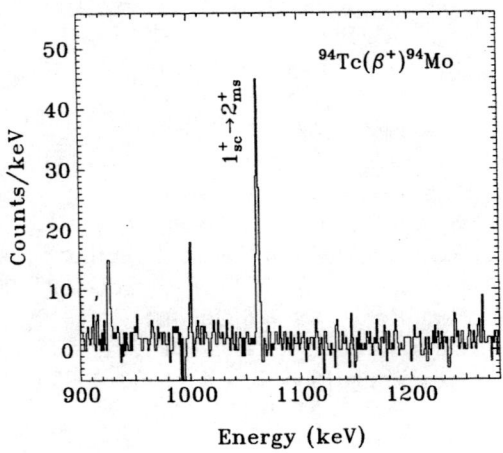

FIGURE 2. Cut spectrum on the decay of the 2_{ms}^+ to the 2_1^+ state showing the first identified γ transition between mixed-symmetry states: We observed a peak resulting from the decay of the 1_{sc}^+ to the 2_{ms}^+ state.

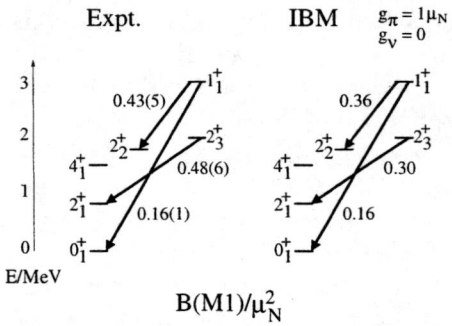

FIGURE 3. Comparison of the $M1$ transition strengths of ^{94}Mo from our experiments (left) and from theoretical calculations with the IBM–2 using the bare orbital g-factors (right). The agreement is reasonable. The enhanced $M1$ transition strengths prove the MS interpretation of the 2_3^+ state and the 1_1^+ state.

We thank D. Belic, A. Fitzler, C. Friessner, A. Nord, I. Schneider, H. Tiesler, and V. Werner for help with the experiments. We acknowledge discussions with T. Otsuka. This work was supported by the DFG under Contracts No. Br 799/8-2/9-1 and No. Kn 154/30.

REFERENCES

1. F. Iachello, Phys. Rev. Lett. **53**, 1427 (1984).
2. N. Lo Iudice and F. Palumbo, Phys. Rev. Lett. **41**, 1532 (1978).
3. D. Bohle, A. Richter *et al.*, Phys. Lett. **B137**, 27 (1984).
4. A. Richter, Prog. Part. Nucl. Phys. **34**, 261 (1995).
5. U. Kneissl, H.H. Pitz, and A. Zilges, Prog. Part. Nucl. Phys. **37**, 349 (1996), and Refs. therein.
6. N. Pietralla *et al.*, Phys. Rev. C **52**, R2317 (1995).
7. P. von Neumann-Cosel *et al.*, Phys. Rev. Lett. **75**, 4178 (1995).
8. N. Pietralla *et al.*, Phys. Rev. C **58**, 184 (1998).
9. J. Enders *et al.*, Phys. Rev. C **59**, R1851 (1999).
10. P. von Brentano *et al.*, Proceedings of the Int. Conf. on Nuclear Structure, Gatlinburg, Tennessee, 1998 (American Institute of Physics, 1998), and Refs. therein.
11. N. Pietralla *et al.*, Phys. Rev. C **58**, 796 (1998).
12. P.E. Garrett *et al.*, Phys. Rev. C **54**, 2259 (1996).
13. N. Pietralla, C. Fransen, *et al.*, Phys. Rev. Lett. **83**, 1303 (1999).
14. N. Pietralla *et al.*, Phys. Rev. C **51**, 1021 (1995).

Terminating bands in ^{98,99,100}Ru and neutron $2d_{5/2}$ - $1g_{7/2}$ energy spacing

J. Timár[1,2], J. Gizon[1], A. Gizon[1], B. M. Nyakó[2], D. Sohler[2],
L. Zolnai[2], A.J. Boston[3], D. T. Joss[4], E. S. Paul[3], A. T. Semple[3],
N. J. O'Brien[5], C. M. Parry[5], and I. Ragnarsson[6]

[1] ISN, IN2P3-CNRS/UJF, Grenoble, France; [2] ATOMKI, Debrecen, Hungary;
[3] University of Liverpool, Liverpool, UK; [4] Staffordshire University, Stoke-on-Trent, UK;
[5] University of York, Heslington, York, UK; [6] University of Lund, Lund, Sweden

Abstract. New high-spin bands have been established in ^{98,99,100}Ru. Some are interpreted as terminating configurations using the Nilsson-Strutinsky cranking formalism. They are observed up to the predicted terminating states which are built from $g_{9/2}$ protons and $N = 3$ proton holes combined with $d_{5/2}$, $g_{7/2}$ and $h_{11/2}$ neutrons relative to a ^{90}Zr core. The observed high-spin states assigned as terminating show systematic behaviour and provide new information on the energy spacing between the $2d_{5/2}$ and $1g_{7/2}$ neutron subshells.

INTRODUCTION

It was predicted recently [1] that in the $A \approx 100$ Pd and Ru nuclei, valence space configurations and core-excited configurations can be followed from the low-spin rotational states up to the termination. In this region terminating high-spin bands have been found in ^{102,103}Pd [2,3] and ^{101,102}Rh [4,5] nuclei. In this contribution we report on the observation and systematic behaviour of terminating high-spin states in the ^{98,99,100}Ru nuclei.

EXPERIMENT

High-spin states in ^{98,99,100}Ru have been populated using the ^{70}Zn(^{36}S,$\alpha x n$) reactions at a bombarding energy of 130 MeV. The beam was provided by the Vivitron accelerator at IReS, Strasbourg. The target was made of two stacked self-supporting foils of Zn, enriched to 70% in ^{70}Zn with a thickness of 440 μg/cm^2 each. γ-rays were detected with the EUROGAM-2 spectrometer. A total of 6×10^8 Compton-suppressed events ($f_s \geq 4$) were recorded. Total and gated $E_{\gamma 1} - E_{\gamma 2} - E_{\gamma 3}$ coincidence cubes were sorted from the events and have been analysed using the

CP495, *Experimental Nuclear Physics in Europe*, edited by B. Rubio et al.
© 1999 American Institute of Physics 1-56396-907-6/99/$15.00

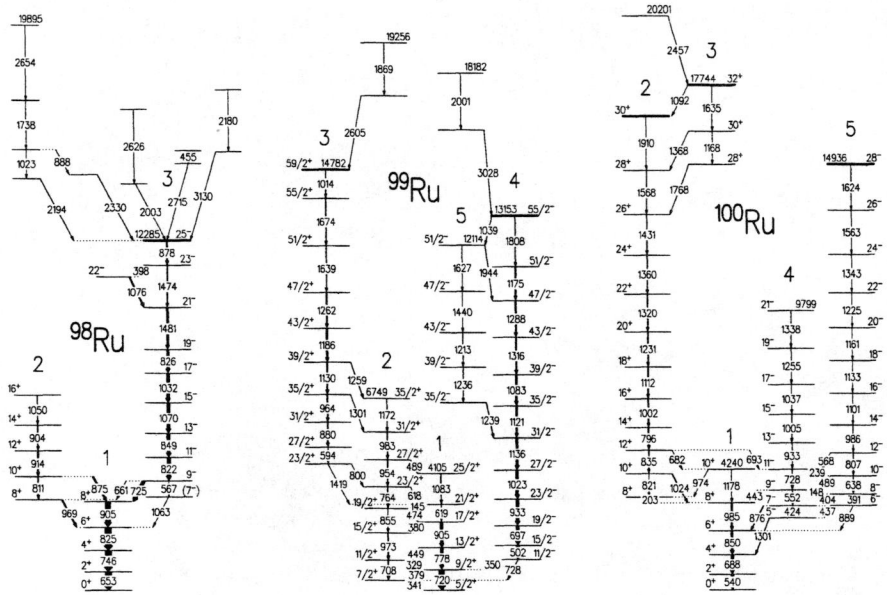

FIGURE 1. Partial level schemes of ^{98,99,100}Ru obtained in the present work.

Radware program package. Spins and parities of levels were deduced from DCO ratios and linear polarization. Partial level schemes of ^{98,99,100}Ru obtained from the present experiment and relevant to the aim of this work are shown in Fig. 1.

DISCUSSION

To interpret our experimental results, calculations have been performed using the cranked Nilsson-Strutinsky formalism. The configurations are labelled by the number of particles in the different j-shells or groups of j-shells relative to the ^{90}Zr core. The number of proton holes p_0 in the $N = 3$ shell, the number of protons p_1 in the $g_{9/2}$ and the number of neutrons n in the $h_{11/2}$ orbitals define the configuration. For the sake of simplicity the shorthand notation $[(p_0)p_1, n]$ is used (p_0 is omitted when 0).

In Fig. 2 the experimental energies of the observed high-spin bands in ^{99}Ru are compared with the corresponding theoretical values, subtracting the same rigid rotor reference from all of them. Only the case of ^{99}Ru is discussed below to illustrate how terminating configurations are established.

Comparing the parities, signatures, relative energies and maximum spins of the observed high spin bands to the corresponding values obtained for the calculated configurations in Fig. 2, a good agreement is found between them. Based on this agreement we assign the [4,1] configuration to band 4 and the [(1)5,0] con-

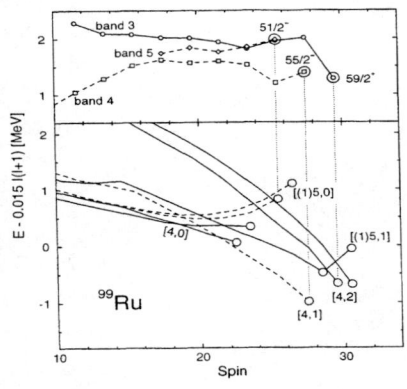

FIGURE 2. Excitation energy relative to a rigid rotor reference as a function of the spin for the experimental (top) and calculated (bottom) bands for ^{99}Ru. Full and open symbols denote $\alpha = 0$ and $\alpha = 1$ signature branches, while solid and dashed lines correspond to positive- and negative-parity bands, respectively. Encircled symbols and open circles at the end of the curves indicate the terminating states.

figuration to band 5. According to these assignments the observed $55/2^-$ and $51/2^-$ states are terminating and they have the $\pi(g_{9/2})^4_{12}\nu(d_{5/2}g_{7/2})^4_{10}(h_{11/2})^1_{5.5}$ and $\pi(N = 3)^{-1}_{2.5}(g_{9/2})^5_{12.5}\nu(d_{5/2}g_{7/2})^5_{10.5}$ single-particle configurations, respectively. We assign band 3 as being the $\alpha = -1/2$ signature branch of the [4,2] configuration at high spins. According to this assignment the observed $59/2^+$ state corresponds to the $\pi(g_{9/2})^4_{12}\nu(d_{5/2}g_{7/2})^3_{7.5}(h_{11/2})^2_{10}$ single-particle configuration. At lower spins band 3 corresponds probably to the [4,0] configuration which is predicted to terminate at spin $47/2\ \hbar$.

In all of the studied Ru nuclei, similarly to the cases of the ^{102,103}Pd and ^{101,102}Rh, [4,1] and [4,2] type terminating configurations have been observed as yrast configurations at high spin. A new feature of the Ru high-spin band structures is, however, that the [(1)5,...] type core excited configurations, in which one proton moves from the $N = 3$ shell to the $g_{9/2}$ subshell, also play an important role at the same spin and excitation-energy region as the above mentioned valence configurations. The [4,1] and [4,2] configurations in the different Ru isotopes were seen up to the terminating states and their terminating spin differs only in the spin contribution of the $d_{5/2}g_{7/2}$ neutrons. The obtained spin contributions are 7.5, 10, and 10.5 $\hbar$ for 3, 4, and 5 positive-parity neutrons, respectively. Results in Refs. [8] and [9] suggest that the energy spacing between the $2d_{5/2}$ and $1g_{7/2}$ neutron orbitals could be larger than it is given by the standard Nilsson parameters for the N=4 shell [6]. Our results have been used to test the size of this spacing assuming that the terminating states correspond to "optimal" single-particle configurations, see, e.g., Refs. [10,11]. The optimal states are obtained from a sloping Fermi surface diagram where the Nilsson single-particle energies are plotted as a function of the m_j quantum number (see Fig. 3). Here we used $\varepsilon_2 = -0.08$ and the standard $\kappa = 0.07$ parameter [6], but varied the μ parameter from 0.31 to the standard value of 0.39. In the figure the solid and dotted lines correspond to the configurations containing one and two $h_{11/2}$ neutrons, respectively. The experimental spin contributions are reproduced

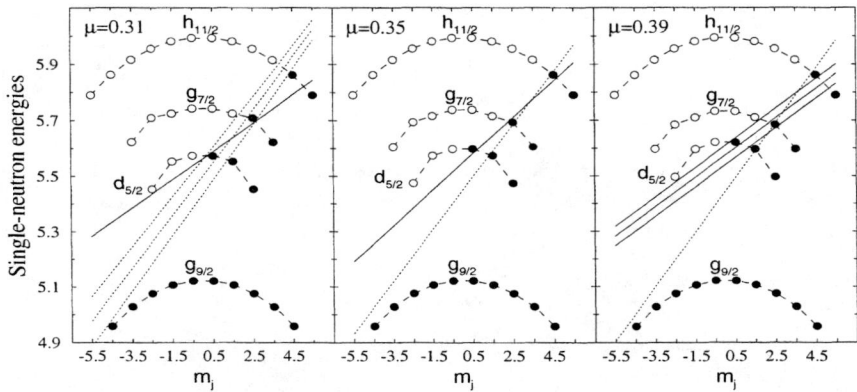

FIGURE 3. Optimal neutron configurations involving one or two $h_{11/2}$ neutrons for different values of the μ parameter. The other parameters used for the diagrams are given in the text.

in the "one $h_{11/2}$ neutron" cases for μ around and larger then 0.35, while in the "two $h_{11/2}$ neutron" cases for μ around and smaller than 0.35, suggesting the $\mu \approx 0.35$ optimal value for the energy spacing.

ACKNOWLEDGEMENTS

This work was supported in part by the exchange programmes between the CNRS and the Hungarian Academy of Sciences, by the Hungarian Scientific Research Fund, OTKA (contract No. 20655) and by the Swedish Natural Science Research Council.

REFERENCES

1. I. Ragnarsson, A.V. Afanasjev, J. Gizon, Z. Phys. A **355**, 383 (1996)
2. J. Gizon et al., Phys. Lett. **B410** (1997) 95.
3. B. M. Nyakó et al., Phys. Rev. C, in press.
4. J. Timár et al., Eur. Phys. J. **A4** (1999) 11.
5. J. Gizon et al., Phys. Rev. **C59** (1999) R570.
6. T. Bengtsson and I. Ragnarsson, Nucl. Phys. **A436** (1985) 14.
7. A.V. Afanasjev and I. Ragnarsson, Nucl. Phys. **A591** (1995) 387.
8. R.E. Azuma et al., Phys. Lett. **86B** (1979) 5.
9. I. Ragnarsson, Atomic Masses and Fundamental Constants 6, eds. J.A. Nolen Jr. and W. Benenson, Plenum Publ. Comp. 1980, p. 87.
10. G. Andersson et al., Nucl. Phys. **A268** (1976) 205.
11. M.J.A. DeVoigt, J. Dudek and Z. Szymanski, Rev. Mod. Phys. **55** (1983) 949.

Nuclear Structure in the Region of ^{100}Sn at N$\simeq$Z

M. Górska[1,2], H. Grawe[1], Z. Hu[1], M. Lipoglavšek[3,4], C. Fahlander[3], A. Axelsson[5], J. Nyberg[5], G. de Angelis[6], A. Gadea[6], E. Roeckl[1]

[1] GSI Darmstadt,D-64291 Darmstadt, Germany,
[2] Institute of Experimental Physics, Warsaw University,
[3] Department of Physiscs, Lund University, Sweden, [4] J. Stefan Institute, Ljubljana, Slovenia,
[5] TSL, Uppsala University, Sweden, [6] INFN Legnaro, Italy

Abstract. With the recent development of efficient γ-arrays with highly selective ancillary detectors, studies of shell structure and residual interaction in exotic nuclei like ^{100}Sn have become feasible. The shell structure at N=Z=50 (^{100}Sn) shows a remarkable similarity to N=Z=28 (^{56}Ni), albeit recent experiments on the ^{100}Sn neighbours ^{102,104}Sn and ^{98}Cd indicate substantial and puzzling deviations in the E2 and E3 polarizability of the magic core [1–3]. New experimental results and their shell model interpretation are discussed.

INTRODUCTION

The region of nuclei close to the heaviest doubly magic nucleus with $N = Z$ – ^{100}Sn – has attracted many experimental as well as theoretical studies. The unique placement of this region in the chart of nuclides is especially important for the crucial tests of the nuclear shell model in the extreme conditions close to the proton drip–line. However, ^{100}Sn is not accessible with respect to γ-ray spectroscopy, and study of the neighbouring nuclei is a considerable challenge for the experimental instrumentation and the present developments in detection techniques. The development of efficient γ-arrays, recoil separators and selective ancillary detectors for evaporation neutrons and charged particles enable in–beam spectroscopy of proton–rich nuclei up to ^{100}Sn with fusion–evaporation reactions and stable beam–target combinations [1,3–5]. The main interest for the γ-ray spectroscopy studies is concentrated on the study of the shell structure, the mean field residual interaction and its influence on the shell closure, on $M1$, $E2$, and $E3$ core excitation. The experimental approach to ^{100}Sn summarized in Ref. [6], is land-marked by the $T_z = 1$ nuclei [1,3] ^{98}Cd and ^{102}Sn in in–beam studies, and by ^{94}Ag [4] and ^{100,101}Sn in β^+/EC decay [5,7,8].

CP495, *Experimental Nuclear Physics in Europe*, edited by B. Rubio et al.
© 1999 American Institute of Physics 1-56396-907-6/99/$15.00

SINGLE PARTICLE STRUCTURE

Because of scarce spectroscopic information on the one– and two–particle neighbours, the single particle structure [9] and the adequacy of various empirical and realistic interactions had to be drawn from a shell model analysis of more remote nuclei. As a result the positions of the $\pi(p_{1/2}, g_{9/2})$ hole and $\nu(d_{5/2}, g_{7/2}, h_{11/2})$ particle levels in ^{100}Sn are determined to an accuracy of about 300 keV, whereas the single–particle energies of the proton particles and the low–spin orbitals $\nu(d_{3/2}, s_{1/2})$ remain floating. The latter, which are of crucial importance for the extraction of the $E2$ polarization for neutrons [10], are accessible to β–decay studies. In the

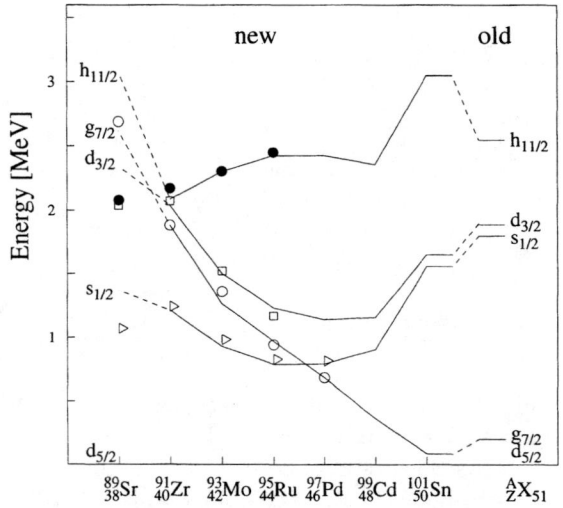

FIGURE 1. Single particle states in N=51 isotones. The points determine the experimental data and lines are result of the shell model calculation.

recent study of the ^{97}Ag β–decay a $\nu s_{1/2}$ single particle state was identified in the $N = 51$ nucleus ^{97}Pd [11]. Based on this information the reevaluation of the experimental data on $N = 51$ isotones was made using the interaction described in Ref. [9]. The results are shown in fig.1 for the evolution of the quasi–neutron energies relative to the $\nu d_{5/2}$ state. The unusual behavior of the $\nu g_{7/2}$ orbital in comparison to the normal parabola like trend is due to the increasing $\pi\nu$ interaction with the filling of the $\pi g_{9/2}$ orbital. The extrapolated value of the $\nu h_{11/2}$ orbital was discarded as only 30% of the experimental $I^\pi = 11/2^-$ wave function contains this configuration, instead the previous value extracted from the tin isotopes [9] is used. It should be noticed that the experimental low–spin data were obtained in β–decay, while information on high spin orbitals is determined by in–beam data.

218

RESIDUAL INTERACTION AND CORE EXCITATION

In the PEX–NORDBALL Pre–EUROBALL campaign at NBI, the ^{100}Sn $T_z = 1$ neighbours ^{98}Cd [3] and ^{102}Sn [1] were identified. As ^{98}Cd represents the two–proton hole spectrum in ^{100}Sn, the N=50 empirical shell model parametrisation in the $\pi p_{1/2}, g_{9/2}$ model space [12] could now be improved to fit the experimental data. The agreement obtained for the nuclei in the upper $\pi g_{9/2}$ shell starts to deterioriate towards mid-shell, as the limited space excludes excitations from the lower $\pi(p, f)$ shell. In the next step the neutron core excitation between $\nu g_{9/2}$ and $\nu d_{5/2}$ were included into the model space by extracting the respective empirical interaction

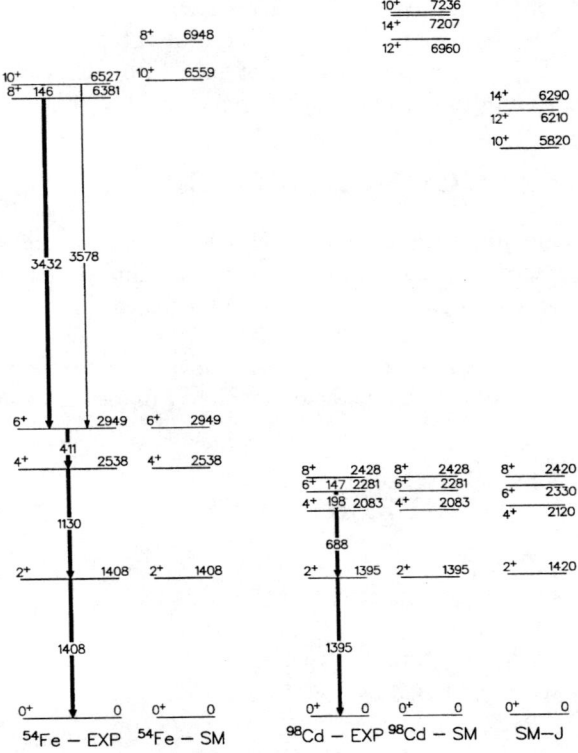

FIGURE 2. Experimental and calculated level schemes of the two–proton hole nuclei ^{54}Fe and ^{98}Cd with respect to the doubly magic cores.

from nuclei in the lower $\pi g_{9/2}$ shell. The result for ^{98}Cd is shown in fig.2.

A similar analysis of the two-body interaction was done for the doubly magic ^{56}Ni region and ^{54}Fe, the two–proton hole nucleus corresponding to ^{98}Cd. The energy of the known 10$^+$ isomer in ^{54}Fe [13] is reproduced fairly well. From the similarity of the shell structure in both regions of nuclei the existence of a high spin, core excited isomer in ^{98}Cd is deduced and confirmed by the full shell model calculation (SM–J) shown in fig.2 [14]. Non-observation of its high energy γ–decay in the previous experiment [3], could significantly influence the conclusion stated in Ref. [3] about the small proton polarization charge of the ^{100}Sn core.

On the contrary, the neutron effective charge deduced from the light tin isotopes ^{102,104}Sn is surprisingly large [3,15]. However, particle–hole excitation of the type $\Delta l = 2$ of an ls–open magic core gives rise to low lying $E2$ strength and large $E2$ polarization charges. The extended calculation including particle–hole excitation of the ^{100}Sn core in the gds space confirms this observation, and indicates only a small residual effective charge $e_\nu \simeq 0.5$, as a remainder from higher order excitation, e.g. the giant quadrupole resonance [15].

The shell structure of ^{100}Sn allows also for $E3$ particle–hole excitation, which can not be observed in the ^{56}Ni region. From the experimental data on ^{104}Sn [2] the significant influence of the octupole vibration of the ^{100}Sn core was deduced from the large $E3$ transition rates observed at high spins.

CONCLUSIONS

The recent developments in the detector technology and in–beam identification techniques have enabled detailed spectroscopy of exotic nuclei at the proton drip–line up to ^{100}Sn. Although the new approach to the neutron single particle energies and first information on the $E2$ and $E3$ polarizability of the closed ^{100}Sn core are a step forward in understanding the nuclear structure in this region, the available information awaits a consistent explanation, which will be the main subject of future experiments.

REFERENCES

1. M. Lipoglavšek et al., Z. Phys. **A356**, 239 (1996).
2. M. Górska et al., Phys. Rev. **C58**, 108 (1998).
3. M. Górska et al., Phys. Rev. Lett. **79**, 2415 (1997).
4. K. Schmidt et al., Z. Phys. **A350**, 99 (1994).
5. Z. Janas et al., Phys. Scr. **T56**, 262 (1995).
6. H. Grawe et al., Z. Phys. **A358**, 185 (1997).
7. R. Schneider et al., Z. Phys. **A348**, 241 (1994).
8. M. Lewitowicz et al., Phys. Lett. **A332**, 20 (1994).
9. H. Grawe et al., Phys. Scr. **T56**, 71 (1995).
10. M. Lipoglavšek et al., Phys. Lett. **B440**, 246 (1998).
11. Z. Hu et al., Phys. Rev. **C60**, 24315 (1999).
12. J. Blomqvist, L. Rydström, Phys. Scr. **31**, 31 (1985).
13. M. Hass et al., Nucl. Phys. **A414**, 316 (1984).
14. I. Johnstone, private communication.
15. R. Schubart et al., Z. Phys. **A352**, 373 (1995).
16. F. Nowacki, private communication.

Confirmation of Magnetic Rotation in the A~110 region

D.G. Jenkins,* R. Wadsworth,* J.A. Cameron,† R.M. Clark,‡ D.B. Fossan,° I.M. Hibbert,*¹ V.P. Janzen,° R. Krücken,‡² G.J. Lane,°³ I.Y. Lee,‡ A.O. Macchiavelli,‡ C.M. Parry,* J.M. Sears,° J.F. Smith,°⁴ S. Frauendorf •⁵

* *Department of Physics, University of York, Heslington, York YO1 5DD, U.K.*
† *Department of Physics and Astronomy, McMaster University, Hamilton, Ontario L85 4M1, Canada*
‡ *Lawrence Berkeley National Laboratory, Berkeley, California 94720*
° *Department of Physics, State University of New York at Stony Brook, Stony Brook, New York 11794-3800*
° *Chalk River Laboratories, AECL Research, Chalk River, Ontario K0J 1J0, Canada*
• *FZ Rossendorf, Postfach 510119, D-01314 Dresden, Germany*

Abstract. Lifetimes of states of a magnetic dipole band in each of the nuclei, ^{106,108}Sn have been obtained using the Doppler Shift Attenuation Method. The deduced B(M1) transition rates show the characteristic behaviour associated with the shears mechanism. A simplified semi-classical analysis yields B(M1) values in qualitative agreement with those expected for previously assigned configurations. The results suggest extremely low deformations for these dipole structures. The ^{106}Sn band appears to be the first example of almost pure magnetic rotation in a spherical nucleus.

Considerable evidence has now been collated for the existence of "rotational-like" magnetic dipole (M1) bands in the lead and tin regions (e.g. [1]). It has been proposed that the mechanism responsible for the existence of such structures in weakly deformed nuclei is the shears mechanism [2]. This mechanism is so-called by analogy with the closing of a pair of shears since nearly all the angular momentum is generated by the gradual alignment of proton and neutron spin vectors ($\mathbf{j}_\pi$ and $\mathbf{j}_\nu$), which are initially coupled perpendicularly at the bandhead, with the total angular momentum vector, $\underline{I}$ (Fig. 1). This behaviour has been discussed

1) present address: Dept. of Physics, University of Liverpool, Liverpool L69 7ZE, U.K.
2) present address: Physics Department, Yale University, New Haven, CT 06520
3) present address: Lawrence Berkeley National Laboratory, Berkeley, California 94720
4) present address: Schuster Laboratory, University of Manchester, Manchester M13 9PL, U.K.
5) present address: Dept. of Physics, University of Notre Dame, Notre Dame, IN 46556

CP495, *Experimental Nuclear Physics in Europe*, edited by B. Rubio et al.
© 1999 American Institute of Physics 1-56396-907-6/99/$15.00

in terms of the Tilted Axis Cranking (TAC) model [3]. A more phenomenological description of the shears mechanism has recently been presented in terms of the coupling of two long vectors, $\mathbf{j}_\nu$ and $\mathbf{j}_\pi$ [4], mediated by an effective quadrupole force attributed to particle-vibration coupling. Both this and the TAC model predict a definitive signature characteristic of the shears mechanism, namely, that the B(M1) transition rates, proportional to the square of the perpendicular component of the magnetic dipole vector (Fig. 1), should decrease markedly with increasing angular momentum.

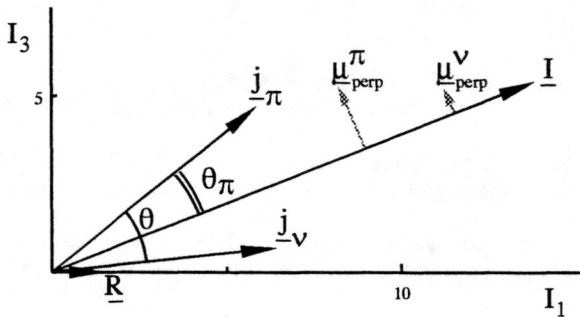

FIGURE 1. Angles associated with the shears mechanism.

The Doppler Shift Attenuation Method (DSAM) has been used to obtain lifetimes for states in two M1 bands, one in each of the nuclei ^{106,108}Sn. B(M1) transition rates have been deduced which clearly demonstrate the expected rapid decrease with increasing angular momentum. High-spin states in the nuclei ^{106,108}Sn were populated using the ^{54}Fe(^{58}Ni,α2p) and ^{54}Fe(^{58}Ni,4p) reactions respectively at a beam energy of 243 MeV. The ^{58}Ni beam was incident on an enriched 600μg/cm^2 ^{54}Fe target, backed with 15.2mg/cm^2 of gold. The resulting γ decay was detected by the Gammasphere array of 95 hyper-pure Ge detectors. Sufficient statistics were available in order to obtain lifetime information for the more intensely populated of the two bands, band 1 of ref [1], in each nucleus.

Lifetimes were extracted by fitting Doppler-broadened lineshapes using the codes of Wells and Johnson [5]. The slowing-down of recoiling nuclei in the target and gold backing was simulated using Monte Carlo methods with shell corrected stopping powers. Lineshapes were fitted simultaneously to spectra from detectors located at forward (31.7°/37.4°), backward (142.6°/148.3°) and transverse directions with respect to the beam, assuming a side-feeding rotational cascade of five transitions, with a moment of inertia of 15$\hbar^2$(MeV)$^{-1}$, feeding into each state, including the topmost state. Typical lineshape fits obtained are shown in Fig. 2.

The effective lifetime of the highest state observed was used as the input parameter to a global fit of the entire cascade. An error analysis was performed by examining the behaviour of the χ^2 value deduced for the fit in the vicinity of the minimum. Quoted errors do not include systematic errors in the stopping powers,

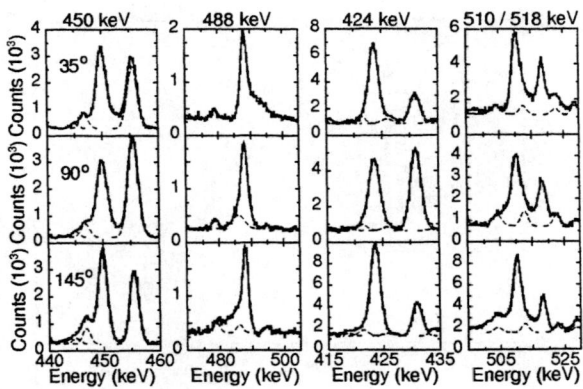

FIGURE 2. Typical lineshape fits.

which may be as large as 20%.

B(M1) values, predicted by the TAC model for the suggested configurations in our original analysis [1] do not decrease as rapidly with increasing spin as the experimental values, indicating that the calculated deformations used previously were too great. Lifetimes obtained in the present work have allowed B(E2) values (or limits) to be determined for the M1 bands in ^{106,108}Sn. A reduction in the Q-Q coupling constant of around 10% is necessary to reproduce the experimental B(E2) values. This adjustment indicated a deformation of around ϵ_2=0.08 for the ^{108}Sn band. The revised deformation of ϵ_2=0.03 for the ^{106}Sn band is much lower than the previous value of ϵ_2=0.11.

TAC calculations with the new deformations, show improved agreement with the experimental B(M1) values (see Fig. 3). The remaining difference between theory and experiment is probably due to treating deformation as a constant within the TAC model, while the experimental B(E2) values for ^{108}Sn clearly decrease with increasing spin.

A semi-classical derivation of B(M1) values in terms of the coupling scheme illustrated in Fig. 1 has been attempted [4]. B(M1) values are derived in terms of the angle, θ_π, which the proton spin vector makes with the total vector, $\mathbf{I}$. It is found that B(M1) $= \frac{3}{8\pi} g_{eff}^2 j_\pi^2 \sin^2\theta_\pi$, where g_{eff} is the difference between the proton and neutron g-factors [4].

The proton spin vector is assumed to originate from the $g_{9/2}$ proton hole, i.e. $|\mathbf{j}_\pi|$ = 4.5$\hbar$, and the length of the remaining vector, $\mathbf{j}_\nu$, is fixed such that it reproduces the length of the total angular momentum vector, $\mathbf{I}$, at the bandhead. Higher spin values are obtained by closing the shears. B(M1) transition rates have been determined for the respective bands using values of g_{eff} obtained from empirical g factors for this mass region. The excellent agreement of these calculations with experimental B(M1) values for both nuclei provides good evidence for the near perpendicular coupling of the proton and neutron configurations at the bandhead.

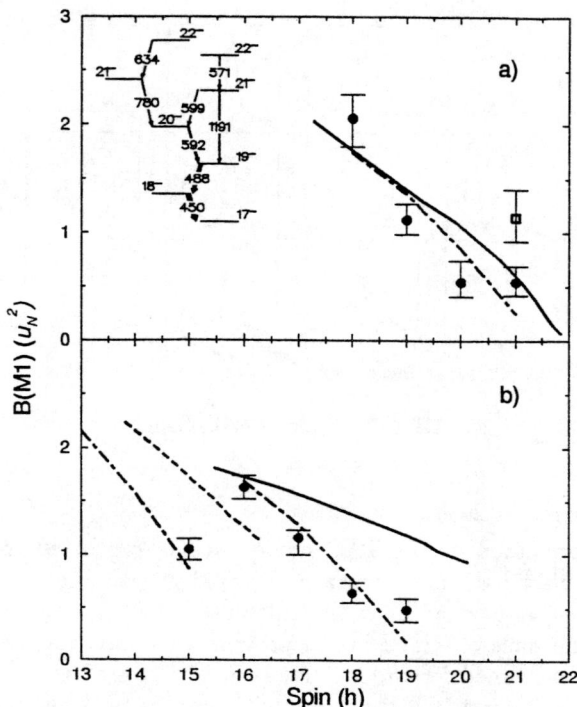

FIGURE 3. Comparison of experimental B(M1) values for ^{106}Sn (top) and ^{108}Sn (bottom). Revised TAC calculations are the solid and dashed lines. Semi-classical calculations are the dot-dashed lines.

Furthermore, the results imply a very low core contribution to the total angular momentum and that the angular momentum is almost entirely generated by the shears mechanism. It could be argued that ^{106}Sn is the first example of almost pure magnetic rotation in a nucleus classified as spherical. This work has been reported elsewhere [6]

REFERENCES

1. D. G. Jenkins *et al.*, Phys. Lett. B428, 23 (1998).
2. S. Frauendorf, J. Meng and J. Reif, in Proceedings of the Conference on Physics from Large γ Ray Detectors, Berkeley, 1994.
3. S. Frauendorf, Nucl. Phys. A557, 259c (1993).
4. A. O. Macchiavelli *et al.*, Phys. Rev. C57, R1073 (1998).
5. J. C. Wells and N. Johnson (private communication) (June 1998)
6. D. G. Jenkins *et al.*, Phys. Rev. Lett. 83, 500 (1999).

Rotation of Highly Excited Nuclei : Mass Dependence of Rotational Damping

B. Million[1], S. Frattini[1], A. Bracco[1], S. Leoni[1], F. Camera[1], N. Blasi[1], G. Lo Bianco[1], M. Pignanelli[1], E. Vigezzi[1], B. Herskind[2], T. Døssing[2], M. Bergström[2], P. Varmette[2], S. Törmänen[2], A. Maj[3], M. Kmiecik[3], D.R. Napoli[4], M. Matsuo[5]

1- Dipartimento di Fisica, Università di Milano, and INFN sez. Milano,
via Celoria, 16, 20133 Milano, Italy
2- The Niels Bohr Institute, Copenhagen, Denmark
3- The Henryk Niewodniczański Institute of Nuclear Physics, 31-342 Kraków, Poland
4- INFN, Laboratori Nazionali di Legnaro, Legnaro, Italy
5- Yukawa Institute of Physics, Kyoto, Japan

Abstract. The γ-decay of the continuum has been measured in two mass regions. The excitation function of the continuum decay as well as spectral shape and fractional Doppler shifts are discussed for both ^{114}Te and ^{164}Yb compound nuclei, and show the typical features of rotational collective motion. Moreover, in both cases an upper limit of Γ_{rot} is given and the number of decay-paths is determined from the fluctuation analysis method. Simulations based on microscopic calculations of the rotational damping model reproduce quite well the experimental findings for both N_{path} and the scaling of Γ_{rot} as a function of the mass number.

INTRODUCTION

In the last years a number of experimental studies aiming at determining the properties of γ-continuum decay have been made for rare earth nuclei. This decay was found to be due to collective transitions between rotational states. Moreover it is well described by the rotational damping model [1] based on microscopic cranking model calculations. This model predicts a mass-dependence of the rotational damping [2]. As it is not easy to have a direct measurement of the rotational damping width, Γ_{rot}, the problem of rotational damping is also adressed by studying the number of nuclear decay paths $N_{path}^{(2)}$ which depends on Γ_{rot} and on the excitation energy U_0 at which the damping sets in. As Γ_{rot} depends also on the deformation parameter and on the spin of the nucleus, it is important to choose nuclei with similar deformation if one wants to disentangle mass effects from other effects.

CP495, Experimental Nuclear Physics in Europe, edited by B. Rubio et al.

Nuclei in the mass region A≈110 are good candidates to be compared with rare earth nuclei A≈160 and to verify the model predictions. In fact, at high excitation energy and spin they were found to have a very similar deformation to that of the rare earth nuclei. We report here on results from both rare earth nuclei, [164]Yb, measured at Eurogam2 [3], [4], and lighter nuclei measured more recently at Euroball, [114]Te [3], [5].

CONTINUUM DECAY PROPERTIES FROM 1D-SPECTRA : COLLECTIVITY AND ROTATIONAL DAMPING

With increasing beam energy the continuous bump measured for both [114]Te and [164]Yb increases in intensity and maximum transition energy, fig. 1, as expected for rotational decay [6]. Moreover, in the case of [114]Te the strength function at the highest beam energy (270 MeV) is practically identical to the 260 MeV one. This indicates the spin limit that the compound nucleus can sustain.

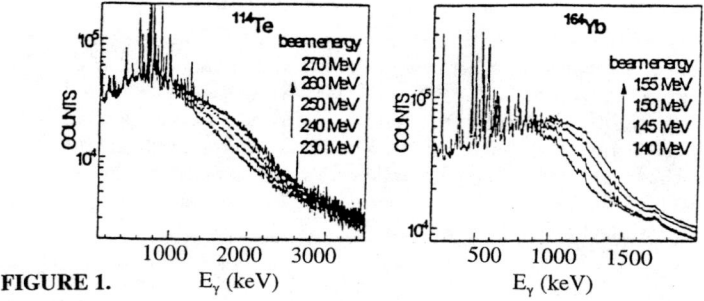

FIGURE 1.

Fractional Doppler shifts, $F(\tau)$, were obtained for the continuum region, fig. 2 and insets, by comparing spectra at forward and backward angles following the technique described in [7]. The experimental values lie around 1.0 for both nuclei, showing that the transitions in the continuum are emitted at a time shorter than 10^{-14}s for the Yb nuclei, [4], and 10^{-12}s for the Te nuclei, [3]) in the decay and have a high degree of collectivity.

An upper limit of the rotational damping width was inferred from the present data. The beam energies have been chosen to populate residual nuclei characterized by approximately two E2-transition difference between two consecutive bombarding energies, fig. 1. By substracting two consecutive spectra, one gets approximately the decay flow from spin I to spin I-4. Such spectra are built up from a convolution of the two decay steps, I→I-2 and I-2→I-4, that, according to the rotational damping model decay both with the damping width Γ_{rot}. Because of the feeding of the decaying states it is not straigthforward to extract a direct value for Γ_{rot}. As a consequence, the width of the curves in figure 3 gives an upper limit of Γ_{rot} which is found to be ≈ 300-400 keV for [164]Yb and 800-1000 keV for [114]Te. The ratio between these numbers is in good agreement with the scaling factor of 2 predicted by the rotational damping model.

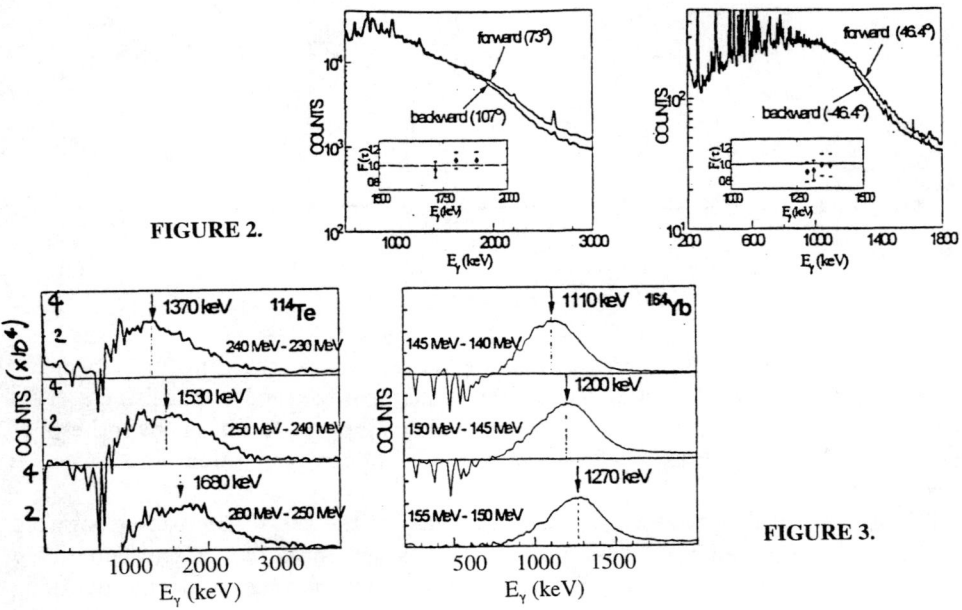

FIGURE 2.

FIGURE 3.

RIDGE STRUCTURES

The rotational decay of the continuum region can be better analyzed through the study of the 2D matrix, $E_{\gamma 1} \times E_{\gamma 2}$. For rotational nuclei, such matrix shows a ridge-valley structure [8], [9] ; two ridges, which correspond to the unresolved regular rotational decay, are built on both sides of a central valley along the main diagonal, corresponding to the rotational damped region.

For the Te nucleus, the ridge-valley structure typical of rotational motion was found although with intensity lower than in the Yb nucleus. In addition, the main valley is wider for [114]Te because of its lower moment of inertia (lower mass).

The fluctuation analysis was applied on both matrices to extract the number of paths $N^{(2)}_{path}$ - i.e. the number of pairs of consecutive transitions - as a function of transition energy [10]. The experimental results are very well reproduced for the ridge structure, see figure 4, and reasonably well reproduced for the valley region by the model predictions, showing that the rotational damping model based on the calculation of microscopic levels describes quite well the interplay of the rotational damping with the level density and the residual interaction in these two mass regions.

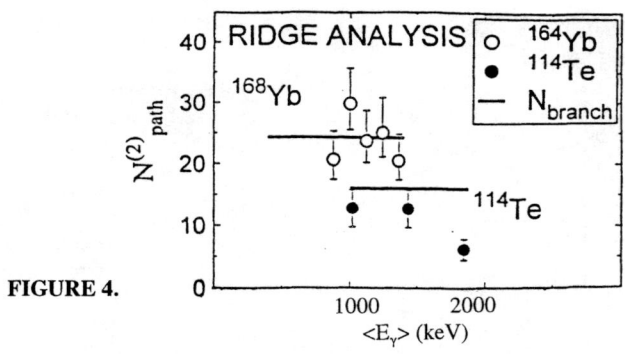

FIGURE 4.

CONCLUSIONS

The experimental results presented here show that also the continuum γ-decay of nuclei with mass A≈110 corresponds to collective rotational transitions. This is the first important result of this study which supports the basic hypothesis of the rotational damping model, that the continuum decay corresponds to collective rotational motion.

The second result is that this continuous decay of the thermally excited rotating nuclei in the damped rotational regime, follows the mass dependence predicted by the model : a scaling factor of 2 has been measured for the rotational damping width Γ_{rot} between the two mass region. Alltogether, the rotational damping model gives a good description of the interplay of the level density and of the damping width in the decay of the excited compound nuclei in the two different mass region studied here, as is seen from the study of the number of paths.

REFERENCES

1. Matsuo M. et al., *Nucl. Phys.* **A617**, 1 (1997).
2. Lauritzen B., Døssing T., Broglia R.A. *Nucl. Phys.* **A457**, 61 (1986).
3. Frattini S., PhD thesis, University of Milan, and to be published.
4. Frattini S. et al.,*Phys. Rev. Lett.* **81**, 2659 (1998).
5. Frattini S. et al., submitted to (*Phys. Rev. Lett.*).
6. *Körner H.J. et al., Phys. Rev. Lett. 43, 490 (1979).*
7. *Million B. et al., Phys. Lett. B415, 321 (1997).*
8. *Deleplanque M.A. et al. Nucl. Phys. A448, 495 (1986).*
9. *Leoni S. et al., Nucl. Phys. A587, 513 (1995).*
10. *Døssing T. et al., Phys. rep. 268, 1 (1996).*

Collective dipole bands in ^{110,112}Te: stability against magnetic rotation

E.S. Paul[1], A.J. Boston[1], C.J. Chiara[2], M. Devlin[3], D.B. Fossan[2],
D.R. LaFosse[3], G.J. Lane[2], I.Y. Lee[4], A.O. Macchiavelli[4],
P.J. Nolan[1], D.G. Sarantites[3], J.M. Sears[2], A.T. Semple[1],
J.F. Smith[2], K. Starosta[2], A.V. Afanasjev[5,6], and I. Ragnarsson[5]

[1] Oliver Lodge Laboratory, University of Liverpool, P.O. Box 147, Liverpool L69 7ZE, UK
[2] Department of Physics and Astronomy, SUNY at Stony Brook, Stony Brook, NY 11794
[3] Department of Chemistry, Washington University, St. Louis, MO 63130
[4] Nuclear Science Division, Lawrence Berkeley National Laboratory, Berkeley, CA 94720
[5] Department of Mathematical Physics, University of Lund, S-22100 Lund, Sweden
[6] Nuclear Research Centre, Latvian Academy of Sciences, Salaspils, Miera str. 31, Latvia

Abstract. Three long, strongly coupled ($\Delta I = 1$) sequences have been identified in ^{110,112}Te by using the GAMMASPHERE array in conjunction with the MICROBALL charged-particle array. These bands are interpreted in terms of deformed proton 1-particle–1-hole bands that reach termination at $I \sim 40\hbar$. This is the first observation of such collective dipole structures in this mass region. In contrast, many shorter dipole sequences have been associated with weakly deformed structures that generate angular momentum by the shears mechanism (magnetic rotation).

Nuclei close to doubly magic ^{100}Sn generate their spin predominantly through the alignment of single-particle angular momenta of the valence nucleons resulting in complex energy spectra. The limited number of valence particles is insufficient to break the spherical symmetry of these nuclei and their structure can be interpreted in terms of a nuclear shell model. However, quadrupole deformation can be induced through 2-particle–2-hole (2p-2h) proton excitations across the $Z = 50$ shell gap. Such structures are able to generate spin through collective nuclear rotation leading to the observation of band structures consisting of regular sequences of $\Delta I = 2$ γ rays. In this region, 1p-1h proton excitations, involving the $g_{9/2}$ high-K proton hole, result in $\Delta I = 1$ bands with little signature splitting. Such structures in $_{50}$Sn [1,2] and $_{51}$Sb [3] isotopes, which are observed to only modest spins, drop off quickly in $M1$ intensity showing shears dynamics (magnetic rotation) related to the perpendicular coupling at low spin of a high-K proton hole to low-K neutron orbitals in a weakly deformed nucleus. Similar dipole bands have also been observed in

CP495, Experimental Nuclear Physics in Europe, edited by B. Rubio et al.
© 1999 American Institute of Physics 1-56396-907-6/99/$15.00

nuclei with $Z < 50$, e.g. ^{108}Cd [4]. These structures have been interpreted through comparison with tilted-axis cranking (TAC) calculations [5], which predict that magnetic rotation should exist around proton or neutron closed shell regions, i.e. regions of weak quadrupole deformation. Magnetic rotation involves the rotation of a large magnetic dipole moment which breaks the spherical symmetry of nuclei, as opposed to the conventional view of collective nuclear rotation, where the symmetry is broken by quadrupole deformation induced by distortion of the electric charge distribution.

In addition to several new high-spin $\Delta I = 2$ collective bands in $^{110,112}_{52}$Te, three strongly coupled ($\Delta I = 1$) bands have been observed up to unusually high spins. Two of the new dipole bands (one in each nucleus) have been connected to the low-spin level schemes thus allowing an estimate of their spins, parities, and excitation energies. In order to assign configurations to the dipole bands in ^{110,112}Te, calculations have been performed using the configuration-dependent Nilsson-Strutinsky approach, but neglecting pairing correlations [7–9]. A lack of signature splitting in the three dipole bands of ^{110,112}Te implies that the configurations involve a single $\pi g_{9/2}$ hole and thus 1p-1h proton excitations (or 3p1h *configurations* relative to $Z = 50$). Indeed, the calculations yield deformed 1p-1h ($\pi h^1_{11/2} g^{-1}_{9/2}$) excitations that are yrast at intermediate spins between low-spin 0p-0h spherical states and high-spin deformed 2p-2h configurations. Furthermore, these deformed 1p-1h configurations terminate smoothly at $I \approx 40\hbar$.

With these new dipole bands in ^{110,112}Te reaching maximum spins for configurations outside ^{100}Sn, while maintaining $M1$ strength, a significant stability against the shears mechanism must be operative. These bands therefore appear to represent the first observation of *deformed* 1p-1h high-spin excitations in this mass region. It is worth pointing out, however, that the new dipole bands in ^{110,112}Te appear to be the only terminating bands in this mass region *not* built on 2p-2h excitations. To date, well over 40 terminating bands are known in this mass region, which are all built on $\pi g^{-2}_{9/2}$ configurations.

This work was supported in part by the US NSF and the UK EPSRC.

REFERENCES

1. Gadea A. et al., *Phys. Rev. C* **55**, R1 (1997).
2. Jenkins D.G. et al., *Phys. Lett.* **428B**, 23 (1998).
3. Jenkins D.G. et al., *Phys. Rev. C* **58**, 2703 (1998).
4. Clark R.M. et al., *Phys. Rev. Lett.* **82**, 3220 (1999).
5. Frauendorf S., *Nucl. Phys.* **A557**, 259c (1993).
6. Lee I.Y., *Nucl. Phys.* **A520**, 641c (1990).
7. Bengtsson T. and Ragnarsson I., *Nucl. Phys.* **A436**, 14 (1985).
8. Ragnarsson I. et al., *Phys. Rev. Lett.* **74**, 3935 (1995).
9. Afanasjev A.V. and Ragnarsson I., *Nucl. Phys.* **A591**, 387 (1995).

DSAM lifetime measurements in ^{119}Xe

H.C. Scraggs[1], E.S. Paul[1], A.J. Boston[1], C.J. Chiara[2], D.B. Fossan[2], C. Fox[1], D.R. LaFosse[2], P.J. Nolan[1], T. Koike[2], K. Starosta[2], A. Walker[1], A.V. Afanasjev[3,4], and I. Ragnarsson[3]

[1] *Oliver Lodge Laboratory, University of Liverpool, P.O. Box 147, Liverpool L69 7ZE, UK*
[2] *Department of Physics and Astronomy, SUNY at Stony Brook, Stony Brook, NY 11794*
[3] *Department of Mathematical Physics, University of Lund, S-22100 Lund, Sweden*
[4] *Nuclear Research Centre, Latvian Academy of Sciences, Salaspils, Miera str. 31, Latvia*

Abstract. Lifetime measurements of states in the yrast band of ^{119}Xe have been performed using a Doppler broadened lineshape analysis. Preliminary results, in the range $27/2^- \leq I^\pi \leq 43/2^-$, indicate a transition quadrupole moment of approximately 2.9 eb, which corresponds to a prolate rotor with deformation $\varepsilon_2 \approx 0.17$.

Nuclei of mass A $\approx$ 110 $-$ 120, with a limited number of valence particles outside the Z = N = 50 doubly magic core, are ideal cases in which to study band-termination effects. In this mass region, these effects occur at experimentally accessible spins with fusion-evaporation reactions (30 $-$ 50 $\hbar$). The high-spin, negative-parity yrast band of ^{119}Xe nucleus has been interpreted in terms of a configuration showing properties of smooth band termination following an experiment at Jyväskylä using the JUROSPHERE array [1]. Theoretical calculations, based on the configuration-dependent Nilsson-Strutinsky approach [2], however yield two possible structures that can describe the high-spin properties of ^{119}Xe. The first structure represents a *valence-space* configuration involving only proton orbitals above the Z = 50 shell gap (4-particle), while the second, more deformed *core-excited* configuration involves two holes in the $\pi g_{9/2}$ orbital that originates from below the shell gap (6-particle 2-hole). Since the predicted quadrupole deformations (both ε_2 and γ) of these structures yield significantly different transition quadrupole moments, it would be possible to experimentally distinguish between the two theoretical configurations by measuring the quadrupole moment at high spin.

To this end, lifetime measurements of high-spin states in ^{119}Xe have been performed using the Doppler Shift Attenuation Method (DSAM). States in ^{119}Xe were populated using the ^{95}Mo(^{28}Si,2p2n)^{119}Xe reaction. The data were collected at the Stony Brook Nuclear Structure Facility using an array of six escape-suppressed HPGe detectors with a fourteen element BGO sum-energy/multiplicity filter. A 131

CP495, *Experimental Nuclear Physics in Europe*, edited by B. Rubio et al.

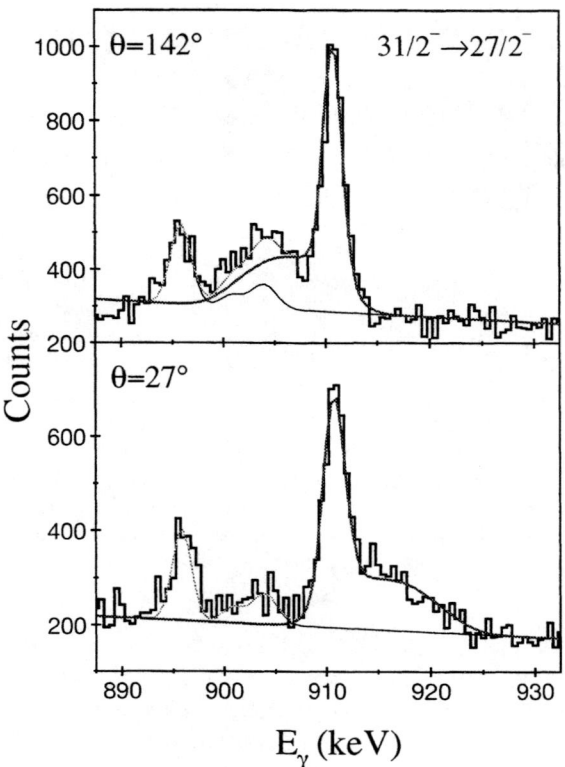

FIGURE 1. Lineshapes generated for the $31/2^- \to 27/2^-$ yrast transition in ^{119}Xe.

MeV beam of ^{28}Si, provided by the SUNY tandem/LINAC accelerator, was used to bombard a 1 mg/cm^2 ^{95}Mo target backed by 15 mg/cm^2 of ^{208}Pb. Approximately 1.5×10^8 $\gamma - \gamma$−BGO events were acquired.

The Doppler shift attenuation method was used to determine the mean level lifetimes of states in ^{119}Xe through a Doppler-broadened lineshape (DBLS) analysis of the data. A γ ray that is emitted while the nucleus is slowing down will possess a broadened lineshape as a consequence of the Doppler effect. The magnitude of the shift depends on the lifetimes of the band member and on the observed and unobserved transitions feeding that state. Monte Carlo techniques were used to model the slowing down history of the recoiling nuclei [3]. The electronic stopping powers of Ziegler [4] have been used, while for low recoil velocities where nuclear stopping is important, the scattering direction was traced. Five thousand recoil histories were generated at a time step of 1 fs and converted into time-dependent velocity profiles for particular angular configurations of detectors. The side-feeding was modelled by a five-state rotational band with a fixed moment of inertia. Spectra were generated by summing gates set on the low-lying, fully stopped transitions of

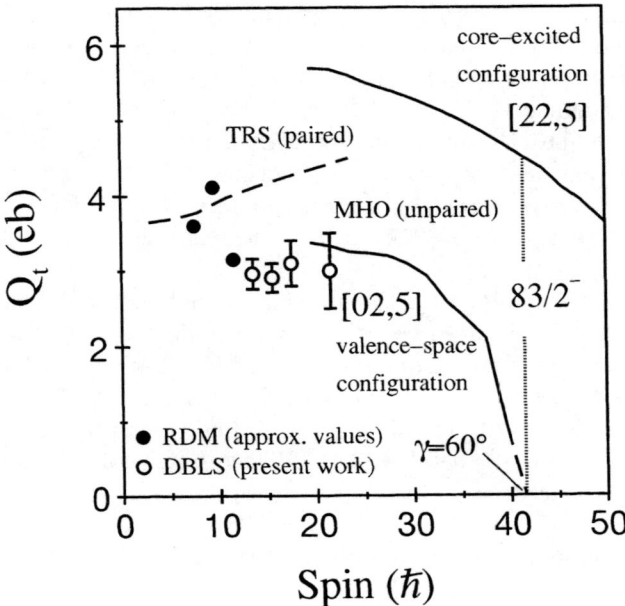

FIGURE 2. Measured and theoretical transition quadrupole moments for the negative-parity yrast band in ^{119}Xe. The approximate values at low spin are taken from Ref. [5] and were obtained from a recoil-distance measurement (RDM).

^{119}Xe. Lineshapes were then generated by simultaneously fitting spectra at forward and backward angles. Known contaminant peaks can also be included in the fitting procedure. An example of such lineshapes is shown in Fig. 1 for the 910 keV $31/2^- \rightarrow 27/2^-$ yrast transition of ^{119}Xe.

The mean lifetime τ of a state can be used to determine the experimental transition quadrupole moment Q_t:

$$\frac{1}{\tau} = 1.22 \times 10^{12} E_\gamma^5 |\langle I\ 0\ 2\ 0\ |I-2\ 0\rangle|^2 Q_t^2. \tag{1}$$

Furthermore, the deformation parameters (ε_2, ε_4, γ) obtained from the Nilsson-Strutinsky calculations can be translated into theoretical Q_t values for comparison:

$$Q_t = Q_{20} \frac{\cos(\gamma + 30°)}{\cos(30°)}, \tag{2}$$

$$Q_{20} = \frac{4}{5} Zer_0^2 A^{2/3} \left[\varepsilon_2 \left(1 + \frac{1}{2}\varepsilon_2 \right) + \frac{25}{33}\varepsilon_4^2 - \varepsilon_2\varepsilon_4 \right]. \tag{3}$$

Preliminary experimental results for the yrast band of ^{119}Xe are shown in Fig. 2 together with theoretical values. Above spin 20$\hbar$, the calculated valence-space and core-excited configurations differ in the magnitude of Q_t by almost a factor of two. Both configurations show decreasing Q_t values with increasing spin which is related to the gradual drift through the γ plane towards termination. The valence-space configuration terminates at $I^\pi = 83/2^-$ while the core-excited configuration terminates well beyond $60\hbar$. Results from Total Routhian Surface (TRS) calculations [6], which include pairing, are also shown at low spin. In this case, the increasing trend of Q_t with spin is caused by a deviation into the negative γ plane below the backbending frequency ($I \sim 20\hbar$); at the backbend the shape returns to axial symmetry ($\gamma = 0°$).

The present results obtained for the low-spin yrast band of ^{119}Xe, in the range $27/2^- \leq I^\pi \leq 43/2^-$, indicate a transition quadrupole moment of approximately 2.9 eb, which, assuming axial symmetry and zero hexadecapole deformation, corresponds to a prolate rotor with quadrupole deformation $\varepsilon_2 \approx 0.17$. Unfortunately, the statistics were too poor to follow the band to higher spin beyond the backbend where the smooth termination takes place. The results, however, would seem to favour the valence-space interpretation.

This work was supported by grants from the UK EPSRC and US NSF.

REFERENCES

1. Scraggs H.C., et al., *Nucl. Phys.* **A640**, 337 (1998).
2. Afanasjev A.V. and Ragnarsson I., *Nucl. Phys.* **A591**, 387 (1995).
3. Wells J.C. and Johnson N.R., *"LINESHAPE: A Computer Program for Doppler Broadened Lineshape Analysis"*, Report No. ORNL-6689 (1991).
4. Ziegler J.F., *The Stopping and Ranges of Ions in Matter* (Pergamon, London, 1985), Vols. 3 and 5.
5. Chaudhury A., et al., *Bull. Am. Phys. Soc.* **30**, 742 (1985).
6. Wyss R., et al., *Phys. Lett.* **B215**, 211 (1988).

Rotational Bands of ^{155}Gd

T. Hayakawa[1], M. Oshima[1], Y. Hatsukawa[1],
J. Katakura[1], H. Iimura[1], M. Matsuda[1], S. Mitarai[2],
Y. R. Shimizu[2], S. -I. Ohtsubo[2], T. Shizuma[3],
M. Sugawara[4], H. Kusakari[5]

1 Japan Atomic Energy Research Institute, Tokai, Ibaraki 319-1195, Japan
2 Kyushu University, Hakozaki, Fukuoka 812-8581, Japan
3 University of Tsukuba, Tsukuba, Ibaraki 305-8577, Japan
4 Chiba Institute of Technology, Narashino, Chiba 275-8588, Japan
5 Chiba University, Inage-ku, Chiba 263-8522, Japan

The nucleus ^{155}Gd ($N = 91$) is located near the edge of the well deformed region, and shows a typical rotational character. The last neutron in ^{155}Gd can occupy various orbitals near the Fermi surface in the single-particle states. The high-spin states have been studied using the Coulomb excitation experiment (1-3) and using ^{150}Nd(^{9}Be, 4n) reaction (4). It is difficult to observe the high spin states using the (HI,xn) reaction because ^{155}Gd is located in the neutron rich region. In order to study a role of interaction between single-particle configurations and collective motion in ^{155}Gd, an in-beam γ-ray spectroscopy has been carried out.

The nucleus ^{155}Gd was produced with the reaction ^{150}Nd (^{12}C, α3n) ^{155}Gd using a 65 MeV ^{12}C beam provided by the tandem accelerator at Japan Atomic Energy Research Institute (JAERI). The target was a self-supporting ^{150}Nd metallic foil enriched to 96.1 % with a thickness of 2 mg/cm^2. Gamma-rays from excited states were detected with an array (5) of 11 HPGe detectors with BGO Compton suppressors; in coincidence with the particles detected by the Si-ball (6) particle filter made up of 20 detector segments. The energy resolutions of HPGe detectors were 2.0−2.3 keV at 1.3 MeV. The efficiencies of typical HPGe detectors were about 40 % relative to 3″ × 3″ NaI detector. The Si-ball was used to identify the nuclei produced by α evaporation channel with a small cross section. The experimental data were recorded when two or more HPGe detectors and at least one segment of the Si-ball were fired. Approximately 2×10^7 γ-γ coincidence events were collected.

Fig. 1 shows a partial level scheme which was constructed from γ - γ coincidence relationships and intensity balances. Three rotational bands in ^{155}Gd are labeled as Band 1, 2 and 3 in Fig. 1, and also the ground state rotational band in ^{156}Gd has been presented. Band 1 with $\nu[651]\frac{3}{2}^+$ configuration and Band 2 with the $\nu[505]\frac{11}{2}^-$ configuration have been extended up

CP495, *Experimental Nuclear Physics in Europe*, edited by B. Rubio et al.
© 1999 American Institute of Physics 1-56396-907-6/99/$15.00

to $(51/2)^+$ and $45/2^-$, and Band 3 has newly been observed.

The backbending and blocking effects are discussed in comparison with the calculation of cranked shell model. The routhians and the $B(M1)/B(E2)$ ratio of the high-Ω rotational band are compared with the calculation by the tilted axis cranking model. Those for Band 1 are nicely reproduced by the usual cranking model, but not for high-K Band 2. The behavior of the $B(M1)/B(E2)$ ratio for Band 2 indicates that the angular momentum vector is tilting appreciably even at high-spin region. The TAC calculation reasonably reproduces both the routhian and the $B(M1)/B(E2)$ ratio of Band 2 simultaneously.

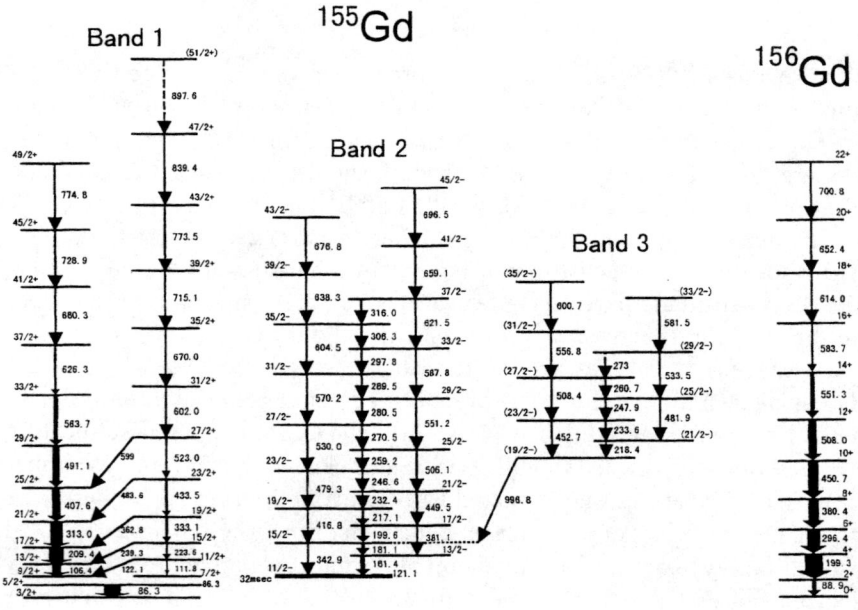

FIGURE.1 The partial level scheme of ^{155}Gd and ^{156}Gd.

REFERENCES

1. Kusakari, H., *et al.*, Phys. Rev. C **46**, 1257 (1992).
2. Kidera, M., *et al.*, J. of Phys. Soc. of Japan **66**, 285 (1997).
3. Stuchbery, A. E., *et al.*, Nucl. Phys. A**642**, 361 (1998).
4. Riley, M. A., *et al.*, Z. Phys. A**345**, 121 (1993).
5. Furuno, K., *et al.*, Nucl. Inst. & Method in Phys. Rese. A **421**, 211 (1999).
6. Mitarai, S., *et al.*, Nucl. Inst. and Method in Phys. Rese. A **277**, 491 (1989).

Complete Spectroscopy and the ^{162}Dy Nucleus

A Aprahamian[a], X Wu[a], S R Lesher[a], D D Warner[b], W Gelletly[c],
Z R Shi[d], F Hoyler[e], H G Borner[e], R F Casten[f] and
K Schreckenbach[g]

(a) University of Notre Dame, South Bend, Indiana 46556 USA
(b) CLRC, Daresbury Laboratory, Warrington, WA4 4AD UK
(c) School of Physics and Chemistry, University of Surrey, Guildford, GU2 5XH UK
(d) Institute of Atomic Energy, Beijing, People's Republic of China
(e) Institut Laue-Langevin, 38042 Grenoble France
(f) Wright Nuclear Structure Laboratory, Yale University, Connecticut 06250 USA
(g) Physik Department, TU-München, D-85748 Garching Germany

I.   Introduction and Experiment

Despite much effort ^{168}Er remains the one nucleus in the rare-earth region where we have comprehensive information about the states up to an excitation energy of ~2MeV. The present work on ^{162}Dy was intended to provide a second, equally well studied case.

To this end we have studied the (n, γ), (n, e⁻) and (α, 2nγ) reactions leading to ^{162}Dy. The first two sets of measurements were carried out at the ILL High Flux Reactor and took advantage of the high precision of the GAMS curved crystal spectrometer and BILL β-spectrometer to study the γ-rays from 30-2000 keV. The (n, γ) spectrum was also studied with Ge detectors and a thermal neutron beam at the Brookhaven HFBR. Complementary studies of the (α, 2nγ) reaction were carried out with the HERA array at the LBL 88" cyclotron. A detailed level scheme for ^{162}Dy has been constructed by combining the results of these measurements with those from an earlier average resonance capture study[1] which serves to guarantee the completeness of the set of low spin states.

II.   Results

Fig.1 shows the complete set of levels observed. They have been arranged into bands according to their decay patterns. Those marked with a full circle were seen

CP495, Experimental Nuclear Physics in Europe, edited by B. Rubio et al.
© 1999 American Institute of Physics 1-56396-907-6/99/$15.00

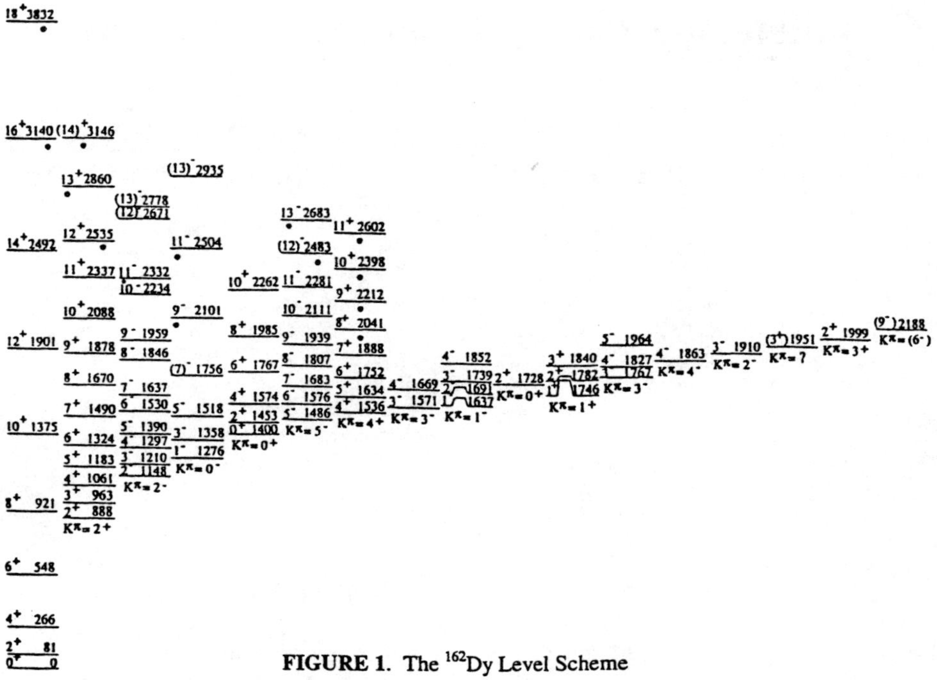

$18^+ 3832$

$16^+ 3140$ $(14)^+ 3146$

$13^+ 2860$ $(13)' 2935$

$(13)' 2778$
$(12)' 2671$

$13^- 2683$

$11^+ 2602$

$14^+ 2492$ $12^+ 2535$ $11^- 2504$ $(12)' 2483$

$10^+ 2398$

$11^+ 2337$ $11^- 2332$ $10^+ 2262$ $11^- 2281$ $9^+ 2212$
$10^- 2234$

$10^+ 2088$ $9^- 2101$ $10^- 2111$ $8^- 2041$

$12^+ 1901$ $9^+ 1878$ $9^- 1959$ $8^+ 1985$ $9^- 1939$ $7^+ 1888$ $5^- 1964$ $2^+ 1999$ $(9^-) 2188$
$8^- 1846$ $7^+ 1888$ $3^- 1910$ $(3^+) 1951$ $K^\pi = 3+$ $K^\pi = (6^-)$

$8^+ 1670$ $(7)^- 1756$ $6^+ 1767$ $8^- 1807$ $4^- 1852$ $3^+ 1840$ $4^- 1827$ $4^- 1863$ $K^\pi = ?$
$7^- 1637$ $7^- 1683$ $2^+ 1728$ $2^+ 1782$ $3^- 1767$ $K^\pi = 2^-$
$7^+ 1490$ $6^- 1530$ $5^- 1518$ $4^+ 1574$ $6^- 1576$ $4^- 1669$ $3^- 1739$ $2^- 1691$ $1^- 1746$ $K^\pi = 4^-$
$5^+ 1634$ $3^- 1571$ $1^- 1637$ $K^\pi = 0+$ $K^\pi = 3^-$
$10^+ 1375$ $5^- 1390$ $5^+ 1518$ $2^+ 1453$ $5^- 1486$ $4^- 1536$ $K^\pi = 1^+$
$6^+ 1324$ $4^- 1297$ $3^- 1358$ $0^+ 1400$ $K^\pi = 5^-$ $K^\pi = 4+$ $K^\pi = 3^-$ $K^\pi = 1^-$
$5^+ 1183$ $3^- 1210$ $1^- 1276$ $K^\pi = 0+$
$4^+ 1061$ $2^+ 1148$ $K^\pi = 0^-$
$8^+ 921$ $3^+ 963$ $K^\pi = 2^-$
$2^+ 888$
$K^\pi = 2+$

$6^+ 548$

$4^+ 266$
$2^+ 81$
$0^+ 0$

FIGURE 1. The ^{162}Dy Level Scheme

for the first time in the (α, 2n) study. The K=4 band at 1536 keV is of some particular interest. To date we have no lifetime for the states in the K = 4 band. If we assume that the intrinsic quadrupole moment is the same as in the ground state band we can deduce the intraband B(E2)s for the K = 4 band from the observed ratios of interband/intraband intensities. These are an order-of-magnitude lower than would be expected for a fully collective, two phonon $\gamma\gamma$-vibration. This is in contrast to the situation in ^{164}Dy which has been interpreted[2] as a two phonon, $\gamma\gamma$-vibration.

REFERENCES

1. Warner, D.D., Casten, R.F., Kane, W.R. and Gelletly, W., *Phys.Rev.* **C27**, 2292 (1983)
2. Carminboeuf, F. et al, *Phys.Rev.* **C56**, R1201 (1997)

Signature-dependent Electromagnetic Transition Rates in the $\pi h_{11/2}$ Rotational Sequence of $^{167}_{73}$Ta

R. Chapman*, K.–M. Spohr*, M.B. Smith*, R.A. Bark†,
G.J. Campbell*, G.B. Hagemann‡, N. Keeley§, D.J. Middleton*,
H. Ryde† and P.O. Tjøm**

*Department of Physics, University of Paisley, Paisley PA1 2BE, UK
†Department of Physics, Lund University, S–223 62 Lund, Sweden
‡Niels Bohr Institute, University of Copenhagen, DK–2100 Copenhagen Ø, Denmark
§Schuster Laboratory, University of Manchester, Manchester M13 9PL, UK
**Department of Physics, University of Oslo, N–0316 Oslo, Norway

Abstract. Excited states in ^{167}Ta, populated in the ^{141}Pr(^{30}Si, 4n)^{167}Ta reaction, have been studied using the NORDBALL Ge detector array. For the $\pi h_{11/2}[514]\frac{9}{2}^{-}$ decay sequence, strong signature-dependent effects in the transition quadrupole moment ratio, $Q_1(I \rightarrow I - 1)/Q_2(I \rightarrow I - 2)$, have been observed over the spin range $\frac{21}{2} \leq I \leq \frac{39}{2}$ which encompasses a BC neutron alignment. This is interpreted as strong evidence for departure from axial symmetry.

In the high-spin regime, the experimental signatures of triaxial nuclear shape in rotational sequences based on high-j orbitals in odd-Z nuclei (e.g. $\pi h_{11/2}$) include (i) signature splitting of the Routhians, (ii) signature splitting of $B(M1)$ values and (iii) signature splitting of $\Delta I = 1$ $B(E2)$ values. When the Fermi level lies low in the $h_{11/2}$ shell, and for triaxiality parameter $\gamma = 0$, signature splitting is expected to be observed in the Routhians and in $B(M1)$ values as a consequence of mixing with the $K = \frac{1}{2}$ component of the $h_{11/2}$ wavefunction. However, the $\Delta I = 1$ $B(E2)$ values are not predicted to exhibit such an effect. High in the $h_{11/2}$ shell (e.g. the $\pi h_{11/2}[514]\frac{9}{2}^{-}$ configuration) the mixing of the wavefunction with low-K components of the wavefunction is negligibly small when $\gamma = 0$ and no signature-dependent effects are expected in Routhians, $B(M1)$ values and $\Delta I = 1$ $B(E2)$ values. For $\gamma \neq 0$ signature-dependent effects are expected for all three quantities. In this respect measurements of $\Delta I = 1$ $B(E2)$ values provide the most definitive evidence for departure from axial symmetry in $h_{11/2}$ proton shell nuclei.

Absolute $B(M1)$ and $B(E2)$ values require the measurement of nuclear lifetimes; in the absence of such data, it is convenient to measure the ratios $B(M1; I \rightarrow$

CP495, *Experimental Nuclear Physics in Europe*, edited by B. Rubio et al.
© 1999 American Institute of Physics 1-56396-907-6/99/$15.00

$I-1)/B(E2; I \to I-2)$ and $B(E2; I \to I-1)/B(E2; I \to I-2)$, which rely on the determination of branching ratios $\lambda\ (= T_\gamma(I \to I-2)/T_\gamma(I \to I-1))$ and mixing ratios δ $(\delta^2 = T_\gamma(E2; I \to I-1)/T_\gamma(M1; I \to I-1))$. Any signature-dependent effects observed in these ratios can be attributed to the $\Delta I = 1$ reduced transition probabilities rather than to the $B(E2; I \to I-2)$ values. No experimental or theoretical evidence exists which suggests that the latter quantities should exhibit significant signature-dependent effects.

Yrast and near-yrast rotational sequences in the odd-Z nucleus ^{167}Ta were populated in the ^{141}Pr(^{30}Si, 4n)^{167}Ta reaction using a ^{30}Si beam of energy 155 MeV delivered by the NBI tandem Van de Graaff. Gamma-rays were detected using the NORDBALL array of 20 escape-suppressed Ge detectors and a 60-element BaF$_2$ inner ball. A total of 500×10^6 coincidence events were recorded. Figure 1 presents the decay sequence for the $\pi h_{11/2}[514]\frac{9}{2}^-$ configuration which resulted from analysis of the coincidence data. The present work has resulted in significant additions to the published level scheme of Theine $et\ al$ [1]. Quasiparticle alignments are shown in figure 2. The alignment of a pair of $i_{13/2}$ neutrons (AB neutrons) takes

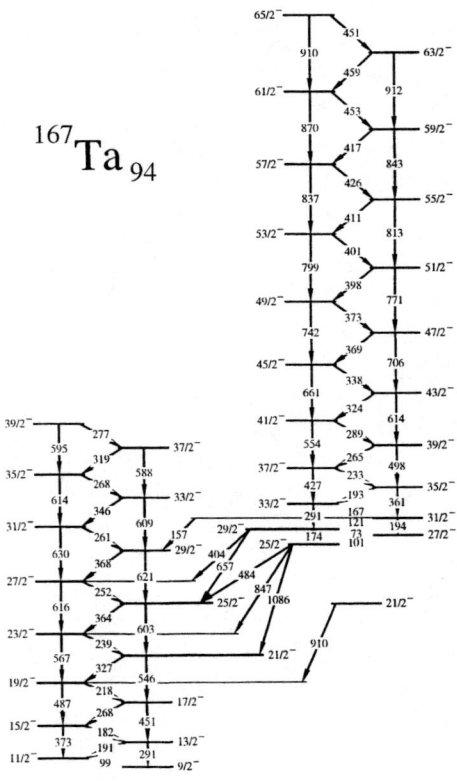

FIGURE 1. Partial level scheme of ^{167}Ta showing the $\frac{9}{2}^-$[514] decay sequence.

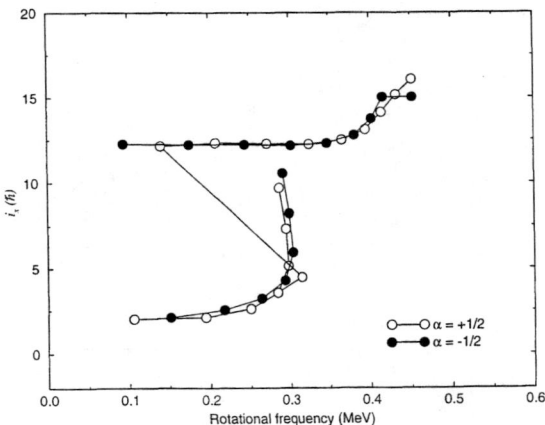

FIGURE 2. Alignment i_x for the $\frac{9}{2}^-[514]$ decay sequence of ^{167}Ta plotted as a function of rotational frequency. A reference band with Harris parameters $\mathcal{J}^{(0)} = 22.5$ MeV$^{-1}\hbar^2$ and $\mathcal{J}^{(1)} = 60$ MeV$^{-3}\hbar^4$ has been subtracted.

place at a rotational frequency of $\hbar\omega = 0.24$ MeV. At $\hbar\omega = 0.4$ MeV the alignment of a second pair of $i_{13/2}$ neutrons (CD) is observed. The extension of the low-spin decay sequence to $I = \frac{39}{2}$ (see figure 1) corresponds, for $I > \frac{23}{2}$, to a three-quasiparticle configuration which involves a pair of BC neutrons (band-crossing frequency $\hbar\omega_c = 0.30$ MeV).

Angle-dependent $\gamma - \gamma$ coincidence matrices were used in a DCO analysis for $E2$ and mixed $M1/E2$ transitions within the $h_{11/2}$ decay sequence. The degree of alignment of states (σ/I) as a function of total angular quantum number I was established from the analysis. σ is the standard deviation of the (assumed Gaussian) distribution of magnetic substates m about $m = 0$. From a knowledge of σ/I versus I and measurements of DCO ratios, it was possible to determine $M1/E2$ mixing ratios for the one- and three-quasiparticle configurations. Measured mixing ratios δ and branching ratios λ were used in a calculation of $B(M1)/B(E2)$ and Q_1/Q_2 ratios using the following expressions (where Q_1 and Q_2 are the transition quadrupole moments for the $\Delta I = 1$ and $\Delta I = 2$ transitions respectively):

$$\frac{B(M1; I \to I-1)}{B(E2; I \to I-2)} = 0.693 \frac{E_\gamma^5(I \to I-2)}{E_\gamma^3(I \to I-1)} \frac{1}{\lambda(1+\delta^2)} (\mu_N/eb)^2 \tag{1}$$

$$\left(\frac{Q_1}{Q_2}\right)^2 = \frac{E_\gamma^5(I \to I-2)}{E_\gamma^5(I \to I-1)} \frac{\delta^2}{\lambda(1+\delta^2)} \frac{<IK20|(I-2)K>^2}{<IK20|(I-1)K>^2}. \tag{2}$$

Figure 3 shows the measured ratios as a function of I. What is especially interesting about these results is the onset of large signature splitting in both ratios corresponding to the spin region where the BC neutrons are aligning. In contrast,

241

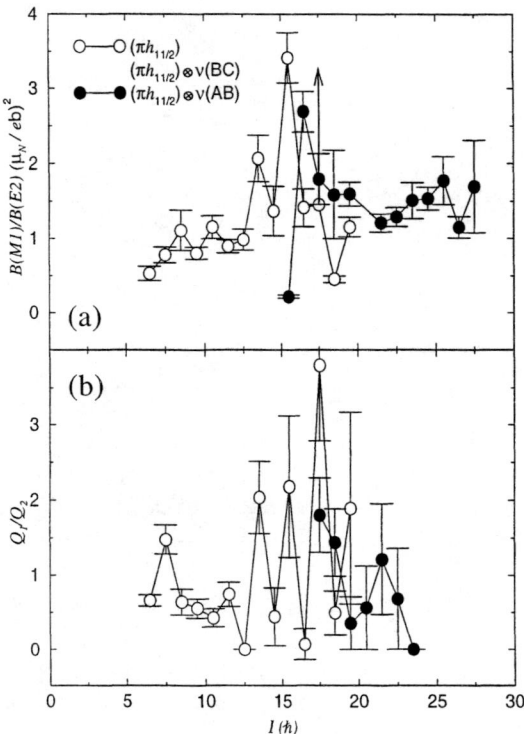

FIGURE 3. (a) $B(M1)/B(E2)$ ratios and (b) Q_1/Q_2 ratios for the $\frac{9}{2}^-[514]$ band in ^{167}Ta plotted as a function of spin I.

for the three-quasiparticle configuration involving AB neutrons there is no evidence of signature splitting. The alignment of a pair of $i_{13/2}$ neutrons (AB) is predicted, within the context of the cranked shell model, to drive the nucleus towards $\gamma = 0$. The cranked shell model predicts that BC neutron alignment should have a similar effect although, in this case, the driving tendency is less pronounced. Calculations based on the particle rotor model [2] appear to be unable to reproduce the observed magnitude of the splitting. The model was also unsuccessful in reproducing the significantly smaller signature-dependent effects in the $\pi h_{11/2}$ decay sequence of ^{157}Ho [3]. We interpret the strong signature-dependent effects as evidence for triaxiality in the three-quasiparticle configuration $\pi h_{11/2} \otimes \nu(\text{BC})$.

REFERENCES

1. K. Theine *et al*, Nucl. Phys. **A536** 418 (1992).
2. A. Bohr and B.R. Mottelson, Nuclear Structure, Vol. II (Benjamin, New York, 1975).
3. D.C. Radford *et al*, Nucl. Phys. **A545** 665 (1992).

Limiting Moments of Inertia in ^{178}W

D.M. Cullen, S.L. King, A.T. Reed, J.A. Sampson.

Oliver Lodge Laboratory, Department of Physics, University of Liverpool, Liverpool L69 7ZE, United Kingdom.

P.M. Walker, C. Wheldon, F. Xu.

Department of Physics, University of Surrey, Guildford GU2 5XH, United Kingdom.

G.D. Dracoulis.

Department of Nuclear Physics, RSPhysSE, Australian National University, Canberra, ACT 0200, Australia.

I.-Y. Lee, A.O. Macchiavelli, R.W. MacLeod.

Lawrence Berkeley National Laboratory, Berkeley, California 94720.

A.N. Wilson[1], C. Barton.

Wright Nuclear Structure Lab., Yale University, 272 Whitney Avenue, New Haven, CT 06511.

Abstract. This contribution focussed on studying nuclear pairing via its effects on the nuclear moment of inertia using K-isomeric states as a probe. The reason for studying the effects on K-isomeric states is that their underlying single-particle configuration can be uniquely defined through a measurement of the in-band to out-of-band branching ratios. The angular momentum limits to populating K-isomers were also addressed.

It has long been established that the reduced moments of inertia observed for rotational bands in deformed nuclei are a consequence of the nuclear pairing force [1–4]. This force is responsible for coupling pairs of nucleons together in an analogous manner to the paired electrons in the BCS theory of superconductors [5].

---

[1] Present Address: Department of Physics, University of York, Heslington, York, YO10 5DD, United Kingdom.

CP495, *Experimental Nuclear Physics in Europe*, edited by B. Rubio et al.
© 1999 American Institute of Physics 1-56396-907-6/99/$15.00

Recently, much evidence has been presented for the demise of nuclear pairing correlations on the basis of increasing angular momentum [6] where the Coriolis force destroys the pairing correlations gradually. In contrast, the demise of pairing correlations based on the number of unpaired nucleons (seniority) is not so well established. As the nuclear pairing force becomes quenched with increasing seniority the nuclear moment of inertia is observed [7] to increase toward that of a classical rigid body, although shell effects may be important [8]. The large number of high-K multi-quasiparticle states identified in ^{178}W [9,10] have offered a unique opportunity to study this effect and the highest-K rotational band, prior to this work, was the $K^\pi = 28^-$ eight-quasiparticle band in ^{178}W [9]. There are, however, predictions [11] that higher-K states should compete to form the yrast line in ^{178}W.

High-spin states were populated in ^{178}W with the ^{170}Er(^{13}C,5n) reaction. The 0.25-pnA, 86 MeV ^{13}C beam was supplied by the 88-inch cyclotron at the Lawrence Berkeley National Laboratory, USA and the data were recorded with the one hundred and two escape-suppressed germanium detectors of the GAMMASPHERE array. The natural beam pulsing frequency of 5.66 MHz resulted in a beam pulse every 177 ns. A thick target of ^{170}Er (enriched to 96.88%) with thickness 4.6mg/cm^2 was used to stop 9.7×10^{10} unpacked triple-coincident events in the focus of the array. The time coincidence information, with respect to the beam pulsing of the cyclotron, enabled various delayed-early (across isomer) histograms to be created. Several new states were observed in the present work and a full report is given in Ref. [12]. In particular, three new high-K multi-quasiparticle intrinsic states, $K^\pi = (29^+)$, $K^\pi = 30^+$ and $K^\pi = (34^+)$, based on ten, eight and ten, unpaired nucleons, respectively were observed. These states lie up to the limit of the angular momentum brought into the compound system in this reaction, and imply that we have not yet reached the angular momentum limits to K-isomeric states.

Figure 1 shows the total aligned angular momentum, I_x, versus rotational frequency, ω for the $K^\pi = 15^+$, $K^\pi = 25^+$, $K^\pi = 28^-$, $K^\pi = 30^+$ and the ground-state-band, $K^\pi = 0^+$, in ^{178}W. In the figure the kinematic and dynamic moments of inertia can easily be deduced by $\Im^{(1)} = \hbar I_x/\omega$ and $\Im^{(2)} = \hbar dI_x/d\omega$. The dynamic moment of inertia for the $K^\pi = 25^+$ band, $\Im^{(2)}=51\hbar^2$MeV^{-1}, is very much greater than that of the ground-state band of 29$\hbar^2$MeV^{-1}. However, this still only represents 60% of the rigid-body value of 85$\hbar^2$MeV^{-1} shown by the full line in Fig. 1. These observations may be reconciled with the fact that the ground-state band (below the first band crossing) has all particles fully paired and therefore, the lowest moment of inertia. In contrast, the eight quasiparticle bands have eight unpaired particles, which are expected to lead to a substantial erosion of the nuclear pairing force, and therefore, a larger moment of inertia. One important feature of this figure is that the $K^\pi = 30^+$ band is the only band which has equality in its dynamic and static moments of inertia, $\Im^{(1)}=55.8\hbar^2$MeV^{-1} and $\Im^{(2)}=55.9\hbar^2$MeV^{-1}. This is precisely the behaviour which would be expected for a rigid body. However, these values are still very much less than the 85$\hbar^2$MeV^{-1} rigid-body value. The $K^\pi = 30^+$ band, therefore, represents a better rotor than the $K^\pi = 25^+$ band, with no apparent alignment, but it still does not show a fully rigid moment of inertia

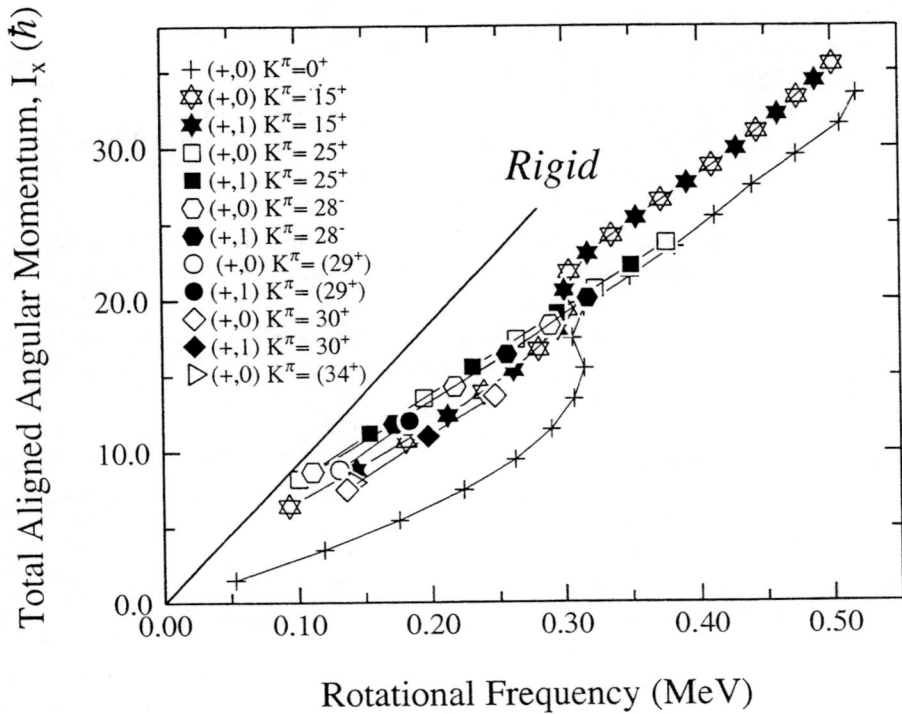

FIGURE 1. Total aligned angular momentum for the $K^\pi = 25^+$, $K^\pi = 28^-$, $K^\pi = 30^+$ and the ground-state band $K^\pi = 0^+$, in ^{178}W as a function of rotational frequency. The solid line represents a rigid rotor with a moment of inertia of 85 $\hbar^2$MeV^{-1}.

even with eight unpaired particles. This is consistent [7] with the substantial pairing energy obtained in Lipkin-Nogami calculations, see Ref. [12], with Δ_n=599 keV and Δ_p=707 keV.

In Ref. [7] Dracoulis *et al.*, argue that if pairing correlations persist for these high-seniority states then despite adding nucleons (not the $h_{9/2}$ proton) the apparent alignment decreases. The implication is that the fall in alignment, caused by the reduction of pairing, is greater than the effect of an underlying increase in the collective moment of inertia, which when using the reference appropriate to the low-seniority states would lead to an apparent increase in alignment. For these new data in the present work the alignments of the 10 quasiparticle $K^\pi = (29^+)$ and $K^\pi = (34^+)$ bands can be compared with their respective underlying configurations, the $K^\pi = 30^+$ and $K^\pi = 25^+$ bands coupled to an additional 2 neutrons ($7/2^-[503]\otimes 1/2^-[521]$) in each case. This comparison is shown in Fig. 1. It can be seen that the alignment indeed reduces for the $K^\pi = (29^+)$ band compared with the $K^\pi = 25^+$ band which seems to indicate that pairing correlations are still having an effect, at least at the 8-quasiparticle level. However, the alignment apparently does not reduce in the $K^\pi = (34^+)$ band (for which there is only a single data point) compared with the $K^\pi = 30^+$ band.

One interpretation of this is that the behaviour of the $K^\pi = (34^+)$ band is just that expected if the pairing is indeed already quenched in the $K^\pi = 30^+$ band, as implied by the equality of its dynamic and static moments of inertia (and no net alignment). The further implication is that those moments-of-inertia (close to $56\ \hbar^2\text{MeV}^{-1}$) correspond to the *underlying unpaired moment of inertia* for ^{178}W. The blocking of two extra neutrons to form the $K^\pi = (34^+)$ configuration from the $K^\pi = 30^+$ configuration can then have no additional effect on the pairing, and therefore on the apparent moment of inertia. The absence in the present results of rotational transitions above the first member of the $K^\pi = (34^+)$ band means, unfortunately, that the dynamic moment of inertia is not defined. A crucial test of this proposition would be provided by such higher data points, which if the proposition is valid, should define the same slope for the $K^\pi = (34^+)$ and $K^\pi = 30^+$ bands in the I_x versus $\hbar\omega$ curves of Fig. 1.

As stated previously, shell effects have been identified [8] as a mechanism which would produce moments of inertia different from the classical rigid-body values, even when pairing is quenched. Furthermore, zero-pairing tilted-axis-cranking calculations, of the type described by Frauendorf [13] have been carried out [14] for the $K^\pi = 30^+$ band in ^{178}W. These calculations are successful in reproducing $\Im^{(1)} = \Im^{(2)} = 56\ \hbar^2\text{MeV}^{-1}$, and lend support to the interpretation proposed here.

REFERENCES

1. A. Bohr, B.R. Mottelson, D. Pines, *Phys. Rev.* **111**, 936 (1958).
2. S.T. Belyaev, *Mat. Fys. Medd. Dan. Vid. Selsk.* **31**, No. 11 (1959).
3. S.T. Belyaev, *Nucl. Phys.* **24**, 322 (1961).
4. A.B. Migdal, *Nucl. Phys.* **13**, 655 (1959).
5. J. Bardeen, L.N. Cooper, J.R. Schrieffer, *Phys. Rev.* **108**, 1175 (1957).
6. Y. Shimizu *et al.*, *Rev. Mod. Phys.* **61**, 131 (1989).
7. G.D. Dracoulis, F.G. Kondev, P.M. Walker, *Phys. Lett. B* **419**, 7 (1998).
8. V.V. Pashkevich and S. Frauendorf, *Sov. J. Nucl. Phys.* **20**, 588 (1975).
9. C.S. Purry, P.M. Walker, G.D. Dracoulis, T. Kibédi, S. Bayer, A.M. Bruce, A.P. Byrne, M. Dasgupta, W. Gelletly, F.G. Kondev, P.H. Regan and C. Thwaites, *Phys. Rev. Lett.* **75**, 406 (1995).
10. C.S. Purry, P.M. Walker, G.D. Dracoulis, T. Kibédi, F.G. Kondev, S. Bayer, A.M. Bruce, A.P. Byrne, W. Gelletly, P.H. Regan, C. Thwaites, O. Burglin, and N. Rowley, *Nucl. Phys.* **A632**, 229 (1998).
11. K. Jain, O. Burglin, G.D. Dracoulis, B. Fabricius, N. Rowley, and P.M. Walker, *Nucl. Phys.* **A591**, 61 (1995).
12. D.M. Cullen *et al.*, Submitted to *Phys. Rev. C*, June 1999.
13. S. Frauendorf, in *Proceedings of the Workshop on Gammasphere physics*, Berkeley, California, 1995, edited by M.A. Deleplanque, I.Y. Lee and A.O. Macchiavelli (World Scientific, Singapore 1996).
14. S. Frauendorf, (private communication).

Rotational Bands in Odd-Odd ^{180}Ir

Y.H. Zhang[1,2], T. Hayakawa[1], M. Oshima[1], J. Katakura[1], Y. Hatsukawa[1],
M. Matsuda[1], H. Kusakari[3], M. Sugawara[4] and T. Komatsubara[5]

[1] Japan Atomic Energy Research Institute (JAERI) , Tokai, Ibaraki 319-1195, Japan
[2] Institute of Modern Physics, Chinese Academy of Sciences, Lanzhou 730000, P.R.China
[3] Chiba University, Inage-ku, Chiba 263-8512, Japan
[4] Chiba Institute of Technology, Narashino, Chiba 275-0023, Japan
[5] Institute of Physics and Tandem Accelerator Center, University of Tsukuba, Ibaraki 305-0006, Japan

In a deformed odd-odd nucleus, a variety of coupling modes between the valence proton and neutron will result in a richness of band structures at moderate and high spins. In order to extend our knowledge about the band structures of odd-odd nuclei, the high-spin states in odd-odd ^{180}Ir have been investigated in our laboratory through in-beam γ-ray spectroscopy via ^{154}Sm(^{31}P,5nγ) ^{180}Ir reaction.

The ^{31}P beam was provided by the JAERI tandem accelerator. The target is an enriched ^{154}Sm metallic foil of 2 mg/cm^2 thickness with 5 mg/cm^2 Au backing. The beam energy of 160 MeV was used during X-γ and γ-γ coincidence measurements. A γ-ray detector array [1] including 12 HPGe's with BGO anti-Compton shields was used. DCO analysis has been made to extract multipolarities of γ-transitions. A level scheme of ^{180}Ir consisting of four new rotational bands has been constructed and shown in Fig. 1. The assignment of these bands to ^{180}Ir is supported by the measurements of γ-ray excitation functions, K X-γ coincidences, and the knowledge of high-spin data of neighboring ^{179}Ir and ^{181}Ir.

The band properties such as rotation alignments, band crossing frequencies, signature splitting, and intra-band B(M1)/B(E2) ratios have been analyzed. These information together with the knowledge of band structures in this mass region and the classification of coupling scheme [2] are used to propose the quasiparticle configurations. It is well known that the nuclei in the lighter Re-Os-Ir mass region are soft and their shapes can depend on the intrinsic quasiparticle configurations. For example, the π5/2$^+$[402] and ν9/2$^-$[514] bands show a gradual increase in alignment at low frequencies with respect to a reference derived from the π1/2$^-$[541] bands [3]. This low-spin anomaly in alignment has been interpreted in terms of β-stretching and/or shape co-existence associated with an $(h_{9/2})^2$ proton pair excitation [3,4]. Thus the intruder orbital π1/2$^-$[541] plays an important role in this β-stretching process. Indeed, the three observed band in ^{180}Ir show normal alignment (roughly constant alignment) at lower frequencies when π1/2$^-$[541] orbital is involved. However, the alignment for the π9/2$^-$[514]+ν7/2$^+$[633] band is increased from 4 $\hbar$ at 0.06 MeV to 11 $\hbar$ at 0.35MeV; this phenomenon is very similar to the behavior of π9/2$^-$[514] band in ^{179}Ir. Total

CP495, Experimental Nuclear Physics in Europe, edited by B. Rubio et al.
© 1999 American Institute of Physics 1-56396-907-6/99/$15.00

routhian surface (TRS) calculations by Bengtsson [4] for the even-even Os isotopes predict that this β-stretching process is very strong for N=100, 102. Therefore, the similar behavior may be observed in the $\pi 9/2^-[514]+vi_{13/2}$ bands of ^{178,180}Ir.

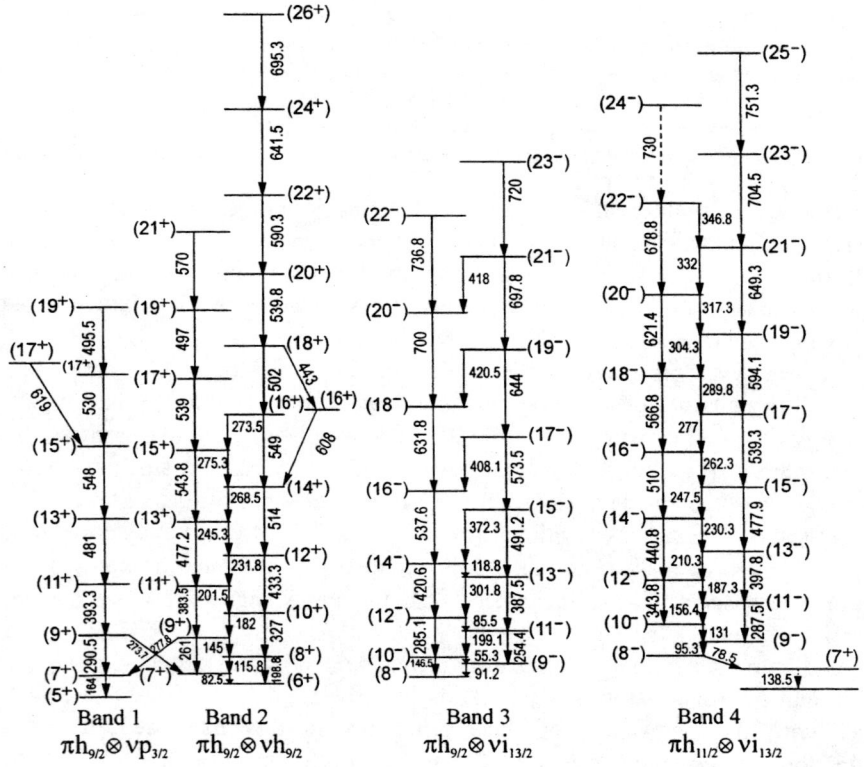

Fig. 1 Proposed level scheme of ^{180}Ir deduced from the present work

REFERENCES

[1] K. Furuno et al., Nucl. Instr. and Meth. A 421 (1999) 211
[2] A. J. Kreiner et al., Phys. Rev. C 36 (1987) 2306
[3] R. Bark, J. Phys. G:17 (1991) 1209
[4] R. Bengtsson, Nucl. Phys. A520 (1990) 201c, and in Proc. of Int. Conf. on High spin physics and gamma-soft nuclei, Pittsburgh, 1990, edited by J. X. Saladin, R. A. Sorensen, and C.M. Vincent (World Scientific, Singapore, 1990), p. 289

Study of Shape Coexistence in the very Neutron-Deficient Nucleus ^{176}Hg

M. Muikku, J.F.C. Cocks, K. Helariutta, P.T. Greenlees, P.M. Jones, R. Julin, S. Juutinen, H. Kankaanpää, H. Kettunen, P. Kuusiniemi, M. Leino, P. Rahkila, A. Savelius, W. Trzaska and J. Uusitalo

University of Jyväskylä, Department of Physics, P.O. Box 35, FIN-40351 Jyväskylä, Finland

Abstract. The ground-state yrast band in ^{176}Hg has been observed up to I = 10$\hbar$ by using the recoil decay tagging (RDT) method. The irregularity of this band indicates that the prolate intruder band, seen in the heavier Hg isotopes near the neutron mid-shell, crosses the nearly spherical ground-state band of ^{176}Hg above I = 6$\hbar$.

The even-mass Hg isotopes with N > 100 are weakly oblate in their ground states. The properties of the ground state band remain rather constant with decreasing neutron number until in ^{188}Hg, where the band is crossed by an intruding prolate deformed band (1,2). The prolate states minimize their energies in ^{182}Hg (3), but they still lie above the ground state (4). The very light even-mass Hg nuclei (N $\leq$ 96) are predicted to be spherical in their ground states. In these nuclei the prolate structure should disappear and give away to a superdeformed structure (2).

In the present work γ-ray transitions in ^{176}Hg have been studied to confirm the tentative assignments of ref. (5) and to probe further the yrast line towards higher spin. Prompt γ rays from ^{176}Hg were resolved from those arising from the dominant background of fission and other reaction products using RDT method (6, 7). The experiment was carried out at the Accelerator Laboratory of the University of Jyväskylä. Excited states in ^{176}Hg were populated via the ^{144}Sm (^{36}Ar, 4n) fusion evaporation channel with a beam energy of 190 MeV and a production cross-section of about 5 μb. Prompt γ rays were detected by the JUROSPHERE array employed in conjunction with the gas-filled recoil separator RITU (8). The events corresponding to the observation of a recoil together with a subsequent α decay at the same position in the Si detector at the RITU focal plane within a maximum time interval of 100 ms were selected in the data analysis (Fig. 1a). In order to construct the level scheme, recoil-gated α-tagged γ-γ coincidence data were required (Fig. 1b).

Our results confirm the earlier tentative assignments by Carpenter et al. (5) up to the 6$^+$ level. The relatively large spacings observed between these three lowest states indicate that the oblate ground state becomes less deformed in very neutron deficient

CP495, *Experimental Nuclear Physics in Europe*, edited by B. Rubio et al.
© 1999 American Institute of Physics 1-56396-907-6/99/$15.00

even-mass Hg nuclei. The energy spacings above the 6^+ state, seen for the first time, reveal that the prolate minimum has not disappeared. The similarity between the observed prolate intruder bands in the even mass Pt, Hg and Pb isotopes close to the neutron mid-shell is well known (9). The comparison of the kinematic moments of inertia derived from the experimental yrast-level energies for the Hg and Pt isotones with N = 96 (10, 11) also supports the idea of the prolate intruder band crossing the ground-state band above I = $6\hbar$. A branch of near-yrast states has also been tentatively observed. These states could be negative-parity states similar to those seen in even-mass Hg isotopes with A ≥ 186 (12-14). Due to the intruding prolate bands these negative-parity states in Hg isotopes close to the neutron mid-shell lie higher above the yrast line and are therefore not observed.

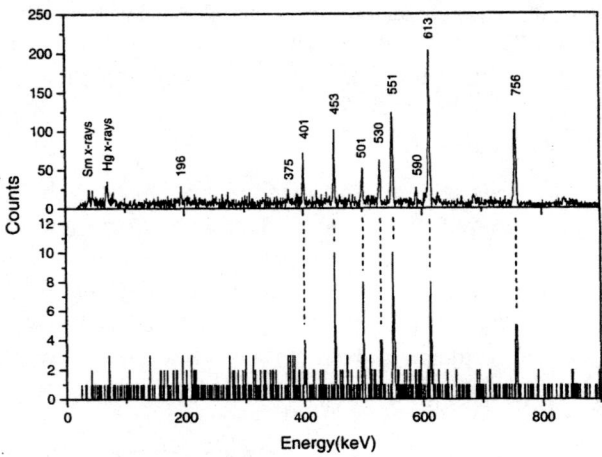

FIGURE 1. a) γ-ray spectrum obtained by gating with fusion evaporation residues and tagging with ^{176}Hg α decays b) Sum of the recoil-gated and α-tagged γ-γ coincidence spectra gated on the seven strongest transitions in the spectrum (b).

REFERENCES

1. Wood, J.L. et al., *Phys. Rep.* **251**, 101 (1992)
2. Nazarewicz, W., *Phys. Lett. B* **305**, 195 (1993)
3. Dracoulis, G.D. et al., *Phys. Lett. B* **208**, 365 (1988)
4. Bindra, K.S. et al., *Phys. Rev C* **51**, 401 (1995)
5. Carpenter, M.P. et al., *Phys. Rev. Lett.* **78**, 3650 (1997)
6. Simon, R.S. et al., *Z. Phys.* **A325**, 197 (1986)
7. Paul, E.S. et al., *Phys. Rev. C* **51**, 78 (1995)
8. Leino, M. et al., *Nucl. Instrum. Meth. B* **99**, 653 (1995)
9. Dracoulis, G.D. et al., *Phys. Rev. C* **49**, 3324 (1994)
10. Muikku, M. et al., *Phys. Rev. C* **58**, R3033 (1998)
11. Dracoulis, G.D. et al., *Phys. Rev. C* **44**, 1246 (1991)
12. Ma, W.C et al., *Phys. Rev. C* **47**, R5 (1993)
13. Hannachi, F. et al., *Nucl. Phys.* **A481**, 135 (1988)
14. Hübel, H. et al., *Nucl. Phys.* **A453**, 316 (1986)

Shape Coexistence in Neutron Deficient Po Nuclei

K. Helariutta[1], J.F.C. Cocks[1], T. Enqvist[1], P.T. Greenlees[1], P. Jones[1], R. Julin[1], S. Juutinen[1], P. Jämsen[1], H. Kankaanpää[1], H. Kettunen[1], P. Kuusiniemi[1], M. Leino[1], M. Muikku[1], M. Piiparinen[1], P. Rahkila[1], A. Savelius[1], W.H. Trzaska[1], S. Törmänen[1], J. Uusitalo[1], R.G. Allatt[2], P.A. Butler[2], R.D. Page[2], M. Kapusta[3]

[1] *Department of Physics, University of Jyväskylä, Finland*
[2] *Oliver Lodge Laboratory, University of Liverpool, England*
[3] *Department of Nuclear Electronics, Soltan Institute for Nuclear Physics, Poland*

Abstract. The excited levels in $^{192-195}$Po have been studied using the recoil-decay tagging method. New levels have been identified. The data are in accordance with the scheme of the coexisting spherical and deformed intruder structures crossing each other with N < 112.

The excited levels in even neutron-deficient polonium nuclei until N=108 have been studied using both in-beam γ-ray and α decay methods. In the resulting energy level systematics, the yrast positive parity level energies stay relatively constant until at neutron number N=114 (^{198}Po) the energies start to decrease. The lowering of the energy levels is regarded as a sign of increased collectivity. Two different explanations has been introduced for the origins of the collectivity in polonium nuclei. According to the first one (1,2) the light polonium nuclei are quadrupole vibrators developing towards more collective anharmonic vibrator due to the opening of the $i_{13/2}$ neutron shell. In the second explanation (3,4) polonium level structure is formed by the states of two coexisting and mixing shapes, almost spherical normal shape and oblate deformed intruder shape. The intruder states are explained to be formed by the proton particle-hole excitations across the Z=82 closed shell.

In order to shed light on the ambiguities concerning the interpretation of the energy levels of light Po nuclei we have performed a study of light odd- and even-mass Po nuclei with N = 108 – 111 (5). The nuclei were produced in several experiments via fusion-evaporation reactions. Due to strong fission competition, the fusion-evaporation reaction channels for populating the light Po nuclei become very weak. Therefore, recoil-decay tagging (6,7) and recoil gating methods were used to resolve the events of interest from the vast background.

CP495, *Experimental Nuclear Physics in Europe*, edited by B. Rubio et al.
© 1999 American Institute of Physics 1-56396-907-6/99/$15.00

The yrast band in ^{192}Po and in ^{194}Po with regularly increasing level spacings up to the 10^+ state can be discussed within the framework of a soft rotor described with a variable moment of inertia (VMI). Similarity between the prolate intruder bands in the even-mass Pt, Hg and Pb isotopes close to the neutron midshell is well known (8). The $J^{(1)}$ values of ^{192,194}Po have been compared with the ones of the intruder bands observed in Pb, Hg and Pt. A difference of about 10 $\hbar^2$/MeV in the $J^{(1)}$ values of the Po nuclei and the prolate bands is observed, revealing a difference in the structure. On the other hand, the $J^{(1)}$ values for the yrast band of ^{198}Rn (9) seem to have structures similar to those in ^{192,194}Po. The structures observed in ^{192,194}Po and ^{198}Rn can represent the oblate minimum which according to calculations (10) should reach the ground state in ^{192}Po. At low spins, irregularities in the smooth behavior of $J^{(1)}$ are seen in ^{192,194}Po. These are supposed to be due to crossing and mixing of two different coexisting structures. Simple two-level mixing calculation reveals that the deformed structure which in ^{194}Po is still slightly above the spherical one has become the ground-state structure in ^{192}Po.

REFERENCES

1. Bernstein, L.A. et al., *Phys. Rev.* **C 52**, 621-627 (1995).
2. Younes, W., Cizewski, J.A., *Phys Rev.* **C 55**, 1218-1226 (1997).
3. Alber, D. et al., *Z. Phys.* **A 339**, 225-229 (1991).
4. Wauters, J. et al., *Phys. Rev.* **C 47**, 1447-1454 (1993).
5. Helariutta, K. et al., *Phys. Rev.* **C 54**, R2799-R2801 (1996) and Helariutta, K. et al., *submitted to Eur. Phys. J.*
6. Simon, R.S. et al., *Z. Phys.* **A 325**, 197-202 (1986).
7. Paul, E. et al., *Phys. Rev.* **C 51**, 78-87 (1995).
8. Dracoulis, G.D., *Phys. Rev.* **C 49**, 3324-3327 (1994).
9. Taylor, R. et al., *Phys. Rev.* **C 59**, 673-681 (1999).
10. May, F.R. et al., *Phys. Lett.* **B 68**, 113-116 (1977).

Indication of Triple-Shape Coexistence in ^{188}Pb

Y. Le Coz[1], F. Becker[1], H. Kankaanpää[2], W. Korten[1], E. Mergel[3],
P.A. Butler[4], J.F. Cocks[2], O. Dorvaux[2], D. Hawcroft[4],
K. Helariutta[2], R.D. Herzberg[4], M. Houry[1], H. Hübel[3], P. Jones[2],
R. Julin[2], S. Juutinen[2], H. Kettunen[2], P. Kuusiniemi[2], M. Leino[2],
R. Lucas[1], M. Muikku[2], P. Nieminen[2], P. Rahkila[2], D. Rossbach[3],
A. Savelius[2] and Ch. Theisen[1]

[1] DAPNIA/SPhN, CEA Saclay, F-91191 Gif-sur-Yvette Cedex, France
[2] JYFL, University of Jyväskylä, Finland
[3] ISKP, Universität Bonn, Nussallee 14-16, D-53115 Bonn, Germany
[4] Dept. of Physics, University of Liverpool, United Kingdom

Abstract. By using conversion electron and γ-ray spectroscopy combined with a recoil tagging method, excited states in ^{188}Pb has been investigated. The existence of two excited low-lying 0^+ states have been confirmed giving a new energy for the 0_3^+ state, which are interpreted as evidence for a triple-shape coexistence in ^{188}Pb.

Nuclei around the Z=82 closed shell provide some of the best examples for the understanding of shape coexistence. In heavy lead isotopes (with N≥106) oblate-deformed structures have been observed at low spin coexisting with spherical ground states while in light ones (down to ^{184}Pb) a coexistence between prolate deformed bands and spherical ground states occurs [1]. The aim of the present experiment was to observe the predicted triple spherical-oblate-prolate shape coexistence in ^{188}Pb with electron-γ spectroscopy.

The experiment was performed at the Jyväskylä laboratory. With the gas-filled separator RITU, evaporation residues from fusion reactions were selected from the overwhelming background of fission products, while 4 Germanium clovers and the ICEMOS Mini-Orange spectrometers were used for in-beam gamma-ray and conversion-electron spectroscopy, respectively. Using the heavy-ion fusion reaction ^{36}Ar+^{156}Gd $\rightarrow$ ^{188}Pb+4n (at 172 MeV), gated by any evaporation residue with RITU, known excited states in ^{188}Pb have been observed up to spin 16$^+$ [2]. The recoil-gated conversion electron spectrum (see Fig. 1) shows two strong lines, corresponding to transition energies of 591(2) and 725(2) keV after correction for

CP495, *Experimental Nuclear Physics in Europe*, edited by B. Rubio et al.
© 1999 American Institute of Physics 1-56396-907-6/99/$15.00

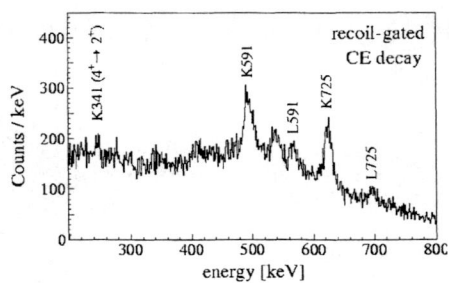

FIGURE 1. Recoil-gated conversion electron (CE) spectrum without Doppler and binding energy corrections; the K and L index stands for the atomic shell from which the CE's originate.

Doppler effect and for a electron binding energy of ~ 88 keV (assuming that these conversion electrons come from the lead atomic K shell). The E0 character of those transitions and then spin and parity of depopulating states, e.g. 0^+ states, have been established [3]. The 591 keV state was previously observed by fine structure measurements in α decay spectroscopy of ^{192}Po [4,5] and interpreted as an oblate state. The 0_3^+ state was also proposed, but at a different energy (767 keV) [5]. The extrapolation of the rotational band (which have been interpreted as prolate deformed due to moment of inertia and systematics considerations [2]) to spin 0 for the band head gives 710 keV. This is in good agreement with the new energy of the 0_3^+ state and suggest a prolate configuration. In addition, the levels systematic of low-spin states in even-even lead isotopes, and also the predicted shape coexistence in ^{188}Pb [7], support this assumption.

In summary, we have observed by electron spectroscopy two excited 0^+ states in ^{188}Pb depopulating via strong E0 transitions. Spin and parity of these states have been therefore unambiguously assigned. These two excited 0^+ states in addition to the spherical ground states are evidence for a triple-shape coexistence. To prove such a shape coexistence, it would be important to identify (i) the non yrast rotational oblate band, and to measure (ii) the collectivity of these bands in a Recoil Distance Method (RDM) experiment.

REFERENCES

1. J.F.C. Cocks et al., EPJ A3, 17-20 (1998) and references therein.
2. J. Heese et al., Phys. Lett. B302, 390-395 (1993).
3. Y. Le Coz et al., submitted to Eur. Phys. J.
4. N. Bijnens et al., Z. Phys. A356, 3-4 (1996).
5. R.G. Allatt et al., Phys. Lett. B437, 29 (1998).
6. A.N. Andreyev et al., J. Phys. G: Nucl. Part. Phys. 25 (1999).
7. W. Nazarewicz, Phys. Lett. B305, 195 (1993)

First Observation of Excited States in [199]At with the Recoil Filter Detector

J. Styczeń[1], W. Męczyński[1], M. Lach[1], P. Bednarczyk[1], R. Chapman[2],
S. Courtin[3], J. Grębosz[1], F. Hannachi[4], P.M. Jones[5], J. Kownacki[6],
A. Lopez-Martens[3], K.H. Maier[7], J.C. Merdinger[3], D. Middleton[2],
M. Palacz[6], N. Schulz[3], M.B. Smith[2], K.-M Spohr[2], M. Wolińska[6],
M. Ziębliński[1]

[1]The Niewodniczański Institute of Nuclear Physics, Kraków, Poland, [2]University of Paisley, Paisley,
Scotland, U. K., [3]Institut de Recherches Subatomiques, Strasbourg, France, [4]Centre de Spectrométrie
Nucléaire et de Spectrométrie de Masse, Orsay, France, [5]University of Jyväskylä, Jyväskylä, Finland,
[6]Heavy Ion Laboratory, University of Warsaw, Warszawa, Poland, [7]Hahn-Meitner Institut, Berlin,
Germany

Abstract. We have deduced for the first time a level scheme for the [199]At nucleus. The angular
distribution data and the conversion electron information have allowed for multipolarity
assignments and spin suggestions to be made. The current data and the level systematics of odd
At isotopes suggest that the first excited state should be isomeric with a lifetime of the order of a
microsecond.

Theoretical predictions [1] point out that a region of stable ground-state
deformation should exist above the $Z = 82$ proton and below the $N = 126$ neutron
shells in nuclei very far from the valley of stability. For Astatine isotopes ($Z = 85$) the
collective features should be pronounced in very light nuclei, and be already strong in
[198]At which is predicted to have a deformed ground state with $|\beta_2| = 0.20$ [2]. No
deformation has been identified experimentally in heavier At nuclei, above A=200 [3].
The intermediate [199]At nucleus should then be a "transition region" nucleus with an
expected small deformation of about $\beta_2 = 0.08$.

Before this work started, nothing was known about its level scheme. The first
hints of some gamma transitions and a very preliminary level scheme we have
presented in the recent work by Spohr [4]. Partially, the results given in the present
contribution were published elsewhere [5]. Moreover, some of the γ-lines have been
also assigned, very recently, to the [199]At nucleus using the RITU separator [6].

In the present experiment, a 141 MeV [28]Si pulsed beam with 430 ns repetition
time from the VIVITRON accelerator (IReS Strasbourg) bombarded a 0.7 mg/cm^2
thick [175]Lu target. Gamma-rays were detected by 14 Compton-suppressed HPGe-
detectors and 1 LEPS, of the GAREL+ set-up, in coincidence with the evaporation
residues separated by the Recoil Filter Detector (RFD) [7].

CP495, *Experimental Nuclear Physics in Europe*, edited by B. Rubio et al.
© 1999 American Institute of Physics 1-56396-907-6/99/$15.00

The γ-recoil coincidence requirement very strongly suppressed the high fission background and gave clearly resolved [199]At and [199]Po γ-lines, invisible in the γγ-projection [5].

The first excited state in [199]At (Fig. 1) is about 200 keV lower than in the more neutron-rich [201-207]At isotopes. And, moreover, we anticipate it to be a long-lived isomeric state with a lifetime of the order of a microsecond since its decay was not observed. The use of the RFD with the necessary thin target makes the measurement of such long lifetimes impossible.

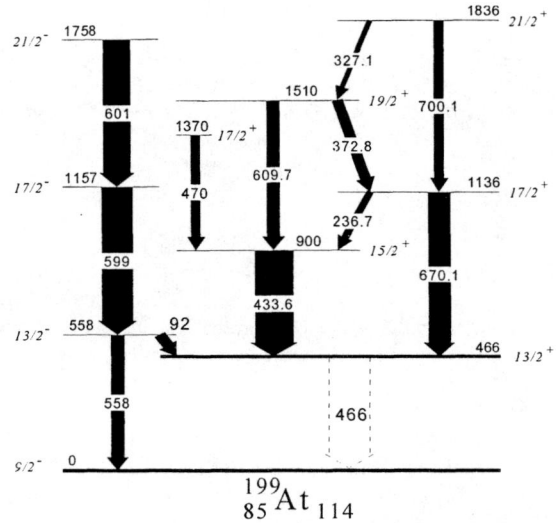

FIGURE 1. The level scheme of low lying states in [199]At

Due to the high selectivity property of the RFD, the recoil-gated singles gamma spectrum becomes a very clean spectrum permitting the identification of weak γ-ray lines. Moreover, the RFD selects the recoil nuclei at a very definite forward direction, therefore, an angular distribution analysis of the de-exciting gamma rays becomes possible. An example is shown in Fig. 2. It is seen that even with the restricted range of the Ge- detector angles, unambiguous multipolarities are possible for the most of the observed γ-rays. The upper-most line corresponds to the 399 keV, E2 transition in [199]Po and is shown for comparison.

The γ-ray intensity ratios in the γγ-recoil-coincidence spectra and the angular distributions of 434, 237 and 670, 700 keV transitions (Fig. 2) suggest their M1, M1/E2, and E2 character, respectively. The 92 keV γ-transition is strongly observed in the coincidence spectrum, thus suggesting its E1 character. The proposed level scheme is based on our experimental data and on the level systematics of the odd-A At isotopes.

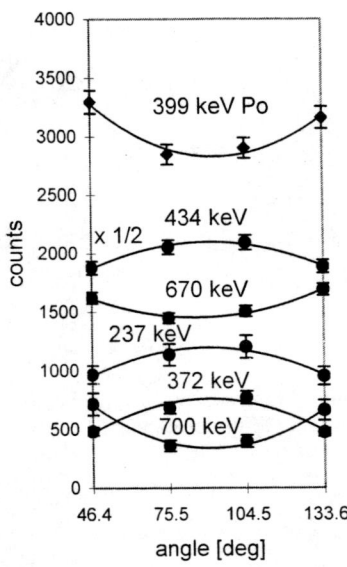

FIGURE 2. The angular distributions of γ-rays of ^{199}At.

At present, it is rather speculative to suggest that there is experimental evidence for the onset of collectivity in this nucleus. The presence of the E2-E2 cascade of 670 and 700 keV above the suggested isomeric state might, however, be considered as a weak evidence for. Nonetheless, the irregular level structure favours, at most, only a rather small deformation, as expected.

The search for the anticipated isomeric state is in progress.

This work was partially supported by the KBN under grant no. *2P03 045 16* and the Polish-French agreement under the *JUMELAGE* and *IN2P3 Conventions.*

References:

1. P. Möller, J.R. Nix, W.D. Myers and W.J. Swiatecki, *At. Data Nucl. Data Tables* **59**, 185 (1995)
2. M.B. Smith *et al., Eur. Phys. J.* A **5**, 43 (1999), see also these proceedings
3. K. Dybdal *et al.,* Phys. Rev. C **28**, 1171 (1983)
4. K. -M. Spohr, *Ph.D. thesis*, Institut für Kernphysik, Jülich-3171, Germany, Report No. ISBN-0944-2952, 1996
5. W. Męczyński *et al., Eur. Phys. J.* A **3**, 31 (1998)
6. R.B.E. Taylor *et al., Phys Rev.* C **59**, 673 (1999)
7. W. Męczyński *et al., Report* No. 1782/PL, 1998, The Niewodniczański Institute of Nuclear Physics, 31-342 Kraków, Poland.

The Onset of Deformation in Neutron-deficient At Nuclei

M.B. Smith*, R. Chapman*, J.F.C. Cocks[†], O. Dorvaux[†],
K. Helariutta[†], P.M. Jones[†], R. Julin[†], S. Juutinen[†],
H. Kankaanpää[†], H. Kettunen[†], P. Kuusiniemi[†], Y. Le Coz[‡],
M. Leino[†], D.J. Middleton*, M. Muikku[†], P. Nieminen[†], P. Rahkila[†],
A. Savelius[†] and K.-M. Spohr*

*Department of Physics, University of Paisley, Paisley PA1 2BE, UK
[†]Department of Physics, University of Jyväskylä, FIN–40351 Jyväskylä, Finland
[‡]DAPNIA/SPhN CEA–Saclay, France

Abstract. Excited states in the [197]At nucleus have been identified for the first time using the recoil-decay-tagging technique. The excitation energy of these states is found to be consistent with the systematics of neutron-deficient At nuclei and with calculations indicating that the nucleus may be deformed in its ground state. A more recent experiment, to study states in [195]At, is discussed.

Theoretical predictions for nuclei far from stability have suggested an extended region of stable ground-state deformation in nuclei above the $Z = 82$ shell gap and below the closure at $N = 126$ (e.g. [1]). Experimentally, this region is largely uninvestigated and little information about the predicted deformed shapes exists. Direct evidence for such deformation would be a rotational sequence of levels based on the ground state, accompanied by significant lowering of the first excited state relative to more neutron-rich spherical nuclei. For At nuclei the onset of deformation is predicted to occur sharply between [199]At and [198]At, with a change in the deformation from $\beta_2 = 0.08$ (near-spherical) to $\beta_2 = -0.20$ (oblate-deformed). Recent work on [199]At [2,3] has shown that the first excited state is lower than for the heavier At nuclei, indicating that [199]At may be deformed. It is this observation that has motivated our study of the [197]At nucleus. A more detailed report on these results can be found in reference [4].

States in [197]At were populated using the ^{165}Ho(^{36}Ar, 4n) reaction at a bombarding energy of 178 MeV. The beam was provided by the $K = 130$ MeV cyclotron at the Accelerator Laboratory of the University of Jyväskylä, Finland. Prompt γ-rays were detected using the SARI array and four TESSA-type Ge detectors were used to detect delayed γ-rays. Recoiling fusion-evaporation products were separated in-

CP495, *Experimental Nuclear Physics in Europe*, edited by B. Rubio et al.
© 1999 American Institute of Physics 1-56396-907-6/99/$15.00

flight using the RITU gas-filled recoil separator and the recoil-decay-tagging (RDT) method was used to select γ-rays associated with ^{197}At. The known α-decay characteristics of ^{197}At [5] were exploited. Two transitions, one prompt ($E_\gamma = 324$ keV) and one delayed ($E_\gamma = 311$ keV, $\tau = 8\pm2$ μs), have been correlated with the ground state α-decay of ^{197}At. Based on energy-level systematics, including the ^{199}At level scheme of reference [2], we originally suggested that these γ-rays correspond to $J^\pi = (\frac{11}{2}^-) \rightarrow \frac{9}{2}^-$ and $J^\pi = (\frac{13}{2}^+) \rightarrow \frac{9}{2}^-$ transitions [4]. However, the new ^{199}At level scheme [3], in which the isomeric $J^\pi = (\frac{13}{2}^+) \rightarrow \frac{9}{2}^-$ transition is proposed, casts considerable doubt on the spin assignment for the prompt transition in ^{197}At. Based on the new level scheme of ^{199}At, it is thought most likely that this γ-ray originates from a state with $J^\pi = \frac{13}{2}^-, \frac{15}{2}^+$ or $\frac{17}{2}^+$. For any of these initial spins, the corresponding state is considerably lower in energy than the same state in heavier At nuclei. The excitation energy of the $(\frac{13}{2}^+)$ state is also seen to decrease significantly for ^{197}At. By comparison with neighbouring even-even Pb and Po isotopes, we believe that the decrease in excitation energy of these levels is associated with an increase in collectivity and an onset of ground-state deformation, in agreement with theoretical predictions.

A further experiment to study excited states in ^{195}At has been performed very recently, again using the RDT method at Jyväskylä. In ^{197}At a weakly-populated low-lying $\frac{1}{2}^+$ intruder state has been observed [4,5]. Recent α-decay studies at Jyväskylä have revealed that this state becomes the ground state in ^{195}At and that the $\frac{9}{2}^-$ level is observed as an excited state [6]. Both the $\frac{9}{2}^-$ and $\frac{1}{2}^+$ states are believed, from the α-decay results, to correspond to oblate-deformed minima in the potential energy surface. Observation of transitions feeding these states, which are populated with comparable intensities, will provide important information concerning the shape of the ^{195}At nucleus. The analysis of these data is in progress.

The SARI project is jointly funded by EPSRC (UK) and CEA/DSM/DAPNIA (France). Support for this work was provided by EPSRC, the Academy of Finland and the Access to Large-scale Facilities programme under the Training and Mobility of Researchers programme of the European Union.

REFERENCES

1. P. Möller, J.R. Nix, W.D. Myers and W.J. Swiatecki, At. Data Nucl. Data Tables **59**, 185 (1995).
2. W. Męczyński et al, Eur. Phys. J. A **3**, 311 (1998).
3. J. Styczeń et al, contribution to this conference.
4. M.B. Smith et al, Eur. Phys. J. A **5**, 43 (1999).
5. E. Coenen, K. Deneffe, M. Huyse, P. Van Duppen and J.L. Wood, Z. Phys. A **324**, 485 (1986).
6. M. Leino et al, to be published.

Excited States in the Heavy Nuclide ^{254}No

H. Kankaanpää[1], M. Leino[1], R.-D. Herzberg[2], A.J. Chewter[2],
F.P. Heßberger[3], Y. Le Coz[4], F. Becker[4], P.A. Butler[2], J.F.C. Cocks[1],
O. Dorvaux[1], K. Eskola[5], J. Gerl[3], P.T. Greenlees[2], K. Helariutta[1],
M. Houry[4], G.D. Jones[2], P. Jones[1], R. Julin[1], S. Juutinen[1], H. Kettunen[1],
T.L. Khoo[6], A. Kleinböhl[3], W. Korten[4], P. Kuusiniemi[1], R. Lucas[4],
M. Muikku[1], P. Nieminen[1], R.D. Page[2], P. Rahkila[1], P. Reiter[7],
A. Savelius[1], Ch. Schlegel[3], Ch. Theisen[4], W.H. Trzaska[1], and
H.-J. Wollersheim[3]

[1] Department of Physics, University of Jyväskylä, P.O. Box 35, FIN-40351 Jyväskylä, Finland
[2] Oliver Lodge Laboratory, University of Liverpool, Liverpool L69 7ZE, UK
[3] Gesellschaft für Schwerionenforschung, Darmstadt, Germany
[4] DAPNIA/SPhN CEA-Saclay, France
[5] Department of Physics, University of Helsinki, Finland
[6] Argonne National Laboratory, Argonne, Illinois, USA
[7] LMU University, Munich, Germany

Abstract. In-beam γ-ray spectroscopy of the excited states in the heavy nuclide ^{254}No have been studied in the reaction ^{208}Pb(^{48}Ca,2n)^{254}No. The techniques of recoil-gating and recoil-decay-tagging were needed due to the dominant fission background. Prompt γ-rays were detected with a Ge detector array, consisting of four clover detectors in close geometry, and a gas-filled recoil separator (RITU) was used for detecting recoils and their α-decays. The observed six γ-rays were associated with E2-transitions in the ground state rotational band of ^{254}No. The value $\beta_2=0.27\pm0.03$ was extracted for the quadrupole deformation from the extrapolated 2^+ excitation energy.

INTRODUCTION

Very recent work speculates an synthesis of the elements up to Z=118. The goals of the study of the heaviest elements are to find their limits of existence end to reach the long-predicted island of spherical super heavy nuclei (1). It is known that the main decay mode of the isotopes of the heaviest elements is α-decay instead of fission. This enhanced stability is due to a deformed shell closure, the most important component being the quadrupole deformation. It is predicted to have a rather constant value of β_2 ~ 0.24 in a large region around ^{256}No (2).

Experimental information on the deformation and the limitations on angular momentum values leading to the formation of heavy isotopes can be achieved by using

CP495, *Experimental Nuclear Physics in Europe*, edited by B. Rubio et al.
© 1999 American Institute of Physics 1-56396-907-6/99/$15.00

the tools of γ-spectroscopy. The Recoil Decay Tagging (RDT) method (3,4) is needed to perform in-beam γ-spectroscopy when the production cross sections are of the order of 1μb or less. The cross sections for neutron evaporation exit channels in fusion reactions decrease rapidly when the proton number of the compound system increases, making in-beam studies of very heavy nuclides generally impossible with presently available techniques. However, anomalously high production cross sections occur in some special cases when using magic nuclei as reaction partners. The reaction $^{208}Pb(^{48}Ca,2n)^{254}No$ leads to the production cross section of ~ 2 μb, and the RDT method together with this reaction led to successful experiments in two laboratories to detect γ-rays from ^{254}No. The first experiment was carried out at Argonne National Laboratory (ANL), and five E2 transitions in the ground state band were observed (5). This observation was confirmed and the ground state band has been extended by one transition in the present work (6) performed at the Department of Physics, University of Jyväskylä (JYFL).

EXPERIMENTAL DETAILS

Prompt γ-rays were observed with the SARI array (Segmented Array at RITU). It consisted of four clover (three of them segmented) detectors placed in close geometry at 50 degrees relative to the beam direction (Fig. 1.). The photo-peak efficiency of the array at 1.33 MeV was 1.7 % when operated in add-back mode. The beam and the dominant fission products are separated from the desired fusion evaporation residues by the RITU dipolemagnet, and focusing fusion products with two quadrupoles to the focal plane where they are implanted in the position sensitive Si-strip detector (7). Signals from clover detectors are delayed by the amount of the flight time of the fusion evaporation residues through RITU, and so prompt γ-rays are in coincidence with the recoils. Final identification is performed by looking forward in time for a possible α-decay at the same position where a possible fusion evaporation residue was implanted. Accepting only those prompt γ-rays together with possible fusion evaporation residues which have exact α-decay properties of ^{254}No gives us high selectivity and final proof that these γ-rays originated from the isotope ^{254}No.

RESULTS

The γ-spectra from the reaction are shown in Fig. 2. The fact that the neutron evaporation exit channels in fusion reactions are especially sharp in this region gives us only one open channel when bombarded at the optimum energy. This can be seen in the γ-ray spectra only gated by possible fusion evaporation residues and α-tagging is only needed to positively identify these transitions to survive and that they originate from ^{254}No. We assumed that the observed six transitions are from an E2 cascade in the ground state band of ^{254}No. The spins of the states between which the observed transitions are can be assigned by making a Variable Moment of Inertia (VMI) fit, where the kinematic moment of inertia is expressed in terms of the Harris parameters. According to the fit we assign the lowest observed transition of 158.9 keV as the transition between 6^+ and 4^+ states. The two lowest transitions (44 keV and 102 keV) cannot be observed due to their high conversion, but their energies can be extrapolated by using the VMI fit.

It is possible to extract the ground state quadrupole deformation from the extrapolated energy of the 2^+ state using global systematics (8). This yields a deformation parameter value $\beta_2 = 0.27\pm0.03$ which is in good agreement with the values calculated using the macroscopic-microscopic method.

Based on this work, the nucleus ^{254}No survives at least up to spin $16\hbar$, we see no indication of the fission barrier leading to a band termination or strong deviations from the rotational pattern or backbending. According to Cranked Shell Model calculations strong neutron interactions are predicted to occur only above transition energies of ~500 keV in ^{254}No, but in ^{252}No at about 60 keV lower in transition energy. In a very recent experiment performed at JYFL (9) the reaction ^{206}Pb(^{48}Ca,2n)^{252}No was used and the result indicate a small upbend in a plot of spin vs transition energy supporting the calculations.

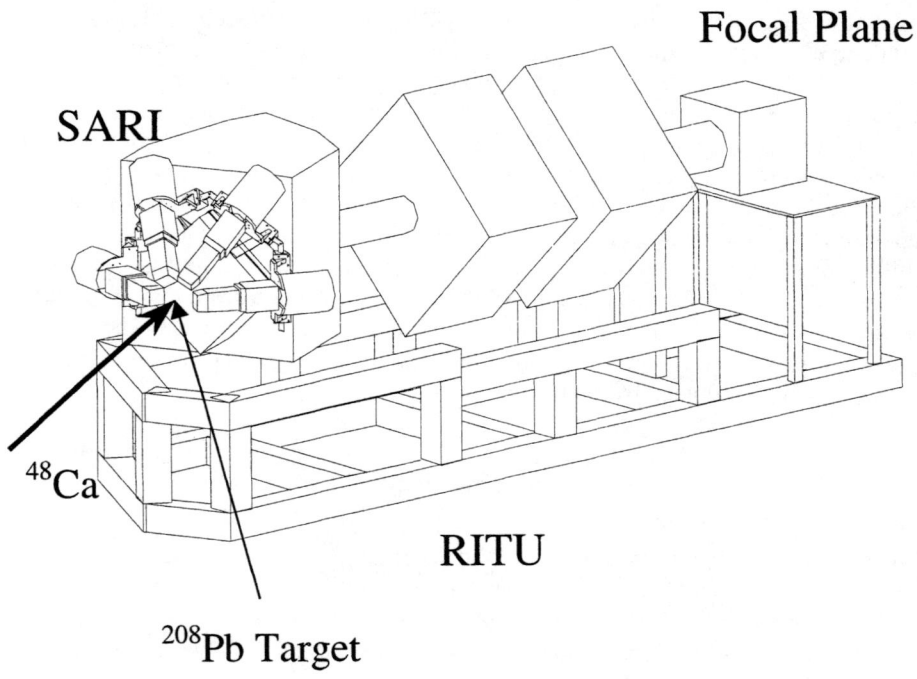

FIGURE 1. SARI-array formed by four clover detectors together with the RITU recoil separator.

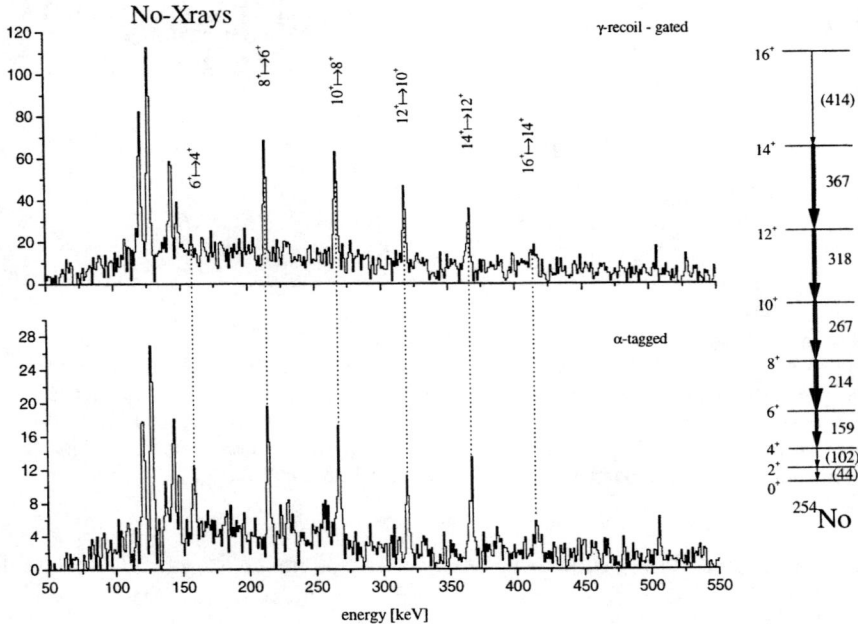

FIGURE 2. γ-spectra from the reaction ^{208}Pb(^{48}Ca,2n)^{254}No when only gated by possible γ-recoil-events and when exact α-decay is required. The proposed ground state band level scheme of ^{254}No is shown on the right.

REFERENCES

1. Seaborg G. T., Loveland W. D., The Elements Beyond Uranium, John Wiley & Sons, 1990.
2. Möller P., Nix. J.R., Myers W. D., Swiatecki W. J., Nuclear Data Tables 59, 185, 1995.
3. Simon R. S. et al., Z. Phys. A 611, 211, 1996.
4. Paul E. S. et al., Phys. Rev. C 51, 78, 1995.
5. Reiter P. et al., Phys. Rev. Lett. 82, 509, 1999.
6. Leino M. et al., Eur. Phys. J. A (in press).
7. Leino M. et al., Nucl. Instr. Meth. B99, 653, 1995.
8. Raman S. et al., Nuclear Data Tables 42, 1, 1989.
9. Herzberg R.-D. et al., to be published.

Euroball Channel Selection Using ISIS and the Neutron Wall

J. Garcés Narro

Dept. of Physics, University of Surrey, Guildford, GU2 5XH, UK

Abstract. A number of methods of channel selection in fusion evaporation reactions studied with the EUROBALL array have been investigated. In particular, a channel selection technique is proposed in the current work whereby the compound nucleus angular momentum (using the gamma-ray fold as measured in the neutron wall) and the entrance excitation energy of the residual system (from measuring the energy of the evaporated charged particles) are used to select individual nuclei.

An experiment to study the near yrast states of nuclei in the vicinity of the N=Z nucleus $^{80}_{40}$Zr was performed at the Legnaro National Laboratory using the fusion-evaporation of a 200 MeV ^{58}Ni beam on a gold backed ^{28}Si target. A high degree of channel selection was required to pick out the weakly populated nuclei of interest which included at least one neutron, from the large flux of gamma rays from the much more intensely populated pure charged particle evaporation channels. Gamma rays were detected in the EUROBALL array in coincidence with both light evaporated charged particles observed using the ISIS, Si-EΔE array [1] and evaporated neutrons were detected with a high-efficiency neutron wall [2] placed at forward angles.

Neutron-γ discrimination was carried out for 15 pseudohexaconical neutron detector units in two rings and a central pentagonal unit placed in the forward direction. Figure 1a presents the results of the discrimination using the zero-crossing (ZCO) and the time-of-flight methods (TOF). Note, there is no overlap of neutron and γ locii. The associated γ-multiplicity for each event could then be inferred by measuring number of γ-ray hits for each event in the neutron wall elements.

Evaporated charged particles were observed using ISIS, a 4-π Si-EΔE array [1], placed inside the Euroball spectrometer. This array consists of 40 telescopes of two silicon detectors each. The two elements in the telescope, ΔE and E have thicknesses of 130 μm and 1000 μm respectively. Figure 1b shows the result of the identification of charge particles according to its energy deposited in the ΔE and E elements of the ISIS array. The energy deposited for every observed charged particled was also measured. This allows the calculation of the entrance excitation energy by the means of Equation 1. Since the energy for the evaporated neutrons

CP495, *Experimental Nuclear Physics in Europe*, edited by B. Rubio et al.
© 1999 American Institute of Physics 1-56396-907-6/99/$15.00

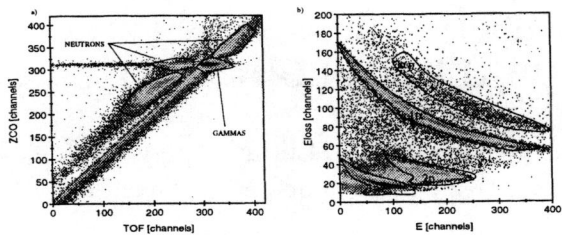

FIGURE 1. a) Example of neutron-γ discrimination using the simultaneous measurement of the zero-crossing time and time-of-flight. b) Identification of light charge partciles using the ISIS array.

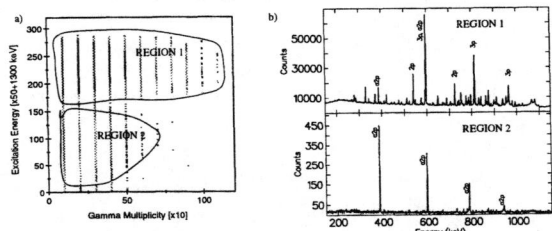

FIGURE 2. a) Excitation energy as a function of the γ-multplicity for the 2p channel evaporated observed in the ISIS ball. b) Energy spectra correponding to Regions 1 and 2, respectively. The α2p-channel is clearly identified in Region 2.

is not measured, the apparent measured excitation energy is generally greater for those neutron evaporating channels with similar numbers of charged particles (eg, 3pn compared to 3p).

$$(E_{ex})_{a\alpha,bp,cn} = (E_{ex})_{CN} - \sum_{i=1}^{a} E(\alpha_i) - \sum_{j=1}^{b} E(p_j) - \sum_{k=1}^{c} E(n_k) \tag{1}$$

Together the γ-multiplicity and the excitation energy of each channel, as shown in Figure 2a for the 2p-ISIS observed channel, were used to identify transitions in a variety of evaporation channels. The gamma energy spectra corresponding to the Region 1 contains the de-excitations correspong to those channels involving at least 2 protons (ie. 2p, 3p, α2p, etc). Region 2 clearly corresponds to those gammas coming from the de-excitation of α2p reaction products. This method presents a way for assigning the reaction channel for weak transitions.

REFERENCES

1. E. Farnea et al., Nuc. Inst. and Meth. **A400** (1997) 87
2. M. Moszynski et al., Nuc. Inst. and Meth. **A421** (1999) 532

Monte Carlo Study of the charged-particle detectors used in NORDBALL and EUROBALL Spectrometers

S. Ashrafi[a], M. Lipoglavšek[b], A. Likar[a], J. Nyberg[c], A. Axelsson[d], H. Grawe[e], T. Vidmar[a], R. Schwengner[f]

[a] Jozef Stefan Institute, Ljubljana, Slovenia
[b] Oak Ridge National Laboratory, Oak Ridge, TN, USA
[c] The Svedberg Laboratory, Uppsala University, Uppsala, Sweden
[d] Department of Radiation Sciences, Uppsala University, Uppsala, Sweden
[e] Gesellschaft für Schwerionenforschung, Darmstadt, Germany
[f] Forschungszentrum Rossendorf, Institute für Kern- u. Hadronenphysik, Dresden, Germany

Abstract. The ancillary charged-particle detector arrays used inside the NORDBALL and the EUROBALL spectrometers were simulated employing the Monte Carlo code GEANT3. The simulated charged-particle spectra are in good agreement with the experimental ones.

In two heavy-ion experiments using the (^{58}Ni+^{50}Cr) [1] and the (^{58}Ni+^{46}Ti) [2] reactions which were carried out with the NORDBALL at NBI, Riso, and an array of large volume HPGe detectors and a EUROBALL cluster detector at MPI, Heidelberg, respectively, an ancillary 4π array of the silicon detectors was used to select the exit channel of the heavy-ion reaction. In the first experiment, a silicon array which consisted of 12 thin (170 μm) pentagonal silicon detectors, the so-called Si-Ball [3], was used inside the NORDBALL spectrometer. The Si-Ball exploits the ΔE technique for proton-alpha discrimination. To increase it's granularity, the detector at 0° (the most forward detector) and the detectors at 37° were divided into 5 and 2 equal segments, respectively. The other silicon array, the so-called RoSiB [2], which consisted of 30 hexagonal and 12 pentagonal 500 μm thick detectors was commissioned in the second experiment and will be used in the EUROBALL spectrometer. It uses pulse shape discrimination technique to distinguish between proton and alpha particles. In this experiment, two pentagons at 0° and 180° and one hexagon at 90° were removed and their positions were used as beam exit, beam entrance and target mounting port, respectively. Also, all of the detectors at 148° and some of the detectors at 117° were replaced with 300 μm thick detectors. To quantitatively resolve the details of the charged particle detection and identification in these two experiments, the response of the Si-Ball and the RoSiB arrays were simulated employing the Monte Carlo code GEANT3 [4]. In the simulation program, the energy losses of the projectile in the target were obtained from stopping power tables. For different energies of the projectile in the target, the fusion reaction cross sections were extracted from the statistical code CASCADE [5]. These cross sections were then used to randomly sample the kinetic energy of the projectile before the reaction as well as the location of the reaction in the target. The relative experimental yields of the reactions were used to randomly sample the multiplicity of the evaporated particles. The charged particles were assumed to evaporate anisotropically in the c.m. system [6] and their kinetic energies were obtained from the CASCADE

CP495, *Experimental Nuclear Physics in Europe*, edited by B. Rubio et al.
© 1999 American Institute of Physics 1-56396-907-6/99/$15.00

code. Using the GEANT3, the charged particles were tracked through different media from their origin in the target to a point at which they stopped or exited the silicon detectors. The total amount of energy deposited by the impinging particles was used to obtain the simulated charged-particle spectra. In Fig.1 the simulated charged-particle spectra are compared with the experimental ones.

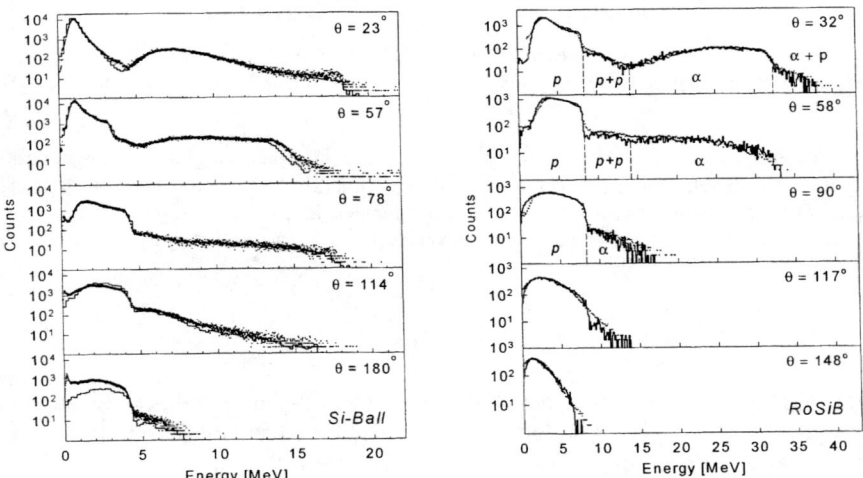

Fig. 1. For different rings of the Si-Ball array (left) and of the RoSiB (right), the simulated (solid lines) charged-particle spectra are compared with the experimental ones (dots). In the right panel, the regions due to single hits of one proton and one alpha-particle are marked with *p* and *α*, and the regions due to double hits of two protons and of one proton and one alpha are marked with *p+p* and *α+p*.

For the two experiments, the average proton and alpha detection efficiencies of the Si-Ball and of the RoSiB arrays were determined from the simulation and values of $\varepsilon_p=59\%$, $\varepsilon_\alpha=37\%$ (Si-Ball) and $\varepsilon_p=60\%$, $\varepsilon_\alpha=44\%$ (RoSiB) were obtained. The corresponding measured efficiencies are $\varepsilon_p=56\%$, $\varepsilon_\alpha=40\%$ (Si-Ball) and $\varepsilon_p=59\%$, $\varepsilon_\alpha=44\%$ (RoSiB). It should be mentioned that due to the temporary failure of the detectors in the backward hemisphere of the RoSiB in this experiment, the obtained efficiencies are much lower than its optimum efficiencies which can be extrapolated from the simulation results [2] as $\varepsilon_p=77\%$, $\varepsilon_\alpha=68\%$.

References:
(1) M. Lipoglavšek et al. Phys. Rev. Lett. 76 (1996) 888.
(2) G. Pausch et al., RoSiB - a 4π silicon ball for charged-particle detection in EUROBALL, submitted to Nucl. Instr. And Meth. A, and C. Borcan et al., Eur. Phys. J. A5 (1999) 243.
(3) T. Kuroyanagi et al., Nucl. Instr. Meth. A316 (1992) 289.
(4) GEANT - Detector Description and Simulation Tool, CERN, Geneva (1993).
(5) F. Pühlhofer, Nucl. Phys. A280 (1977) 267.
(6) C. Beck et al. Phys. Rev. C 39 (1989) 2202.

Direct Mathematical Calculation of the Photopeak Efficiency for Gamma Rays in Cylindrical NaI(Tl) Detectors

Mahmoud I. Abbas[1] and M. Bassiouni[+]

Physics Department, Faculty of Science, Alexandria University, EGYPT.
+) Basic & Applied Sciences Department, Faculty of Engineering & Technology, AAST & MT, Alexandria, EGYPT.

Abstract. A direct mathematical formalism for the determination of the photopeak (full energy peak) efficiency and the photofraction (peak to total ratio) of cylindrical (2RxL) NaI(Tl) scintillation detectors is deduced. The results have been compared with previous computational treatments. The comparison of our calculated data with the published experimental values shows a very satisfactory agreement in most of the practical energy region.

Introduction

Determination of total efficiency, full energy peak efficiency and peak to total ratio for gamma rays in NaI(Tl) scintillation detector is required and plays an important role in gamma rays spectroscopy. For example, to determine the gamma decay scheme of excited nuclei or study particle-gamma reactions.

The total efficiency of the source-detector system is defined as the number of gamma rays interacting with the detector active medium divided by the number of gamma rays generated by the source (say, N_t). The full energy peak efficiency of the source-detector system is defined as the number of gamma rays interacting with the detector active medium and there losing all the primary energy divided by N_t. The peak to total ratio is defined as the ratio between the full energy peak- and the total efficiencies, Belluscio et al.[1].

Mathematical Viewpoint

Several authors have treated the efficiencies of cylindrical (2RxL) NaI(Tl) scintillation detectors for gamma rays. The calculations of the total efficiency of right circular cylindrical NaI(Tl) detectors using isotropic radiating (point-disk) sources are described in previous papers, Selim and Abbas [2,3] and Selim et al.[4].

Treatments of peak to total ratio (photofraction) and full energy peak efficiency (photopeak) for isotropic radiating axial point sources have been given using Monte Carlo calculations by Weitkamp [5], Giannini et al.[6] and Belluscio et al.[1]; experimentally by Heath [7], Green et al.[8], Chinaglia et al.[9], Mishra et al.[10], Grosswendt et al.[11] and Sudarshan et al.[12] and finally using semi-empirical formulae by Mitchell et al.[13].

---

[1] Corresponding author

CP495, *Experimental Nuclear Physics in Europe*, edited by B. Rubio et al.

Recently Selim and Abbas using spherical coordinates derived four direct mathematical expressions for the efficiency of cylindrical (2RxL) NaI(Tl) detectors (in the general case of a point source displaced by a lateral distance ρ less than the detector's radius R) with the appropriate gamma coefficient of the detector's material, Selim et al.[4]. The four mathematical expressions are in the integral form; a computer program has been set for those expressions. For speedy direct calculations, an approximated from for the efficiencies is as follows:

$$\varepsilon = \frac{1}{8}\Big[\ 2\exp(-\mu\ t).[1 - \exp(-\mu\ L)](1 - \beta_1)$$
$$+ \exp(-\mu' t / \beta_1).[1 - \exp(-\mu\ L / \beta_1)](2 - \beta_2 - \beta_3)$$
$$+ \exp(-\mu' t / \beta_1).[1 - \exp(-2\mu\ \rho / \alpha_1 - \mu L / \beta_1)](\beta_3 - \beta_2)$$
$$+ \exp(-\mu' t / \beta_2).[1 - \exp(-2\mu\ \rho / \alpha_2)](\beta_1 - \beta_4)\ \Big]\quad (1)$$

with,

$$\alpha_i = \sin\theta_i\ ;\quad \beta_i = \cos\theta_i\quad and\quad i = 1,2,3\ and\ 4.$$

$$\theta_1 = atan\left(\frac{R - \rho}{h + L}\right); \theta_2 = atan\left(\frac{R - \rho}{h}\right); \theta_3 = atan\left(\frac{R + \rho}{h + L}\right) and\ \theta_4 = atan\left(\frac{R + \rho}{h}\right).\quad (2)$$

where h is the source-to-detector distance, μ is the photopeak (total attenuation) coefficient of the scintillator, looking for photopeak (total) efficiency of the detector at the corresponding energy, Hubbell and Seltzer [14]. Finally, μ' is the total attenuation coefficient of the cap material and t its thickness. This approximated form gave a good representation to the four integrals in the familiar energy range. Setting the lateral distance $\rho = 0$ in the previous equations, we obtain the efficiency of cylindrical (2RxL) NaI(Tl) detectors for isotropic axial-point source.

Systematic calculations of both the full energy peak efficiency (photopeak) and peak to total ratio (photofraction) of different cylindrical NaI(Tl) crystal sizes for an axial point source at different heights (h = 10 and 15 cm) as a function of the gamma energy were calculated and represented in figures (1-7). The overall errors in the published experimental values vary between 7% and 10%.

Conclusion

In this paper, it becomes easier using a compact mathematical expression to calculate the total efficiency, the photopeak efficiency and consequently photofraction of cylindrical (2RxL) NaI(Tl) scintillation detectors.

Acknowledgment

The authors would like to thank Prof. Younis S. Selim for his fruitful discussion and also, Prof. John H. Hubbell, NIST, Gaithersburg, MD, USA, for kindly providing us with comprehensive accurate tables of the attenuation coefficients.

Figures Captions

FIGURE 1: Variations of photofraction for a $2''x2''$ NaI(Tl) crystal with the photon energy.

FIGURE 2: Variations of photofraction for a $3''x3''$ NaI(Tl) crystal with the photon energy.

FIGURE 3: Variations of photofraction for a $5''x4''$ NaI(Tl) crystal with the photon energy.

FIGURE 4: Variations of photofraction for a $3''x1''$ NaI(Tl) crystal with the photon energy.

FIGURE 5: Variations of photofraction for a $2.5''x2.5''$ NaI(Tl) crystal with the photon energy.

FIGURE 6: Variations of the intrinsic photopeak efficiency for a $3''x3''$ NaI(Tl) crystal with the photon energy.

FIGURE 7: Variations of the intrinsic photopeak efficiency for a $5''x4''$ NaI(Tl) crystal with the photon energy.

References

1- Belluscio M., De Leo R., Pantaleo A. and Vox A. Nucl.Instrum. Meth. **118** (1974) 553.

2- Selim Y.S. and Abbas M.I. Egypt. J. Phys. **26**, No. 1-2(1995) 79.

3- Selim Y.S. and Abbas M.I. Radiat. Phys. Chem. **48**(1996) 23.

4- Selim Y.S., Abbas M.I. and Fawzy M.A. Radiat. Phys. Chem. **53**(1998) 589.

5- Weitkamp C. Nucl. Instrum. Meth. **23**(1963) 13.

6- Giannini M., Oliva P.R. and Ramorino M.C. Nucl. Instrum. Meth. **81**(1970) 104.

7- Heath R.L. (1964) AEC Report-IDO 16880.

8- Green R.M. and Finn R.J. Nucl. Instrum. Meth. **34**(1965) 72.

9- Chinaglia B. and Malvano R. Nucl. Instrum. Meth. **45** (1966) 125.

10- Mishra U.C. and Sadasivan S. Nucl. Instrum. Meth. **69**(1969) 330.

11- Grosswendt B. and Waibel E. Nucl. Instrum. Meth. **133**(1976) 25.

12- Sudarshan M., Joseph J. and Singh R. J. Phys. D: Appl. Phys. **25**(1992) 1561.

13- Mitchell D.G., Sanger H.M. and Marlow K.W. Nucl. Instrum. Meth. **A 276**(1989) 547.

14- Hubbell J. H. and Seltzer S. M. (1995) NISTIR 5632, USA.

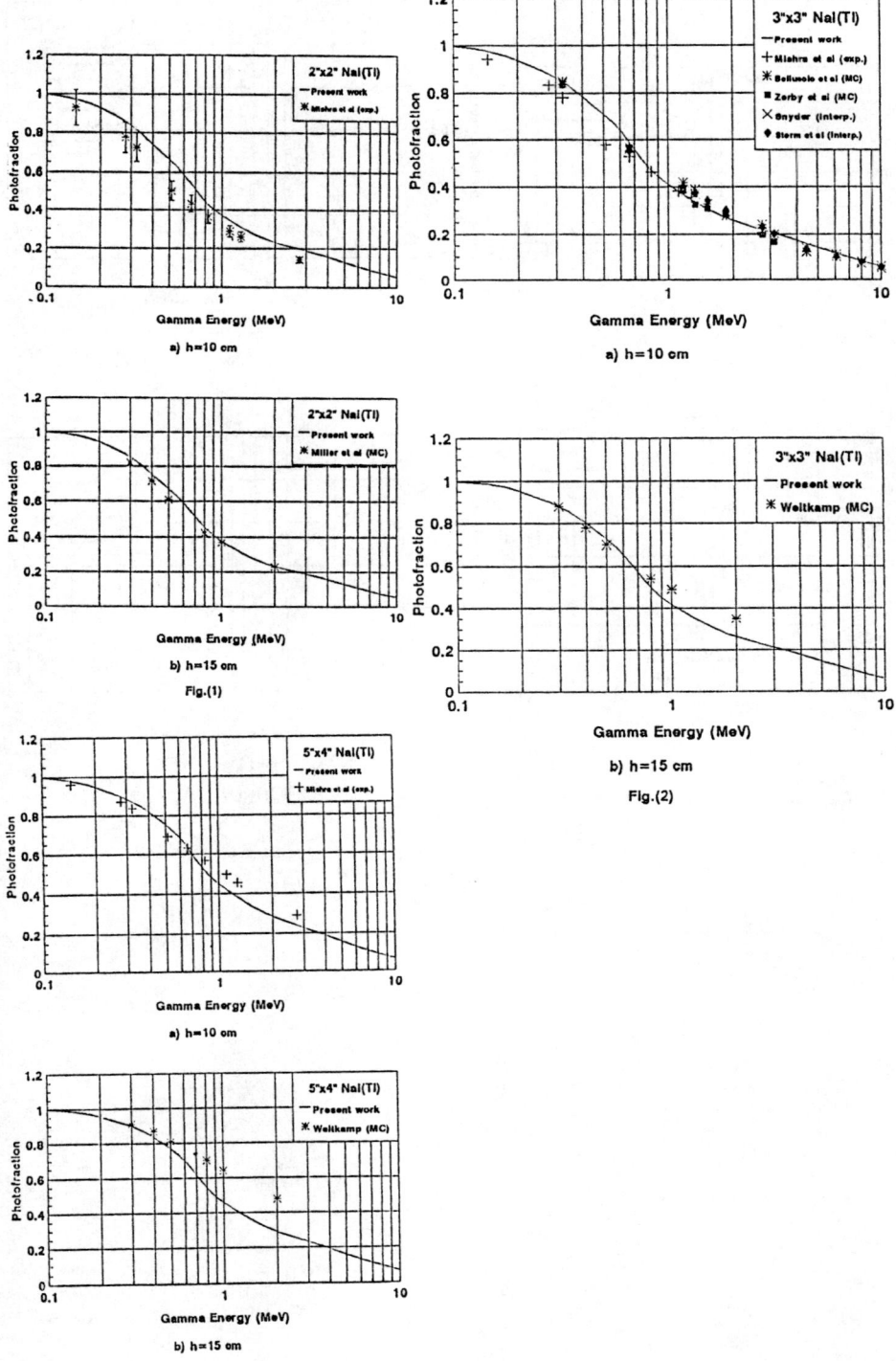

a) h=10 cm

b) h=15 cm

Fig.(1)

a) h=10 cm

a) h=10 cm

b) h=15 cm

Fig.(2)

b) h=15 cm

271

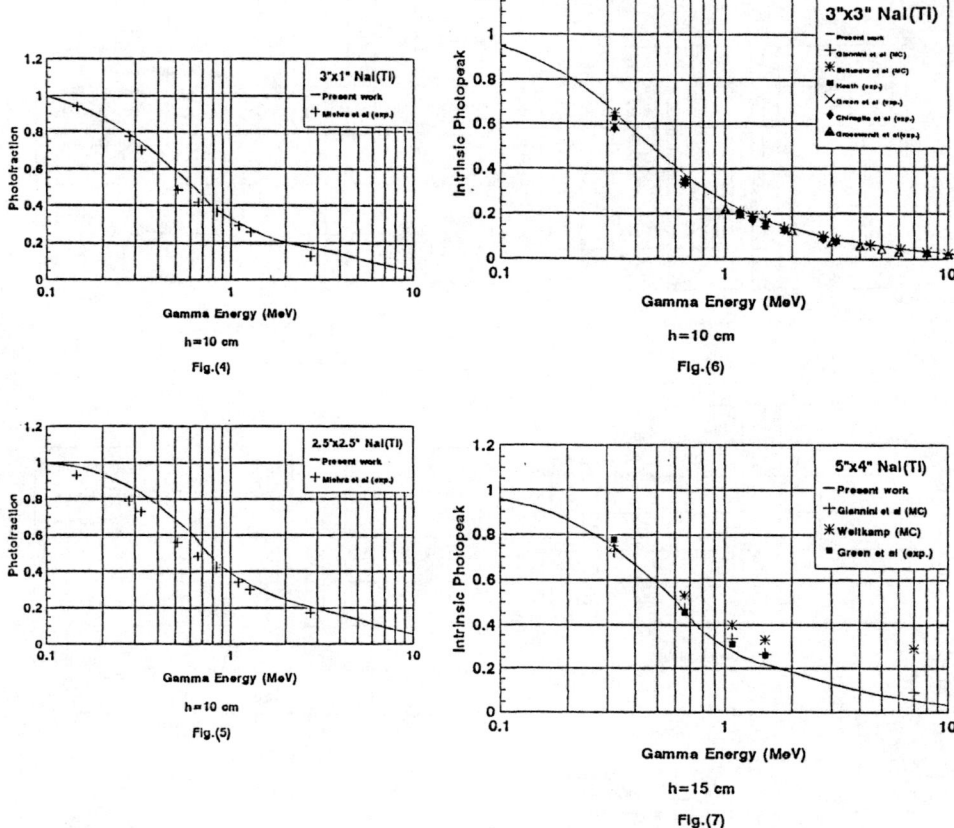

Fig.(4)

Fig.(5)

Fig.(6)

Fig.(7)

REACTIONS AND RADIOACTIVE
NUCLEAR BEAMS

Isospin and Spin-Isospin Modes in Charge-Exchange Reactions

M. N. Harakeh[a], H. Akimune[b,c], A. M. van den Berg[a],
S. Brandenburg[a], M. Fujiwara[b,c], H. Laurent[d],
A. Willis[d], and R. G. T. Zegers[a]

[a] Kernfysisch Versneller Instituut, Zernikelaan 25, 9747AA Groningen, The Netherlands
[b] Research Center for Nuclear Physics, Osaka University, Mihogaoka 10-1 Ibaraki, Osaka 657-0047, Japan
[c] Advanced Science Research Center, Japan Atomic Energy Research Institute, Tokai, Ibaraki 319-1195, Japan
[d] Institut de Physique Nucléaire, IN2P3, Université de Paris-Sud, Orsay Cedex, France

Abstract. The microscopic structure of the Gamow-Teller resonance (GTR) and spin-dipole resonance (SDR) in ^{208}Bi has been investigated in the ^{208}Pb(^{3}He,tp)^{207}Pb reaction at E(^{3}He) = 450 MeV and very forward scattering angles. The partial and total branching ratios and the escape widths for GTR and SDR decay to the residual neutron-hole states in ^{207}Pb were deduced. These are found to be in good agreement with recent theoretical estimates. The (^{3}He,tp) reaction on Pb at E(^{3}He) = 177 MeV was also studied in order to locate isovector monopole strength corresponding to $2\hbar\omega$ transitions. Monopole strength at excitation energies above 25 MeV was discovered and compared to calculated strength due to the isovector giant monopole resonance and the spin-flip isovector monopole resonance. Calculations in a normal-mode framework show that all isovector monopole strength can be accounted for if the branching ratio for decay by proton emission is 20%.

INTRODUCTION

The microscopic structure of isospin and spin-isopsin modes can be well studied through charge-exchange reactions followed by particle decay. The (p,n) reaction is not very useful in this respect because of efficiency and resolution problems in detecting the outgoing neutron. On the other hand, the (^{3}He,t) reaction is very well suited for this purpose because the outgoing tritons can be detected with very high resolution and essentially 100% efficiency using magnetic spectrometers. Furthermore, the decay protons can be detected in solid-state detectors (SSDs) allowing a good total energy resolution and thus possible discrimination of decay to various final states in the residual nucleus. This is imperative for disentangling the microscopic structure of the investigated resonances.

CP495, *Experimental Nuclear Physics in Europe*, edited by B. Rubio et al.
© 1999 American Institute of Physics 1-56396-907-6/99/$15.00

The study of spin-flip charge-exchange (isovector) giant resonances, and in particular their microscopic structure, received a boost recently with the availability of magnetic spectrometers that allowed the investigation of the (^{3}He,t) charge-exchange reaction at intermediate energies and very forward scattering angles including 0°. It was possible, for instance, to deduce reliably the microscopic structure of the Gamow-Teller resonance (GTR) in ^{208}Bi via the ^{208}Pb(^{3}He,tp)^{207}Pb reaction at E(^{3}He)=450 MeV [1,2]. These experimental results helped to solve a long-standing problem concerning a discrepancy between the experimental total width and branching ratios for proton decay from the GTR in ^{208}Bi measured earlier [3] and those calculated in different theoretical approaches [4–6].

The study of particle decay can in addition be useful in identifying high-lying resonances such as the isovector giant monopole resonance (IVGMR) and its spin-flip partner, the spin-flip isovector giant monopole resonance (IVSMR). The experimental investigation of these resonances is seriously hampered due to their large widths and a high underlying non-resonant continuum background. This continuum background consists of contributions of quasi-free knock-on charge-exchange process, and pickup-breakup and breakup-pickup processes. All of these processes lead in the case of the (^{3}He,t) reaction to "high-energy" protons emitted in the forward direction in coincidence with the ejectile tritons. If a coincidence with backward emitted protons is required this continuum background is largely suppressed. This experimental procedure has been used to locate monopole strength (IVGMR and IVSMR) in ^{208}Bi.

The IVGMR has been observed previously in the (n,p)-type (π^-, π^0) reaction, but for the (p,n)-type (π^+, π^0) reaction the evidence is less clear [7–9]. Indications for the IVGMR in the (n,p) and (^{13}C,^{13}N) charge-exchange reactions has also been reported [10–13]. Furthermore, indications for the IVSMR have been found in the ^{90}Zr(^{3}He,t) reaction at bombarding energies of 600-900 MeV [14,15], and in the (p,n) reaction at 795 MeV [16] and 295 MeV [17].

SPIN-FLIP DIPOLE RESONANCE

The total and partial proton-decay branching ratios of the isobaric analog state (IAS) and the GTR, and a preliminary result for the total proton branching ratio for the isovector spin-flip dipole resonance (SDR) in ^{208}Bi, have been reported earlier [2]. Here, the experimental results for the total and partial branching ratios of proton decay from the SDR in ^{208}Bi will be reported and compared to recent calculations [18].

The ^{208}Pb(^{3}He,tp)^{207}Pb experiment was carried out at the Research Center for Nuclear Physics (RCNP) using the high-resolution magnetic spectrometer "Grand Raiden" [19] to detect the tritons. Decay protons leading to final states in ^{207}Pb were measured in SSDs in coincidence with the tritons. The detection system of Grand Raiden allowed to reconstruct the scattering angle at the target via ray-tracing techniques. This facilitated the projection of singles and coincidence spectra gated at scattering angles of $0°$ and $1°$ at which the GTR+IAS and SDR peak, respectively. The experimental technique is described in detail in [2].

Scatter plots of triton energy versus proton-decay energy (not shown here), gated on events with scattering angles centered at $\theta \approx 0°$ and $1°$, were generated for the prompt peak in the timing spectrum. The loci for decay of the IAS, GTR, and SDR to the ground state and low-lying neutron-hole states in ^{207}Pb (i.e. $3p_{1/2}$, $2f_{5/2}$, $3p_{3/2}$, $1i_{13/2}$, $2f_{7/2}$) could be observed along lines for which $E_t + E_p = constant$. The total energy resolution obtained for $E_t + E_p$ of 400-580 keV was not sufficient to completely resolve the decay to the first and second excited states of ^{207}Pb at E_x = 570 keV and 898 keV, respectively. The final neutron-hole-state spectra in ^{207}Pb obtained by projecting the coincidence events onto the axis of excitation energy for ^{207}Pb, gated on the excitation-energy regions of the GTR+IAS and SDR in ^{208}Bi and on the angles $\theta \approx 0°$ and $1°$ for the GTR+IAS and SDR, respectively, showed that the population pattern of the various neutron-hole states for the SDR is quite different from that for the GTR+IAS [20]. The $1i_{13/2}$ and $2f_{7/2}$ are more strongly and the $3p_{1/2}$ more weakly populated in the case of proton decay from the SDR. This can be understood qualitatively by considering the increase of the proton energy available above the Coulomb barrier and also the average decrease of the centrifugal barrier for the decay protons as compared to the GTR+IAS. There is also a clear indication for proton decay from the SDR to the broad $1h_{9/2}$ deep-neutron-hole state in ^{207}Pb. These qualitative arguments are confirmed by the detailed analysis of the data.

In Fig. 1, the relative branching ratios for proton decay from the IAS, GTR and SDR regions to the various neutron-hole states in ^{207}Pb are shown as a function of excitation energy. These were obtained by generating the final-state spectra for 1 MeV excitation-energy bins. These final-state spectra were then fitted to obtain the contributions to decay to the neutron-hole states in ^{207}Pb. From this detailed comparison of relative branching ratios to neutron-hole states, it can be easily seen that decay to the high-spin neutron-hole states $1i_{13/2}$ and $1h_{9/2}$ from the "low-spin" SDR gain in importance for higher excitation energies because of the increase in proton decay energy.

In Table 1, the deduced branching ratios and partial escape widths for the summed SDR strength composed of 0^-, 1^-, and 2^-, are given in columns 5 and 6, respectively. The partial escape widths have been determined as follows. First, the singles cross section for the SDR has been obtained from fitting the singles spectra. The partial double-differential cross sections for the SDR have been determined from fitting the final-state spectra (not shown here) to obtain the cross sections for populating the ground state and low-lying neutron-hole states in ^{207}Pb (i.e. $3p_{1/2}$, $2f_{5/2}$, $3p_{3/2}$, $1i_{13/2}$, $2f_{7/2}$, $1h_{9/2}$). These were then integrated over the full solid angle taking the angular correlation into account. The ratio of the integrated double-differential cross section to the singles cross section yields the branching ratio. Note that in this ratio, the target thickness, collected charge and spectrometer (triton) solid angle cancel out.

It was not possible in the present experiment to disentangle the various spin components (2^-, 1^- and 0^-) of the SDR. Possible weak contributions from the 1^- non-spin-flip isovector giant dipole resonance could also not be identified. The total proton decay branching ratio for the SDR has been determined from the coincidence spectra to be $13.4 \pm 3.9\%$. If the SDR is assumed to be one single resonance with the total width of 8.4 ± 1.7 MeV as determined from fitting the singles spectra, a total proton escape width of 1.12 ± 0.32 MeV is deduced. Even if one assumes that the observed total proton escape width should be divided among the three components, the resulting proton escape width per component is still considerably larger than that for the GTR of $4.9 \pm 1.3\%$ [2].

Results of recent theoretical calculations by Moukhai *et al.* [18] for the branching ratios and the partial escape widths to the various neutron-hole states in ^{207}Pb are listed in columns 3 and 4 of Table 1, respectively. These calculations were

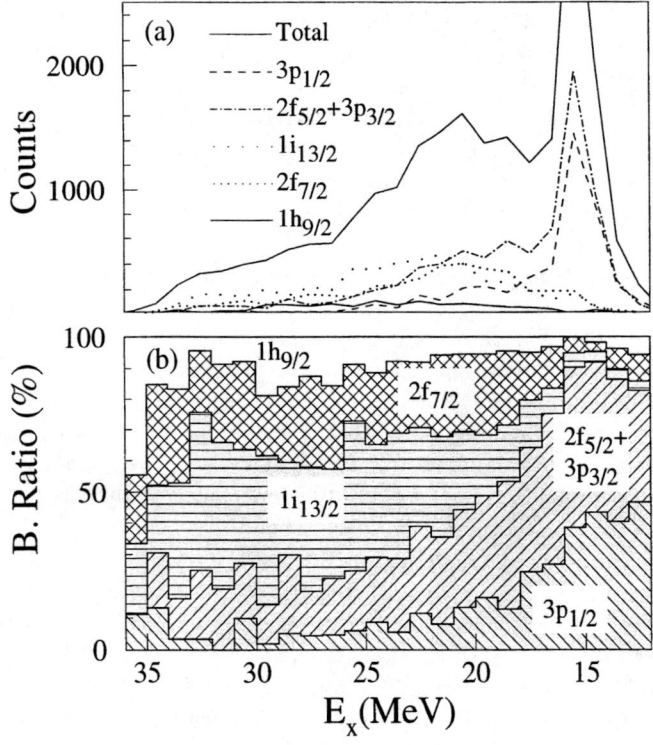

FIGURE 1. (a) Total decay spectrum (relative units) measured at $1°$, and decomposition into contributions for proton decay to the various low-lying neutron-hole states in ^{207}Pb (i.e. $3p_{1/2}$, $2f_{5/2}$, $3p_{3/2}$, $1i_{13/2}$, $2f_{7/2}$, $1h_{9/2}$), and (b) relative branching ratios for the decays to these neutron-hole states as a function of excitation energy in steps of 1 MeV. The spectrum overlaps with the resonances (IAS at 15.165 MeV, GTR at 15.6 MeV, and SDR at 21.1 MeV).

TABLE 1. Theoretical and experimental partial escape widths and branching ratios for the decay of SDR in ^{208}Bi into the neutron-hole states in ^{207}Pb. The branching ratios are given in % and the widths in keV.

Decay channel	E_x [a] (keV)	Theory[b] $\Gamma_{p_i}^\uparrow/\Gamma$	Theory[b] $\Gamma_{p_i}^\uparrow$	Exp.[c] $\Gamma_{p_i}^\uparrow/\Gamma$	Exp.[c] $\Gamma_{p_i}^\uparrow$
$3p_{1/2}$	0	1.05	88.4	0.95±0.28	79.8±23.5
$2f_{5/2}$	570	1.94	163.2	2.10±0.61	176.4±51.2
$3p_{3/2}$	898	2.18	183.1	2.79±0.81	234.4±68.0
$1i_{13/2}$	1633	3.80	318.8	3.41±0.98	286.4±82.3
$2f_{7/2}$	2340	4.02	337.3	3.14±0.91	263.8±76.4
$1h_{9/2}$	3413	1.17	98.6	0.97±0.27	81.5±22.7
$\sum_i \Gamma_{pi}^\uparrow/\Gamma$		14.2	1190	13.4±3.9	1122±324
Γ			8400[d]		8400[d]

[a] Nuclear Data Sheet.
[b] Moukhai *et al.* [18]. A small contribution to deep-hole states of 1.94% is not listed.
[c] present experiment.
[d] Total width is taken from the present experimental results.

performed in a semi-microscopic approach based on continuum RPA (CRPA), i.e. with coupling to the continuum, and with the strengths of the Landau-Migdal force f' and g' chosen to be 1.0 (to reproduce the experimental difference of the neutron and proton separation energies in ^{208}Pb) and 0.76 (to describe the GTR excitation energy in ^{208}Bi), respectively. Furthermore, the mean-field depth has been increased slightly to describe the nucleon separation energies in ^{208}Pb better. With these adjustments of the parameters to reproduce the global properties of single particles in ^{208}Pb and GTR in ^{208}Bi, the theoretical results are found to reproduce the experimental partial escape widths very well. The new calculations of Moukhai *et al.* [18] reproduce also the experimental branching ratios and partial escape widths of the IAS and GTR published earlier [2].

ISOVECTOR MONOPOLE STRENGTH

The experiment for the IVGMR and IVSMR was carried out using the Big-Bite Spectrometer (BBS) [21] at the Kernfysisch Versneller Instituut (KVI). The 177 MeV ^{3}He^{2+} beam was delivered by the AGOR cyclotron. A 7.8 mg/cm^2 thick ^{nat}Pb target was used. The BBS was set at -1°. In order to obtain a very accurate angle definition in the excitation-energy range of interest for the IVGMR and IVSMR, a special vertical-angle-defining slit was used with openings in the vertical direction chosen to coincide with the extrema of the IVSMR and IVGMR (-3°, 0° and 3°)

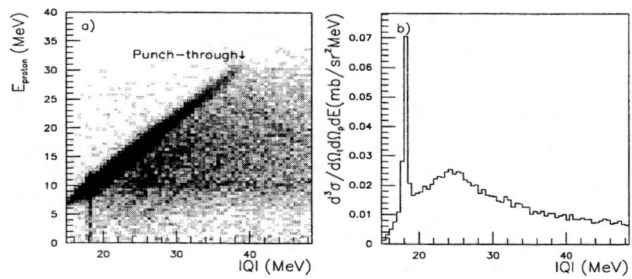

FIGURE 2. The two-dimensional coincidence spectrum of proton energy versus reaction $|Q|$-value (a) and its projection on the Q-axis (b) for the Pb(^{3}He,tp) reaction at 177 MeV and $\theta = 1°$.

Other experimental details are similar to those of the experiment at RCNP and are given in [10].

In Fig. 2 the coincidence spectra are shown. In the 2-dimensional spectrum, Fig. 2a, the proton energy is plotted against the $|Q|$-value of the reaction. Random events are subtracted. Around $|Q| \approx 38$ MeV the highest-energy protons begin to punch through the detector and to deposit less energy in the SSDs. The most intense band in the spectrum corresponds to direct decay to low-lying states in the final nucleus, mixed with coincidences stemming from quasi-free processes. The two contributions can not be distinguished kinematically since they both populate single-neutron hole states. In Fig. 2b the projection of Fig. 2a on the Q-axis is shown.

In order to enhance monopole strength at high $|Q|$-values, the difference spectrum between 0° and 3° was studied. The results are shown in Fig. 3. In Fig. 3a the spectra at 0° and 3° are shown; Fig. 3b shows the difference. Excess cross section is present. It was confirmed that this excess is isotropically distributed over the proton detectors as is expected for monopole excitations.

An estimate for the combined cross section due to the IVGMR and IVSMR has been obtained by performing distorted-wave Born approximation (DWBA) calculations. Wave functions were constructed in the normal-mode framework. This procedure gives an upper limit for the total transition strength since all 1p-1h configurations that contribute to a specific resonance are added coherently. Calculations indicate that even at this relatively low projectile energy, the expected cross section for the IVSMR is three times higher than the cross section of the IVGMR. The cross sections for the IVGMR and IVSMR were performed for this $|Q|$-value range and the fractions of the measured cross sections relative to these were determined. The results, expressed in terms of fraction of the non-energy-weighted sum rule (NEWSR) calculated in the normal-mode framework, for the IVSMR and IVGMR are shown in Fig. 3c. Since the calculated cross section drops strongly

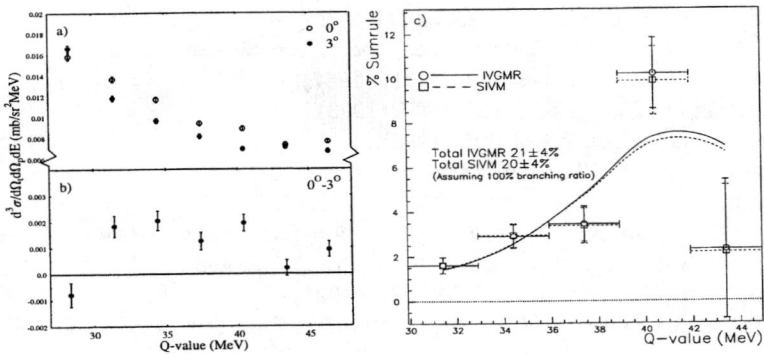

FIGURE 3. Coincidence spectra at 0° and 3° (a) and their difference (b). (c) same as (b) after unfolding the dependence of the excitation probability of monopole strength as a function of $|Q|$-value for the separated IVGMR and IVSMR contributions. The curves and fractions of NEWSR-strength exhausted are explained in the text.

as a function of $|Q|$, because of the increase in momentum transfer, results above $|Q| = 45$ MeV contain large error bars and are not included in the figure. The Lorentzian fits in the figures are merely to guide the eye. It can be concluded that if the branching ratio for decay by proton emission from the IVSMR and IVGMR is 20%, the full strength of both resonances can be accounted for. However, RPA correlations will lead to a quenching of the strength with respect to the normal-mode calculations. Therefore, a higher total branching ratio must be expected in order to make up for 100% of the NEWSR.

The research was performed as part of the research program of the "Stichting voor Fundamenteel Onderzoek der Materie" (FOM) with financial support from the "Nederlandse Organisatie voor Wetenschappelijk Onderzoek" (NWO).

REFERENCES

1. Akimune, H., *et al.*, *Phys. Lett.* **B323**, 107 (1994).
2. Akimune, H., *et al.*, *Phys. Rev.* C **52**, 604 (1995).
3. Gaarde, C., Larsen, J. S., Drentje, A. G., Harakeh, M. N., and van der Werf, S. Y., *Phys. Rev. Lett.* **46**, 902 (1981).
4. Colò, G., Nguyen, Van Giai, Bortignon, P. F., and Broglia, R. A., *Phys. Rev.* C **50**, 1496 (1994).
5. Knobles, D. P., Stotts, S. A., and Udagawa, T., *Phys. Rev.* C **52**, 2257 (1995).
6. Chekomazov, G. A., Muraviev, S. E., and Urin, M. H., *Nucl. Phys.* **A599**, 259c (1996).
7. Erell, A., *et al.*, *Phys. Rev.* C **34**, 1822 (1986).
8. Erell, A., *et al.*, *Phys. Rev. Lett.* **52**, 2134 (1984).

9. Irom, F., *et al.*, *Phys. Rev.* C **34**, 2231 (1986).
10. Zegers, R. G. T., *et al.*, to be published, and references therein.
11. Ford, T. D., *et al.*, *Phys. Lett.* **195B**, 311 (1987).
12. Bérat, C., *et al.*, *Nucl. Phys.* **A555**, 455 (1993).
13. Lhenry, I., *Nucl. Phys.* **A599**, 245 (1996).
14. Ellegaard, C., *et al.*, *Phys. Rev. Lett.* **50**, 1745 (1983).
15. Auerbach, N., *et al.*, *Phys. Lett.* **219B**, 184 (1989).
16. Prout, D. L., *et al.*, in *Proc. of the Eighth International Symposium on Polarization Phenomena in Nuclear Physics*, AIP Conf. Proc. **339**, 458 (1995).
17. Wakasa, T., *et al.*, *Phys. Rev.* C **55**, 2909 (1997).
18. Moukhai, E. A., Rodin, V. A., and Urin, M. H., *Phys. Lett.* **B447**, 8 (1999).
19. Fujiwara, M., *et al.*, *Nucl. Instrum. Meth. Phys. Res.* **A422**, 484 (1999).
20. Akimune, H., *et al.*, to be published, and references therein.
21. van den Berg, A. M., *Nucl. Instr. Meth.* **B99**, 637 (1995).

Nuclear Structure Physics at GSI – Challenges and Perspectives

G. Münzenberg

*Gesellschaft für Schwerionenforschung mbH, Planckstr. 1,
D64291 Darmstadt, Germany and Johannes Gutenberg - Universität Mainz*

Abstract. The perspectives at GSI for the exploration of exotic regions with new high-current accelerators and recent technical developments including single-atom decay studies, ultra sensitive γ-spectroscopy, new set-ups for reaction studies at relativistic energies in complete kinematics, heavy ion storage rings, and low-energy heavy-ion-electron colliders will be discussed.

I   INTRODUCTION

In recent years significant experimental progress has been made to access the limits of nuclear stability [1–3]. New phenomena were discovered such as the nuclear halo [2,4]: the spreading out of weakly bound nuclear matter far away from the nuclear core, and the extension of the number of chemical elements by shell stabilisation [5]. New experimental tools are: single-atom techniques employing separation in flight to investigate short lived systems far-off stability [6], 4π Ge-arrays for efficient γ-spectroscopy, detectors for break-up studies of energetic nuclei in complete kinematics, and storage rings for high precision experiments including direct measurement of nuclear masses, and decay studies of exotic atomic nuclei. Key issues are: the discovery of the proton radioactivity from the ground state [7] marking for the first time the limits of stability at the proton rich side, the first observation of the doubly magic nuclei [9,10] ^{100}Sn and ^{78}Ni, and the discoveries of the heaviest elements [8] with Z = 107 to 112. Working groups have been established to evaluate future perspectives and developments of nuclear structure research at GSI [11].

II   THE GSI PROGRAMME

GSI has the possibility of using heavy-ion beams of all elements and energies ranging from near Coulomb-barrier to 1 AGeV. Consequently exotic nuclei can be produced by complete heavy ion fusion as well as by projectile fragmentation and

CP495, *Experimental Nuclear Physics in Europe*, edited by B. Rubio et al.

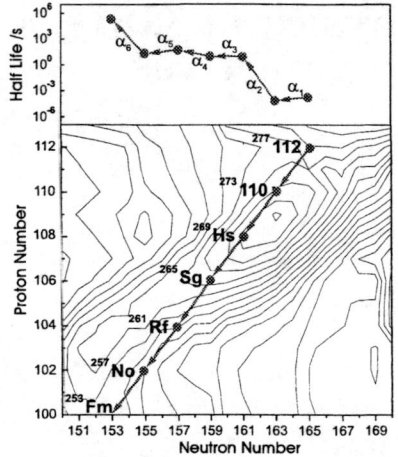

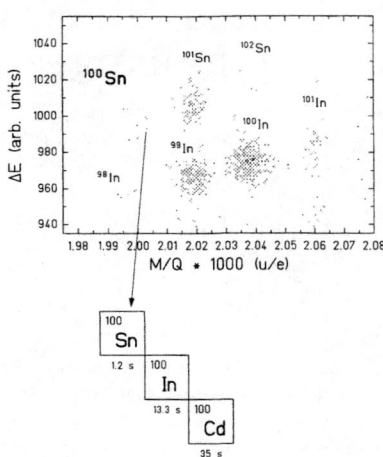

FIGURE 1. The calculated microsopic corrections of the heaviest elements [13]. The path of the decay chains from elements 112 are indicated. The upper panel displays the measured correlation times for one of the chains.

FIGURE 2. The in-flight identification of ^{100}Sn and the example of a aingle-atom β-decay sequence.

fission in-flight. Using these unique possibilities the GSI nuclear structure research program covers:

- Physics and chemistry of heavy and superheavy elements
- Gross properties of nuclei at the limits of β-stability
- Reaction studies of skin and halo nuclei, and fission
- Precision experiments with stored and cooled exotic nuclei and for the future
- Reaction studies of highest precision in a heavy ion cooler ring
- Electron scattering on exotic nuclei in a heavy-ion-electron collider

A   Heavy and Superheavy Elements

The exciting discovery in recent heavy-element research is the enhanced stability of the elements beyond rutherfordium ($Z = 104$) against fission due to shell stabilisation [12]. Theoretical models explain this in terms of a deformed shell region centred at $Z = 108$ and $N = 162$, created by a hexadecapole deformation. The heaviest element presently identified unambiguously [8] is element 112. It was discovered on the basis of two atoms followed by long α-decay chains to known isotopes of the elements seaborgium, rutherfordium, and nobelium. The α-chains pass by the centre of the shell region as displayed in Fig. 1 in the landscape of calculated microscopic corrections [13]. In the upper panel the experimental time distances between the α-decays observed for one of the chains clearly show the acceleration of the decays when passing the N = 162 shell. Recently results on the synthesis of first elements in the superheavy region Z = 114 and even Z = 118 have been reported. The still open question is the location of the spherical superheavy shell closure [14,15]. Self consistent calculations predict the location of he proton shell at Z = 120, or even Z = 126. A prerequisite for the exploration of the region near and beyond element 114 are accelerators with high beam intensities up to the order of 10^{14} projectiles per second.

The new generation of highly efficient 4π γ-arrays in combination with recoil separators for in-beam and decay spectroscopy will give access to the level structure of this new species of shell nuclei at and above the ground state. First experiments with nobelium have been carried out at Argonne (USA) [16] and Jyvaskyla (Finland). The long half-lives of the heaviest elements allow the application of new experimental methods such as trapping techniques to investigate nuclear, atomic, and chemical properties. The production and properties of the heaviest elements are described in more detail in the contribution by S. Hofmann to this conference.

B   Groundstate Properties far-off Stability

Exotic nuclei beyond uranium can be produced by projectile fragmentation at relativistic energies [3]. Eventwise in-flight identification and implantation into silicon detectors allows for single-atom decay studies in close analogy to the decay studies of superheavy elements. Fig. 2 displays the example of the investigation of ^{100}Sn by in-flight identification and a single-atom β-decay chain. Isomer spectroscopy has proven an extremely sensitive tool for structure investigation far-off stability [17]. Interesting regions are specifically around the doubly magic ^{78}Ni, ^{132}Sn, and the neutron rich region around ^{208}Pb. In flight trapping of projectile fragments e. g. at the fragment separator is under preparation. Fragments, once they are trapped, can be re-accelerated to any desired energy.

C   Nuclear Halos and Skins

The specific feature of halo nuclei is that the nuclear core and the halo are separable. So wave-functions of specific nucleons in well defined states become accessible. Break-up reactions at energies above the Fermi domain are a principal source of information on the halo structure.

- In peripheral collisions halo nucleons are removed by nucleon-nucleon collisions. As the collision times are short as compared to the motion of the nucleons structure effects can be observed in the exit channel.

- Excitation by the electromagnetic or nuclear mean field allows to study continuum effects.

Kinematic complete experiments are the key to reaction studies at GSI to investigate nuclear structure at and beyond the driplines. They are carried out with

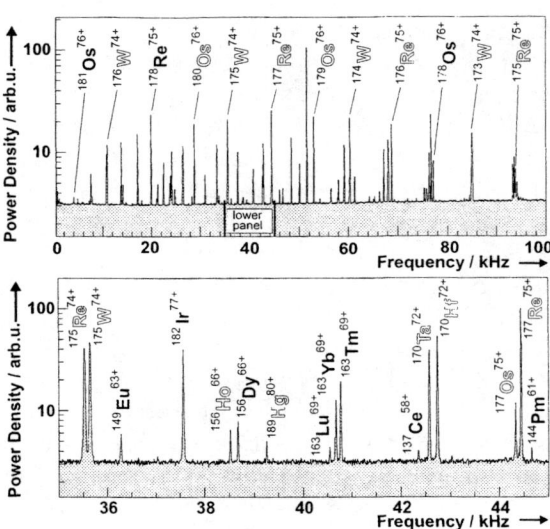

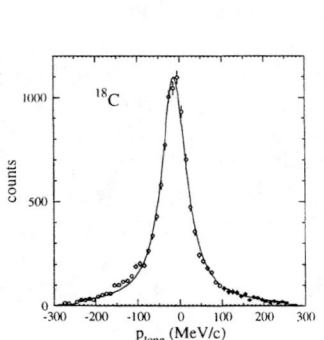

FIGURE 3. Momentum distribution [20] of ^{18}C after one-neutron removal from ^{19}C.

FIGURE 4. Mass spectrum of stored and cooled bismuth fragments in various ionic charge states, lower panel: zoomed part of the upper spectrum.

the combination of the Heidelberg CRYSTAL-ball for γ-spectroscopy, the ALADIN magnet for charged particle identification, and the LAND large area neutron detector. The possibilities of this method are still not fully explored. Improved detector systems with increased momentum resolution in combination with enhanced beam intensities would allow studies at high resolution, accompanying gamma radiation and to proceed to heavier drip-line systems

The heaviest halo nucleus known so far is ^{19}C. The neutron separation energy is 242 ± 95 keV. Fig. 3 shows the momentum distribution of the ^{18}C core after one-neutron removal, measured at the fragment separator with high momentum resolution. The longitudinal momentum spread is 69 ± 3 MeV/c and a factor of three more narrow than expected from the Goldhaber independent particle model for tightly bound nucleons [20] which is a strong indication for an extended neutron halo.

D   Precision Experiments in Storage Rings

Heavy ion cooler- and storage-rings open completely new possibilities for precision experiments with heavy ions, and specifically with exotic nuclei [21]. The advantages of storage rings as compared to traps are the large acceptance and the possibility to store nuclei with high energies appropriate for nuclear reactions.

First generation experiments are direct mass measurements. A an example of a mass spectrum of exotic nuclei produced by fragmentation of ^{209}Bi and stored and electron-cooled in the GSI experimental storage ring ESR is displayed in Fig 4 [23]. The low-resolution frequency spectrum (upper part) shows the large number of isotopes of elements osmium, tungsten and rhenium stored in the ring. The lower spectrum shows isobaric doublets and triplets of various ionic charge states, bare, hydrogen-, or helium-like. The precision of the mass determination is about 100 keV. Large mass surfaces can be surveyed at high precision: in two experiments 105 new masses of proton rich isotopes in the region from lanthanum to polonium could be determined [22,23]. Presently cooling time limits the application of this method to nuclei with half-lives of the order of tens of seconds. Fast cooling schemes are under development.

Short-lived nuclei at the very limits of stability can be measured with the storage ring operated in the isochronuous mode where the ring optics is chosen so that all nuclei of the same mass circulate with the same frequency irrespective of their velocity. First tests were successful [24].

In forthcoming experiments shell regions around ^{100}Sn and ^{78}Ni will be mapped. Of special interest are the nuclei along the r-process path. A long term goal is the investigation of the N = 126 shell region and the end of the r-process path. Hyperfine structure and isotope-shifts can be studied with laser methods. The new generation of power laser systems will significantly enhance the sensitivity so that laser spectroscopy with few atoms circulating inside the ring will become possible.

Next generation experiments will be nuclear reaction studies inside the storage ring using cooled fragment beams and thin internal targets with a thickness of $10^{12}/cm^2$ to $10^{14}/cm^2$ where atomic interactions such as energy loss and scattering play practically no role.

A low-energy electron-heavy-ion collider system would be a completely new experimental tool for structure studies e.g. to determine charge radii or electromagnetic excitation at GSI facility [21].

III   THE NEXT GENERATION FACILITY

A future facility for in-flight studies as GSI should comprise a powerful low-energy branch for studies of fusion products, preferably heavy and superheavy elements, and a high energy branch providing beams up to uranium at energies of up to 1 AGeV. An improved in-flight separator system of large acceptance will separate projectile fragments and fission products for decay studies, for reaction-experiments with a new setup for complete kinematics and high momentum resolution, or for injection into the storage ring system comprising an accumulator with fast stochastic cooling, a small electron collider of 400 to 800 MeV collision energy, and a high power laser system.

REFERENCES

1. A. Richter, Nucl. Phys. A553 (1993) 417c
2. P.G. Hansen, and B. Jonson, Europhys. Lett.4 (1987) 409 45 (1995) 591
3. H. Geissel, G. Münzenberg, and K. Riisager, Ann. Rev. Nucl. Part. Sci. 45 (1995) 163
4. I. Tanihata et al., Phys. Lett. 160B (1985) 380
5. S. Hofmann, Rep. Prog. Phys. 61 (1998) 639
6. G. Münzenberg in: Handbook of Nuclear Decay Modes, Eds. D. N. Poenaru and W. Greiner, W. De Gruyter, Berlin, 1997
7. S. Hofmann et al., Z. Phys. A305 (1982) 111
8. S. Hofmann et al., Z. Phys. A350 (1995) 277, Z. Phys. A350 (1995) 281, Z. Phys. A354 (1996) 229
9. R. Schneider et al., Z. Phys A348 (1994) 241
10. W. Schwab et al., Z. Phys. A350 (1995) 284
11. GSI working group: Nuclear Strucuture Phyics With Radioactive Beams, conveners: D. Habs, H. Lenske, P. Ring, G. Mñzenberg
12. G. Münzenberg, S,. Hofmann, H. Folger, F. P. Heßberger et al., Z. Phys. A322 (1985) 227
13. P. Möller, J. R. Nix, W. D. Myers, and W. J. Swiatecki Atomic and Nuclear Data tables 59 (1995) 185
14. S. Cwiok et al., Nucl. Phys. A611 (1996) 211
15. K. Rutz et al., Phys. Rev. C56 (1997) 238
16. P.Reiter, T. Khoo, C. J. Lister et al., Phys. Rev. C82 (1999) 509
17. R. Grywacz et al., Phys. Rev. C55(1997)1126, P. Regan, priv. Comm 1999
18. T. Aumann, L. V. Chulkov, V,. N. Pribora, and M. H. Smedberg, submitted to Nucl. Phys. A
19. D. Aleksandrow, T. Aumann, L. Axelsson, T. Baumann et al., submitted to Nucl. Phys. A
20. T. Baumann, H. Geissel, H. Lenske, K. Markenroth et al., submitted to Phys. Lett. B
21. G. Münzenberg, J. Friese, H. Geissel, I. Meshkov, G. Schrieder, and E. Syresin, Nucl. Phys. A626 (1997) 249c,
    I. Meshkov et al., Nucl Instr. Meth. A (1997)
22. H. Irnich et al., Phys. Rev. Lett. 75 (1995) 4182
23. T. Radon, H. Geissel, G. Münzenberg, B. Franzke et al., submitted to Phys. Rev. C
24. H. Geissel, priv comm, 1999

Status of the present radioactive beam facilities and perspectives for second generation installations

Daniel GUERREAU

GANIL, (IN2P3/CNRS, DSM/CEA)
BP 5027, 14021 Caen Cedex, France

Abstract. Radioactive nuclear beams offer very new opportunities in the domain of nuclear physics for the investigation of the isospin degree of freedom very far away from the β-stability line. The present situation with respect to the first generation Radioactive Beam Facilities (RIB) will be described. Intensity limitations of such facilities have led to the development of several projects all aiming at the increase by several orders of magnitude the RIB intensities. Innovative solutions are being explored and very ambitious projects are now discussed in Japan, USA and Europe. The main trends and goals of these projects will be reviewed.

INTRODUCTION

The understanding of the structure of nuclei and related properties of nuclear matter requires exploring extreme states of nuclear matter, especially along the isospin degree of freedom. The later is connected with the production and detailed studies of the properties of exotic nuclei far beyond the β-stability line and the present proceedings indeed stress the very intense undergoing activity in this domain.

One of the main tools for such nuclear structure studies, which will improve significantly our understanding of nuclear matter, is the development of dedicated radioactive ion beam (RIB) facilities. With such facilities, one is willing to focus on a number of fundamental questions: What is the limit for the existence of nuclei? What is the behaviour of nuclear matter with respect to the strong interaction near the limits of stability? The structure of exotic nuclei may contradict existing standard nuclear models. How to understand this structure far from the β-stability line? What can we learn in connection with the nucleosynthesis problem? An overview of the main scientific directions as well as the various possible solutions can be found in several reports. The reader may refer for instance to the NuPECC report (1) or the OECD Megascience Forum on Nuclear physics (2).

This paper intends to present the current status of the first generation Radioactive Beam Facilities (RIB), the detector developments as well as the undergoing R&D for developing innovative solutions for future second generation RIB facilities. Giving the large number of existing or planned facilities, it is excluded to give an exhaustive

CP495, *Experimental Nuclear Physics in Europe*, edited by B. Rubio et al.
© 1999 American Institute of Physics 1-56396-907-6/99/$15.00

overview of the present situation. The reader may also refer to ref. (3)(4)(5) for details on the physics as well as on the facilities. Special emphasis will be given here to the developments in Europe and to the NuPECC initiatives.

PRODUCTION TECHNIQUES: ISOL AND IN-FLIGHT METHODS; EXISTING FACILITIES

Two methods are heavily used for the production of radioactive ion beams, which are quite complementary, the ISOL and the In-Flight techniques.

The ISOL method (Isotopic Separation On-Line) has been extensively used especially in Europe, the most well known facility being ISOLDE at CERN (6). With such a method, the radioactive ions are produced in a thick target where the primary beam (light charged particles or heavy ions) as well as usually the reaction products are stopped. The radioactive species are then extracted, ionised and the needed mass is obtained trough an electromagnetic separation before to be possibly post-accelerated in a dedicated accelerator. The advantage of the ISOL method is that various drivers can be used (light particles as well as heavy ions) and the use of a post accelerator allows to deliver low energy RIB of high optical qualities as good as with stable beams. On the other hand, selective ion sources should be used and the access to very short lived species (below several msec) is somewhat limited to the time release from the production target except maybe with the IGISOL method used at Jyväskylä.

The main existing facilities of this type without post-acceleration are ISOLDE (6), IGISOL (7) and TISOL at TRIUMF (8). Facilities using a post accelerator are Louvain-La-Neuve (LLN) and HRIBF at Oak Ridge. At LLN, the pioneer post accelerated RIB facility (9), the driver is a high intensity (170µA) 30 MeV proton machine. RIB are produced by transfer reactions and post accelerated in a cyclotron up to a few MeV/u. Light beams as ^{13}N are produced with good intensities (10^5-10^7 pps). A new cyclotron, dedicated to astrophysics is presently under commissioning.

In-flight facilities rely on the use of medium and high-energy heavy-ion driver accelerators. The RIB are essentially produced in peripheral collisions by projectile fragmentation, transfer reactions and even projectile-fission. These products are very forward focused and the use of an on-line fragment separator as LISEIII (10) at GANIL allows to get a purified beam. The production of in-flight RIB gives easily access to very short periods and the acceptance is quite high due to the forward focusing of the products. Giving the high energy, those products can be easily identified. However, the optical quality might be limited and it is difficult to get access to low energies (E<30 MeV/u), except for some specific applications using a storage ring as in GSI.

The main operating facilities (all coupled with performing fragment separators) are GANIL (SISSI and LISE), MSU (A1200), RIKEN (RIPS), Dubna (Acculina and Combas) Lanzhou (RIBLL) at medium energies and GSI (FRS) at high energy.

FIRST GENERATION FACILITIES: FACILITY UNDER CONSTRUCTION, UPGRADES AND PROJECTS

Several projects of post-accelerated RIB based on the ISOL concept are presently under construction.

SPIRAL at GANIL (11) will be using the present GANIL cyclotrons as a driver. A vigorous R&D program (SIRA) on the production and ionization of the radioactive species produced by projectile fragmentation has been pursued since several years. This has led to the concept of a universal specially designed carbon target and to the development of performing ECR sources. Highly charged RIB will be injected in a new cyclotron CIME to be post-accelerated in the energy range 1.7-25 A.MeV. Beam tests with stable beams have been successfully achieved. For example, typical expected beam intensities for Ar isotopes are ranging from 2.10^8 pps for ^{35}Ar, 10^5 for ^{33}Ar and 10^3 for ^{32}Ar. The facility is ready to operate and waiting for authorization from the safety authorities. At the same time, the primary beam intensities have been increased up to 10^{13} pps and will finally reach a beam power close to 6KW.

At the same time, very performing experimental equipments are under construction trough large European collaborations, EXOGAM (12), a large γ-array detector and a large acceptance spectrometer VAMOS (13).

Prospects for a second phase of SPIRAL are underway. This second phase could be based on the use of a deuteron beam followed by a Be converter for the production of a high neutron flux which then induces fission in a thick U target. Intense fission fragments beams are expected to be produced (for instance 10^9 pps ^{91}Kr and 2.10^5 pps ^{78}Ni for a 6KW d beam at 200 MeV)[14]. A European Union RTD program[15] has been settled to experimentally test this concept of a deuteron based system and the first results are very promising (16)(GANIL, IPN Orsay, Jyväskylä, Groningen, Louvain-La-Neuve). This idea of using fast neutrons was first suggested by Argonne (17) and similar projects based on the use of thermal neutrons from reactors have been planned (the former PIAFE project and now the MAFF project near the FRM-II reactor at Munich (18).

Another facility is under construction at CERN. This is the REX-ISOLDE project (19) which will use the RIB produced by the existing ISOLDE facility. These RIB will be then accumulated, cooled and bunched using a Penning trap, ionized with an EBIS source and post-accelerated up to 2 A.MeV. The installation should be completed by the middle of 2000. At the same time, a new high efficiency γ-array (MINIBALL) will be operating.

Other ISOL projects presently under construction are ISAC (8) at TRIUMF (500 MeV p driver up 100μA and a post-accelerator up to 1.5 A.MeV) and EXCYT at LNS Catania (20) (heavy-ion superconducting cyclotron driver and a tandem post-accelerator).

Several other projects are being discussed but not founded yet. Among them, one may mention:

- The two projects at Argonne and Oak Ridge (see next section on second generation facilities)

- The Dubna project (possible use of two different drivers, heavy-ions for light RIB production and 25 MeV electrons for the production (by photofission) of intense fission fragments beams (21).
- SIRIUS at Rutherford Laboratory (22), a 800 MeV p driver up to 100μa followed by a linear post-accelerator up to 10 A.MeV. For example, intensities for ^{132}Sn should be close to a few 10^8 pps.
- Legnaro (high intensity 200 MeV d driver and a linear post-accelerator) (23).
- E-arena project at KEK (24) a 3GeV p driver up to 10 μA and a linac post-accelerator up o 6.5 A.MeV; expected intensities reach 10^9 at ±5 units from the stability).

As far as the In-Flight facilities are concerned, two main installations are being upgraded. The NSCL K1200 superconducting cyclotron will be coupled to the K500 cyclotron. Primary beam intensities up to 1pμA will be available with energies of 200 A.MeV for light beams and 100 A.MeV for Uranium. Gains of several orders of magnitude in the RIB intensities are expected to be achieved (25). At the same time, the A1200 fragment separator will be replaced by an improved system, the A1900 beam analysis system with a higher bending power more adapted for very n-rich beams. The project is scheduled to be completed in 2001.

The GSI complex will be also significantly improved; the UNILAC injector will be reconstructed and the SIS synchrotron will be filled up to its incoherent space charge limit (26). Intensities as high as 10^{10}pps for ^{238}U will then be achieved. This upgrade is expected to be operational in 2000.

TOWARDS SECOND GENERATION FACILITIES

If one is willing to ultimately answer the question about the nuclear stability and the origin of the elements, an exploration of the nuclear landscape as close as possible from the neutron and proton drip lines is required. In that context, the main limitation of the first generation radioactive beam facilities is the beam luminosity. As mentioned by the OECD Megascience report (2), there is a worldwide interest in the construction of advanced radioactive beam facilities. True second-generation facilities are at least aiming at a 10^3 intensity increase with respect to the present facilities (or those under construction), a necessary condition for a real breakthrough with respect to the present facilities. This is a real challenge for the next millennium.

Two projects in United States and Japan fulfill this condition (the situation in Europe will be discussed later).

The most advanced project is the project of a Radioactive Beam Factory at RIKEN (27). The new accelerator system will consist, in addition to the existing ring cyclotron, of a 4 sector ring cyclotron (IRC) and a 6 sector superconducting cyclotron (SRC) in order to boost the energy into the range 150-400 A.MeV up to uranium and reach primary beam intensities as high as 1particle μA (corresponding to a beam power of 100KW). A new beam separator (RIPS II) will be constructed. With such a system, nuclei up to N/Z=3 should be available. This first phase of the project is accepted and the IRC cyclotron is under construction. First beam from the SRC cyclotron is scheduled for 2003. The second phase of the project will consist of the MUSES complex (a booster synchrotron up to 1.4 A.GeV, an accumulator cooler, double storage ring for electron-RIB collisions). 2.5 GeV electrons will be provided by a LINAC. This very ambitious project is aiming at the production of more than 2000 nuclei with at least 10 pps. Preliminary schedule intends to achieve the whole project by 2006.

In United States, the high priority has been given by the NSAC to the definition of an advanced RIB facility. The main physics opportunities have been defined in the so-called Columbus white paper (28) and the present stage of the reflection with respect to a new project is the following (29): this is a projectile fragmentation based ISOL concept with a multiple heavy ion beam driver. The driver would be a 200 A.MeV heavy ion linear accelerator with an intensity of 1 particle μA (50-100KW-beam power). The radioactive products created by the in-flight method are then slowed down after passing through a fragment separator and stopped in a gas cell acting as an ion guide where they come to rest as 1^+ ions. The extraction is then possible in a few msec with a high efficiency independent of the chemical properties of the radioactive species. A linear post-accelerator is then boosting the ions up to a maximum energy of 15 A.MeV. This project is obviously at a very preliminary stage and may be modified in the following months.

What is the present situation en Europe? NuPECC has clearly considered that the existence of performing RIB facilities was a key-tool for the future of Nuclear Physics. A study group (first chaired by R.Siemssen and now by B.Jonson) has been settled in order to investigate the main options for two types of second generation RIB facilities in Europe (In-Flight and ISOL).

As far as the ISOL type is concerned, a large European Collaboration, supported by the EU 5th Research program, has been settled which intends to define a preliminary design study of the next generation European ISOL RIB facility (EURISOL collaboration: GANIL, Chalmers Univ., K.U.leuven, GSI, INFN, IPN Orsay, ISOLDE, Jyväskylä, RAL, Saclay). The first step is to investigate scientific and technical challenges, identify the required R&D, establish a cost estimate and look at possible synergies with other major European projects. This period should be followed by a vigorous R&D program on key technologies before to reach the point where the decision to build such a European Facility can be made.

Possible upgrade towards second generation facilities is under discussion. At GSI, one possibility has been studied implying the construction of a high intensity, high-energy synchrotron (with BR=50Tm) with an accumulator ring (with BR=18Tm). Intensities up to 10^{12} pps for 1A.GeV Uranium could be obtained. This facility could be also well suited for plasma physics studies.

At GANIL, it is presently thought to propose to add a new high intensity proton driver taking advantage of the existing R&D IFHI project of a linear proton accelerator for different purposes, in particular the developments of hybrid reactors for nuclear waste incineration. Discussions about the use of a 100mA, 30 MeV proton machine are underway which could at the same time favor crossdisciplinary developments.

All these ambitious projects require strong needs for technological developments. This concerns accelerator R&D, high power target, high intensity sources, highly efficient beam handling including cooling, traps, fragment separator, beam storage and instrumentation.

Many ongoing R&D programs have started in Europe, which should be supported and may be more coordinated. The OECD recommendations should be followed: it would be very positive to decide a world initiative of the funding agencies to exchange information and define opportunities and partnership for international cooperation.

There is a need in Europe for the next millennium to have two performing second-generation facilities based on ISOL and In-Flight methods. A positive decision for such very ambitious projects will be achieved only providing there is an agreement between the main laboratories and the funding agencies.

REFERENCES

1. *Nuclear Physics in Europe : highlights and opportunities* (NuPECC report, december 1997)
2. *OECD Mega Science Forum on Nuclear Physics* (Final report, january 1999)
3. A.C.Mueller, *Proceedings of ENAM98 : Exotic Nuclei and Atomic Masses*, edited by B.M.Sherill, D .J.Morissey and C.N.Davids, 1998, American Institute of Physics, p 933
4. I .Tanihata, id ref 3, p 943
5. J.A.Nolen, id ref 3, p 952
6. H.L. Ravn, *Nucl. Inst. Methods* B70, 107-112(1992)
7. M.Huhta et al, *Nucl. Inst. Methods* B126,201(1996)
8. TRIUMF, see www.triumf.ca/isac/lothar/isac.html
9. J.Vervier, Nucl .Phys. A616, 97c (1997)
10. A.C.Mueller and R.Anne, *Nucl.Inst.Methods* B70 107-112(1992)
11. M.P. Bourgarel et al, *Proceedings of the International Conference on Cyclotrons and their Applications* (Caen, 1998), IOP publishing, p 311.
12. F.Azaiez et al, Nouvelles du GANIL, 60(1997)4
13 . H.Savajols et al, Nouvelles du Ganil, 65(1999)
14. D.Ridikas and W.Mittig, *GANIL Preprint* P98-22
15. The SPIRAL Phase II project, http//www.ganil.fr
16. F.Clapier et al, Exotic Beams Produced by Fast Neutrons, *Phys.Rev. Special Topics* AB vol 1,1998
17. J.A.Nolen, *Proceedings of the 3rd Int. Conf. On Radioactive Nuclear Beams*, 1993, East Lansing, USA, ed. by D.J.Morissey (Ed. Frontières, 1993) p 111-115
18. D.Habs et al, *Nucl. Phys.*, A616, 39c(1997)
19. D.Habs et al, *Nucl.Inst.Methods*, B126,218(1996)
20. P.Finocchiaro and D.Vinciguerra, Nucl.Phys.News, vol.8, 4 (1998)
21. Y.Oganessian, private communication
22. The SIRIUS Project, CLRC, may 1999
23. L.Tecchio, private communication
24. H.Sakurai et al, *Nucl. Phys.*, A616, 311c(1997)
25. For more details see http://www.nscl.msu.edu
26. GSI web site : www.gsi.de
27. I.Tanihata, *J.Phys.G : Nuc. Part. Phys.* 24(1998)1311
28. www.er.doe.gov/production/henp/isolpaper.pdf
29. H.Grunder, ISOL Task Force Update to NSAC, April, 1999

Reaction mechanism and structure interplay for proton elastic scattering from halo nuclei

R. Crespo * and R.C. Johnson [†]

* Departamento de Física, Instituto Superior Técnico, Lisboa, Portugal
† Department of Physics, University of Surrey, GU2 5XH, United Kingdom

Abstract.
The aim of this work is to clarify what properties of the projectile w.f. are relevant to describe elastic scattering of halo nuclei from stable nuclei. In particular, we examine how far elastic scattering observables probe correlation effects among projectile nucleons.

Our treatment is based on a multiple scattering expansion of the proton-projectile transition amplitude in a form which is well adapted to the weakly bound cluster picture of halo nuclei. In the specific case of ^{11}Li scattering from protons at 800 MeV/u we show that because core recoil effects are significant, scattering crosssections can not, in general, be deduced from knowledge of the total matter density alone. We advocate that the optical potential concept for the scattering of halo nuclei on protons should be avoided and that the multiple scattering series for the full transition amplitude should be used instead.

This work concerns the scattering of light halo nuclei from a proton target within a framework in which only the degrees of freedom of the loosely bound valence nucleons orbiting around a relatively tightly bound core are taken into account [1].

We develop a multiple scattering expansion of the nucleon-projectile transition amplitude (MST) for the scattering of a weakly bound few body system from a proton target. The transition amplitude, T, can be written as a multiple scattering expansion in the transition amplitudes $\hat{t}_\mathcal{I}$ for proton scattering from each projectile sub-system $\mathcal{I}$ [2].

We contrast this with other works [3,4] based on a multiple scattering expansion of the optical model operator (MSO) which treats the ground and excited states of the projectile on a different footing. The leading term of this expansion envolves the total halo matter density distribution, and higher order terms are express in terms of the projectile correlations. The present approach is more appropriate for few-body projectiles at high projectile energy.

Here we address the question of whether elastic scattering from halo nuclei is determined by the total matter density of the projectile alone [5,6].

CP495, *Experimental Nuclear Physics in Europe*, edited by B. Rubio et al.
© 1999 American Institute of Physics 1-56396-907-6/99/$15.00

We have applied our formalism to the scattering of ^{11}Li from a proton target at 800MeV/nucleon. We have shown that effects associated with correlations between the nucleons in the double scattering term are small in this case.

We also show that to a good approximation the scattering involves the halo wavefunction in two distinct ways: through the halo density function $\rho_v(\Delta)$ and through the density distribution for the motion of the core center of mass, $\rho_{cm}(\Delta)$, [1]. Both of these density functions are defined in terms of the 2-body halo density

$$\rho_2(\vec{\Delta}_1, \vec{\Delta}_2) = \int d\vec{Q}_1 d\vec{Q}_2\, \varphi_{nn}^*(\vec{Q}_1, \vec{Q}_2)\varphi_{nn}(\vec{Q}_1 + \vec{\Delta}_1, \vec{Q}_2 + \vec{\Delta}_2) \ , \tag{1}$$

by

$$\rho_v(\vec{\Delta}) = \rho_2\left(\tfrac{m_3}{M_{23}}\vec{\Delta}, \tfrac{m_4}{M_{234}}\vec{\Delta}\right) \ , \quad \text{and} \quad \rho_{cm} = \rho_2\left(0, \tfrac{M_{23}}{M_{234}}\vec{\Delta}\right) \ , \tag{2}$$

where $M_{23} = m_2 + m_3, M_{234} = m_2 + m_3 + m_4$, etc. In eq.(1) $\varphi_{nn}(\vec{Q}_1, \vec{Q}_2)$ is the Fourier transform of wave function of the two body valence system relative to the core $\varphi_{nn}(\vec{r}, \vec{R})$. The departure of $\rho_2(0, \tfrac{M_{23}}{M_{234}}\vec{\Delta})$ from unity arises from core recoil effects. Because core recoil effects are important, halo structure information associated with $\rho_v(\Delta)$ and $\rho_2(0, \tfrac{M_{23}}{M_{234}}\vec{\Delta})$ does not contribute to the scattering simply combined as a total matter density.

Thus, a proper treatment of the reaction mechanism for halo nuclei elastic scattering need necessarily to incorporate structure information features that go beyond knowledge of the total halo matter density distribution. We advocate that in microscopic theories of proton scattering from light nuclei such as halo nuclei, at intermediate and high energies the multiple scattering series for the full transition amplitude should be used and that the optical potential concept should be avoided.

Acknowledgements: This work is supported by Fundação para a Ciência e Tecnologia (Portugal) through grant No. PRAXISXXI/PCEX/P/FIS/4/96, and by EPSRC(UK) through Grant No. GR/J95867. We would like to thank Professor Ian Thompson for providing us with the ^{11}Li wave functions in a convenient form.

REFERENCES

1. M.V. Zhukov, D.V. Fedorov, B.V. Danilin, J.S. Vaagen, J.M. Bang and I.J. Thompson, Nucl. Phys, **A552**, 353 (1993).
2. K.M. Watson, Phys. Rev. **105**, 1338 (1957); M.L. Goldberger and K.M. Watson, **Collision Theory** (John Wiley and Sos, New York, 1964).
3. R. Crespo, R.C. Johnson and J.A. Tostevin, Phys. Rev. **C44**, R1735 (1991), *ibid.* Phys. Rev. **C46**, 279 (1992).
4. R. Crespo, J.A. Tostevin and I.J. Thompson, Phys. Rev. **C54**, 1867 (1996).
5. J.S. Al-Khalili *et al.*, Phys. Rev. **C54**, 1843 (1996),
6. G.D. Alkhazov *et al.*, Phys. Rev. Lett. **78**, 2313 (1997).

Nuclear reactions involving weakly bound nuclei

R. C. Johnson, J. S. Al-Khalili, N. K. Timofeyuk, N. Summers

Department of Physics, University of Surrey, Guildford, Surrey GU2 5XH

Abstract. The adiabatic approximation in the theory of reactions involving halo nuclei is reviewed. This approximation gives new insights into the role of the halo as well as suggesting practical ways of including the effects of strong coupling to continuum states of the halo system. The application of these techniques to total reaction cross section and elastic scattering calculations is reviewed and new results for transfer reactions are compared with recent GANIL data. Corrections to the adiabatic approximation are briefly discussed.

INTRODUCTION

Halo nuclei are often described by few-body models which identify halo and core degrees of freedom and treat them on a different footing. In theories of nuclear reactions involving these nuclei and stable targets additional approximations of unknown validity are usually made. In particular the use of the DWBA must be considered suspect because the weak binding inherent in the halo nucleus concept suggests that strong coupling to excitations of the halo degree of freedom will be important. The fact that these excitations include the continuum complicates the situation even further.

THE ADIABATIC APPROXIMATION

The 'adiabatic' approximation (some authors refer to the same approximation as 'sudden') is consistent with a strong coupling situation and permits a significant simplification of the few body scattering problem. The motion of the halo degrees of freedom is assumed to be much slower than the projectile's centre-of-mass motion. The adiabatic approximation has been used for reactions induced by deuterons and other light halo systems by the Surrey group [1–8]. It is of importance to understand the range of validity of the adiabatic approximation and how to incorporate leading corrections into a practical theory.

CP495, *Experimental Nuclear Physics in Europe*, edited by B. Rubio et al.
© 1999 American Institute of Physics 1-56396-907-6/99/$15.00

Application to total reaction cross section calculations

The adiabatic approximation underlies Glauber's [9] microscopic eikonal theory of the scattering of composite systems. In practical applications additional approximations are often made. In particular the static density, or 'optical limit' approximation, is frequently used. This approximation has been shown [3,4] to have serious shortcomings when loosely bound neutron halo nuclei are involved. When applied to the interpretation of fragmentation cross sections the improved calculations of [3,4] required halo radii which were significantly larger than those deduced using the static density approximation. This effect has recently been shown to be a very general consequence of the static density approximation [10].

Applications to reactions to specific channels

There now exist procedures for calculating the scattering wavefunction corresponding to projectiles with 2 [11,12] or 3 [5] constituents whose relative coordinates are treated adiabatically but whose remaining degrees of freedom are treated quantum mechanically. Using eikonal methods as an additional approximation, the case of 5 constituents (^{8}He) has also been evaluated using the best available projectile wavefunctions [13,14]. These methods have been successfully applied to the analysis of the elastic scattering and reaction cross sections of halo nuclei.

It was shown in [1] that when the interaction between the halo particles and the target is absent the adiabatic scattering wavefunction has a simple form. This result has led to new insights into the role of the halo in elastic scattering [1] and also been used as the basis of a new non-perturbative quantum mechanical theory of the Coulomb break-up of neutron halo nuclei [6–8].

The same idea has recently been used in a new approach to the theory of transfer reactions involving a halo nucleus in initial and final channels [15]. Here we compare the new theory with recent data from GANIL [16] for the reaction ^{11}Be(p,d)^{10}Be leading to the 0^+ ground state and the 1st excited 2^+ state of ^{10}Be. The theory of [15] takes into account break-up of the deuteron by ^{10}Be and excitation and break-up of ^{11}Be by the interaction of the incident proton with the ^{10}Be core of ^{11}Be (REB effect). An important feature of this new approach is that information about p-^{11}Be elastic and inelastic scattering and d-^{10}Be does not play a role in the analysis of the transfer reaction. For the evaluation of the REB effect elastic and inelastic p-^{10}Be data are the crucial phenomenological input.

Fig.1 shows that the REB effect produces changes in differential cross section in both magnitude and angular dependence. Deuteron breakup is included in all curves in Fig. 1. For other reactions in the same mass region and energy, this effect is crucial for obtaining good agreement between theoretical and experimental angular distributions. The spectroscopic factors extracted with the chosen geometry of the optical potentials and form factors as in [15] are the following: $S(0^+) = 0.35$

± 0.04 and $S(2^+) = 0.34 \pm 0.05$ without REB and $S(0^+) = 0.19 \pm 0.02$ and $S(2^+)$ $= 0.24 \pm 0.04$ with REB effects taken into account. The numbers quoted include an estimate of the uncertainty with which we are able to normalise the theoretical curves to the experimental ones. The value we deduce for the ratio $S(2^+)/S(0^+)$ is therefore 0.96 ± 0.26 and 1.28 ± 0.35 without and with REB effects.

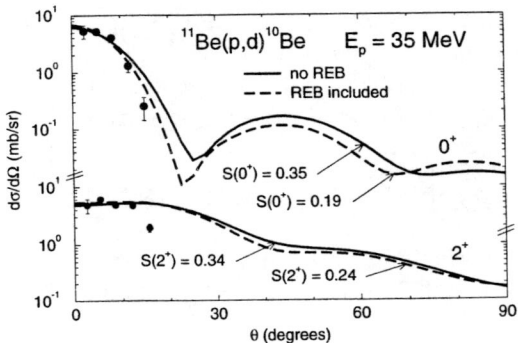

FIGURE 1. Cross sections for the ^{11}Be(p,d)^{10}Be reactions calculated with and without REB effects and normalised to the experimental data [16]

CORRECTIONS TO THE ADIABATIC APPROXIMATION

Although the conditions for the validity of the adiabatic approximation are well known qualitatively, the evaluation of leading corrections to scattering observables is difficult. For the case of elastic scattering of a projectile consisting of a core C and a set of valence nucleons v the leading correction to the T-matrix is

$$\Delta T^{el} = \langle \Psi^{ad(-)}_{\vec{K}'} \mid H_{vC} + \varepsilon_0 \mid \Psi^{ad(+)}_{\vec{K}} \rangle, \tag{1}$$

where the $\Psi^{ad(\pm)}$ are scattering states calculated in the adiabatic approximation, H_{vC} is the Hamiltonian describing the relative motion of the valence particles and the core, and $-\varepsilon_0$ is its ground state eigenvalue. Because of the long range of the operator $H_{vC} + \varepsilon_0$ and the oscillatory behaviour of the scattering states this expression must be handled with care. We write the scattering states in the form

$$\Psi^{ad(+)}_{\vec{K}} = \phi_0(\vec{r})\chi^{(+)}_{\vec{K}}(\vec{R}, \vec{r}), \tag{2}$$

where ϕ_0 is the ground state of H_{vC} and the notation for co-ordinates is standard. Note the dependence of $\chi^{(+)}_{\vec{K}}$ on both co-ordinates. Eq.(1) can now be written more conveniently as

$$\Delta T^{el} = \frac{\hbar^2}{2\mu_{vC}} \langle (\nabla_{\vec{r}} \chi_{\vec{K}'}^{(-)}) \phi_0 \mid \phi_0 (\nabla_{\vec{r}} \chi_{\vec{K}}^{(+)}) \rangle, \tag{3}$$

where μ_{vC} is the valence-core reduced mass.

This expression has been evaluated under the following assumptions: (i) The eikonal approximation is used for the $\chi^{(\pm)}$. (ii) The valence-target interaction is neglected. For the elastic scattering of ^{11}Be from ^{12}C at 50 MeV/A, the case discussed in [1], assumption (i) is a good one. We obtain negligible corrections to the angular distributions calculated in the adiabatic approximation and shown in [1]. However these calculations ignore excitations of the projectile produced by the valence-target interaction and these are likely to generate much larger non-adiabatic corrections. Calculations which include these effects are in progress.

ACKNOWLEDGEMENTS

The support of the U.K. Engineering and Physical Science Research Council in the form of Grant no. GR/J95867 and a Research Studentship for NS is gratefully acknowledged.

REFERENCES

1. R. C. Johnson, J. S. Al-Khalili and J. A. Tostevin, Phys. Rev. Lett **79**, 2771 (1997).
2. J. S. Al-Khalili and R. C. Johnson, Nucl. Phys. A**546**, 622 (1992).
3. J. S. Al-Khalili and J. A. Tostevin, Phys. Rev. Lett **76**, 3903 (1996).
4. J. S. Al-Khalili, I. J. Thompson and J. A. Tostevin, Phys. Rev. C**54**, 1843 (1996).
5. J. A. Christley, J. S. Al-Khalili, J. A. Tostevin and R. C. Johnson, Nucl. Phys. A **624**, 275 (1997).
6. J. A. Tostevin, S. Rugmai and R. C. Johnson, Phys. Rev. C**57**, 3225 (1998).
7. P. Banerjee, I. J. Thompson, and J. A. Tostevin, Phys. Rev. C**58**, 1042 (1998).
8. P. Banerjee, J. A. Tostevin, and I. J. Thompson, Phys. Rev. C**57**, 1337 (1998).
9. R. J. Glauber, in *Lectures in Theoretical Physics*, edited by W. E. Brittin (Interscience, New York, 1959), Vol. 1, pp. 315-414.
10. R. C. Johnson, in preparation, July 1999.
11. H. Amakawa, S. Yamaji, A. Mori, and K. Yazaki, Phys. Lett. B**82**, 13 (1979).
12. I. J. Thompson, Computer Programme ADIA, Daresbury Laboratory Report, 1984, unpublished.
13. J. A. Tostevin, J. S. Al-Khalili, M. Zahar, M. Belbot, J. J. Kolata, K. Lamkin, D. J. Morrissey, B. M. Sherrill, M. Lewitowicz, A. H. Wuosmaa, Phys. Rev. C**56**, R2929 (1997).
14. J. S. Al-Khalili, and J. A. Tostevin, Phys. Rev. C**57**, 1846 (1998).
15. N. K. Timofeyuk and R. C. Johnson, Phys. Rev. C**59**, 1337 (1999).
16. S. Fortier, *et al*, Phys. Letts., in the press. Preprint IPNO DRE-99-17, Orsay, 1999.

Breakup Reactions of ^{11}Li within a Three–body Model

E. Garrido*, D.V. Fedorov† and A.S. Jensen†

*Instituto de Estructura de la Materia, CSIC, E-28006 Madrid, Spain
†Institute of Physics and Astronomy, DK-8000 Aarhus C, Denmark

Abstract. Breakup reactions of ^{11}Li ($n+n+^9$Li) on a light target are investigated. The projectile–target interactions are described by phenomenological optical potentials. The model predicts dependence on beam energy and target, differences between longitudinal and transverse momentum distributions, and provides absolute values for all computed differential cross sections. Good agreement with the experimental data is obtained with a relative neutron–^{9}Li p–wave content of about 40%.

The collision between a spatially extended three–body halo and a small target is studied assuming that the probability than more than one of the constituents interacts with the target is small. The differential cross section is then written as a sum of three terms, each describing the independent contribution to the process from the interaction between the target and each of the constituents in the projectile. Under this model, and assuming that the halo constituent i that interacts with the target has spin 0 or 1/2, the differential cross section takes the form

$$\frac{d^9\sigma_{el}^{(i)}(\boldsymbol{P}',\boldsymbol{p}'_{jk},\boldsymbol{p}'_{0i})}{d\boldsymbol{P}'d\boldsymbol{p}'_{jk}d\boldsymbol{p}'_{0i}} = \frac{d^3\sigma_{el}^{(0i)}(\boldsymbol{p}_{0i}\to\boldsymbol{p}'_{0i})}{d\boldsymbol{p}'_{0i}} |M_s(\boldsymbol{p}_{i,jk},\boldsymbol{p}'_{jk})|^2 \tag{1}$$

$$\frac{d^6\sigma_{abs}^{(i)}(\boldsymbol{P}',\boldsymbol{p}'_{jk})}{d\boldsymbol{P}'d\boldsymbol{p}'_{jk}} = \sigma_{abs}^{(0i)}(p_{0i}) |M_s(\boldsymbol{p}_{i,jk},\boldsymbol{p}'_{jk})|^2 \, , \tag{2}$$

where eqs.(1) and (2) refer to the case of elastic scattering and absorption of the halo constituent i by the target. The momenta in the previous expressions refer to the different relative momenta between the particles in the final state [1]. The differential cross sections have factorized in two terms. The first of them corresponds to the individual differential cross section for the interaction between the constituent i and the target (elastic scattering or absorption), and the second term ($|M_s(\boldsymbol{p}_{i,jk},\boldsymbol{p}'_{jk})|^2$) is the square of the overlap between the three-body halo projectile wave function and the wave function of the spectators in the final state. This second term is precisely the cross section used in the sudden approximation [2].

CP495, *Experimental Nuclear Physics in Europe*, edited by B. Rubio et al.
© 1999 American Institute of Physics 1-56396-907-6/99/$15.00

To apply the model to fragmentation of ^{11}Li on carbon several interactions must be specified. The three–body halo projectile wave function requires the neutron-neutron interaction (chosen as in [2]) and the neutron-^{9}Li interaction, for which three different potentials with 40%, 30%, and 20% of p-wave content have been constructed. These potentials are labelled as I, II, and III, respectively, and together with the neutron-neutron interaction reproduce the main experimental features in ^{11}Li. For the neutron-carbon interaction we have chosen a non-relativistic optical potential valid for neutron energies ranging from 30 MeV to 1 GeV [1].

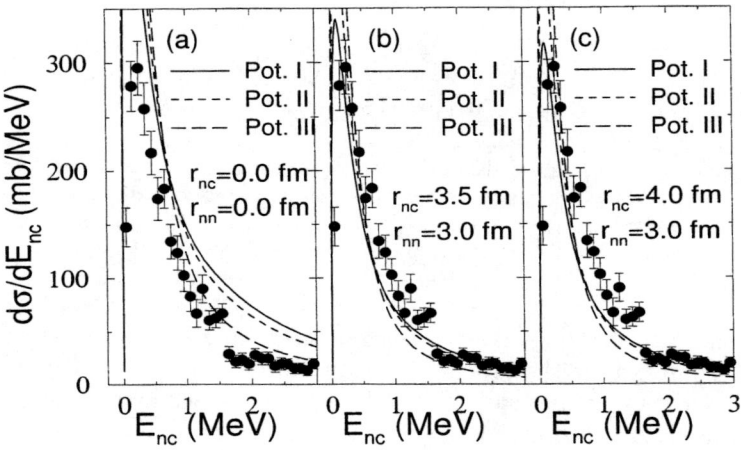

FIGURE 1. Invariant mass spectrum of ^{10}Li after fragmentation of ^{11}Li on C at 280 MeV/u.

The last ingredient to be considered is the so called shadowing. The fact that two of the constituents in the halo projectile are not affected by the target, together with the finite extension of the particles involved in the reaction, makes that configurations where the spectators pass too close to the target are excluded. This is done by killing the halo wave function if the spectator constituents j and k are closer to the participant i than a certain distance r_{ij} and r_{ik}, respectively.

Application of the model to ^{11}Li on carbon gives a good agreement with the experiment for a large variety of experimental data. As an illustration, in fig.1 we show the invariant mass spectrum of ^{10}Li after fragmentation of ^{11}Li on a carbon target at 280 MeV/u. Calculations with the three n-^{9}Li potentials and with different shadowing parameters are shown. It is clearly seen that potential I (40% p-wave) together with shadowing parameters $r_{nn} = 3$ fm and $r_{nc} = 4$ fm reproduces not only the experimental shape of the distribution but also the absolute value.

REFERENCES

1. Garrido, E., Fedorov, D.V., and Jensen. A.S., *Phys. Rev.* **C59**, 1272 (1999).
2. Garrido, E., Fedorov, D.V., and Jensen. A.S., *Phys. Rev.* **C55**, 1327 (1997).

Structure Studies of ^{11}Be and ^{12}Be: Observation of Molecular Rotational Bands

H.G. Bohlen*, A. Blažević*, B. Gebauer*, S.M. Grimes†, R. Kalpakchieva‡, T.N. Massey†, W. von Oertzen*, S. Thummerer*, and M. Wilpert*

*Hahn-Meitner-Institut, Glienicker Str. 100, D-14109 Berlin, Germany[1]
†Dept. of Physics and Astronomy, Ohio University, Athens, Ohio 45701-2979, USA[2]
‡Flerov Laboratory of Nuclear Reactions, JINR, 141980 Dubna, Russia,
and Bulgarian Academy of Science, Sofia, Bulgaria

Abstract. Excited states of ^{11}Be have been studied with several transfer reactions. Nine states between 3.96 MeV and 25.0 MeV excitation energy show the characteristic energy dependence of a rotational band. The deduced large moment-of-inertia of this band is consistent with a two-α structure with large deformation. For ^{12}Be four high lying states at 7.30 MeV, 10.7 MeV, 14.6 MeV and 21.7 MeV, which were observed in the ^{9}Be(^{15}N,^{12}N)^{12}Be reaction, also form a rotational band with almost the same moment-of-inertia as for ^{11}Be, using the tentative spin assignments of 2^+ - 8^+.

INTRODUCTION

Neutron-rich Be-isotopes have attracted strong interest since many years for structure studies due to the possibility to form pronounced α-cluster structures and even molecular states with two α-particles as cores. It is known from earlier studies that the *ground states* of Be-isotopes with A≥10 have a more compact shape. Molecular states can be localized at higher excitation energy, as discussed by von Oertzen [1]. For ^{11}Be he assigned the states at 3.96 MeV, 5.25 MeV, 6.72 MeV and 8.82 MeV, known from (t,p)-reactions on ^{9}Be [2,3], as a molecular rotational band with a $3/2^-$ band head. Kanada-En'yo et al. [4] obtained in AMD-calculations pronounced molecular structures of excited states of ^{10}Be, whereas the ground state is less deformed. Exotic cluster structures of ^{12}Be have been searched for by Korsheninnikov et al. [5] using the $^{12}Be+p$ inelastic scattering, and, recently Freer et al. [6] found an extremely deformed molecular band using

---

[1] Supported by NATO grant No. 960375.
[2] Supported by DOE grant No. DE-FG02-88ER40387.

CP495, *Experimental Nuclear Physics in Europe,* edited by B. Rubio et al.
© 1999 American Institute of Physics 1-56396-907-6/99/$15.00

the fragmentation of ^{12}Be into $^6He+^6He$ by coincident detection of the correlated ^6He fragments.

A MOLECULAR ROTATIONAL BAND IN ^{11}BE

We have investigated the structure of ^{11}Be using two-neutron transfer reactions on ^9Be, and also the $(^{14}N,^{13}N)$ one-neutron transfer reaction on a radioactive ^{10}Be-target. In the latter reaction only the single particle states were populated, up to the $3/2^-$ state at 2.69 MeV, above which no other state was observed. This behaviour changes dramatically using ^9Be as the target nucleus and two-neutron transfer reactions: states up to 25 MeV excitation energy were observed in the $^9Be(^{13}C,^{11}C)^{11}Be$ reaction at 379 MeV. Five new states were found at 10.8 MeV, 13.8 MeV, 18.6 MeV, 21.6 MeV and 25.0 MeV [7] (Fig. 1, upper panel). When these new states were included in the K=3/2$^-$ molecular band starting at 3.96 MeV exci-

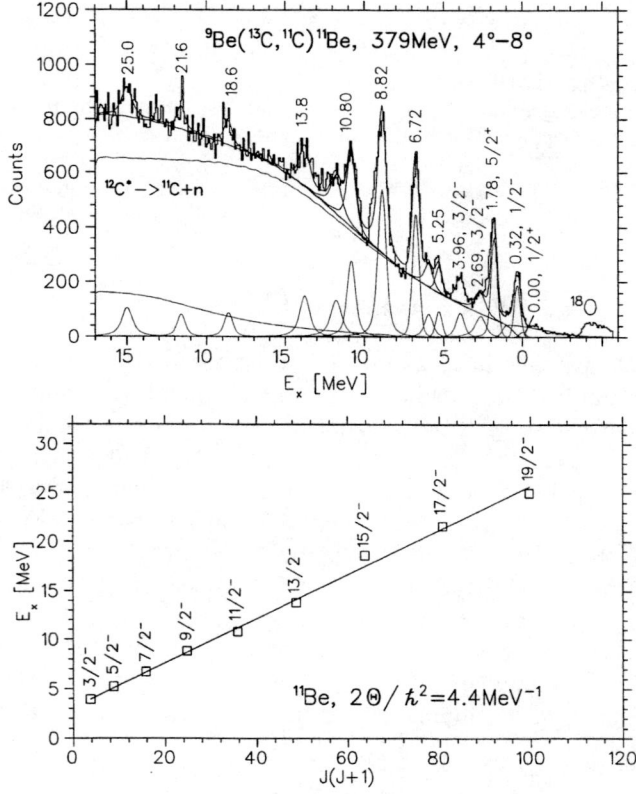

FIGURE 1. Spectrum of the $^9Be(^{13}C,^{11}C)^{11}Be$ reaction (upper panel), and plot of the excitation energies versus J(J+1) for the K=3/2$^-$ molecular rotational band of ^{11}Be (lower panel).

tation energy, a rather good linear dependence was obtained in a plot of excitation energies versus J(J+1) (Fig. 1, lower panel). The spins of the band members run up to a maximum spin of $19/2^-$ for the highest lying state. Such a high spin is still compatible with angular momenta available from single particle motion: the ^8Be-core can be excited up to the 4^+ state, the two transferred neutrons are placed into the 1d5/2 shell and couple up to a maximum spin of 4^+, and the single neutron of ^9Be adds with $j^\pi=3/2^-$ to the total spin. A large moment of inertia of 230 keV is found for this band, which supports the interpretation as a molecular structure. The observed widths of the nine band members are rather small (<15 keV [2] for the $3/2^-$ band head, increasing to 0.7 MeV for the $(19/2^-)$ state at 25.0 MeV excitation energy), which is consistent with a shape isomeric state and high spins.

We have also measured the $^9Be(^{15}N,^{13}N)^{11}Be$ reaction, $E_{Lab}=$ 240 MeV, where we observe again states at 10.8 MeV, 13.8 MeV and 21.6 MeV (not shown here).

MOLECULAR STRUCTURE STUDIES OF ^{12}BE

Low-lying states of ^{12}Be have been measured by Fortune et al. [8] using the (t,p) reaction on ^{10}Be. The formation of molecular structures in a transfer reaction depends sensitively on the structure conditions in the initial channel and the reaction mechanism. As we have seen for ^{11}Be, the molecular band could be well populated starting with ^9Be as the target, which has in its ground state already a strongly deformed two-α structure, whereas this was not the case using a ^{10}Be target. A similar selectivity is observed for molecular states of ^{12}Be. In Fig. 2, left panel, a spectrum of the $^{12}C(^{14}C,^{14}O)^{12}Be$ reaction, 335.9 MeV, is shown [9]: no state

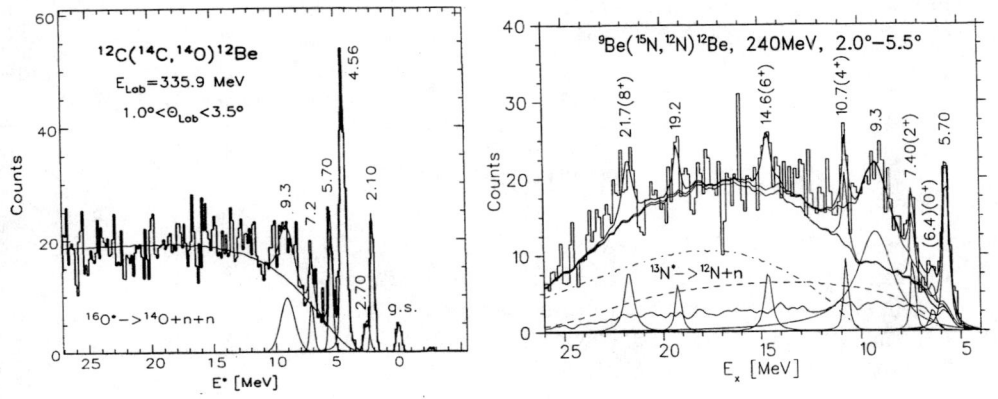

FIGURE 2. Spectra of reactions, which populate final states of ^{12}Be: the double charge exchange reaction $(^{14}C,^{14}O)$ on ^{12}C (left panel), and the three-neutron transfer reaction $^9Be(^{15}N,^{12}N)^{12}Be$ (right panel). The broad distributions correspond to sequential decay contributions.

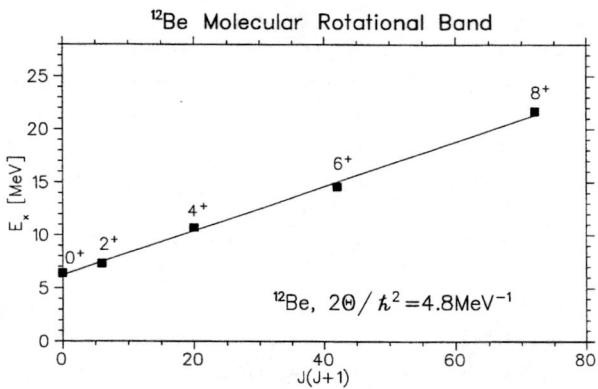

FIGURE 3. Plot of excitation energies versus J(J+1) for states of ^{12}Be, which are assumed to be members of a molecular rotational band. The moment-of-inertia is indicated in the figure.

could be observed above 9.3 MeV within the statistical significance. However, using a ^9Be target and the three-neutron transfer reaction $^9Be(^{15}N,^{12}N)^{12}Be$, further states at 10.7 MeV, 14.6 MeV, 19.2 MeV and 21.7 MeV excitation energy were observed (Fig. 2, right panel) with a significance $> 4\sigma$. In ^{12}Be a molecular rotational band can be formed in a similar way as for ^{11}Be using the same configurations: a ^8Be core (0^+ - 4^+); two neutrons transferred to the 1d5/2 orbit, which couple to a maximum spin of 4^+; the spin of the $3/2^-$ neutron from the ^9Be target, however, is now compensated by the third transferred neutron. For this case a maximum spin of 8^+ can be obtained. Using the tentative assignment of 8^+ for the state at 21.7 MeV, which is consistent with the results of Freer *et al.*, a molecular rotational band is found with the members 14.6 MeV (6^+), 10.7 MeV (4^+), and 7.3 MeV (2^+) (Fig. 3), with a moment-of-inertia of 210 keV very similar to ^{11}Be. The 0^+ band head is not directly observed due to the low cross section for a 0^+ state. In the plot of excitation energies versus J(J+1) it falls at an excitation energy of 6.4 MeV.

REFERENCES

1. Oertzen, W. von, Z. Physik A**357** (1997) 355.
2. Ajzenberg-Selove, F., *et al.*, Phys. Lett. **40B** (1972) 205.
3. Liu, G.B., and Fortune, H.T., Phys. Rev. C**42** (1990) 167.
4. Kanada-En'yo, C., *et al.*, J. Phys. G: Nucl. Part. Phys. **24** (1998) 1499.
5. Korsheninnikov, A.A., *et al.*, Phys. Lett. **82** (1995) 1383.
6. Freer, M., *et al.*, Phys. Rev. Lett. **82** (1999) 1383.
7. Bohlen, H.G., *et al.*, Proc. SNEC98 Conf., Padua, Nuovo Cim. **111**A (1998) 841.
8. Fortune, H.T., *et al.*, Phys. Rev. C**54** (1994) 1355.
9. Oertzen, W. von, Bohlen, H.G., *et al.*, Nucl. Phys. A**588** (1995) 129c.

On the breakup of low energy ^{8}B

F. M. Nunes[1,2] and I.J.Thompson[3]

1) Universidade Fernando Pessoa, Porto, Portugal
2) CENTRA, Inst. Sup. Técnico, Lisboa, Portugal
3) University of Surrey, Guildford, UK

Abstract. Coulomb dissociation has been proposed on many occasions as a means of determining the interaction between fragments at low relative energies. For the case of ^{8}B the main motivation has been the determination of the S_{17}. This method is briefly reviewed from a theoretical point of view, and the validity of certain approximations is discussed, focussing on the sub-Coulomb Notre Dame (ND) regime.

The reaction rates in stellar conditions happen at low relative energy and in most cases the cross section drops down steeply as $E \to 0$ due to the Coulomb repulsion. Even though great experimental progress has been made, the energy reached in laboratories is still not low enough. If the structure of the bound states and continuum is well known, a theoretical extrapolation $E \to 0$ can be made from higher energy data, given that the capture reaction mechanism is well established. However, often there is disagreement between the normalisation of independent measurements. Fortunately, due to the direct relation between $\sigma(a+X \to b+c+X)$ and $\sigma(b+c \to a)$, Coulomb Dissociation may provide information on astrophysical nuclear capture reactions [1,2].

Typically, in Coulomb Dissociation experiments, the cross section is always contaminated with E2 (whereas the low energy S-factors are usually just E1), nuclear effects may be important (introducing uncertainties due to the choice of the optical potentials) and the possibility of more complicated multistep reaction mechanisms has to be considered.

The ^{8}B breakup to extract the S_{17} has been the object of recent experiments: the high energy GSI experiment on ^{208}Pb [4], the intermediate energy RIKEN experiment on ^{208}Pb [5], and the low energy ND experiment on ^{58}Ni [6]. This last experiment measured the cross section at larger angles with the intention of extracting information on the E2. The object of our work has been the study of the reaction mechanisms for the low energy ^{8}B breakup.

The basic assumption in the semiclassical theory [3] is that the Coulomb field is much stronger than the nuclear field, as long as the impact parameter is larger than the sum of the radii of the projectile and the target ($b > R_p + R_T$). Under this premise, the projectile is classically scattered and the relative motion between the projectile fragments is treated quantum mechanically. In [8] we show that it is not possible to understand the ND data within this framework.

CP495, *Experimental Nuclear Physics in Europe*, edited by B. Rubio et al.
© 1999 American Institute of Physics 1-56396-907-6/99/$15.00

It has become very clear that for loosely bound projectiles the above assumption is not valid. Within a 1-step DWBA we demonstate that the finite size of the projectile reduces the pure Coulomb differential cross section by a significant amount [9]. The 1-step DWBA calculations also showed that there is a strong nuclear peak for $\theta \simeq 70°$ but even for angles as low as $\theta = 20°$, nuclear effects start to become important, in agreement with [10]. We find that there is destructive interference, reducing the Coulomb tail and the nuclear peak as well as Coulomb-Coulomb interference between the various multipoles. The sensitivity to the optical potentials is not large in the region where ND data was taken ($\theta = 30° - 50°$) but becomes very strong for backward angles.

Recently we have analysed this reaction using the continuum discretised coupled channels (CDCC) method [7]. As shown in [11], the continuum-continuum couplings are so strong that the usual Born series does not converge (the use of *Padé accelerants* enables multistep convergence). The inclusion of ^{8}B breakup channels reduces the cross section. Although the Coulomb component is reduced by about 10% through the inclusion of continuum-continuum couplings, the strong nuclear peak seen in 1-step calculations virtually disappears. Nevertheless there is an interplay between Coulomb and nuclear in such a way that the nuclear interaction should not by any means be neglected, over the whole range where experimental data is available. Extracting the E2 can only be done in a model dependent way, given that even when the nuclear effects are reduced, E1 and E2 components are entangled (we find the main destructive interference comes from 2-step E1 cancelling E2).

Finally, and most importantly, the differential cross section for the ND experiment is sensitive to the ^{8}B structure: by changing the single particle model for ^{8}B, the differential cross section is renormalised while the shape remains essentially the same. In this way, if other uncertainties can be isolated, this experiment will provide additional information on the S_{17}.

REFERENCES

1. A.C. Shotter et al., Phys. Rev. Letts. **53** (1984) 1539
2. G. Baur and H. Rebel, J. Phys. G **20** (1994) 1
3. K. Alder and A. Winther, *Electromagnetic Excitation*
4. N. Isawa et al., to appear in Phys. Rev. C.
5. T. Motobayashi et al., Phys. Rev. Letts. **73** (1994) 2680; T. Kikuchi et al., Phys. Letts. **B391** (1997) 261
6. Johannes von Schwarzenberg et al., Phys. Rev. **C53** (1996) R2598
7. Y. Sakuragi, M. Yahiro and M. Kamimura, Prog. Theo. Phys. Suppl.**89** (1986) 136
8. F. M. Nunes, R. Shyam, I.J. Thompson, J. Phys. **G 24** (1998) 1575
9. F.M. Nunes and I.J. Thompson, Phys. Rev. **C57** (1998) R2818
10. C.H. Dasso, S.M. Lenzi and A. Vitturi, Nucl. Phys. **A639** (1998) 635
11. F.M. Nunes and I.J. Thompson, Phys. Rev. **C59** (1999) 2652

Single-particle structure of radioactive beams from one-nucleon knockout reactions

A. Navin[a,b], T. Aumann[a,1], D. Bazin[a], B.A. Brown[a,c],

T. Glasmacher[a,c], P.G. Hansen[a,c], R.W. Ibbotson[a,2]

V. Maddalena[a,c], B.M. Sherrill[a,c], J.A. Tostevin[d], J. Yurkon[a]

[a] NSCL, Michigan State University, East Lansing, MI 48824.
[b] Nuclear Physics Division, Bhabha Atomic Research Centre, Trombay, Mumbai 400 085, India.
[c] Department of Physics and Astronomy, Michigan State University, East Lansing, MI 48824.
[d] Department of Physics, University of Surrey, Guildford, Surrey, GU2 5XH, U. K.

Abstract. Studies of the single-particle structure of radioactive beams produced in fragmentation reactions are described. The experiments are based on observing the individual final states of the projectile residues produced in one-nucleon knockout reactions. The measured partial cross sections to the various final states of the projectile residue and the shape of the corresponding longitudinal momentum distributions reflect the single particle properties, spectroscopic factors and the angular momentum l of the removed nucleon. Applications to ^{26,27}P and ^{15}C are discussed.

Investigations of the single particle structure of exotic nuclei are important for understanding the evolution of nuclear structure towards the drip lines. Transfer reactions using low energy stable beams were extensively used in the past to measure the single particle structure of nuclei near the valley of stability. Presently secondary beams of nuclei far from the valley of stability can be produced by projectile fragmentation. Typical energies of these exotic nuclei are around 50 MeV/u and higher. At these energies, the reaction processes tend to become simpler while the cross section for the transfer process decreases rapidly. On the other hand one-nucleon knockout reactions have large cross sections which are almost independent of energy [1]. Exclusive measurements of cross sections and parallel momentum distributions of individual final states populated in the projectile residue after one-nucleon knockout reactions on a light target have been shown to be a powerful way to extract spectroscopic factors and angular momentum assignments [2]. Thick targets and high detection efficiencies involved in these experiments push the sen-

---

1) Present address: G.S.I., Planckstr.1, 64291 Darmstadt, Germany.
2) Present address: Brookhaven National Laboratory, Upton, NY 11973-5000, U.S.A.

CP495, *Experimental Nuclear Physics in Europe*, edited by B. Rubio et al.
© 1999 American Institute of Physics 1-56396-907-6/99/$15.00

sitivity of this method to beam intensities of as low as 1 particle/sec.

In this contribution we briefly describe knockout reactions as a tool for spectroscopy of radioactive beams and illustrate the sensitivity and selectivity of this technique with a few examples.

The experiments were performed at the National Superconducting Cyclotron Laboratory (NSCL) at Michigan State University. The required secondary beams obtained from projectile fragmentation of primary beams from the K1200 cyclotron on a thick Be target were purified using the A1200 fragment separator and transmitted to the S800 spectrograph [3] operated in a dispersion matched mode. Intensities

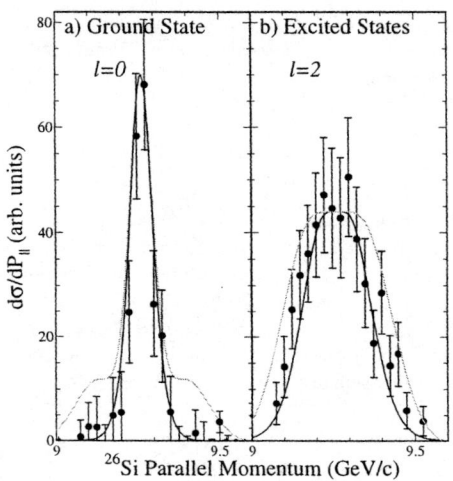

FIGURE 1. Components of the measured parallel momentum distributions after a one-proton knockout from ^{27}P on a Be target. The dark curves correspond to calculations taking into account the effect of shadowing [6] and the grey curves are those obtained neglecting this effect.

of the secondary beams ranged from $\simeq$ 50 particles/sec (^{26}P) to 10^5/sec (^{15}C). The secondary beams with energies of $\simeq$ 65 MeV/u and momentum spreads of 0.5% were incident on thick Be targets. Fragments produced in one-nucleon knockout reactions of the beam were identified and detected using time of flight and the focal plane detectors of the S800. The parallel momentum distributions of the fragments were reconstructed from their measured positions in two x-y position sensitive drift chambers and the known profile of the magnetic fields. Coincident γ rays from the excited fragments were detected using a position sensitive NaI(Tl) array [4] surrounding the target. The γ ray spectrum in the projectile frame was determined from the measured energies and positions in the array. From the above observables the total and partial cross sections after a one-nucleon knockout reaction populating various final states in the projectile residue were obtained.

The deconvolution of the measured momentum distribution into its components corresponding to ground and excited states was made in two different ways. In the case of ^{27}P the method described in [2] has been used and the results are shown in Fig. 1. For ^{15}C in addition to the above method the momentum distribution for the 1$^-$ excited state at 6.09 MeV was obtained by gating on the deexcitation γ rays and correcting for indirect feeding and the background. The fit to the measured coincidence gamma ray spectrum (Fig. 2a) was obtained by a simulation for the 6.09 MeV γ ray using GEANT [5] and an exponential background. The shape and magnitude of the background were measured independently in knockout reactions where no high energy γ rays were expected (^{12}Be,^{16}C on a Be target). The reliability of the simulation was confirmed by being able to reproduce source intensities to within 5% and also the line shapes. A similar method was used for the 0$^-$ state. The ground state momentum distribution is obtained from a subtraction of the excited states distribution from the inclusive momentum distribution. The present analysis yielded consistent results with those using the method in [2]. Shown in

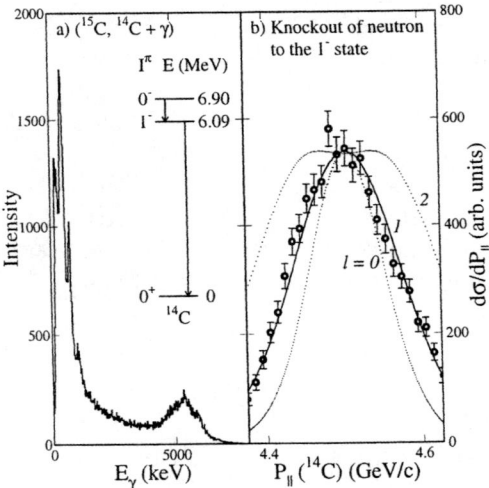

FIGURE 2. a) Doppler corrected γ ray spectrum in coincidence with ^{14}C fragments. b) Parallel momentum distributions corresponding to population of the 1$^-$ state. The calculated momentum distributions for various l values are shown.

Fig. 2b is the momentum distribution corresponding to the 1$^-$ excited state in ^{14}C. The predictions for the momentum distribution assuming various values of l for the knocked out nucleon obtained from an eikonal model [6] are also plotted. As can be seen from the figure the momentum distribution is characteristic of an $l=1$ nucleon which corresponds to the removal from the deeply bound p orbit in ^{14}C. This part contributes to 25% of the total one-neutron knockout cross section. Such a removal from a core leaving the halo nucleon intact was found to have a 15% contribution to the measured total one neutron removal cross section in the case of ^{11}Be on a Be target [7]. The measured ground state distribution is characteristic of a $l=0$ nucleon knockout.

The measured cross sections to various final states in the projectile residue are compared with a model [2] which incorporates single particle removal cross-sections (σ_{sp}) calculated using a modified eikonal model [1] and spectroscopic factors (C^2S) obtained from the shell model. The cross section to a given final state c in the core is expressed as $\sigma(c) = \sum_j C^2S(c, nlj)\sigma_{sp}(l_j)$. The measured cross sections for the various states in ^{14}C populated in one neutron knockout reactions are in good agreement with these calculations. The structure of ^{15}C has been well studied using (d,p) reactions and the derived spectroscopic factors in the present work are in good agreement with the literature [8] and a recent shell model calculation. Inclusive measurements of ^{15}C have been reported earlier [9]. Detailed results for ^{15}C on Be and various heavier targets will be presented elsewhere [10]. Comparing the experimental results for the total and partial cross sections using this method we have been able to reaffirm the tentative spin assignments of 3^+ and $\frac{1}{2}^+$ for ^{26,27}P. Assuming a spin 1^+ for ^{26}P would give a smaller cross section and be inconsistent with the measured ground state $s_{1/2}$ spectroscopic factors. The significance of the large s-component in the ground state of the neutron deficient phosphorus isotopes is an indication of a proton halo and has been discussed elsewhere [2,11].

The present results and those obtained for ^{11}Be [7], ^{12}Be and ^{16,17,19}C [12] indicate the versatility of this method. Applications at different beam energies for ^{15}C and ^{17}O (studied by complementary methods) will further test the validity of this method. New generation gamma ray detectors and higher beam intensities available after the NSCL upgrade will give us an opportunity to probe the structure near the drip lines for heavier nuclei.

REFERENCES

1. J. Tostevin, to be published; J. Phys. G **25**, (1998) 735.
2. A. Navin *et al.*, Phys. Rev. Lett. **81** (1998) 5089.
3. B.M. Sherrill *et al.*, to be published.
4. H. Scheit *et al.*, Nucl. Instr. Meth. A **422** (1999) 124.
5. GEANT, Cern Library Long Writeup W5013 (1994).
6. P.G. Hansen, Phys. Rev. Lett. **77**, (1996) 1016.
7. T. Aumann *et al.*, to be published.
8. J.D. Gross *et al.*, Phys. Rev. **C** 12 (1975) 1730.
9. D. Bazin *et al.*, Phys. Rev. C **57** (1998) 2156.
10. A. Navin *et al.*, to be published.
11. A. Navin *et al.*, in *Proc. ENAM98*, ed. by B.M. Sherrill *et al.*, AIP Conf. Proc. No. 455 (New York, 1998), p209.
12. A. Navin *et al.*, to be published, V. Maddalena *et al.*, to be published.

Formation and decay of ^{24}Mg in the ^{13}N+^{11}B collision.

P.Figuera[a], F.Amorini[a], W.Bradfield-Smith[c], M.Cabibbo[a], G.Cardella[b], T.Davinson[c], A.Di Pietro[c], W.Galster[d], P.Leleux[d], A.Musumarra[d], A.Ninane[d], M.Papa[b], G.Pappalardo[a], F.Rizzo[a], A.C.Shotter[c], C.Sukosd[e], S.Tudisco[a], P.J.Woods[c].

[a]*INFN-Laboratori Nazionali del Sud, Via S.Sofia 44, 95123 Catania Italy*
[b]*INFN-Sezione di Catania, Corso Italia 57, 95129 Catania Italy*
[c]*Department of Physics and Astronomy University of Edinburgh. Edinburgh U.K.*
[d]*Institut de Physique Nucléaire Universitè Catholique de Louvain. Louvain la Neuve Belgium.*
[e]*Department of Physics University of Budapest. Budapest Hungary.*

Abstract. Different aspects of the formation and decay of ^{24}Mg in the collision ^{13}N+^{11}B have been studied using a large solid angle and highly segmented Silicon strip detector. Results concerning the fusion cross section, the 6 α decay of ^{24}Mg and the GDR gamma ray emission are discussed.

The study of nuclear reactions with post accelerated radioactive beams allows the formation of the same intermediate systems studied with stable beams but with additional entrance channels. We have studied different aspects of the ^{24}Mg formation and decay in the reaction ^{13}N+^{11}B at E_{lab}=45.0 and 29.5 MeV. The experiment was performed at the radioactive beam facility of Louvain la Neuve. A large solid angle and highly segmented (224 strips) silicon strip array "Leda+Lampshade", sketched in figure 1, was used to detect charged particles which have been identified with the standard TOF technique. Three BaF$_2$ clusters covering the backwards angle were used to detect high energy gamma rays (E_γ>4 MeV).

The energy spectra of the evaporation residue are well reproduced by statistical model Montecarlo calculations. The fusion cross section obtained integrating the angular distributions, is in agreement, within the experimental error, with the fusion excitation functions of similar systems (^{14}N+^{10}B, ^{12}C+^{12}C) leading to the same compound nucleus. Therefore we do not observe strong structure effects on the fusion process due to the presence of the weakly bound proton (S$_p$=1.9 MeV) in ^{13}N.

The ^{24}Mg at E(^{13}N)=45 MeV is populated with $E^*\approx$47 MeV. In this E^* region of ^{24}Mg several resonances have been observed with ^{12}C+^{12}C scattering [1,2]. In particular a broad resonance was originally observed in [1] and attributed to the 6α linear chain configuration in ^{24}Mg. We studied the decay of the ^{24}Mg intermediate system into two ^{12}C* each of them decaying into ^{8}Be+α. Events with 6α in the final channel have been selected with the help of E-TOF identification spectra. In addition we constructed the Q value spectrum Q=Σ(E$_i$)-E$_{inc}$ and put a gate around the value Q=-500 KeV corresponding to the ^{13}N+^{11}B→6α reaction. The two ^{8}Be were selected

CP495, *Experimental Nuclear Physics in Europe*, edited by B. Rubio et al.
© 1999 American Institute of Physics 1-56396-907-6/99/$15.00

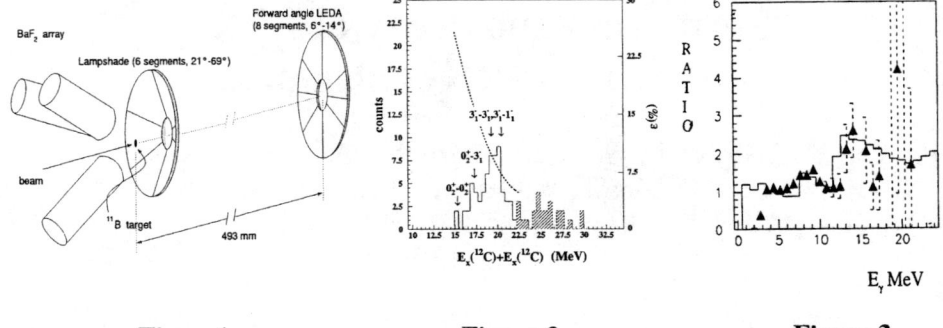

Figure1
Detection system

Figure 2
Excitation energy spectrum

Figure 3
γ spectra ratio

by looking at the relative energy ($E_{rel} \approx 90$ KeV) of α particles detected in adjacent strips. By measuring energies and angles of these two ^{8}Be and α particles, the double excitation energy spectrum for the two primary ^{12}C* nuclei has been reconstructed and is shown in figure 2. The calculated efficiency for the setup is shown in the same figure (dashed line). Although our data set is much smaller, there is a reasonable consistency between the present ^{13}N+^{11}B results and what observed by other authors in ^{12}C+^{12}C inelastic scattering. Bearing in mind the single particle structure of the ^{13}N+^{11}B channel compared to the α cluster nature of ^{12}C+^{12}C, this finding seems surprising.

The GDR gamma decay from the reaction ^{13}N+^{11}B (total IsopsinT=0,1 in the entrance channel) has been studied and compared with the one for the collision ^{14}N+^{10}B (T=0) leading to ^{24}Mg at the same excitation energy $E^* \approx 47$ MeV. Due to the E1 decay selection rules, a GDR gamma decay from an initial state T=0 will have to populate one of the less numerous T=1 states and the GDR yield will be suppressed. Due to isospin mixing T=1 states can also be populated in a reaction with T=0 in the entrance channel, therefore the GDR yield will be sensitive to the degree of isospin purity/mixing in the formed CN. Using radioactive and stable beams we are now forming the same CN with two very similar entrance channels characterized by T=0,1 and T=0. The calculated (Cascade-histogram) and experimental (symbols) ratios between gamma spectra (Eγ-(^{13}N+^{11}B)/Eγ-(^{14}N+^{10}B)) are shown in figure 3. The isospin suppression of the GDR yield in the ^{14}N+^{10}B reaction is clearly seen. Within the experimental error the data can be reproduced using a mixing coefficient α^2=0.1. This appears to be coherent with previous studies [3] on the same C.N. performed producing different isotopes of Mg in stable beam induced reactions.

REFERENCES

1. A.H. Wuosmaa et al., Z.Phys. A349, 249, (1994)
2 R.A. Le Marechal et al., Phys. Rev. C 55, 1881, (1997)
3 M.N. Harakeh et al., Phys. Lett. B176, 297, (1986).

Complete and incomplete fusion in the $^{32}S +^{12}C$ reaction at $E(^{32}S)$=20 MeV/A

S.Aiello[1], N.Arena[1,3], Seb.Cavallaro[1,3], E.Geraci[2,3], G.Lanzalone[1,4], S.Pirrone[1], G.Politi[1], F.Porto[2,3], S.Sambataro[1,3]

[1] INFN, Sezione di Catania, [2] INFN, Laboratori Nazionali del Sud, [3] Dipartimento di Fisica, Università di Catania, [4] CSFNSM, Catania

Abstract. Velocity distribution of mass identified evaporation residues produced in the $^{32}S +^{12}C$ reaction at $E(^{32}S)$=20 MeV/A have been measured using time-of-flight techniques. These distributions were used to separate the complete and incomplete fusion components. The complete fusion cross section and the deduced critical angular momenta are compared with other experimental data and the predictions of existing models.

INTRODUCTION

We study of the $^{32}S +^{12}C$ reaction at $E(^{32}S)$= 20 MeV/A [1] to investigate about the still interesting controversial question if the fusion cross section is either limited by the properties of the compound nucleus or by entrance channel effects.

EXPERIMENTAL PROCEDURE AND RESULTS

The experiments were performed at the Cyclotron accelerator facilities of the Laboratori Nazionali del Sud (LNS). The experimental system [2] used consists of a sliding seal scattering chamber rigidly connected to $\Delta E - E$ multianode ionization chamber with a 300 μm silicon detector at the end. A time of flight telescope consisting of two MCPs is coupled to the ionization chamber. The flight path is about 1 m. Example of charge resolution is shown in fig.1.

The separation of the different reaction mechanism from the inclusive velocity spectra is possible with complex deconvolution techniques, as you can see in fig.2. In this figure are reported the invariant velocity spectra of evaporation residues for the $^{32}S +^{12}C$ reaction at $E(^{32}S)$=20 MeV/A. The curves centered at $V_{CN}cos\vartheta_L$ (vertical arrow) represent the contributions of complete fusion. The dot and dash dot curves represent respectively the incomplete fusion and the direct reactions contributions.

CP495, *Experimental Nuclear Physics in Europe*, edited by B. Rubio et al.
© 1999 American Institute of Physics 1-56396-907-6/99/$15.00

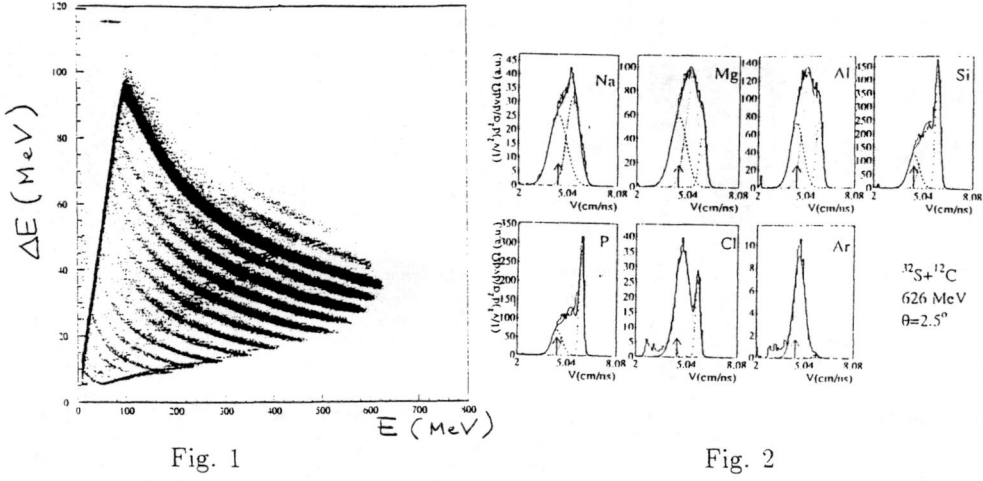

Fig. 1   Fig. 2

The experimental excitation function for $^{32}S + ^{12}C$, constructed by our data and those available in the literature [3] is shown in fig.3 with the theoretical predictions derived in the frame of Bass [4] and Matsuse [5] models.

Finally the fig.4 shows the critical angular momenta extracted from the complete fusion cross section at different excitation energies, for several reactions leading to the ^{44}Ti compound nuclei. The solid line curve corresponds to the statistical yrast line [6] . The dashed line indicates the angular momenta at which the fission barrier vanishes as predicted by the Sierk model [7].

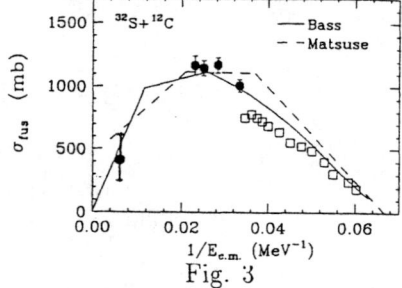

Fig. 3

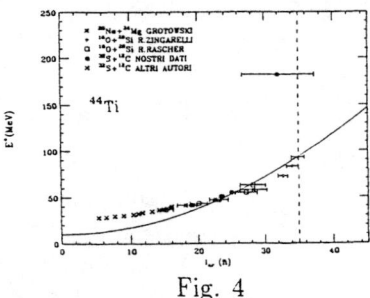

Fig. 4

The trend of the experimental points seems to show a saturation at high energy which could be consistent with the hypothesis that the limitation of fusion cross section is due to compound nucleus effects.

REFERENCES

1. S.Pirrone et al., *Atti del Congresso* **SIF 98**, 28 (1998).
2. P.Figuera, S.Pirrone et al., *Il Nuovo Cimento* **A104**, 251 (1991).
3. N.Arena et al., *Phys. Rev.* **C44**, 1947 (1991).
4. R.Bass, *Phys. Rev. Lett.* **39**, 265 (1977).
5. T.Matsuse et al., *Phys. Rev.* **C26**, 2338(1982).
6. S.M.Lee et al., *Phys. Rev. Lett.* **45**, 165 (1980).
7. Arnold J.Sierk et al., *Phys. Rev.* **C33**, 2039 (1986).

Yields of Neutron Rich Isomers Produced via Deep Inelastic Collisions

M.Ogawa, I.Hossain, T.Ishii*, M.Asai*, A.Makishima*, S.Ichikawa*,
M.Itoh, P.Kleinheinz and M.Ishii*

*Research Laboratory for Nuclear Reactors, Tokyo Institute of Technology,
Ohokayama, Meguro-ku, Tokyo 152-8550 Japan*
*Japan Atomic Energy Research Institute, Tokai, Ibaraki, 319-1195, Japan

Abstract. Neutron rich isomers near ^{68}Ni have been investigated via deep-inelastic collisions as a function of projectile mass and incident energy. Gamma-rays from the ns- μs isomers were measured in coincidence with projectile-like fragments. Yields of 26 isomers have been derived from three experiments with different projectile conditions.

EXPERIMENTAL RESULTS

Deep-inelastic collisions (DIC) are feasible for study of neutron-rich nuclei (1). In particular, spectroscopic data on neutron rich nuclei including the doubly magic ^{68}Ni and ^{78}Ni are required to examine the nuclear shell structures for nuclei toward neutron drip line. Isomers contain fruitful information on nuclear shell structures. We have expected isomers of E2 type based on $(g_{9/2})^2$ configuration and of M2 type resulting from $g_{9/2}$-$f_{5/2}$ spin sequence in the mass region of A=60 to 90 (2).

We have performed three DIC experiments at the JAERI tandem-booster facility; [1] ^{76}Ge (635 MeV), [2] ^{76}Ge (550 MeV) and [3] ^{74}Ge (625 MeV) projectiles with a ^{198}Pt target. An isomer scope (3) consisting of a γ-ray shield, an annular type of Si detector of 10 cm in diameter and four Ge detectors was employed to intensify the γ rays emitted from the isomers. Projectile-like fragments (PLF) ejected to the grazing angle were stopped with the Si detector where flight time of PLFs was about 1 ns. The kinetic energy of PLF measured by the Si detector was roughly proportional to the number of transferred mass, i.e., $|Z| + |N|$ (4). Gamma rays in coincidence with the PLF signals were recorded event by event. Procedure of isomer identification, i.e., assignment of mass and atomic number is explained in our previous works (2-4).

About 40 isomers were observed and 26 of them were identified. Table 1 lists relative yield ratios of I(^{74}Ge)/I(^{76}Ge) where isomer yields of I(^{76}Ge) and I(^{74}Ge) were deduced from experiments 1 and 3, respectively. Less-neutron-rich isomers have the larger yields with the lighter projectile of ^{74}Ge for Ni and Cu isotopes. The similar trend

CP495, *Experimental Nuclear Physics in Europe,* edited by B. Rubio et al.

was also observed for As isotopes. Figure 1 shows dependence of isomer yields on projectile energy resulting from experiments 1 and 2. Yield ratios of I(550 MeV)/I(635 MeV) are plotted as a function of isomer mass. Note that the ratios are separated into two groups taking an abrupt jump at the mass of projectile. The mechanism of this jump is not understood. However, the present data indicate that the lower beam energy leads to the higher yields of heavier isomers in the energy region of 7 to 8 MeV/u.

REFERENCES

1. R.Broda et al., Phys. Rev. Lett. **7 4**, 868-871 (1995)
2. T.Ishii et al., Phys. Rev. Lett. **8 1**, 4100-4103 (1998)
3. T.Ishii et al., Nucl. Instr. Meth. **A395**, 210-216 (1997)
4. I. Hossain et al., Phys. Rev. **C58**, 1318-1320 (1998)

TABLE 1. List of relative yield ratios of I(^{74}Ge)/I(^{76}Ge). Yield I(^{74}Ge) for ^{65}Ni was not obtained.

^{61}Cu	^{64}Co	^{63}Ni	^{65}Ni	^{66}Ni	^{67}Ni	^{64}Cu	^{66}Cu	^{67}Cu
90	150	317	NA	139	80	280	120	133
^{69}Cu	^{71}Cu	^{67}Zn	^{68}Ga	^{70}Ga	^{70}Ge	^{72}As	^{73}As	^{74}As
42	22	164	165	218	218	131	77	158
^{76}As	^{79}As	^{81}Br	^{83}Br	^{84}Kr	^{87}Rb	^{86}Sr	^{90}Zr	
89	17	74	37	32	119	47	119	

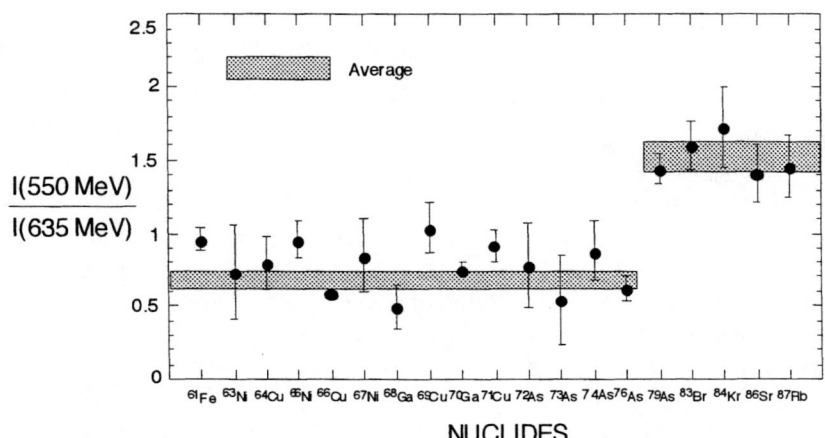

FIGURE 1. Ratios of isomer yields observed with ^{76}Ge projectile energies of 550 and 635 MeV for different isomers.

Two-neutron interferometry in low- and intermediate-energy heavy-ion reactions

G. Tagliente[a], R. Ghetti[b], N. Colonna[a], E. De Filippo[c]
for the ARGOS-CHIC-RIPEN collaboration

[a] INFN and Dip. Fisica, V. Amendola 173, 70126 Bari, Italy
[b] Department of Physics, Lund Univ., Box 118, SE-221 00 Lund, Sweden
[c] INFN and Dip. Fisica, C.so Italia 57, 95129 Catania, Italy

Abstract. Two-nucleon interferometry can be used in low- and intermediate-energy heavy ion reactions to probe the space-time evolution of hot and possibly compressed nuclear systems. At low energy, angle-gated correlation functions have allowed to succesfully extract the lifetime of moderately excited compound nuclei. At intermediate energy, two-neutron interferometry has been used to obtain information on the contribution of different mechanisms to the reaction dynamics.

INTRODUCTION

The relative wave function of identical nucleons is influenced by the Quantum-Statistical Interference (QSI), the short range Final State Nuclear Interaction (FSI) and, for charged pairs, the long-range repulsive Coulumb Interaction (CI). Since the magnitude of these effects depends on the space-time separation between particles, multi-dimensional two-nucleon interferometry can be used to obtain information on the radius and lifetime of a decaying system [1].

In low energy heavy ion reactions, both two-proton and two-neutron interferometry can be applied to study the lifetime of compound nuclei [3,4]. Two-neutron correlations are easier to interpret, as compared to pp ones, since they are not distorted by Coulomb effects. However, several experimental problems, and in particular detector cross-talk, have to be addressed [2].

At intermediate energy, where the dynamical evolution of the reaction is more complicated, two-neutron interferometry can be applied to verify processes, such as expansion or the formation of hot fireball-like regions. In this energy region, however, because of the complicated dynamical evolution, a selection and characterization of the emitting sources is mandatory.

In this talk, we present the results of a study of two-neutron interferometry at Tandem energy, together with an analysis of reactions in the Fermi energy domain.

CP495, *Experimental Nuclear Physics in Europe*, edited by B. Rubio et al.

COMPOUND NUCLEUS LIFETIME AT LOW ENERGY

A study of two-neutron interferometry at low energy was performed at the Tandem Accelerator of Laboratori Nazionali di Legnaro, Italy. Correlation funtions were measured for neutrons emitted in the 130 MeV ^{18}O + ^{26}Mg reaction, which leads to the formation of ^{44}Ca compound nucleus at 100 MeV excitation energy. The presence of a unique source with well defined velocity is fundamental in the analysis of angle-gated correlation functions. The experimental apparatus consisted of 22 neutron detectors, BC501A liquid scintillator cell 12.7x12.7 cm, mounted in a close-packed configuration, with angular separation between detectors of $\approx$ 6 degrees. The cluster was positioned at 45 degrees, at distance of 2 m from target.

When studying two-neutron correlaton functions, particular attention has to be paid to the problem of cross-talk, that is a neutron scattering from one detector into a neighboring one, which might produce large distortions of the measured correlation functions. Several methods can be used to identify and reject cross-talk events. The one used in this analysis, described in more details in [2], relies purely on the measured time-of-flight.

Experimental angle-gated correlation functions for neutrons emitted from the ^{44}Ca compound nucleus reaction are shown in figure 1. The solid and empty symbols in the figure represent respectively logitudinal and transverse correlation functions, constructed by gating on the angle ψ between the relative and the total momentum of the neutron pair, the later evaluated in the rest frame of the ^{44}Ca compound nucleus. As expected, transverse correlation functions are suppressed relative to the longitudinal ones, because of the fermionic anticorrelation from the attractive final state interaction. The magnitude of both correlation functions were exploited to obtain information on the source dimension and on the neutron emission time-scale, by comparison with theoretical predictions. The minimum value of χ^2 indicate a radius of 4.4 fm and an emission scale time of $\approx$ 1100 fm/c [4].

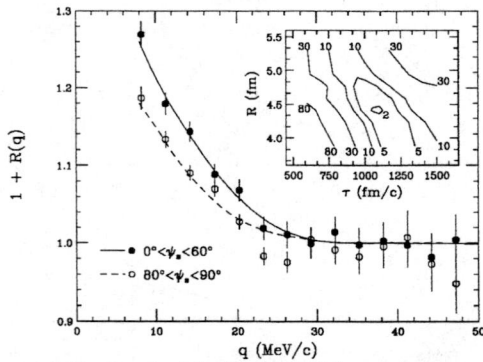

FIGURE 1. Experimental longitudinal (solid symbols) and transverse (open symbols) correlation functions. The solid and dashed curves depict calculations with Koonin-Pratt formalism (see text for details). The panel in the figure shows the reduced χ^2 between the experimental and theoretical correlation functions.

INTERFEROMETRY IN FERMI ENERGY DOMAIN

While at low energy the reaction picture is sufficently clear, at intermediate energy the evolution of the reaction is far from being understood. Some aspects of the reaction dynamic in the Fermi-energy domain might be clarified with the help of two-nucleon interferometry. At this energy, particle emission is characterized by a fast pre-equilibrium stage followed by a slower emission from equilibrated system. A measurement of the fragment decay channel, in coincidence with the nucleon pair may help to correlate the results of interferometry studies to the properties of the emission sources, in particular to their excitation energies.

With the aim of studing the space-time evolution of the reaction zone in intermediate energy heavy ion reactions, an European collaboration (Argos-Chic-Ripen) has undertaken an experimental investigation of 45 MeV/u ^{58}Ni-induced reactions at the Superconducting Cyclotron of Laboratori Nazionali del Sud, Catania, Italy. A complex apparatus has been put together in order to simultaneously measure nn, np and pp correlation functions in coincidence with fragments. Neutrons were detected in 48 BC501A cells mounted in 4 cluster at 2.7 m from the target at polar angles of 90° and 45° on one side of the beam, 45° and 25° on the other side. Proton detection was provided by the EMRIC CsI(Tl) array, while the ARGOS forward wall, made of 32 high-resolution phoswich detectors, was used to allow the measurement of fragments (see ref. [5] for more details).

Figure 2 shows two-neutron correlation function measured in the 45 Mev/u ^{58}Ni + ^{27}Al reaction, in coincidence with at least one fragment in the forward wall. To extract information from the experimental results, a comparison with the prediction of different theoretical models has been performed. These are obtained by convoluting the phase-space distribution, generated with any source model, with the two-neutron relative wave function, with account of antisymmetrization and final state interactions. The calculations include experimental effects, in particular the cross-talk rejection. Fig. 2a shows results from a statistical and from BUU dynamical transport models. The two models are completary, since the first lacks of pre-equilibrium emission, while the second does not treat properly the late evaporation. The results from the evaporative calculation with two sources separated by ≈6 fm is in agreement with experimental data for the most peripherical collisions. The BUU model, instead, produces a too strong correlation. Fig 2b shows the increase in the correlation function strenght observed when only particle pairs with total momentum in the source frame larger 270 MeV/c are selected (open square). A better agreement with BUU prediction (solid line) in this case indicates that high momentum pair originate preferably in the early stages of the reaction.

To characterize the space and time extent of the emitting source, a third model was used. This model, originally derived for an expanding fireball formed at high bombarding energies and modified for low energy by including evaporation and cooling [6], describes a statistical particle emission parametrized with a few global parameters. With this model, utilizing one source, we are able to reproduce the energy spectra at all angles and the total integrated correlation function as well

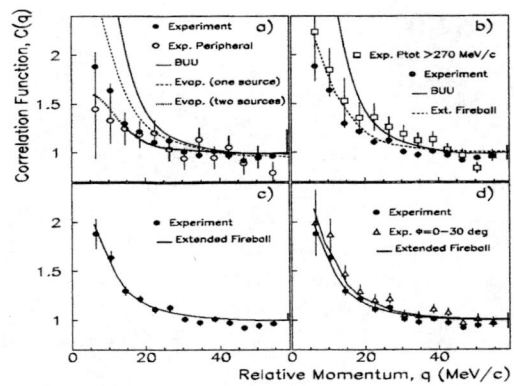

FIGURE 2. Comparison of experimental nn correlation functions (symbols) with the prediction of theoretical models (lines). See text for explanation.

as the longitudinal correlation function (triangle symbol in fig. 2d), obtained by gating on the angle ψ between the relative and the total momentum of the neutron pair in the source reference frame.

CONCLUSIONS

Two-nucleon interferometry can be used in low- intermediate-energy heavy ion reactions to study the space-time properties of excited nuclear matter. At low energy multidimensional two-neutron correlation functions have been succesfully used to extract the lifetime of a moderately excited compound nucleus. An intermediate energy, two-neutron correlation functions measured in coincidence with fragments indicate emission from a source at normal density on a fast timescale (≈ 100 fm/c). Comparisons to theoretical calculations suggest that careful analysis of both single particle spectra and multidimensional correlation functions, with accurate event identification, is needed in order to disentangle the different contributions of mecahnism such pre-equilibrium, flow, evaporation, impact parameter averaging etc. The analysis of identical and mixed-pairs correlation functions might allow to extract more reliable information on these aspects of the reaction dynamics.

REFERENCES

1. D.H. Boal et al., Rev. Mod. Phys. **62**, 553 (1990)
2. N. Colonna et al., Nucl. Instr.and Meth. A **381**,
3. A. Elmani et al., Phys. Rev. C **49**, 284 (1994). 472 (1996).
4. N. Colonna et al., Phys. Rev. Lett. **75**, 4190 (1995).
5. R. Ghetti et al., Nucl. Phys. A, in press
6. J. Helgesson et al., Phys. Rev. C **56**, 2626 (1997).

Evidence of pre-equilibrium γ-ray emission in heavy ion collisions at intermediate incident energies

S.Tudisco[a,c], F.Amorini[d], G.Cardella[b], A.Di Pietro[a,c], P.Figuera[a],
G.Lanzalone[b,c], A.Musumarra[a], M.Papa[b], G.Pappalardo[a,c], S.Pirrone[b,c],
F.Rizzo[a,c]

[a]INFN-Laboratori Nazionali del Sud, Via S.Sofia 44, 95123 Catania Italy
[b]INFN-Sezione di Catania, Corso Italia 57, 95129 Catania Italy
[c]Dipartimento di Fisica, Università di Catania, Corso Italia 57, 95129 Catania Italy
[d]Institute de Physique Nucleaire, IN2P3-CNRS, Orsay, France

Abstract. The experimental results of $^{40}Ca + {}^{48}Ca,{}^{40}Ca,{}^{46}Ti$ reactions are reported. The comparison between γ-ray spectra measured in coincidence with fusion evaporation residues for the three colliding systems shows a clear evidence of pre-equilibrium γ-rays emission in the region around 10 MeV. BNV simulations also predict this emission. The saturation of GDR strength with temperature has been found with some dependence on the colliding system.

INTRODUCTION

The study of the Giant Dipole Resonance (GDR) has offered, during the past years, a powerful method to investigate the nuclear shape and its evolution with temperature and angular momentum (1). More recently the attention has been focused on the role of the reaction dynamics on the γ-ray emission. In particular some experimental works have been performed to investigate the effect of mass and charge asymmetry in the entrance channel.

For instance, Flibotte et al. (2) found different GDR strengths selecting fusion events from reactions having different charge asymmetry between projectile and target. This was interpreted as resulting from giant dipole phonon excited at the moment of the collision in N/Z asymmetric reaction.

Similar effects (even if with a more peaked strength distribution around 10 MeV), have been observed comparing the γ-ray spectra measured in coincidence with charged particles emitted in deep inelastic(3) and incomplete fusion(4) reactions, respect to the calculated ones using a statistical code (like CASCADE). For such systems also quantitative theoretical estimations, based on Boltzman Noretheim Vlasov (BNV) simulations, have been performed(5). Within this approach the pre-equilibrium γ-ray emission is associated to the excitation of a dipole mode generated by the charge equilibration dynamics.

CP495, *Experimental Nuclear Physics in Europe*, edited by B. Rubio et al.
© 1999 American Institute of Physics 1-56396-907-6/99/$15.00

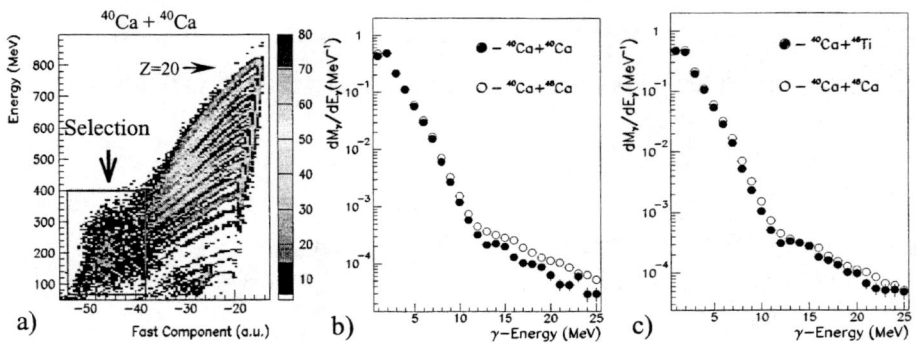

FIGURE 1.a) Identification matrix from pulse-shape analysis; the arrow indicates the selected events. b) Comparison between γ-ray spectra from ^{40}Ca+ ^{48}Ca,^{40}Ca. c) Comparison between γ-ray spectra from ^{40}Ca+ ^{48}Ca, ^{46}Ti.

The above mentioned works (2-4) have been in the incident energy range up to 10 MeV*A. It seems then interesting to follow the evolution of these pre-equilibrium effects at much higher beam energy where no data are available and where new dynamical aspects should arise. For this reason we decided to measure γ-rays in coincidence with heavy fragments produced in reactions induced by ^{40}Ca beam at 25 MeV*A. We used three different targets ^{48}Ca, ^{40}Ca and ^{46}Ti. We chose the ^{48}Ca target, which presents a larger N/Z ratio (N/Z=1.4), in order to enhance as much as possible the visibility of the pre-equilibrium γ emission connected with the charge asymmetry. Our goal was to compare the γ spectra measured using the ^{48}Ca target with the ones measured using the other two targets. In fact these last two present practically the same N/Z ratios (N/Z(^{40}Ca)=1, N/Z(^{46}Ti)=1.09), moreover ^{46}Ti was chosen because its mass is closer to ^{48}Ca. This direct comparison overcomes the problems of statistical codes that, at these higher energies, don't take into account the complex fragment emission.

EXPERIMENTAL SET-UP

The experiment was performed using the ^{40}Ca beam delivered by the Superconducting Cyclotron of the Laboratori Nazionali del Sud in Catania. Self-supporting targets of ^{48}Ca, ^{40}Ca, ^{46}Ti, 2 mg/cm^2 thick and respectively 95%,99%,95% isotopically enriched, were used. The targets were stored and mounted under vacuum in order to avoid oxygen contamination. We used the TRASMA multidetector system(6) to measure γ-rays in coincidence with charged particles. It consists of 63 BaF$_2$ detectors for the γ detection and of a forward angle ΔE-E hodoscope (Si-CsI) for charged particle identification. In order to remove the contamination of neutrons from the γ spectra, time of flight technique and analysis of fast-slow components have been used.

Using a pulse shape analysis the heavy residues stopped in the first stage of the hodoscope (annular silicon strip detector, 300 μm thick) were identified. We looked at the γ-ray spectra in coincidence with charged particles coming from complete fusion (CF) and incomplete fusion (ICF) events (Figure 1-a).

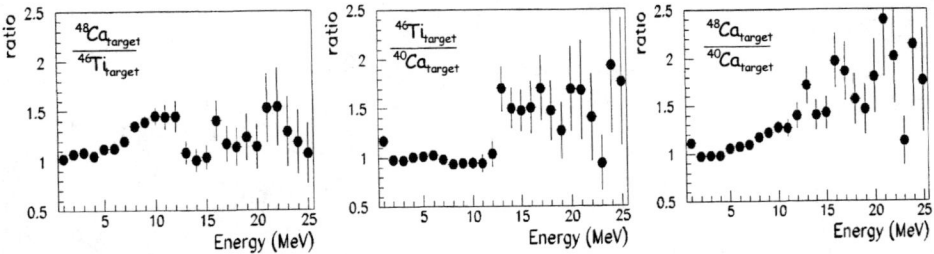

FIGURE 2. Experimental ratios of γ-spectra from the three reactions ^{40}Ca + ^{48}Ca, ^{40}Ca, ^{46}Ti.

EXPERIMENTAL RESULTS

Figure 1-b and 1-c shows the γ-multiplicity spectra measured for the three different reactions. The data of ^{40}Ca+^{48}Ca reaction, where we expect a contribution due to the charge equilibration dynamics, are compared with the spectra obtained using the other two targets. The first consideration that can be done is related to the rather low multiplicity of the GDR region observed for the ^{40}Ca+^{40}Ca system respect to the ^{40}Ca+^{48}Ca one. From that we could wrongly conclude that the N/Z difference between projectile and target produces an enhancement in the whole GDR range. Nevertheless such conclusion falls when we look at the comparison between ^{48}Ca and ^{46}Ti targets. In this case the γ-multiplicity in the GDR region for both systems is rather similar; moreover an extra-strength around 10 MeV is present in the reaction with ^{48}Ca target. This is more evident if we plot the ratios between γ-rays spectra measured for the three reactions (Figure 2). Looking at ^{48}Ca/^{46}Ti ratio we find an extra strength around 10 MeV. From the ^{46}Ti/^{40}Ca ratio, where we don't aspect any effect originating from charge asymmetry, we observe an enhancement in the whole GDR region. Both these contributions appear mixed in the ^{48}Ca/^{40}Ca ratio. It seems then clear that the effect of N/Z asymmetry in the entrance channel is essentially related to the extra strength observed in the region around 10 MeV of the γ spectra. This experimental finding is confirmed by a BNV simulations. The calculated pre-equilibrium strength, at 2 fm of impact parameter, is plotted in Figure 3a. As a consequence of the collisional damping (which characterizes, at this energies, the dynamics) a broad distribution is found. This result is in agreement with the experimental evidence. We find in fact a larger strength distribution respect to the ones observation in previous experimental works realized at lower energy (3,4).

A further interesting point is connected to the observed GDR strength. In the last decade the study of the evolution of the GDR parameters with excitation energy has shown, at temperature around 5 MeV, the saturation effect of GDR strength (7). This has been interpreted in terms of a much larger particle emission probability (8,9) and, however, is strictly related to loss of collectivity of the system.

In our selection we populate three similar compound systems a temperature that we have roughly estimate about 7 MeV. This estimation is in agreement with the results of other published data (^{40}Ca+^{48}Ti at 25 MeV*A (10)). Then, even if some small differences in excitation energies or in masses are presents, due to the saturation effect

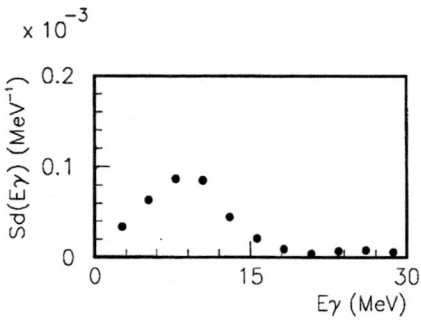 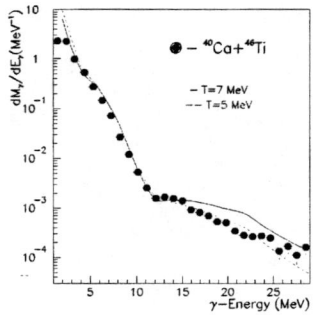

FIGURE 3.a)BNV strength distribution of pre-equilibrium γ emission (b=2 fm). b)Comparison between CASCADE calculations and experimental data from $^{40}Ca+^{46}Ti$. Solid line: E*=290(T≅7) MeV. Dashed line: E*=180(T≅5) MeV. For both systems A=61,Z=30 E_{GDR}=18MeV, Γ_{GDR}=15MeV, S=1, a=A/10.

we aspect to observe the same GDR yield for the three studied colliding systems. In Figure 2b the results of two CASCADE calculations (using an emitting source like in ref.(10)) are plotted. The comparison with the experimental data shows that, in order to reproduce the GDR region, we need to reduce the temperature to about 5 MeV. Then we confirm the presence of the GDR saturation strength, but (as is clear from ratios) we observe a dependence of GDR yield from the colliding system. The stronger evidence comes from $^{40}Ca+^{40}Ca$ system, consequently such difference can not be due to the N/Z asymmetry. Also Bremsstrahlung effect can be excluded, as a consequence of the same yield measured in the high energy region of the γ spectra. A simple explanation could be related to the peculiar characteristic of the $^{40}Ca+^{40}Ca$ system. In particular a more pronounced light particle production (Z=1 and Z=2) is found comparing previously published data on $^{40}Ca+^{40}Ca$ (11) respect to the $^{40}Ar+^{45}Sc$ (12) at E_{inc}=35 MeV*A. This could be connected to the lower observed GDR yield even if a more quantitative analysis is needed.

REFERENCES

1. J.J.Gaardhoje; Ann.Rev.Nucl.Part.Sci 42(1992)483.
2. S.Flibotte et al., Phys.Rev.Lett. 77(1996)1448
3. L.Campajola et al., Z.Phys. A352(1995)421
4. F.Amorini et al., Phys. Rev. C58(1998)987
5. M.Papa et al., Eur.Phys Journ. A4(1999)69
6. A.Musumarra et al., NIM A370(1996)558
7. J.J.Gaardhoje, Phys. Rev. Lett 59(1987)1409
8. P.F.Bortignon et al., Phys.Rev.Lett. 67(1991)3360
9. P.Chomaz, Nucl. Phys A569(1994)203
10. T.M.V. Bootsma et al. Z. Phys. A359 (1997)391
11. P.Pawlowski et al. Z.Phys A357 (1997)387
12. T.Li et al. Phys. Rev. C 49(1994)1630

Mass Measurements of Stored Exotic Nuclei at Relativistic Energies

H. Geissel[1,2], F. Attallah[1], K. Beckert[1], F. Bosch[1], H. Eickhoff[1],
M. Falch[3], B. Franzke[1], M. Hausmann[1], F. Herfurth[1],
Th. Kerscher[3], O. Klepper[1], H.-J. Kluge[1], C. Kozhuharov[1],
Yu. A. Litvinov[4], K. E. G. Löbner[3], G. Münzenberg[1], F. Nolden[1],
Yu. N. Novikov[4], Z. Patyk[5], T. Radon[1] H. Schatz[1],
C. Scheidenberger[1], J. Stadlmann[2], B. Schlitt[1], M. Steck[1],
K. Sümmerer[1], H. Weick[1,2], H. Wollnik[2]

[1] *Gesellschaft für Schwerionenforschung, Planckstrasse 1, D-64291 Darmstadt, Germany*
[2] *II. Physikalisches Institut, Universität Giessen, Heinrich -Buff-Ring 16, D-35392 Giessen,*
Germany
[3] *Sektion Physik, Ludwig-Maximilians-Universität München, Am Coulombwall, D-85748*
Garching, Germany
[4] *St. Petersburg Nuclear Physics Institute, Gatchina 188350, Russia*
[5] *Soltan Institute for Nuclear Physics, Warsaw, Poland*

Abstract. Relativistic exotic nuclei produced via projectile fragmentation were separated in flight by the fragment separator FRS and injected into the storage ring ESR for precise mass measurements using Schottky spectrometry. Nuclei with half-lives shorter than the time required for electron cooling have been investigated by time-of-flight measurements with the ESR being operated in the isochronous mode. This novel experimental technique gives access to all nuclei with half-lives down to the μs-range.

I INTRODUCTION

Research of nuclear structure is extended by new secondary beam facilities, storage rings and trap systems. Special effort has been devoted to experiments using radioactive nuclei at energies above the Coulomb barrier [1]. An advantage of exotic nuclear beams at relativistic energies is that the reaction products are bare or populate only few-electron states. This condition allows clean monoisotopic separation in flight and favorable resolution of particle tracking and identification detectors. Indeed, the selection of the projectile energy can be used to separate and store the fragments in desired charge states to study astrophysical problems under conditions which prevail in stellar plasmas [2]. In this contribution we will

CP495, *Experimental Nuclear Physics in Europe*, edited by B. Rubio et al.

concentrate on mass measurements of relativistic projectile fragments performed with the unique combination of the FRS [3] and the experimental storage ring ESR [4].

II MASS SPECTROMETRY OF STORED RELATIVISTIC IONS AT THE SIS-FRS-ESR FACILITY

The masses of stable nuclei and their neighborhood are well known, whereas near the drip lines the data are scarce and difficult to obtain due to the small production cross sections and short lifetimes. However, this information is necessary for a crucial check and improvements of nuclear models.

Stable beams of relativistic heavy ions provided by the synchrotron SIS with a maximum magnetic rigidity (Bρ) of 18 Tm are converted into exotic nuclei by nuclear collisions in the production target at the entrance of the spectrometer FRS. Projectile fragmentation and projectile fission combined with in-flight separation are rich sources for experiments with relativistic exotic nuclei. The FRS separates the fragments within a few hundred ns and injects them into the ESR for precise mass determination, performed by measuring the revolution frequency of the stored ions. The ESR is equipped with an electron cooler [5] and can store ions in the range of (0.5$\leq$ Bρ $\leq$10)Tm. The phase-space density of the stored ions can be drastically increased by electron cooling, e.g., the relative velocity spread of a low-intensity cooled beam can be less than 10^{-6}.

Precise mass measurements of stored ions circulating in the ESR can be performed with two experimental methods: 1.Mass spectrometry using cooled ion beams. 2.Mass spectrometry of hot fragments operating the ESR in the isochronous mode. Both principles can be easily understood by

$$\frac{\Delta m}{m} = -\gamma_t^2 \frac{\Delta f}{f} + \left(\gamma_t^2 - \gamma^2\right) \frac{\Delta v}{v}, \tag{1}$$

the first-order relation between the mass m, the revolution frequency f, and the velocity v. The mass resolution is given by the precision achieved for the frequency and velocity determination. The formation of the exotic nuclei always causes a large velocity spread, which is inherent due to the nuclear reaction dynamics and depends on the selected fragments. γ is the relativistic Lorentz-factor and γ_t represents the transition energy which characterizes the ion-optical mode of the ring. From this formula it is obvious that either cooling $\frac{\Delta v}{v} \to 0$ or the isochronous condition, $\gamma \to \gamma_t$, are the basis of precise mass measurements.

Schottky Mass Measurements with Cooled Fragments

Schottky spectroscopy is widely used for beam diagnosis in circular accelerators and storage rings. The induced signals of the stored circulating ions in nondestructive probes are recorded and analyzed. Already in our pilot experiments with cooled projectile fragments we have applied Schottky diagnostics and since then this technique has been gradually developed for the requirements of precision mass spectrometry [6–8]. The experimental procedure and progress of our Schottky mass spectrometry (SMS) are described in reference [8] and a forthcoming paper in detail [9].

The analysis of the data profited from the feature of SMS that a large number of known and unknown masses are simultaneously measured. The reliability of the method was first carefully investigated in the region of known masses before a large unknown territory of about 100 neutron-deficient isotopes in the element region ($57 \leq Z \leq 84$) was studied. A maximum likelihood method has been applied for the evaluation of masses and their errors. In the analysis it was assumed that the dependence of mass-to-charge ratio on the frequency is a polynomial in first order. For reference masses only isotopes with no isomeric states according to Ref. [10] have been used. The achieved accuracy was about 100 keV and the mass resolving power $6.5 \cdot 10^5$ (FWHM).

The systematic mass measurements over a large unknown territory present e.g. a mapping of the proton dripline, i.e., to follow the one-proton separation along the isotopes of the different elements, see figure 1.

The knowledge of the one-proton separation energies is an important nuclear structure information, particular for the investigation of proton radioactivity. The data clearly map and even cross the proton dripline for $Z \geq 83$. Furthermore, the shell closure at lead can be seen by the systematics of the S_p values of the different elements. More information on the existence of the magic shell gap of Pb far off stability and the onset of nuclear shape changes can be investigated from our mass values by deriving the difference of the two-proton separation energies for the vicinity of $Z=82$ as a function of the neutron number, see figure 2:

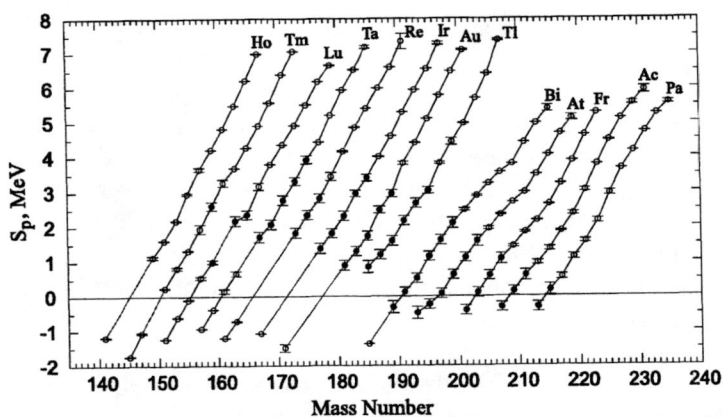

FIGURE 1. One-proton separation energies for odd-A and odd-Z isotopes of the elements from holmium to protactinium measured in this experiment (bold circles) and taken from ref. [10].

$$2G_p = S_{2p}(Z, N) - S_{2p}(Z + 2, N) = (M(Z - 2, N) - 2M(Z, N) + M(Z + 2, N))c^2 \tag{2}$$

The figure illustrates that the large $2G_p$ values of about 7 MeV at the doubly magic nucleus ^{208}Pb decrease towards the proton dripline to about 2 MeV. First microscopic calculations indicate that the change in deformation of the compared isotopes of Po, Pa, and Hg can explain the observation [11].

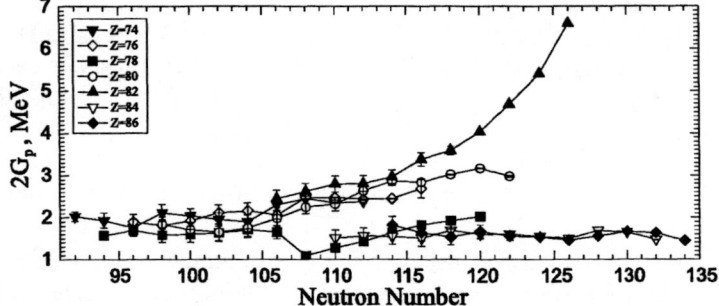

FIGURE 2. Difference of the two-proton separation energies for different elements in the vicinity of Z=82 as a function of the neutron number.

Isochronous Mass Measurements of Stored Short-lived Fragments

Masses of short-lived exotic nuclei with half-lives shorter than the cooling time can be investigated by the time-of-flight technique where the ESR is operated in the isochronous mode [12–14]. In this case, the magnetic fields of the ESR quadrupole and hexapole magnets are set such that the revolution frequency of an isotope with fixed charge state becomes independent of its velocity, see equation (1). A time-of-flight detector, based on secondary electrons released from a thin foil, has been installed in the ESR to record the revolution frequency of the stored fragments [14]. This novel experimental technique has been successfully applied in first measurements with chromium fragments. The ESR lattice was tuned to γ_t=1.37, corresponding to 345 A·MeV, to fulfill the isochronicity condition.

A measured spectrum of the revolution periods of chromium fragments recorded in the isochronous mode is shown in fig. 3. Although the momentum acceptance inside the ESR is strongly reduced in the isochronous mode, as compared to the standard operation, the measured m/q acceptance still reached 4.5% due to the large velocity spread of the fragments. Already in the development phase of the isochronous mode we achieved a remarkable mass resolving power of m/Δm = $1.5 \cdot 10^5$ (FWHM) and an agreement with known masses within several tens of keV [13].

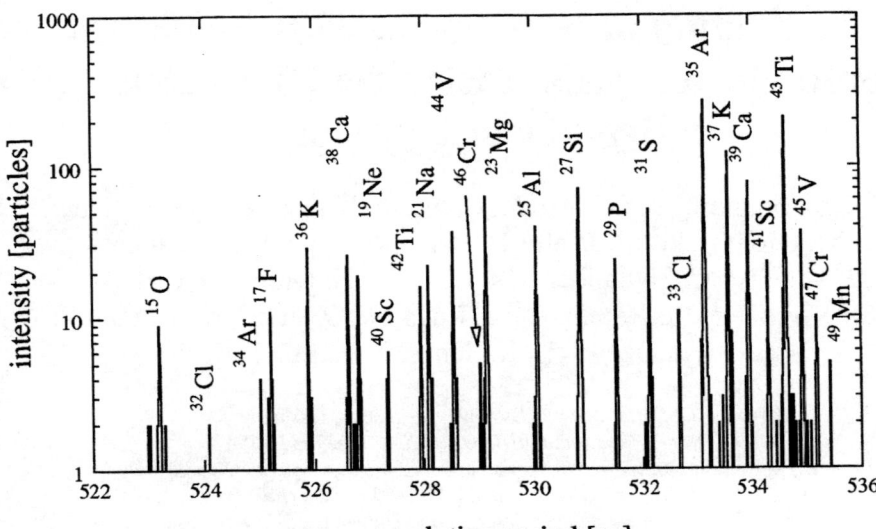

FIGURE 3. Mass-frequency spectra of hot fragments recorded in the isochronous mode of the ESR.

In conclusion we can state that with Schottky and Isochronous Mass Spectroscopy we have powerful experimental tools to improve the knowledge of nuclear structure physics.

REFERENCES

1. H. Geissel, G. Münzenberg, K. Riisager, Ann. Rev. Nucl. Part. Sci. 45 (1995) 163.
2. F. Bosch, et al., Phys. Rev. Lett. 77 (1996) 5190
3. H. Geissel, et al., Nucl. Instr. and Meth. B70 (1992) 286.
4. B. Franzke et al., Nucl. Instr. and Meth. B24/25 (1987) 18.
5. M. Steck, K. Beckert, H. Eickhoff, et al., Phys. Rev. Lett. 77 (1996) 3803.
6. B. Franzke et al., Physica Scripta T59 (1995) 176.
7. T. Radon, Th. Kerscher, B. Schlitt, et al., Phys. Rev. Lett. 78 (1997) 4701.
8. H. Geissel, T. Radon, et al., AIP conference proceedings 455 Edts. B.M. Sherrill, D.J. Morrissey, C.N. Davids (1998) 11.
9. T. Radon, H. Geissel, G. Münzenberg, et al., subm. to Nucl. Phys. A 1999.
10. G. Audi, O. Bersillon, J. Blachot and A.H. Wapstra. Nucl.Phys., A624, 1997, 1.
11. J. Maruhn, T. Bürvenich, T. Cornelius, private communication (1999).
12. H. Wollnik, et al., GSI-Report 86-1 (1986) 372.
13. M. Hausmann, et al., Proc. of the 6th EPAC, Stockholm 1998 Bristol, Institute of Pysics Publishing.
14. M. Hausmann, PHD thesis, university Giessen 1999.

Production of N≅Z nuclei with A≈100 via complete heavy-ion fusion reactions followed by cluster emission

M. La Commara[$], J. Gómez del Campo[#], A. D'Onofrio[^], A. Gadea[§],
M. Glogowski[$], P. Jarillo-Herrero[$], N. Belcari[§], R. Borcea[$],
G. de Angelis[§], C. Fahlander[▽], M. Górska[$], H. Grawe[$], M. Hellström[▽],
R. Kirchner[$], M. Rejmund[$], V. Roca[^], E. Roeckl[$], M. Romano[^],
K. Rykaczewski[#], K. Schmidt[$] and F. Terrasi[^]

[$]GSI, Planckstr. 1, D-64291, Darmstadt, Germany
[#]Physics Division, ORNL, Oak Ridge, TN 37831, USA
[^]INFN Napoli, Complesso Universitario M. S. Angelo, I-80126, Napoli, Italy
[§]INFN Legnaro, via Romea 4, I-35020, Legnaro (Pd), Italy
[▽]Division of Cosmic and Subatomic Physics, Lund University, S-22100, Lund, Sweden

Abstract. The production of isotopes near ^{100}Sn has been investigated by using on-line mass separation of evaporation residues produced by heavy-ion induced complete fusion reactions. We measured β-delayed protons and γ-rays, deduced the mass-separated beam intensities for ^{99}Cd, ^{100}In, ^{101}Sn and ^{102}In, and determined the corresponding production cross-sections σ for ^{58}Ni+^{58}Ni reactions followed by emission of nuclear clusters, and for ^{58}Ni+^{50}Cr reactions accompanied by emission of protons, neutrons and α particles. The results indicate that the σ values for reactions followed by emission of clusters are comparable with those for reactions proceeding through light particle emission. Due to recoil-loss effects in on-line mass separation, the former mechanism is less suitable than the latter one for producing such neutron-deficient nuclei in this kind of experiments.

Due to the double shell closure, several interesting nuclear structure phenomena arise in the region of ^{100}Sn. Among these, one of the most exciting ones is the possible occurrence of a cluster-radioactivity island, that is particle with Z>2 decay, beyond the already known proton- and α- ones lying "north-east" of ^{100}Sn. We investigated the population of nuclei close to ^{100}Sn by using heavy-ion induced fusion-evaporation reactions, and studied in particular those followed by emission of nuclear clusters. The latter reaction mechanism was recently studied in an in-beam experiment, which has shown that the ^{58}Ni+^{58}Ni reaction at beam energies above 5 A MeV results in high production rates of nuclei below, although not very close to ^{100}Sn (1). In order to measure the small production cross-sections σ closer to ^{100}Sn, i. e. for ^{99}Cd, ^{100}In, ^{101}Sn and ^{102}In, we used isotope separation on-line (ISOL). We studied the reaction ^{58}Ni(380÷430 MeV) + ^{58}Ni followed by cluster emission, and the reaction ^{58}Ni(320÷380 MeV) + ^{50}Cr followed by emission of protons, neutrons and α particles. The experiment was performed by using the ISOL at the heavy-ion accelerator UNILAC of GSI. The isotopes were produced as reaction recoils, extracted from the separator ion source, and separated according to their mass in a

CP495, Experimental Nuclear Physics in Europe, edited by B. Rubio et al.
© 1999 American Institute of Physics 1-56396-907-6/99/$15.00

dipole magnet. Their β-delayed proton (βp) and β-delayed γ (βγ) activities were measured with silicon and HP-Ge detectors, respectively. The σ values for individual isotopes were deduced from mass-separated beam intensities, which were determined on the basis of the experimental βp and βγ activities. The branching ratios for these radiations were taken from previous experiments in the case of ^{99}Cd and ^{102}In, whereas for ^{100}In and ^{101}Sn only an experimental lower limit (2) and a theoretical prediction (3), respectively, were available. Due to the more pronounced angular spreading of the residues after cluster emission, the σ values corresponding to the ^{58}Ni+^{58}Ni reaction had to be corrected for losses occurring in the collection of the reaction recoils in the catcher placed inside the ion source. The experimental results show, on the one hand, that the ^{58}Ni+^{58}Ni and ^{58}Ni+^{50}Cr reactions yield about the same σ value for ^{99}Cd while cluster emission is less favourable for the production of ^{100}In than the "standard" light-particle one. On the other hand, ^{58}Ni+^{58}Ni reaction yields somewhat higher σ values, if compared to the ^{58}Ni+^{50}Cr one, for ^{101}Sn and ^{102}In. As an example, the results for ^{99}Cd and ^{101}Sn are shown in Fig. 1.

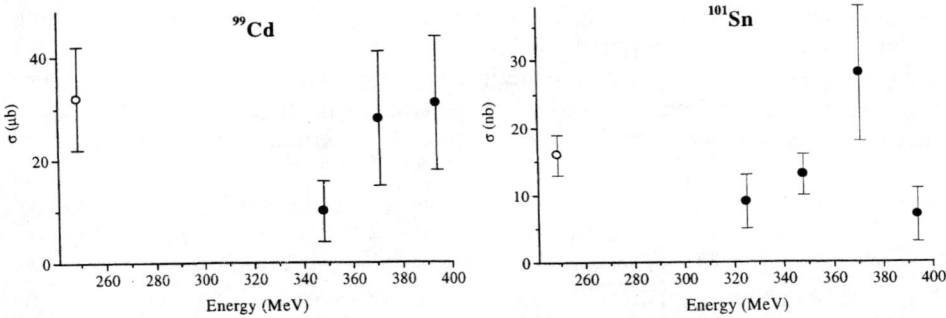

FIGURE 1. Experimental σ values for the isotopes ^{99}Cd and ^{101}Sn as a function of the ^{58}Ni-beam energy in the middle of targets. The data were deduced from βp activities for the ^{58}Ni+^{50}Cr reaction (open circles) and the ^{58}Ni+^{58}Ni one (solid circles).

All in all, the two reaction mechanisms yield about the same experimental σ values for nuclei near ^{100}Sn. However, because of the low recoil-collection efficiency of the ISOL method for reactions accompanied by cluster emission, the latter mechanism does not seem to be very suitable for investigating very neutron-deficient isotopes close to ^{100}Sn in ISOL experiments. The comparison of the data with statistical model calculations is reported elsewhere (4).

REFERENCES

(1) J. Gómez del Campo et al., Phys. Rev. **C 57**, 2 (1998).
(2) Z. Janas et al., Phys. Scripta **56**, 262 (1995).
(3) J. Szerypo et al., Nucl. Phys. A **584**, 221 (1995).
(4) M. La Commara et al., proceedings of the 7th Int. Conf. on "Clustering Aspects of Nuclear Structure and Dynamics", June 14-19, 1999, Rab, Island of Rab, Croatia.

Comprehensive Studies of Heavy Ion Reactions

M. Cavinato, E. Fabrici, E. Gadioli, E. Gadioli Erba and D. Leoni

Dipartimento di Fisica, Università di Milano and I N F N, Sezione di Milano, Italia

Recently the Milano-Wits-NAC-Stellenbosch collaboration has shown that it is possible to provide a comprehensive description of the reactions which are induced by ^{12}C on A$\approx$100 nuclei up to incident energies of about 35 MeV/Amu [1-5]. This study showed that the accuracy obtained in reproducing the excitation functions for production of nuclei spanning a very large range of masses and charges and the spectra of the emitted α particles (the yield of which is especially large in ^{12}C induced reactions at these incident energies) is comparable to that reached in the analysis of nucleon induced reactions [6] in spite of the much greater complexity of ^{12}C induced reactions which include complete and incomplete fusions and nucleon transfers.

We show now that our model (which is discussed in the above quoted papers) may equally well reproduce the reactions induced by ^{12}C on sizeably lighter nuclei such as ^{63,65}Cu at notably larger energies (45 MeV/Amu). The data which we have analysed are the production cross-sections and the average forward ranges of a large number of radioactive residues measured by the nuclear activation technique [7].

The results of our analysis are summarized in the following figure.

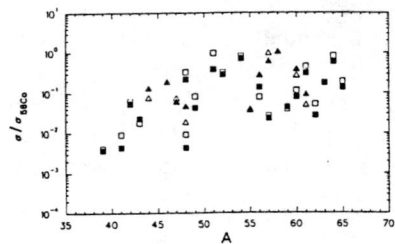 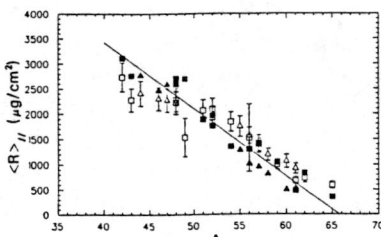

In the left part the comparison of the experimental (open symbols) and calculated (black symbols) production cross sections is shown. To reduce systematic uncertainties, the experimental cross sections which we have reproduced are relative to the cross section for production of the most abundant residue which has been observed (^{58}Co). In the right part of the figure the experimental average forward ranges (open symbols) are compared to the calculated ones (black symbols and the χ^2 straight line drawn through the theoretical values).

In the next figure it is shown the calculated isobar yield distribution which includes also a large number of unobserved residues. The production of target-like residues appears to be especially large.

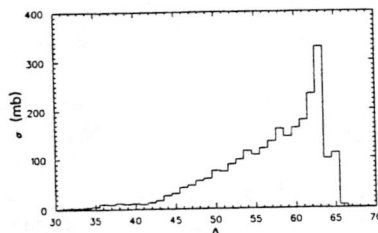

These results together with those of previous analyses [1-5] suggest that, at this energy, ^{12}C transfers, on the average, a very small fraction of its energy and linear momentum to the target nucleus. A much larger fraction of the ^{12}C energy and linear momentum is given to projectile's fragments and pre-equilibrium ejectiles which are emitted before the production of equilibrated nuclear matter.

REFERENCES

1. C Birattari et al., Phys. Rev. C **54**, 3051 (1996)
2. E Gadioli et al., Phys. Lett. B **394**, 29 (1997)
3. E Gadioli et al., Nucl. Phys. A **641**, 271 (1998)
4. E Gadioli et al., Heavy Ion Phys. **7**, 275 (1998)
5. E Gadioli et al., Nucl. Phys. A *in course of publication*.
6. E Gadioli and P E Hodgson, *Pre-equilibrium Nuclear Reactions*, (Clarendon Press, Oxford, 1992).
7. S Y Cho, N T Porile and D J Morissey, Phys. Rev. C **39**, 2227 (1989)

Reaction Spectroscopy on ^{140}Nd

Teruo Suehiro*, Kengo Ogawa† and Yoseo Iwasaki*

*Tohoku Institute of Technology, Sendai 982-8577, Japan
†Faculty of Science, Chiba University, Chiba 263-8522, Japan

Abstract. Experimental data on ^{142}Nd(p, t)^{140}Nd reaction at 40 MeV has been analyzed in terms of the shell-model and the finite-range DWBA theory. Overall agreement is quite satisfactory. A significant discrepancy has been observed, however, for the $L = 0$ ground state transition. An alternative method for obtaining the transition form factor has been proposed which may resolve this discrepancy.

INTRODUCTION

The experiment on ^{142}Nd(p, t)^{140}Nd reaction was done with 40 MeV protons at Kernfysisch Versneller Instituut, Rijksuniversiteit Groningen [1]. The experimental energy resolution of 30 keV was good enough so that more than 30 peaks observed below an excitation energy of about 3.5 MeV were well resolved and found to correspond to levels of ^{140}Nd with $J^\pi = 0^+$, 2^+, 3^-, 4^+, 5^-, 6^+ and 7^- . In the present paper the data have been analyzed in the framework of the microscopic shell model and the finite-range distorted wave Born approximation.

SHELL-MODEL AND DWBA CALCULATIONS

The shell-model hamiltonian H is written as $H = H_p + H_n + V_{pn}$. Ten protons are distributed in the five orbitals $1d_{3/2}$, $2s_{1/2}$, $0h_{11/2}$, $1d_{5/2}$ and $0g_{7/2}$. The neutrons are assumed to make closed shells in ^{142}Nd. After solving the eigenvalue problem for the proton part H_p, proton motions are coupled with two-neutron holes to describe states in ^{140}Nd. All the two-neutron-hole states generated from the $(1d_{3/2}, 2s_{1/2}, 0h_{11/2}, 1d_{5/2}, 0g_{7/2})^{-2}$ configurations are taken into account. We use the surface-delta force for the neutron-neutron interaction and the quadrupole-quadrupole force for V_{pn} . The single neutron-hole energies are basically those given in [2]. Distorted wave Born approximation calculations using the transition amplitudes obtained by the shell-model calculation have been done with the exact finite range code [3]. The transfer form factor comprizes radial wave functions of the two transferred neutrons, which are most commonly abtained by solving motions of neutrons in a Wood-Saxon well with $r_0 = 1.25$ fm and $a_0 = 0.65$ fm. The cal-

CP495, *Experimental Nuclear Physics in Europe*, edited by B. Rubio et al.
© 1999 American Institute of Physics 1-56396-907-6/99/$15.00

TABLE 1. Ratios of experimental and calculated peak cross sections summed over levels with the same J^π, $\sum \sigma_{\text{calc}}(J^\pi)/\sum \sigma_{\text{exp}}(J^\pi)$.

J^π	0^+	2^+	3^-	4^+	5^-	6^+	7^-
Standard F. F.	0.349	0.795	0.982	0.951	0.993	0.772	0.971
Adjusted F.F.	0.623	1.33	0.865	1.54	0.876	0.740	1.07

culated cross section for the predicted lowest 7^- state has been normalized to the experimental one for the lowest 7^- state at 2.22 MeV. The number and the position of the strong transitions have been well reproduced for each J^π. The results of the calculation with the standard form factors are summarized in Table 1.

RESULTS AND DISCUSSIONS

The overall agreement with the experimental data is satisfactory in general, except for the $L = 0$ ground state transition, for which the theory gives only 33% of the experimental cross section. This has been a long-standing puzzle in the interpretation of the two-neutron tranfer reaction in terms of the standard model. A preliminary calculation has shown that the enhancement of the ground state transition by the factor 3 is hardly plausible by simply expanding the shell model space. The Wood-Saxon potential used above with the standard parameters predicts, however, an ordering of the single particle orbits which is inconsistent with the assumption made in the shell-model calculation. The standard potential gives the $0h_{11/2}$ orbit far above the $2s_{1/2}$ orbit. An alternative binding potential has been obtained by using a potential search code which produces the ordering more consistent with the shell-model calculation. The binding potential [4] is a superposition of a volume part ($r_0 = 1.10$ fm and $a_0 = 0.50$ fm) and a surface part ($r_D = 1.16$ fm and $a_D = 0.65$ fm). Results obtained by this potential is referred to as "Adjusted form factor (F.F.)" in Table 1. Reasonable agreement with the experimental angular distributions has been obtained by using the adjusted form factors. It is very noticeable that the adjusted form factors enhance largely the transitions with lower angular momentum transfers. In this case the ratio $\sigma(0_1^+)/\sigma(7_1^-)$ becomes almost twice as large and the calculated ground state transition strength amounts 62% of the experimental value.

REFERENCES

1. Iwasaki, Y., Bhowmik, R. K., Harakeh, M. N., Janssens, R. V. F., Put, L. W. and Sakai, H., Experimental data of (p,t) reactions on Nd isotopes (unpublished).
2. Heyde, K. and Brussaard, P. J., *Z. Physik* **259**, 15 (1973).
3. Kunz, P. D., Computer code DWUCK5 for PC (unpublished).
4. Suehiro, T., "Does the core melt?", in *Proceedings of International Conference on Nuclear Data for Science and Technology*, 1997, pp. 796 - 798.

Temperature dependence of friction from pre- and post-scission neutron multiplicities

Aleksandra Kelić [1,2] and Gérard Rudolf [1]

[1] *IReS, UMR7500, IN2P3-CNRS, BP 28, F-67037 Strasbourg Cedex (France)*
[2] *INN "Vinča", lab. (010), 11000 Belgrade - Yugoslavia*

Abstract. The concept of varying friction is introduced in the frame of the Bass model. The temperature of transition is extracted from all published pre- and post-scission neutron multiplicities.

The concept of temperature dependence of friction has been introduced in the analysis of γ rays emitted from the Giant Dipole Resonance [1]. There, it was suggested that friction is small below approximately T=1 MeV, and large above. These studies allow also to extract the delay time for fission, since one can distinguish the γ rays emitted by the GDR of the compound nucleus and those emitted by the fission fragments.

The neutron multiplicities have been recently reanalysed in the frame of the Bass model [2]. In this method, the key parameter is the excitation energy left at the saddle or at the scission point, and not the delay time. Simple analytical expressions were obtained which reproduce nicely almost all published pre- and post-scission neutron multiplicity data. It was shown that very different conclusions can be drawn when compared to the usual statistical model approach, especially when a temperature dependent friction is introduced. To compare the results of this model with the existing experimental data, we have supposed that above a given temperature - the temperature of transition (T_{tran}) - the friction is large and therefore the motion of the system is slow. During this step pre-scission neutrons are emitted. When the excitation energy falls below the one corresponding to T_{tran}, friction becomes small and the subsequent motion towards scission fast. So, all the available excitation energy in this second step is spent on the post-scission emission. By fitting all published pre- and post-scission neutron multiplicities we have extracted the temperature T_{tran}. For systems for which an excitation function has been measured, it was found that one single value of T_{tran} is valid over a broad bombarding energy range. This holds even for high bombarding energies but then the excitation energy has to be compared not to the fission barrier, but to the

CP495, *Experimental Nuclear Physics in Europe*, edited by B. Rubio et al.
© 1999 American Institute of Physics 1-56396-907-6/99/$15.00

Bass barrier in the exit channel. This transition from one to another barrier is supported with the results of ref. [3], where it was shown that the fission barrier is not fixed, but varies with the number of neutrons emitted inside the saddle. In figure 1. is shown the temperature T_{tran} as a function of total mass of the systems. The experimental data are from refs. [4–8].

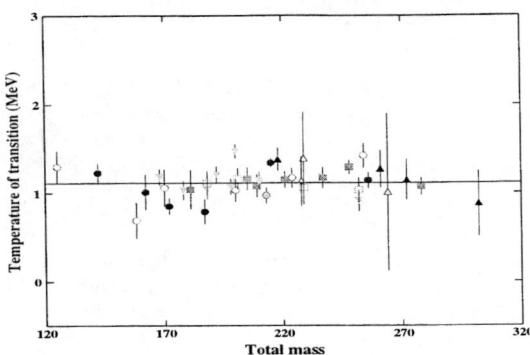

FIGURE 1. The dependance of T_{tran} on the total mass of the system. The full line represents the mean value of T_{tran}.

A mean value of $T_{tran} = 1.11 \pm 0.05 MeV$ has been obtained by taking into account some 60 systems ranging from $^{16}O+^{144}Sm$ to $^{64}Ni+^{238}U$ and bombarding energies ranging from the Coulomb barrier to 18 MeV/u.

REFERENCES

1. P. Paul and M. Thoennessen, *Annu. Rev. Nucl. Part. Sci.* **44**, 65(1994).
2. G. Rudolf, *Proc. RIKEN Symposium on Dynamics in Hot Nuclei*, Tokyo 13-14 March (1998), not published.
3. K. Pomorski et al., *Nucl. Phys.* **A605**, 87 (1996).
4. J.O. Newton et al., *Nucl. Phys.* **A483**, 126 (1998).
5. D.J. Hinde et al., *Phys. Rev.* **C39**, 2268(1989).
6. E.Mordhorst et al., *Phys. Rev.* **C43**, 716 (1991).
7. D.J. Hinde et al, *Phys. Rev* **C45**, 1229 (1992).
8. K. Knoche et al., *Phys. Rev.* **C51**, 1908 (1995).

Fusion Barrier Distribution in ^{9}Be+^{209}Bi

C.Signorini[1], Z.H.Liu[2,5], Z.C.Li[1,5], K.E.G.Löbner[3], L.Müller[1],
M.Ruan[2], K.Rudolph[3], F.Soramel[4], C.Zotti[3], A.Andrighetto[1], L.Stroe[1],
A.Vitturi[1], H.Q.Zhang[5]

[1]Physics Department of University and INFN, via Marzolo 8, 35131 Padova, Italy, [2]INFN, Legnaro
National Laboratories, Legnaro, Padova, Italy, [3]Sektion Physics, Ludwig-Maximlians-University
Munich, 85748 Garching, Germany, [4]University and INFN, via delle Scienze 208, 33100 Udine,
Italy, [5]China Institute of Atomic Energy, P. O. Box 275 (10), Beijing 102413, P. R. China

Abstract. The fusion barrier distribution obtained for the system ^{9}Be+^{209}Bi from fusion
and elastic scattering cross sections do not show any specific signature of the ^{9}Be break-
up process which is however significant above the Coulomb barrier .

The main goal of this study was to search for possible break-up effects of the
loosely bound ^{9}Be nucleus (S_n=1.67 MeV) into the fusion barrier distributions of the
^{9}Be+^{209}Bi system in the ^{9}Be energy range 36-50 MeV; both these distributions were
extracted from excitation functions of the fusion cross section and of the elastic
scattering at backward angles. The experiments [1,2], done at the Munich Tandem
accelerator, were mainly focused and optimized for precise total fusion cross sections
determination of ^{209}Bi with both ^{9}Be and ^{8}Be (originating from ^{9}Be break-up); anyhow
the detector geometry adopted allowed to observe also elastic scattering events at
rather backward angles but with large solid angle acceptance (±9°, Ω=85 msr).
    The barrier distribution from fusion cross section data was calculated according
to the well established formula [3]:

$$D_{fus}(E) = \frac{1}{\pi R_f^2} \frac{d^2}{dE^2}(E\sigma_{fus}(E))$$

were E is the bombarding energy. The quantity πR_f^2 was extracted from the fusion
data with the equation $E\sigma_{fus}(E) = \pi R_f^2(E-V_b)$ with V_b Coulomb barrier. Fig.1 top
panel left shows the experimental fusion cross sections and the theoretical predictions,
also separately, for the two different fusion channels originating from ^{9}Be (^{218}Fr) and
^{8}Be (^{217}Fr). Fig.1 top panel right shows the relative barrier distributions. The
theoretical calculations were done according to the Dasso and Vitturi approach [4]
with the inclusion of only one break-up channel, namely ^{9}Be→^{8}Be+n. In these
calculation the potential was of Christensen and Winther type; its well depth was
adjusted to reproduce as well as possible in the subbarrier region the well known
system ^{16}O+^{208}Pb [5] where break-up effects are very unlikely to occur; these results
are show in Fig.2. The obtained depth was then scaled for the ^{9}Be + ^{209}Bi geometry.

CP495, *Experimental Nuclear Physics in Europe*, edited by B. Rubio et al.

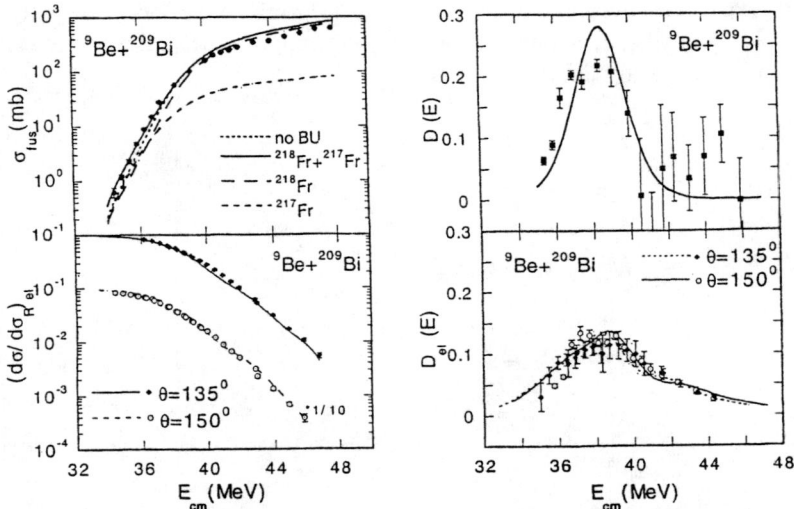

FIGURE 1. Fusion excitation functions experimental and theoretical (top left) and relative barrier distributions (top right). Elastic scattering excitation functions experimental and theoretical (bottom left) and related barrier distributions (bottom right)

This is a very crude approximation but good enough to take into account the most relevant first order vibration effects, expected to be similar in both systems.

The elastic scattering cross sections were measured in the same energy range 36-50 MeV at 135° and partly at 150° and 160° in two different runs. From these data the barrier distributions were extracted according to the formula [6]:

$$D_{el} = -\frac{d}{dE}\left(\frac{d\sigma_{el}}{d\sigma_R}(E)\right)^{\frac{1}{2}}$$

valid for 180° elastic scattering. The distributions were anyhow extracted for 135° and 150°scattering angles (the 160° data were normalized to the 150° scattering angle) with a point difference formula averaging over 3 consecutive points. Support for the utilization of the above formula for the large detectors opening angles which had to be adopted comes from the positive results of ref. [7]. The relative theoretical elastic scattering calculations were done with the code PTOLEMY, with the parameters of a Woods-Saxon type potential obtained from an elastic scattering measurement we recently performed. The results are shown in the Fig.1, lower panels; the energy scale was corrected by the centrifugal energy E_{cent}:

$$E_{cent} = E_{cm}\frac{\sin^{-1}(\theta_{cm}/2)-1}{\sin^{-1}(\theta_{cm}/2)+1}$$

with θ_{cm} detection angle in the center of mass system.

Since the barrier distribution results are rather sensitive to the step sizes adopted in the various point difference formulas the same step sizes were taken for all the data:

fusion elastic, experimental and theoretical. Namely: 1.0 MeV in the low energy region and 1.5 MeV in the high energy one. The adopted formulas automatically normalize to a total strength of 1.0 both barrier distributions.

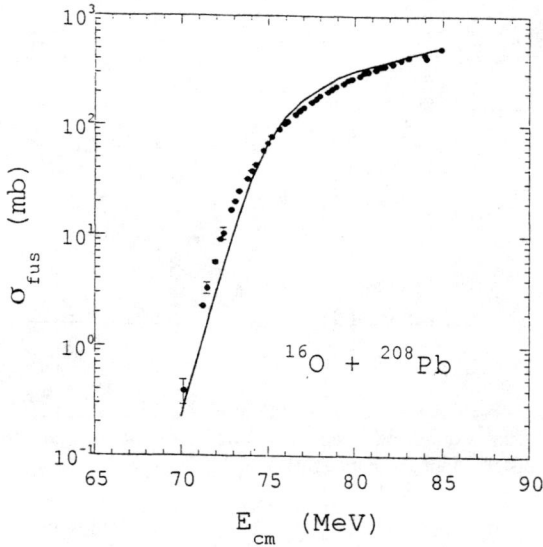

FIGURE 2. Fusion cross sections. In the theoretical calculations (continuos line) the potential depth has been adjusted in order to reproduce as well as possible the subbarrier region.

In principle we can expect that the ^{9}Be BU could influence the barrier distributions if the projectile breaks at a certain point, while approaching the ^{209}Bi target, into $\alpha + \alpha + n$ and then fuses as two separate α plus eventually one neutron; in this case the two α will feel a very different barrier. But the analysis of the barrier distribution in Fig. 1 shows basically one broad peak i.e. one barrier. Consequently there are no evident BU effects into the barrier distribution even if the fusion cross section above the barrier is much lower than theoretically predicted (see fig.2 top left) which is definitely a BU signature.

REFERENCES

1. Signorini C. et al., *Eur. Phys. J.* **2**, 227 (1998).
2. Signorini C. et al., *Eur. Phys. J* **5**, 7 (1999).
3. Leigh J.L. et al., *Phys. Rev.* **C52**, 3151 (1995).
4. Dasso C.H., A. Vitturi, *Phys. Rev.* **C50**, R12 (1994) .
5. Morton C.R. et al., *Phys. Rev.* **C52**, 243 (1995).
6. Rowley N. et al., *Phys. Lett.* **B373**, 23 (1996).
7. Zhang H.Q. et al., *Phys. Rev.* **C57**, R1047 (1998)

A New Interpretation for the Even-Odd Effect in Fission-Fragments Yields

F. Farget[1,2], A. V. Ignatyuk[3], A. R. Junghans[2], K.-H. Schmidt[2]

[1] *Institut de Physique Nucléaire, 91406 Orsay, France*
[2] *Gesllshaft für SchwerIonen Forschung, Planckstraße 1, 64291 Darmstadt, Germany*
[3] *Institute for Physics and Power Engineering, Bondaresko Sq. 1, 249020 Dubna, Russia*

Abstract.
   Calculations which describe the probability to break a pair on the way from saddle to scission point when the nucleus undergoes fission are presented. It is shown that they reproduce the main characteristics of the measured data on even-odd structure in fission-fragment yields. In particular, the strongly differing even-odd structures in proton and neutron numbers are explained within the assumption of thermal equilibrium of intrinsic excitations at scission.

   In the fission of an even-even nucleus, the production of fragments with an odd number of protons (or neutrons) is observed, even if the fissioning nucleus passes the fission barrier with an excitation energy below the pairing gap. That means that a certain amount of heat is gained by the nucleus on its way from saddle to scission point in order to break a pair and to fission into 2 odd fragments. A model, based on the statistical properties of a superfluid nucleus is presented. This model relates the magnitude of the even-odd effect in fission fragment yields to the excitation energy at the scission point.

   The description of the even-odd effect in fission fragment yields is based on the assumption that, if a certain amount of heat is gained by the nucleus on the way to scission, there is a certain probability that the excitation energy is distributed only in the subsystems of protons (or neutrons), while the other subsystem of nucleons remains completely paired. The probability to find the proton subsytem in a completely paired configuration at a given excitation energy can be defined from the level densities of proton and neutron quasi-particle excitation:

$$P_0^Z(U) = \sum_{n_N} \rho_{n_Z=0,n_N}(U) / \sum_{n_N,n_Z} \rho_{n_Z,n_N}(U) \tag{1}$$

   The description of the level density of n_Z-proton and n_N-neutron quasi-particle excitations as a function of the excitation energy U is given in Ref. [1,2]. Parameters of the level density have been adapted to the scission point [3]. The dependence of

CP495, *Experimental Nuclear Physics in Europe*, edited by B. Rubio et al.
© 1999 American Institute of Physics 1-56396-907-6/99/$15.00

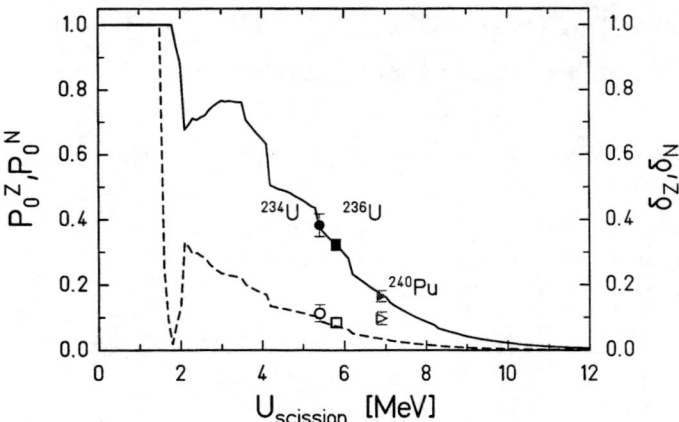

FIGURE 1. Dependence of the survival probabilities of the completely paired configurations for protons P_0^Z (full line) and neutrons P_0^N (dashed line) on the excitation energy at the scission point. The experimental data on the proton and neutron even-odd effects δ_Z and δ_N are shown for the fissioning nuclei ^{234}U, ^{236}U and ^{240}Pu at high kinetic energies of the light fission fragments (which do not allow for neutron evaporation).

the probability to find one of the subsystems (protons or neutrons) in a completely paired configuration on the excitation energy at the scission point is shown in Figure 1. Experimental data [4–6] on the even-odd effect for proton (full symbols) are set on the theoretical curve. The excitation energy of the fissioning system at scission is thus deduced. A remarkable agreement is observed in comparing the experimental values of the even-odd effect in neutron numbers (open symbols) to the theoretical curve.

The model presented here is able to quantitatively describe the different magnitudes of the even-odd effect observed in neutron and proton numbers. This difference, which was not understood up to now, is explained as the result of the difference in the phase space for the nuclear excitations of both subsystems.

REFERENCES

1. Ignatyuk A. V., Sokolov Yu. V., *Sov. J. Nucl. Phys.* **17**, 376 (1973)
2. Ignatyuk A. V., *Statistical properties of excited atomic nuclei (Russian)*, Energoatomizdat, Moscow, 1983. Translated as Report INDC(CCP)-233/L.IAEA,Vienna,1985
3. Farget F., Ignatyuk A. V., Junghans A. R., Schmidt K.-H, to be published in Nucl. Phys. A
4. Lang W., et al., *Nucl. Phys. A* **345**, 34 (1980)
5. Schmitt C., et al., *Nucl. Phys. A* **430**, 21 (1984)
6. Quade U., et al., *Nucl. Phys. A* **487**, 1 (1988)

Study of Complete Fusion Reactions Leading to the Production of Heavy and Superheavy Nuclei

Roman N. Sagaidak

Flerov Laboratory of Nuclear Reactions, JINR, 141980 Dubna, Moscow reg., Russia,
E-mail: sagaidak@sunvas.jinr.ru

Abstract. Cross section values for heavy evaporation residues (ER) produced in complete fusion reactions induced by heavy ions on spherical and deformed target nuclei are analyzed in the framework of barrier penetration and statistical model approximations. For the reactions leading to Rn–Pa nuclei, a strong influence of the entrance channel on the measured cross section values is observed for nearly symmetric projectile-target combinations. In order to reproduce the observed excitation functions in such combinations we had to introduce the quantity of fusion probability. Considering the asymmetric reactions leading to the heaviest nuclei we also had to use the fusion probability to reproduce the cross section values obtained for cold fusion reactions induced by ^{50}Ti and heavier projectiles on the Pb and Bi target nuclei, and also the values obtained for hot fusion reactions induced by ^{34}S on actinide target nuclei. The scaling of fusion probabilities derived for both the reactions allowed us to predict the values of cross sections for superheavy elements (SHE) produced in the ^{48}Ca induced reactions on actinide target nuclei and in the cold fusion reactions induced by the Zn and heavier projectiles.

Production of heavy ER has been considered in different models. The models based on the dinuclear system (DNS)-concept (1) are among of them. In the calculation of the cross sections for the deformed SHE produced in cold fusion reactions good agreement with the experimental data (2) was achieved when assuming that the ratio of the partial widths for neutron emission and fission is constant (3). More sophisticated calculations in the part relating the decay of the compound nucleus (CN) were performed later (4). They also showed very good agreement with the experiments, in despite of the difference in the approaches (3, 4) to the consideration of the CN decay. Production of the deformed SHE in the cold fusion reactions was successfully considered within a very simple model treating the formation of a CN as a quantal tunneling through a one-dimensional potential barrier and the following statistical decay of the CN by the one neutron emission in competition with fission (5).

The aim of present work is a systematical analysis of the large body of the data on the production of ER in different complete fusion reactions with the use of the well-known barrier penetration complete fusion model (BPCFM) and standard statistical model (SSM) approximations realized in the HIVAP-code (6).

We used mainly the data obtained with modern recoil separators: SHIP (2, 7–9), DGFRS (10) and VASSILISSA (8, 11–15). These systems separate extremely rarefied flows of atoms of ER from intensive heavy ion beam and background particles. It is achieved due to the difference in the electrical and/or magnetic rigidity for ER and other particles. The focusing systems of these apparatuses form a flow of ER, which

CP495, *Experimental Nuclear Physics in Europe*, edited by B. Rubio et al.
© 1999 American Institute of Physics 1-56396-907-6/99/$15.00

are stopped in a silicon detector array. The α-decay and spontaneous fission of the implanted ER as well as their time-of-flight and kinetic energy are detected by similar detection systems of the separators. The normalization of the observed yields for the identified ER to the beam flux, transmission for ER, target thickness and detection efficiency allows one to obtain reliable values of the production cross sections.

The main features of the HIVAP-code are considered in (6). Our approach to the analysis of the data and the choice of the model parameters is presented in (9). We note that the calculated production cross sections for ER at energies well above the fusion barrier are weakly sensitive to the form of the nuclear potential (7, 9). At these energies, they are determined by the SSM parameters describing a CN de-excitation. The complete fusion cross sections were calculated in the framework of BPCFM using the nuclear exponential potential and the fusion barrier fluctuations expressed via the radius-parameter percentage. Transmission coefficients were obtained in the framework of the WKB approximation.

The numerous data on heavy ER produced in nearly mass-symmetric massive systems were obtained and analyzed earlier (7, 8). The most salient feature of such systems is considerable deficit of fusion above the expected fusion barrier. This hindrance to fusion was attributed to the dynamical evolution of the composite system which undergoes immediate reseparation (quasi-fission). The experimental data reveal the fact that the hindrance to fusion is strongly influenced by the nuclear structure of reactants (7). Our data on the ^{86}Kr+^{136}Xe reaction obtained later (9) confirm the influence of the structure on the production cross sections. At the same time, the cross section values in the ^{86}Kr+^{130}Xe reaction are about 500 times smaller than those observed in ^{86}Kr+^{136}Xe. This observation has so far no quantitative explanation.

The measured cross section values for the observed ER produced via the ^{222}Th* CN formed in the ^{48}Ca+^{174}Yb (15) and ^{86}Kr+^{136}Xe (9) reactions agree well with each other, at least it is so for the 3n–6n channels. Differences for the 1n and 2n channels could be explained by differences in the values of the fusion barrier. Calculations of the excitation functions for both the reactions were performed with identical parameters describing the fusion and de-excitation of the CN. From the comparison we conclude that the fusionability of both the considered systems is about the same, or that for the ^{86}Kr+^{136}Xe reaction the fusion probability $P_f = 1.0$. The situation becomes quite different when we consider the asymmetric and nearly symmetric combinations leading to the spherical ^{216}Th* CN. The measured cross section values for the ER produced in the ^{40}Ar+^{176}Hf reaction (see ref. in (8)) are reproduced rather well. For the data corresponding to the nearly symmetric ^{86}Kr+^{130}Xe (9) and ^{124}Sn+^{92}Zr (see ref. in (8)) combinations, we had to introduce the fusion probability values in order to reproduce the data with the use of the same parameters describing the ^{216}Th* CN decay.

We extended our analysis using the available data in the region of Rn–Pa compound nuclei produced in the "asymmetric" and "symmetric" combinations. The data allowed us to deduce the P_f values in the same way as in the study of systems leading to the ^{222}Th* and ^{216}Th*. The less asymmetric combinations with the ^{51}V and ^{58}Fe projectiles were also included in the analysis. We have not found a good scaling for the derived P_f values covering the range of 0.01 (^{86}Kr+^{130}Xe) to 0.3 (^{58}Fe+^{158}Sm).

We applied the same approach to the analysis of the data on the production of the heaviest ER. We fitted the calculated excitation functions to the available experimental data for complete fusion reactions induced by heavy ions on deformed and spherical target nuclei. For the reactions with the deformed nuclei we obtained a good fit to the data using the liquid-drop fission barriers which are by 10–30% higher than the nominal ones, with the exception of the reactions induced by [34]S. The strong reduction of these values (up to zero) was insufficient to reproduce the obtained cross sections values. In the case of reactions induced by [16]O to [48]Ca projectiles on spherical target nuclei, we obtained a good fit to the data with the barrier which are by 10–30% lower than the nominal ones. For the reactions leading to heavier compound nuclei, we obtained a strong overestimate of the calculated cross sections for ER. We normalized the experimental data to the calculated ones using the nominal values of liquid-drop fission barriers and fixed parameters of the nuclear potential to derive the values of the fusion probability for these reactions. Searching for regularities in the behavior of P_f, we have plotted it as a function of the mean arithmetic fissility, x_{ma} (Fig. 1). Different atomic masses (16–19) were used in the calculations for $Z_{CN} \geq 108$ keeping in mind the neglected values of the liquid-drop fission barriers for those nuclei and the fact that real fission barriers were determined by the shell corrections (the difference between empirical or calculated and liquid-drop masses of nuclei).

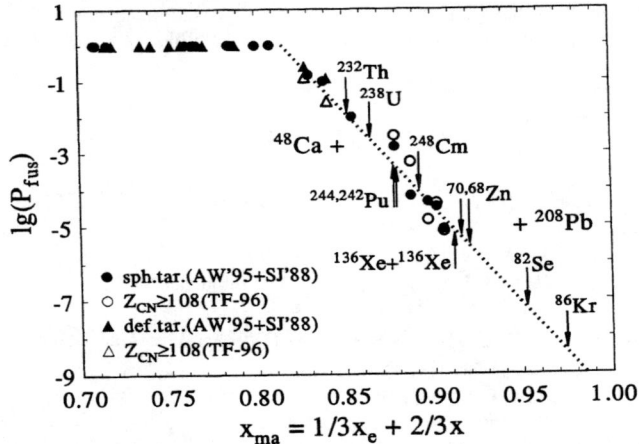

FIGURE 1. The fusion probability values derived in accordance with the analysis of the production of the heaviest ER in the reactions induced on spherical and deformed target nuclei. The P_f values for the reactions induced by [48]Ca on actinide target nuclei and cold fusion reactions induced by Zn and heavier projectiles are marked by arrows (at the corresponding x_{ma} values) near the line of approximation.

Comparing different mass tables (17–19) with the empirical values (AW'95) (16) for the heaviest nuclei (from Db to element 110) we have found that the masses (17) (SJ'88) have the minimal values of the root mean square deviations (RMSD) compared with the other considered mass tables. Further, we considered the production of SHE using those tables; the masses based on the Thomas-Fermi model (TF-96) (18) and the finite-range droplet-model masses (DM-95) (19). The TF-96 masses give the values of RMSD comparable with the SJ'88 ones, whereas DM-95 ones underestimate the values which leads to the corresponding overestimate of the absolute values of shell corrections. The results of calculations (the values of the excitation function maxima of the most probable evaporation channels, i.e., 3n for the ^{48}Ca induced reactions and 1n for the cold fusion reactions at $P_f = 1.0$) are shown in Fig. 2 together with the fusion probabilities derived from the calculations, approximation and extrapolation. The increasing cross section values (at $P_f = 1.0$) for the ER from $Z = 110$ to $Z = 114$ produced in the ^{48}Ca induced reactions and for the ER from $Z = 108$ to $Z = 116$ produced in the cold fusion reactions are the result of the growing shell corrections for these nuclei.

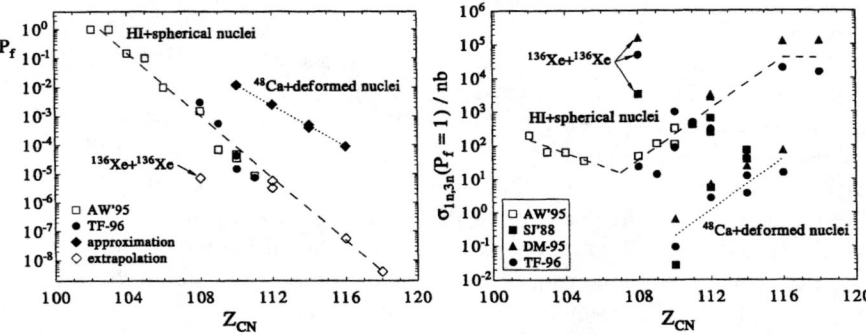

FIGURE 2. Fusion probabilities (left panel) and survivability (right panel) in the ^{48}Ca and cold fusion reactions leading to the production of very heavy and superheavy nuclei in the 3n and 1n evaporation channels, respectively.

REFERENCES

1. Antonenko A. V., *et al.*, *Phys. Lett.* B **319**, 425–430 (1993); *Phys. Rev. C* **51** 2635–2645 (1995).
2. Hofmann S., *Rep. Prog. Phys.* **61**, 639–689 (1998).
3. Adamian G. G., *et al.*, *Nucl. Phys. A* **633**, 409–420 (1998).
4. Cherepanov E. A., *PRAMANA J. Phys.* **23** (1999), to be published.
5. Smolańczuk R., *Phys. Rev. C* **59**, 2634–2639 (1999).
6. Reisdorf W. and Schädel M., *Z. Phys. A* **343**, 47–57 (1992) and references therein.
7. Quint A.B., *et al.*, *Z. Phys. A* **346**, 119–131 (1993).
8. Schmidt K. -H. and Morawek W., *Rep. Prog. Phys.* **54**, 949–1003 (1991).
9. Sagaidak R.N., *et al.*, "Evaporation residues production in the ^{16}O+^{208}Pb and ^{86}Kr+^{130,136}Xe reactions near the fusion barrier", In *Proceedings of the VI International School-Seminar on Heavy Ion Physics, Dubna, Russia, 22–27 September 1997*, World Scientific, Singapore, 1998, pp. 323–330.
10. Utyonkov V. K., *et al.*, "Discovery of enhanced nuclear stability near the shell closures N=162 and Z=108", ibid., pp. 400–408.
11. Yeremin A. V., *et al.*, *Nucl. Instr. Meth. in Phys. Res. A* **350**, 608–617 (1994).
12. Oganessian Yu. Ts., *et al.*, *Eur. Phys. J. A* **5**, 63–68 (1999).
13. Oganessian Yu. Ts., *et al.*, *Nature* **400**, 242–245 (1999).
14. Yeremin A. V., *et al.*, *JINR Rapid Communication* **6[92]-98**, 21–34 (1998).
15. Sagaidak R. N., *et al.*, *FLNR JINR Scientific Report 1997–1998*, Dubna, 1999, to be published.
16. Audi G. and Wapstra A. H., *Nucl. Phys. A* **595**, 409–480 (1995).
17. Spanier L. and Johansson S. A. E., *At. Data Nucl. Data Tables* **39**, 259–264 (1988).
18. Myers W. D. and Swiatecki W. J., *Nucl. Phys. A* **601**, 141–167 (1996).
19. Möller P. *et al.*, *At. Data Nucl. Data Tables* **59**, 185–381 (1995).

Radioactive Ion Beams at ISOLDE/CERN Recent Developments and Perspectives

U. Georg[a], J.R.J. Bennett[b], U.C. Bergmann[c], R. Catherall[a],
P. Drumm[b], V.N. Fedoseyev[d], T. Giles[a], O.C. Jonsson[a],
A.R. Junghans[e], U. Köster[a], E. Kugler[a], J. Lettry[a], V.I. Mishin[d],
T. Nilsson[a], H. Ravn[a], K.-H. Schmidt[e], H. Simon[a], C. Tamburella[a]
and the ISOLDE Collaboration

[a] EP Division, CERN, Geneva, Switzerland
[b] CLRC, Rutherford Appleton Laboratory, Chilton, Didcot, Oxon, UK
[c] Institut for Fysik og Astronomi, Aarhus Universitet, Aarhus Denmark
[d] Institute of Spectroscopy, Russian Academy of Science Troisk, Russia
[e] GSI, Darmstadt, Germany

Abstract. Since the move of ISOLDE from CERN's synchrocyclotron (SC) to the Proton Synchrotron Booster (PSB) in 1992 extensive work has been devoted to the development of new beams, i.e. the production of new isotopes, beams of higher intensity and the ionization of further elements. Most of these developments were driven by the particular needs of the physics community proposing new experiments.

The main achievements were the adaption of liquid metal targets to the pulsed proton beam to prevent shockwaves and splashing inside the target container [1] and systematic studies on the time structure of the release of the isotopes from the target [2]. Furthermore the work on laser ion-sources already started at ISOLDE-2 was continued, the so-called RIST target was developed, and most recently first tests on the isotope production while increasing the proton energy from 1 GeV to 1.4 GeV were done. The latter topics are discussed in this paper.

INTRODUCTION

At the on-line mass separator ISOLDE ion beams of radioactive isotopes are produced by nuclear reactions induced by bombardment of thick targets with a 1 GeV, and now also 1.4 GeV, pulsed proton beam. With the use of a large variety of target materials and ion sources over 600 isotopes of more than 60 elements (Z=2 to 88) have been produced up to now.

Physics at ISOLDE is pursued in several directions, e.g. (i) the systematic study of atomic and nuclear properties, (ii) exotic decays far from the line of stability,

CP495, *Experimental Nuclear Physics in Europe*, edited by B. Rubio et al.
© 1999 American Institute of Physics 1-56396-907-6/99/$15.00

(iii) astrophysics, (iv) weak-interaction physics, (v) solid-state physics and (vi) bio-medical studies using radioactive isotopes for diagnosis and therapy.

The demands from these different experiments have led to the development of the chemically selective laser ion-source producing more intense and purified beams. Annother technical improvement is a target with a geometry optimized for small diffusion path and fast effusion delivering very high numbers of the short lived neutron halo isotope ^{11}Li. Recently CERN's PSB has been upgraded in proton beam energy from 1 GeV to 1.4 GeV, offering now the possibility to compare the production yields at both energies.

LASER IONIZATION

At ISOLDE stepwise resonant laser excitation in hot cavities is used for the production of radioactive ion beams. The 2 or 3 laser beams are sent collinearly into a high temperature metallic cavity where the atoms which effused out from the target container are ionized. The laser set-up consists of copper vapour laser pumped dye-lasers. For those elements with UV transitions BBO crystals are used to obtain the desired frequency by frequency doubling or tripling. The copper vapour lasers are pulsed at a frequency of 10 kHz. The chemical selectivity of the resonant laser ionization is perturbed by the surface ionization of other elements inside the hot cavity. This effect can be partially suppressed by choosing a low work function material like Nb for the hot cavity and by a time selective extraction from the LIS [3].

The development on the chemically selective laser ion-source led to the production of intense, purified beams of Be, Mn, Ni, Cu, Zn, Ag and Cd. The measured ionization efficiencies for these elements are compiled in table 1.

These beams were used to obtain several highly interesting results. These are the half life of the waiting point nucleus ^{129}Ag [4]; the beta decay studies of ^{58}Zn [5] by nuclear spectroscopy; the magnetic moments of ^{7}Be and ^{11}Be by collinear laser spectroscopy and β-NMR [6]; and the magnetic moments of ^{67}Ni and ^{67}Cu by low temperature nuclear orientation [7].

The LIS itself gives also direct access to the measurement of atomic and nuclear properties if the hyperfine splitting (HFS) exceeds the line width of the laser (typically 5 GHz). Furthermore the isotopic shift (IS), providing information on the mean square nuclear charge radii, can be measured by monitoring the radioisotope production with respect to the scanned laser frequency.

TABLE 1. Offline measured LIS ionization efficiencies of those elements for which radioactive ion beams have beam produced at ISOLDE except Sn.

Be	Mn	Ni	Cu	Zn	Ag	Cd	Sn
7% [8]	19.2% [9]	0.8% [10]	6.6% [11]	4.9% [11]	14% [3]	8.8% [11]	8.5% [12]

RIST TARGET

The Radioactive Ion Source Test (RIST) project was originally started to study and design a high power tantalum target for use with proton currents of up to 100 μA [13]. Currently this development programme is continued in order to improve the production yield of very short-lived isotopes using ^{11}Li ($T_{1/2}$=8.7 ms) as a first study case, for which the release time from the target is much longer than its half-life.

For the first time a target was constructed based on theoretical modelling studies [14] of diffusion and effusion mechanisms inside a target. Using 199 tantalum foils with a thickness of only 2 μm, 1 cm high and 15 cm long mounted inside a standard ISOLDE tantalum target container connected to a tungsten surface ionizer, an increased yield of 7076 ^{11}Li per μC was observed [15]. For comparison with other tantalum foil targets at ISOLDE see table 2. Further tests are planned on the production on other very short-lived isotopes like neutron rich beryllium and sodium.

TABLE 2. Production yields of ^{11}Li from RIST and ISOLDE tantalum targets.

Target	foil thickness [μm]	target thickness [g/cm^2]	yield [ions/μC]
standard rolls	20	111	500
RIST discs	100	149	2500
RIST strips	2	10	7076

1.4 GEV PRODUCTION TEST

After CERN's PS-Booster was up-graded from 1 GeV to 1.4 GeV in 1998 it has become possible to use the higher energy proton beam for isotope production at ISOLDE. First tests were done using a uranium carbide target with a plasma ion source and a cooled transfer line [16] and a niobium foil target with tungsten surface ionizer. The ratio $\mathcal{R} = Yield(1.4 \text{ GeV})/Yield(1 \text{ GeV})$ of the production of the mainly neutron-rich isotopes of He, Ne, Ar, Kr, Xe, Rn and Na, Rb respectively has been measured, using the ISOLDE tape system with a beta counter. For those isotopes decaying via the emission of neutrons a highly efficient neutron counter was used, as described in [17].

For the production by fragmentation, i.e. helium, neon and sodium, a considerable increase has been found (see table 3). For spallation, the production of radon isotopes is increased at 1.4 GeV by more than a factor of 2, while the rubidium yield remained unchanged. The fission products krypton and xenon are slightly better produced with the higher energy.

Calculations using a two-step reaction model, consisting of an intranuclear cascade code [18], and a evaporation code [19], reproduce the measured trends in the yield ratio $\mathcal{R}$ reasonably well. The isotopes produced by fission seem to be more or less unaffected varying the kinetic energy of the impinging protons wheras the yields of light nuclei are considerably enhanced in the calculation.

Further tests with other targets and other isotopes are currently in progress and will be published after their evaluation [20].

TABLE 3. Ratio $\mathcal{R}$ of the production yields from a uranium carbide and a niobium target respectively. (The sequence of ratios at the bottom corresponds directly to the mass numbers in the second line.)

Element	Uranium Carbide						Niobium	
	He	Ne	Ar	Kr	Xe	Rn	Na	Rb
Isotope	6, 8	23-28	49	90, 93	117, 137-146	220	20, 26	75, 80
$\mathcal{R}$	1.8, 2.4	5.5-2.5	1.6	1.5, 1.6	0.4, 0.9-1.5	2.5	3.9, 4.5	≈ 1.0

REFERENCES

1. J. Lettry et al., Nucl. Instr. and Meth. **B 126**, 170-175, (1997).
2. J. Lettry et al., Nucl. Instr. and Meth. **B 126**, 130-134, (1997).
3. Y. Jading et al., "Production of radioactive Ag ion beams with a chemically selective laser ion source", Nucl. Instr. and Meth. **B 126**, 76-80, (1997).
4. Kratz K.-L., et al., "Laser isotope and isomer separation of heavy Ag nuclides: Half-life of the r-process waiting point isotope [129]Ag and structure of neutron-rich Cd nuclides", presented at Int. Conf. on Fission and Properties of Neutron-Rich Nuclei, Sanibel Island, Florida, December 1997.
5. Jokinen A., et al., Eur. Phys. J. **A 3**, 271-276 (1998).
6. Kappertz, S., et al., 2nd Int. Conf. on Exotic Nuclei and Atomic Masses ENAM98, AIP conf. Proc. **455**, 110-113, Ed. B.M. Sherill, D.J. Morrissey and C.N. Davids, AIP, Woodbury, NewYork, (1998).
7. T. Giles et al., to be published.
8. V.N. Fedoseyev, CAARI 98, Denton AIP conf. Proc., 1998, to be published.
9. V.N. Fedoseyev et al., Nucl. Instr. and Meth. **B 126**, 88, (1997).
10. A. Jokinen et al., Nucl. Instr. and Meth. **B 126**, 98, (1997).
11. J. Lettry, et al., Rev. Sci. Instr. **69**, 761, (1997).
12. V.N. Fedoseyev, Res. Ion. Spectr. AIP conf Proc. **329**, 465, Ed. H.J. Kluge, J.E. Parks and K. Wendt, AIP, Woodbury, New York, (1994).
13. J.R.J. Bennet et al., Proceedings of 4th EPAC 94, 1415, (1994).
14. J.R.J. Bennet, 8th international Conference on Heavy Ion Accelerator Technology, Argonne National Laboratory, AIP conf Proc. **473**, 490, Ed. K.W. Shepard, AIP, Woodbury, New York, 1999.
15. J.R.J. Bennet et al., Letter to the editor, accepted for publication in Nucl. Instr. Meth. **B**, 1999.
16. J. Lettry et al., submitted to **HFI**.
17. U.C. Bergmann et al., "β-delayed Neutron Emission from [12,14]Be", this conference.
18. J. Cugnon, et al., Nucl. Phys. **A620**, 457, (1997).
19. A.R. Junghans, et al., Nucl. Phys. **A 629**, 635, (1998).
20. U. Georg et al., to be published.

The Chimera Facility at LNS.

S.Aiello[b] , A.Anzalone[a] , M.Bartolucci[d] , M.G.Campisi[a] , G.Cardella[b] ,
Sl.Cavallaro[a] , E.De Filippo[b] , E.Geraci[a] , F.Giustolisi[a,c] , P.Guazzoni[d] ,
C.M.Iacono-Manno[a,e] , G.Lanzalone[b,c,e] , G.Manfredi[d] , G.Lanzanò[b] , S.Lo
Nigro[b,c,e] , A.Pagano[b] , M.Papa[b] , S.Pirrone[b] , G.Politi[b] , F.Porto[a,c] ,
S.Sambataro[b,c] , M.L.Sperduto[a,c] , C.M. Sutera[b,c] , L.Zetta[d]

[a]INFN-Laboratori Nazionali del Sud, Via S.Sofia 44, 95123 Catania Italy
[b]INFN-Sezione di Catania, Corso Italia 57, 95129 Catania Italy
[c]Dipartimento di Fisica, Università di Catania, Corso Italia 57, 95129 Catania Italy
[d]Dipartimento di Fisica dell'Università and INFN, v. Celoria 16,20133 Milano, Italy
[e]CSFNSM, Corso Italia 57, 95129 Catania Italy

Abstract. CHIMERA is a 4π detector for charged particles whose construction at LNS is in progress and is coming at end. The main features, some performances and the status of the project is presented.

The CHIMERA facility was designed [1,2,3] in order to detect and identify charged particles and fragments emitted in heavy ion nuclear reactions to investigate many topics of the heavy ion physics at intermediate energy, as for instance: i) the properties of nuclei under extreme conditions of density and temperature; ii) the deexcitation of hot and compressed systems formed in the early stages of the collision; iii) the appearance and the amount of the multifragmentation.

The construction of the multidetector is in progress and is coming at end. It is made of 1192 telescopes arranged in 35 rings in a cylindrical geometry around the beam axis. The forward 18 rings are assembled in 9 wheels covering polar angles between 1° and 30° and are placed at distances from the target variable from 350 to 100 cm. The remaining 17 rings covering the angular range 30°-176° are assembled in a sphere of 40 cm in radius. An assonometric view of the mechanical structure supporting the telescopes is given in fig.1. The shape and dimensions of CHIMERA make it suitable for TOF techniques measurements and enable a mass identification of the fragments stopped in the first stage of the telescope.

Each telescope is made of one silicon detector 300 microns thick and one CsI(Tl) crystal variable in thickness from 12 to 3 cm. Several tests have been performed [4,5,6] in order to select the optimum characteristics of the detectors making the telescopes. The total overall covering of the solid angle is about 94%.

The figs. 2, 3 and 4 respectively, show the results of some tests on the charge and mass identification of the detected fragments carried out by using standard CHIMERA telescopes and on a LCP identification obtained by using a pulse shape analysis of the signals coming from the CsI crystals read out by photodiodes.

Recently, the 9 wheels of CHIMERA have been assembled in a vacuum chamber at the LNS and soon will be used to get experimental data on the multifragmentation and

CP495, *Experimental Nuclear Physics in Europe*, edited by B. Rubio et al.
© 1999 American Institute of Physics 1-56396-907-6/99/$15.00

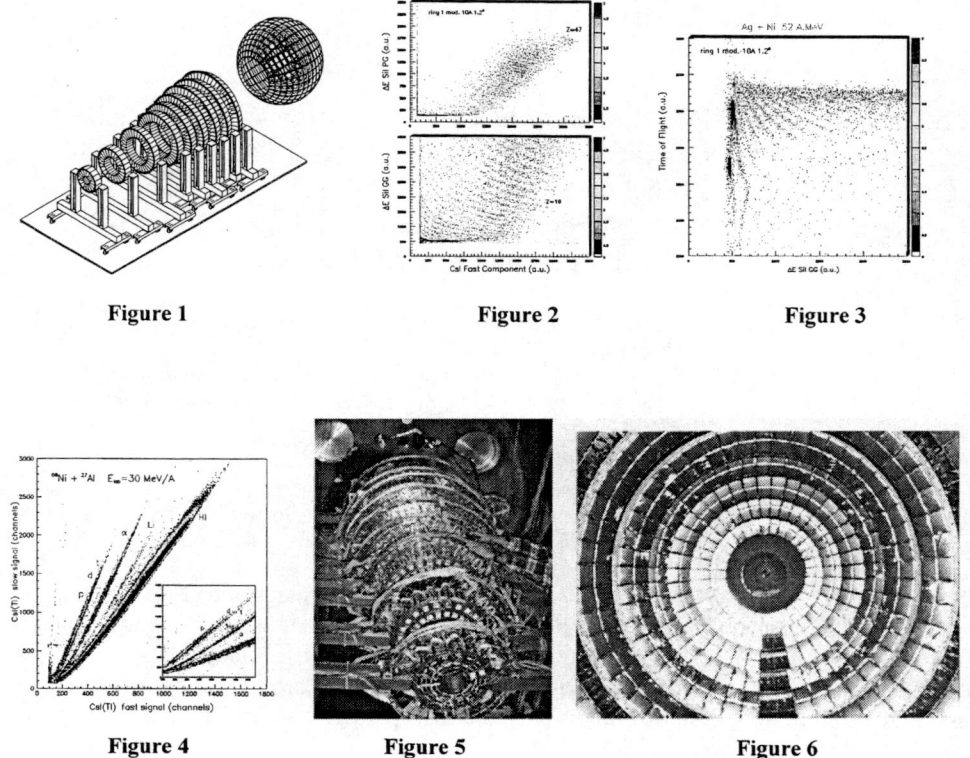

Figure 1   Figure 2   Figure 3

Figure 4   Figure 5   Figure 6

on the effects of clustering and isospin on this mechanism in some reactions in reverse kinematics.

The figures 5 and 6 show the outer side, with the wiring on, and the inner side views, respectively, of the 9 wheels during the assembling.

The mechanical structure of the spherical part is by now ready and the production of the 500 telescopes is starting in summer '99. Considering that several months are necessary for the final assembling, the whole detector should be ready to run fully at LNS in the first year of the next millenium.

REFERENCES

1. S.Aiello et al., Nucl.Phys.A583(1995)461
2. A.Pagano et al., Proc.of 2nd INFN-RIKEN Symposium, Riken, Japan, May 1995, pg.119
3. A.Pagano et al., Proc.of Int.Research Workshop Poiana-Brasov, Romania, Oct. 1996,pg.129
4. S.Aiello et al., Nucl.Instr.and Meth.A369(1996)50
5. S.Aiello et al., Nucl.Instr.and Meth.A385(1997)306
6. S.Aiello et al., Nucl.Instr.and Meth.A427(1999)510

NUCLEAR ASTROPHYSICS

Measurement of the $^3He(^3He,2p)^4He$ cross section down to the lower edge of the solar Gamow peak

THE LUNA COLLABORATION

R. Bonetti[1], C. Broggini[2], L. Campajola[3], P. Corvisiero[4],
A. D'Alessandro[5], M. Dessalvi[4], A. D'Onofrio[3], A. Fubini[6],
G. Gervino[7], U. Greife[8], A. Guglielmetti[1], C. Gustavino[5],
M. Junker[5], C. Marino[7], P.Prati[4], V. Roca[3], C. Rolfs[8],
M. Romano[3], F. Schiemann[8], F. Strieder[8], F. Terrasi[3],
H.P. Trautvetter[8] and S. Zavatarelli[4].

1) Universita' di Milano, Dipartimento di Fisica and INFN Milano
2) INFN Padova
3) Universita' di Napoli, Dipartimento di Fisica and INFN Napoli
4) Universita' di Genova , Dipartimento di Fisica and INFN Genova
5) Laboratori Nazionali del Gran Sasso
6) ENEA Frascati and INFN Torino
7) Universita' di Torino, Dipartimento di Fisica and INFN Torino
8) Institut fur Experimentalphysik III, Rhur-Universitat Bochum

Abstract. The LUNA Collaboration has provided the first cross section measurements of the key reaction $^3He(^3He,2p)^4He$ of the proton-proton chain at the thermal energy of the Sun. This successful project has shown that, by going underground and by using the typical techniques of the low background physics, it is possible to measure down to the energy of the nucleosynthesis inside the stars.

For the first time the cross section of the key reaction $^3He(^3He,2p)^4He$ of the proton-proton chain, that affects strongly the neutrino luminosity from the Sun, has been measured in almost all the energy range covered by the Gamow peak. The Gamow peak region is far below the Coulomb barrier E_C (approximately $E_0/E_C=0.01$) and the reaction cross section drops nearly exponentially with the decreasing energy according to [1]:

$$\sigma(E) = [S(E)/E]\exp(-2\pi\eta)$$

where $S(E)$ is the astrophysical factor and η is the Sommerfeld parameter, given by $2\pi\eta = 31.29\ Z_1 Z_2 (\mu/E)^{0.5}$. Z_1 and Z_2 are the nuclear charges of

CP495, *Experimental Nuclear Physics in Europe*, edited by B. Rubio et al.

the interacting nuclei, μ is their reduced mass in AMU, E is the center of mass energy in keV. Because of the extremely low value of the cross section within the Gamow peak (we are dealing with values of the order of the picobarn), all the measurements where carried out at the Gran Sasso underground laboratories, where the cosmic ray background in our sensible energy region is highly suppressed (around a factor 10^3). The beam and the target used in experiment are usually made of ions and neutral atoms respectively, and the electron clouds sorroundings the interacting nuclei act as a screening potential that reduced the strength of the Coulomb barrier, leading to a higher cross section $\sigma_S(E)$ with respect to the case for the bare nuclei inside the solar plasma. We can link the bare nuclei cross section $\sigma_b(E)$ to the one measured in our experimental conditions by the factor:

$$f_{lab}(E) = \sigma_S(E)/\sigma_b(E) = \exp(\pi\eta U_e/E)$$

where U_e is the electron screening potential energy. It should be stressed that the screening effect has to be measured and taken into account in order to evaluate the bare nuclei cross section which is the input data for the models of stellar nucleosynthesis. The experimental apparatus is described in ref. [2]. In Tab. 1 we give the new results which conclude the LUNA measurements of $^3He(^3He,2p)^4He$ reaction. For these measurements we can state that the $^3He(^3He,2p)^4He$ cross section increases at the thermal energy of the Sun as expected, but does not show any evidence for a narrow resonace. Consequently the astrophysical solution for the solar neutrino problem based on the existence of such a narrow resonace is ruled out by our results.

Tab.1   The LUNA new results

Energy (keV)	Events	S(E) (MeV b)	$\Delta S_{stat}(E)$	$\Delta S_{sys}(E)$
16.50	1	7.70	7.70	0.49
16.99	7	13.15	4.98	0.83
17.46	1	5.26	5.26	0.33
18.46	7	7.86	2.97	0.47
18.98	13	8.25	2.29	0.48

REFERENCES.
[1] C. Arpesella et al. Phys. Lett. B389(1996)452.
[2] U. Greife et al. Nucl. Instr. and Meth. A350(1994)327.

The ^{7}Li(p,γ)^{8}Be reaction
at astrophysical energies

J. M. Sampaio*, A. M. Eiró* and I. J. Thompson[†]

* Centro de Física Nuclear e Departamento de Física da Universidade de Lisboa, Portugal
[†] Department of Physics, University of Surrey, Guildford GU2 5XH, UK

Abstract. We calculate the (p,γ) capture to the 0^+ ground state of ^{8}Be by means of the E1 and M1 multipoles. Scattering states are obtained in a single-particle model using potentials that reproduce the elastic phase-shift of the ^{7}Li+p system at low energies, and the ^{8}Be g.s. is considered to be a pure single-particle $p_{3/2}$ state outside a ^{7}Li core. Describing the M1 resonance with a spectroscopic factor α^2_{M1}, we obtain for the extrapolated astrophysical S-factor and for the M1 contribution at $E_{lab} = 0$, values of $S(0) = 0.27$ keV-barn and $\sigma_{M1}/\sigma_{TOT} = 3\%$.

In the last few years there has been considerable interest in the study of the ^{7}Li(p,γ)^{8}Be reaction, especially concerning the relative importance of E1 and M1 multipoles ($E_{LAB} < 1$ MeV) and the implications for the extrapolation of the astrophysical S-factor. To the p-wave final state, M1 proceeds via p-wave capture and, due to the centrifugal barrier, this decreases the extrapolated astrophysical S-factor to that from the E1 contribution, arising from s-wave capture. An early measurement [1] of ^{7}Li($\vec{p},\gamma$)^{8}Be gave a significant A_y, implying a significant p-wave (M1) admixture. This p-wave capture is influenced by the 441 and 1030 keV 1^+ resonances, but this mechanism is still unclear, since both experimental and theoretical results give inconsistent results (see Tab. 1) [1–5].

The parameters for the ^{7}Li+p potential of volume type were obtained by fitting the s-wave phase-shifts (see Fig. 1 left, dotted line). In order to describe the narrow ($\Gamma_{LAB} \approx 12$ keV) 441 1^+ resonance of ^{8}Be, we needed a surface repulsive potential (SRP) contribution. A quite good fit was obtained by varying the SRP and the spin-orbit strength (Fig. 1 left, dashed line).

From the E1 matrix element for (p,γ) direct capture, we obtain a total cross section with a good description of the energy dependence for the background (Fig. 1 right, dotted line). The M1 matrix element generates a total cross section with a resonant peak at the correct 441 keV energy, and with a good shape and width, but larger than the experimental data (see Fig. 1 right, dashed line). A spectroscopic factor of α^2_{M1}=0.03 is needed to reproduce (by the long dashed line) the experimental peak. Exclusion of the SRP gives the "M1 background" (dot-dashed line)

CP495, *Experimental Nuclear Physics in Europe*, edited by B. Rubio et al.
© 1999 American Institute of Physics 1-56396-907-6/99/$15.00

FIGURE 1. Left: Elastic phase-shift analysis with the code FRESCO [6]. Right: (p,γ) S-factor.

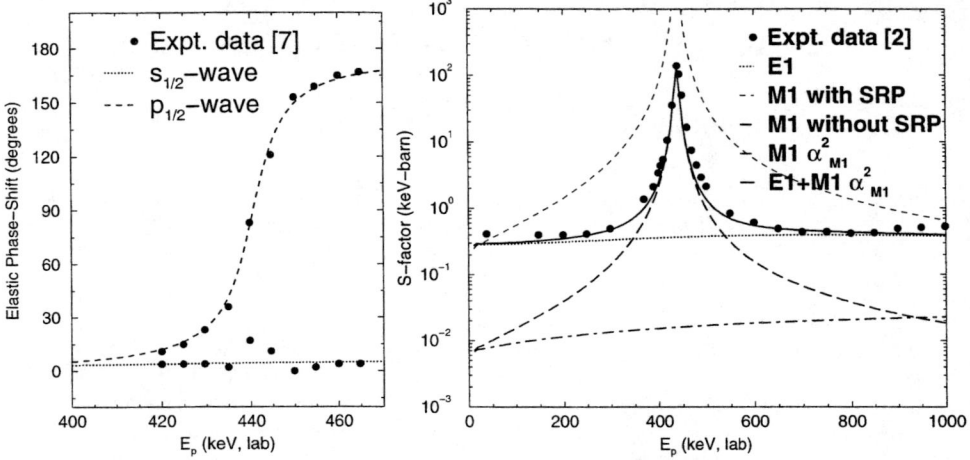

TABLE 1. Estimates of (p,γ) S-factors (keV-barn) and M1 contributions.

Source	Ref.		$S(0)$	$S(70)$	σ_{M1}/σ_{TOT}	$A_y(90)$
Cecil	[8]	Expt. & s-wave DC	0.25 ± 0.05	0.25 ± 0.05	-	-
TUNL94	[1]	Expt.	-	-	18-95 %	0.35
Zahnow	[2]	Expt. & DC+BW	0.40 ± 0.03	0.40 ± 0.03	-	-
TUNL96	[5]	Expt. & DC+BW	0.24	0.25	6-10 %	0.35
Barker	[3]	R-Matrix fit	-	-	9.1%	-
Csótó	[4]	RGM	-	-	3.5 %	-
Our work			0.27	0.30	3 %	-

about 100 times smaller, as would normally be expected.

We conclude that the single-particle model describes the non-resonant scattering channels, but is not completely adequate to describe the relevant 1^+ resonances in ^{8}Be. Improved structure information is needed from new calculations to give the correct resonant scattering wave functions in ^{8}Be.

REFERENCES

1. R. Chasteler, H. Weller et al. *Phys. Rev. Lett.* **72**, 3949 (1994).
2. D. Zahnow et al. *Z. Phys.* **A351**, 229 (1995).
3. F. Barker *Aust. J. Phys.* **48**, 813 (1995).
4. A. Csótó et al. *Nucl. Phys.* **A607**, 62 (1996).
5. M. Godwin it et al. *Phys. Rev.* **C53**, R1 (1996) and **C56**, N3 (1997).
6. I. Thompson *Comput. Phys. Reports* **7** (1988);
7. L. Brown et al. *Nucl. Phys.* **A206**, 353 (1973).
8. F. Cecil et al. *Nucl. Phys.* **A539**, 75(1992).

Determination of $S_{17}(0)$ from Transfer Reactions

R.E. Tribble[a], A. Azhari[a], V. Burjan[c], F. Carstoiu[b] J. Cejpek[c],
H.L. Clark[a], C.A. Gagliardi[a], V. Kroha[c], Y.-W. Lui[a],
A.M. Mukhamedzhanov[a], Š. Piskoř[c], A. Sattarov[a], L. Trache[a],
J. Vincour[c]

[a] Cyclotron Institute, Texas A&M University, College Station, Texas 77843, [b] Institute of Atomic
Physics, Bucharest, Romania, [c] Institute for Nuclear Physics, Czech Academy of Sciences,
Prague-Řež, Czech Republic

Abstract. The S factor for the direct capture reaction $^7Be(p, \gamma)^8B$ can be found at
astrophysical energies from the asymptotic normalization coefficients which provide the
normalization of the tails of the overlap functions for $^8B \to {}^7Be + p$. Peripheral transfer
reactions offer a technique to determine these asymptotic normalization coefficients.
Using this technique, the $^{10}B(^7Be,{}^8B)^9Be$ and $^{14}N(^7Be,{}^8B)^{13}C$ reactions have been
used to measure the asymptotic normalization coefficient for $^7Be(p, \gamma)^8B$. These results
provide an indirect determination of $S_{17}(0)$.

Nuclear capture reactions such as (p, γ) and (α, γ), play a major role in defining
our universe. Until recently, the only reliable method to determine a reaction rate
that is dominated by direct capture has been to measure it at laboratory energies
with a low energy particle beam and extrapolate the result to astrophysical energies.
Often the reaction of interest involves a radioactive target which makes measurements quite difficult. Hence it is important to develop alternative techniques to
determine rates. Direct capture reactions of astrophysical interest usually involve
systems where the binding energy of the captured proton is low. Hence at stellar
energies, the capture proceeds through the tail of the nuclear overlap wave function. The shape of the overlap integral in this tail region is completely determined
by the Coulomb interaction, so its amplitude alone dictates the rate of the capture
reaction. The $^7Be(p, \gamma)^8B$ reaction is an excellent example of such a direct capture
process where new measurements are still needed, as was underscored in a recent
review of stellar reaction rates [1] which includes a detailed discussion of the uncertainties in our present knowledge of $S_{17}(0)$ and its importance to the solar neutrino
problem. Below we describe a new indirect technique for these measurements.

The asymptotic normalization coefficient (ANC) C for $A + p \leftrightarrow B$ specifies the
amplitude of the overlap function for nucleus B when the core A and the proton

CP495, *Experimental Nuclear Physics in Europe*, edited by B. Rubio et al.
© 1999 American Institute of Physics 1-56396-907-6/99/$15.00

are separated by a distance large compared to the nuclear radius. In previous communications [2,3], we have pointed out that astrophysical S factors for peripheral direct radiative capture reactions can be determined through measurements of ANC's using traditional nuclear reactions such as peripheral nucleon transfer. Of course it is extremely important to test the reliability of the technique in order to know the precision with which it can be applied. Determining the S factors for $^{16}O(p,\gamma)^{17}F$ from its ANC's has been recognized as a suitable test for this method [1] because the results can be compared to existing direct measurements of the cross sections [4,5]. Furthermore, the $^{16}O(p,\gamma)^{17}F$ reaction has substantial similarities to the $^{7}Be(p,\gamma)^{8}B$ reaction.

For a peripheral transfer reaction, ANC's are extracted from the angular distributions by comparison to a DWBA calculation. Consider the proton transfer reaction $a + A \rightarrow c + B$, where $a = c + p$ and $B = A + p$. The experimental cross section is related to the DWBA according to

$$\frac{d\sigma}{d\Omega} = \sum (C_{Ap}^{B})^{2}(C_{cp}^{a})^{2}\frac{\tilde{\sigma}^{DW}}{b_{Ap}^{2}b_{cp}^{2}}. \tag{1}$$

$\tilde{\sigma}$ is the calculated DWBA cross section and the b's are the asymptotic normalization constants for the single particle orbitals used in the DWBA. The sum in Eq. (1) is taken over the allowed angular momentum couplings, and the C's are the ANC's for $B \rightarrow A + p$ and $a \rightarrow c + p$. The normalization of the DWBA cross section by the ANC's for the single particle orbitals makes the extraction of the ANC for $B \rightarrow A + p$ essentially independent of the parameters used in the single particle potential wells. See [6] for additional details.

The relation of the ANC's to the direct capture rate at low energies is straightforward. The cross section for the direct capture reaction $A + p \rightarrow B + \gamma$ can be written as

$$\sigma \approx \lambda |\langle C\frac{W_{-\eta,l+1/2}(2\kappa r)}{r}\mid \hat{O}(\mathbf{r})\mid \psi_{i}^{(+)}(\mathbf{r})\rangle |^{2}, \tag{2}$$

where λ contains kinematical factors, $\hat{O}$ is the electromagnetic transition operator, $\psi_{i}^{(+)}$ is the incident scattering wave, W is the Whittaker function, η is the Coulomb parameter for the bound state $B = A + p$, κ is the bound state wave number and C is the ANC. This equation is valid if the dominant contribution to the matrix element comes from outside the nuclear radius. In this case, the ANC's defined in Eq. (1) determine the capture cross section and thus the S factor.

ANC's for $^{17}F \rightarrow {}^{16}O + p$, corresponding to ground and first excited state transitions, were obtained from the $^{16}O(^{3}He,d)^{17}F$ reaction which we measured at 29.75 MeV. Using the ANC's, the S factors describing $^{16}O(p,\gamma)^{17}F$ to both states were calculated, with no additional normalization constants. The agreement between the measured S-factors and those calculated from our $^{17}F \rightarrow {}^{16}O + p$ ANC's is quite good indicating that the technique works to better than 9%. Details of the experiment and S-factor determination can be found in [7].

We have measured the (^{7}Be,^{8}B) reaction on ^{10}B and ^{14}N targets in order to extract the ANC for ^{8}B $\rightarrow$ ^{7}Be + p. The radioactive ^{7}Be beam was produced at 12 MeV/A by filtering reaction products from the ^{1}H(^{7}Li,^{7}Be)n reaction in the recoil spectrometer MARS. The reaction was initiated by a primary ^{7}Li beam at 18.6 MeV/A from the TAMU K500 cyclotron incident on an H_2 cryogenic gas target cooled by LN_2. Reaction products were measured by 5 cm × 5 cm Si detector telescopes consisting of a 100 μm ΔE strip detector, with 16 position sensitive strips, followed by a 1000 μm E counter. A single 1000 μm Si strip detector was used for initial beam tuning. This detector, which was inserted at the target location, allowed us to optimize the beam shape and to normalize the ^{7}Be flux relative to a Faraday cup that measured the intensity of the primary ^{7}Li beam.

Figure 1 shows the results for the elastic scattering angular distributions from the two targets. A Monte Carlo simulation was used to generate the solid angle efficiency for each angular bin. The absolute cross section is fixed by the target thickness, number of incident ^{7}Be, the yield in each bin, and the solid angle. The curves are from optical model calculations. The solid curve includes smoothing appropriate to the angular resolution in the experiment. No separate normalization factor was used for either target to match the data. The angular distributions for the transfer reactions are shown in Fig. 2 along with DWBA predictions from the full finite range code PTOLEMY [8]. Only the dominant components in the transfer reactions are shown, but all are included in the fits. The optical model parameters used in the DWBA calculations were obtained from folding model predictions using the JLM interaction. Details can be found in [9–11].

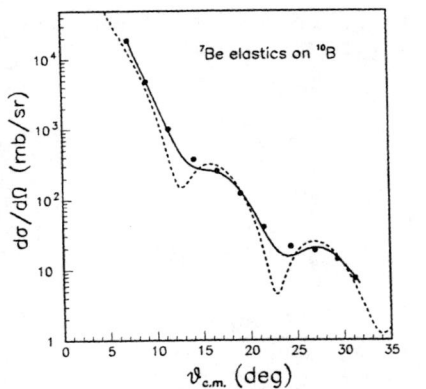
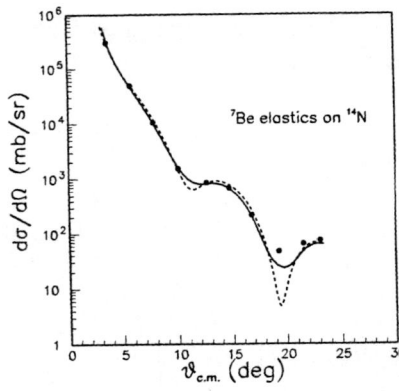

FIGURE 1. Elastic scattering angular distributions and optical model calculations. No scale factor has been applied to the calculations to match the data.

The ANC for ^{8}B $\rightarrow$ ^{7}Be + p was extracted based on the fits to the angular distributions and the ANC's for ^{10}B $\rightarrow$ ^{9}Be + p [6] and ^{14}N $\rightarrow$ ^{13}C + p [12]. These ANC's represent one of the major uncertainties (9% for ^{10}B and 6.4% for ^{14}N) in the extraction of the ANC for ^{8}B $\rightarrow$ ^{7}Be + p. The other major uncertainty in the extracted ANC is due to the optical model parameters (10%) for both targets. The S factor for ^{7}Be $(p, \gamma)^8$B has been determined from the ANC's. The values that we find for $S_{17}(0)$ are 17.8 ± 2.8 eV·b and 16.6 ± 2.2 eV·b for the ^{10}B and ^{14}N targets, respectively. Both values are in good agreement with the recommended value of 19^{+4}_{-2} eV·b [1]. Combining the two values requires a careful consideration of the correlation in the results due to the optical model parameters. This will be the subject of a future publication.

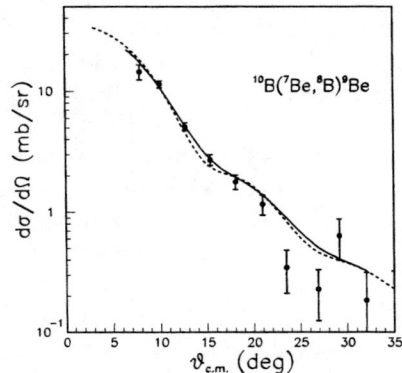 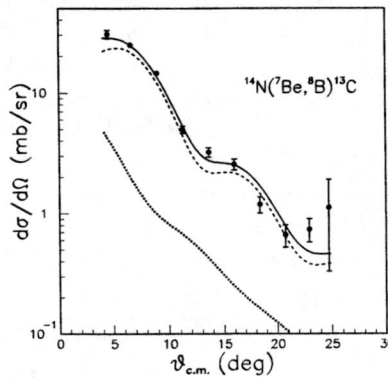

FIGURE 2. Angular distributions for the transfer reactions. Only the dominant components in the reaction on the ^{14}N target have been shown. DWBA calculations have been smoothed to match the angular resolution in the data.

This work was supported in part by the U.S. Department of Energy under Grant number DE-FG03-93ER40773 and by the Robert A. Welch Foundation.

REFERENCES

1. E.G. Adelberger *et al.*, Rev. Mod. Phys.**70(4)**, 1265 (1998).
2. H.M. Xu *et al.*, Phys. Rev. Lett.**73**, 2027 (1994).
3. A.M. Mukhamedzhanov and N.K. Timofeyuk, JETP Lett. **51**, 282 (1990).
4. R. Morlock *et al.*, Phys. Rev. Lett. **79**, 3837 (1997).
5. H.C. Chow, G.M. Griffith and T.H. Hall, Can. J. Phys. **53**, 1672 (1975).
6. A.M. Mukhamedzhanov *et al.*, Phys. Rev. C **56**, 1302 (1997).
7. C.A. Gagliardi *et al.*, Phys. Rev. Lett. **C59**, 1149 (1999).
8. M. Rhoades-Brown, M. McFarlane and S. Pieper, Phys. Rev. C **21**, 2417 (1980); Phys. Rev. C **21**, 2436 (1980).
9. A. Azhari *et al.*, Phys. Rev. Lett. **82**, 3960 (1999).
10. A. Azhari *et al.*, Phys. Rev. C (submitted).
11. L. Trache *et al.*, Phys. Rev. C (submitted).
12. L. Trache *et al.*, Phys. Rev. C **58**, 2715 (1998).

NACRE: A European Compilation of Reaction Rates for Astrophysics[1]

Carmen Angulo
for the NACRE Collaboration

Institut de Physique Nucléaire, Université catholique de Louvain
Chemin du cyclotron, 2, B-1348 Louvain-la-Neuve, Belgium
email: angulo@fynu.ucl.ac.be

Abstract. We report on the program and results of the NACRE network (Nuclear Astrophysics Compilation of REaction rates). We have compiled low-energy cross section data for 86 charged-particle induced reactions involving light ($1 \leq Z \leq 14$) nuclei. The corresponding Maxwellian-averaged thermonuclear reactions rates are calculated in the temperature range from 10^6 K to 10^{10} K. The web site "http://pntpm.ulb.ac.be/nacre.htm", including the cross section data base and the reaction rates, allows users to browse electronically all the information on the reactions studied in this compilation.

Nuclear astrophysics plays a crucial role in the understanding of nucleosynthesis in the Universe [1]. Nuclear reactions determine stellar evolution and represent the major energy source in stars. Current modelizations of stellar systems require a very large amount of data, which remains a challenge for nuclear physicists. Due to the impressive skill and growth in experimental and theoretical nuclear physics in the last years, nuclear astrophysics faces a new and difficult task: the organization and accessibility to the huge amount of nuclear data needed to stellar modelisations. This goal has been the principal motivation for the establishment in 1993 of a network of 10 nuclear physics and astrophysics laboratories from several European countries.

The main task of this collaboration has been the building-up of a detailed compilation of evaluated cross section data of charged-particle-induced reactions on stable targets up to Silicon and the calculation of the corresponding reaction rates for temperatures up to 10^{10} K [2]. This work is meant to supersede the widely used compilation of such reactions issued by W.A. Fowler and collaborators [3]. NACRE comprises a total of 86 reactions (see Table 1) involved in the "cold" pp-chain, CNO cycle, NeNa cycle and MgAl chain, the first two burning modes being

---

[1] This work has been supported by the European Commission under the Human Capital and Mobility network contract ERBCHRXCT930339.

TABLE 1. Reactions included in NACRE

^{1}H(p,νe^{+})^{2}H	^{7}Li(α,n)^{10}B	^{13}C(α,n)^{16}O	^{18}O(p,α)^{15}N	^{23}Na(p,α)^{20}Ne
^{2}H(p,γ)^{3}He	^{7}Be(p,γ)^{8}B	^{13}N(p,γ)^{14}O	^{18}O(α,γ)^{22}Ne	^{23}Na(α,n)^{26}Al^{g}
^{2}H(d,γ)^{4}He	^{7}Be(α,γ)^{11}C	^{14}N(p,γ)^{15}O	^{18}O(α,n)^{21}Ne	^{23}Na(α,n)^{26}Al^{m}
^{2}H(d,n)^{3}He	^{9}Be(p,γ)^{10}B	^{14}N(p,n)^{14}O	^{19}F(p,γ)^{20}Ne	^{23}Na(α,n)^{26}Al^{t}
^{2}H(d,p)^{3}H	^{9}Be(p,n)^{9}B	^{14}N(p,α)^{11}C	^{19}F(p,n)^{19}Ne	^{24}Mg(p,γ)^{25}Al
^{2}H(α,γ)^{6}Li	^{9}Be(p,d)^{8}Be	^{14}N(α,γ)^{18}F	^{19}F(p,α)^{16}O	^{24}Mg(p,α)^{21}Na
^{3}H(d,n)^{4}He	^{9}Be(p,α)^{6}Li	^{14}N(α,n)^{17}F	^{20}Ne(p,γ)^{21}Na	^{25}Mg(p,γ)^{26}Al^{g}
^{3}H(α,γ)^{7}Li	^{9}Be(α,n)^{12}C	^{15}N(p,γ)^{16}O	^{20}Ne(p,α)^{17}F	^{25}Mg(p,γ)^{26}Al^{m}
^{3}He(^{3}He,2p)^{4}He	^{10}B(p,γ)^{11}C	^{15}N(p,n)^{15}O	^{20}Ne(α,γ)^{24}Mg	^{25}Mg(p,γ)^{26}Al^{t}
^{3}He(α,γ)^{7}Be	^{10}B(p,α)^{7}Be	^{15}N(p,α)^{12}C	^{21}Ne(p,γ)^{22}Na	^{25}Mg(α,n)^{28}Si
^{4}He(αn,γ)^{9}Be	^{11}B(p,γ)^{12}C	^{15}N(α,γ)^{19}F	^{21}Ne(α,n)^{24}Mg	^{26}Mg(p,γ)^{27}Al
^{4}He($\alpha\alpha$,γ)^{12}C	^{11}B(p,n)^{11}C	^{16}O(p,γ)^{17}F	^{22}Ne(p,γ)^{23}Na	^{26}Mg(α,n)^{29}Si
^{6}Li(p,γ)^{7}Be	^{11}B(p,α)^{8}Be	^{16}O(α,γ)^{20}Ne	^{22}Ne(α,γ)^{26}Mg	^{26}Al^{g}(p,γ)^{27}Si
^{6}Li(p,α)^{3}He	^{12}C(p,γ)^{13}N	^{17}O(p,γ)^{18}F	^{22}Ne(α,n)^{25}Mg	^{26}Al^{m}(p,γ)^{27}Si
^{7}Li(p,γ)^{8}Be	^{12}C(α,γ)^{16}O	^{17}O(p,α)^{14}N	^{22}Na(p,γ)^{23}Mg	^{27}Al(p,γ)^{28}Si
^{7}Li(p,α)^{4}He	^{13}C(p,γ)^{14}N	^{17}O(α,n)^{20}Ne	^{23}Na(p,γ)^{24}Mg	^{27}Al(p,α)^{24}Mg
^{7}Li(α,γ)^{11}B	^{13}C(p,n)^{13}N	^{18}O(p,γ)^{19}F	^{23}Na(p,n)^{23}Mg	^{27}Al(α,n)^{30}P
				^{28}Si(p,γ)^{29}P

essential energy producers, all four being important nucleosynthesis agents. It also includes the main reactions involved in non-explosive helium burning.

The final manuscript [2] comprises an introductory text containing a general formalism for the data evaluation and the calculation of the rates. Each reaction is presented individually with some comments about the adopted experimental data. If necessary a table of resonances is given containing the energies and strengths of an ensemble of low energy resonances important for the rates, providing the reference source. Figures of the astrophysical S-factor versus energy are also given. The calculated ground state rates are presented in tabular form as a function of temperature. The temperature steps are the same as in ref. [3]. The recommended rates are complemented with lower and upper uncertainty bounds extracted from the experimental errors and theoretical model uncertainties. We have also evaluated the effect of the thermal excitation of the target nuclei in the framework of the Hauser-Feschbach model, in order to derive the thermalized rates of direct astrophysical relevance. Analytical approximations to the adopted rates, as well as to the inverse/direct rate ratios, are provided.

The present updated NACRE rates, needed for stellar models, are expected to be used by an extended community of astrophysicists.

REFERENCES

1. Clayton, D.D. *Principles of Stellar Evolution and Nucleosynthesis*, Chicago: Chicago Press, 1983.
2. Angulo, C. et al., *Nucl. Phys. A*, in press.
3. Caughlan, C.G., and Fowler, W.A., *At. Data and Nucl. Data Tables* **40**, 283 (1988).

A Study of the ^{15}O(α, γ) reaction via the α-decay of ^{19}Ne[*]

A.M. Laird[1], A.N. Ostrowski[1], T. Davinson[1], A. Di Pietro[1],
D. Groombridge[1], A.C. Shotter[1], S. Cherubini[2],
W. Galster[2], J.S. Graulich[2], J. Hinnefeld[2], P. Leleux[2],
L. Michel[2], A. Musumarra[2], A. Ninane[2], J. Vervier[2],
M. Aliotta[3], F. Cappuzzello[3], A. Cunsolo[3], P. Figuera[3],
C. Spitaleri[3], A. Tumino[3]

[1] Department of Physics and Astronomy,
University of Edinburgh, UK
[2] Institut de Physique Nucléaire,
Université Catholique de Louvain, Belgium
[3] Dipartimenti di Fisica,
Universitá di Catania, Italy

Abstract. Post-accelerated radioactive ion beams produced at Louvain-la-Neuve, Belgium have been used to investigate the hot-CNO cycle breakout reaction ^{15}O$(\alpha, \gamma)^{19}$Ne. This reaction not only leads to the synthesis of proton-rich nuclei via the rp-process but is also thought to initiate the massive release of energy which can trigger stellar explosions such as novae, supernovae and x-ray bursters. The properties of this reaction have been investigated via a study of the relevant decays of excited states in ^{19}Ne. These resonances have been populated using an inverse ^{18}Ne(d,p) reaction on a deuterated polyethylene target. Experiments have been carried out using an especially designed high granularity large solid angle silicon strip detector array consisting of 320 elements. The method, the experimental set-up, as well as results on the population and decay of states of astrophysical interest will be presented.

MOTIVATION

Nuclear astrophysics is one of the most studied topics in current nuclear physics. While it has been very successful in explaining galactic isotopic abundances, much has still to be measured and understood. One of the most important advancements experimentally has been the introduction of radioactive ion beams. Many important reactions and processes in nuclear astrophysics involve radioactive nuclei. Therefore the possibility of inducing reactions with radioactive projectiles has

CP495, *Experimental Nuclear Physics in Europe*, edited by B. Rubio et al.
© 1999 American Institute of Physics 1-56396-907-6/99/$15.00

opened up new ways of measuring key quantities of astrophysical interest. One particular area of research to benefit from radioactive ion beams is that of explosive hydrogen burning. Explosive hydrogen burning is a process thought to be the main energy source in cataclysmic scenarios such as, e.g. novae, X-ray bursters and supernovae, where the temperatures and time scales are such that the reaction pathways are often far from stability. For temperatures up to 0.2×10^9 K, the hot-CNO cycle [1].

$$^{12}C(p,\gamma)^{13}N(p,\gamma)^{14}O(e^+\nu)^{14}N(p,\gamma)^{15}O(e^+\nu)^{14}N(p,\alpha)^{12}C \tag{1}$$

is expected to dominate energy production. Due to the long half-lives for the β-decay of the two oxygen isotopes, most of the nuclear material accumulates as ^{14}O and ^{15}O. At higher temperatures the ^{14}O bottleneck is bypassed by the chain

$$^{14}O(\alpha,p)^{17}F(p,\gamma)^{18}Ne(e^+\nu)^{18}F(p,\alpha)^{15}O \tag{2}$$

further enhancing the amount of ^{15}O. However, if high enough temperatures and densities are reached, the β-decay of ^{15}O will be superseded by radiative α-capture to ^{19}Ne [2]. Once this occurs, material is processed by a sequence of proton capture reactions (rp-process) and possibly alpha capture reactions (αp-process) to higher masses. In addition to proton-rich nucleosynthesis, the rp-process is responsible for a massive increase in energy production. Estimates suggest that the energy generation rate is increased by a factor of 100 over the hot-CNO cycle. However, network calculations predicting the nucleosynthetic pathway and energy production rely on accurate information on the breakout reaction, $^{15}O(\alpha,\gamma)$.

Until recently, data on this reaction was based on nuclear systematics and a knowledge of the mirror nucleus [3]. However, a paper by de Oliviera [4] claims that these estimates are only accurate to within one order of magnitude. A direct measurement using a ^{15}O beam has recently been proposed at the new radioactive ion facility, ISAC, at TRIUMF, Canada [5], though it may be a few years before the results are known. The reaction rate for a narrow resonance can be calculated from

$$<\sigma v> = (\frac{2\pi}{\mu kT})^{3/2}\hbar^2 \omega\gamma exp\frac{-E_R}{kT} \tag{3}$$

where $\omega = \frac{2J+1}{(2J_a+1)(2J_b+1)}$ is the statistical spin factor and $\gamma = \frac{\Gamma_\alpha\Gamma_\gamma}{\Gamma} = \Gamma_\alpha(1-B_\alpha)$. Consequently, given information on the total width, the α-branching ratio is sufficient to calculate the reaction rates for known resonances. An indirect technique has already been utilised with some success by Magnus et al. [6] using the reaction $^{19}F(^3He,t)^{19}Ne^*$ and measuring the α-decay of the excited states in ^{19}Ne. Despite the success of this method for higher lying states, for the states of astrophysical interest no new information was obtained. Consequently, for this technique to work another populating mechanism is needed.

EXPERIMENTAL TECHNIQUE

These experiments were undertaken at the radioactive ion beam facility at Louvain-la-Neuve, Belgium. Excited states in ^{19}Ne were populated via a ^{18}Ne induced neutron pickup reaction on a deuterated polyethylene target. The recoiling proton tagged the populated state, and the subsequent decay into the α-^{15}O channel measured. The incident beam was 54.3 MeV ^{18}Ne in the 3+ charge state with a typical intensity of 5×10^5 pps. The experiment ran for 30 hours on the production target. The detector set-up consisted of three LEDA (*Louvain Edinburgh Detector Array*) type detectors shown schematically in Figure 2. The LEDA is a silicon strip detector made up of 8 sectors, each having 16 annular strips [7]. In total there were 320 detector elements each giving energy and timing signals. One LEDA was positioned upstream from the target to measure the tagging protons and two LEDAs were placed downstream to detect the α and ^{15}O from the decay.

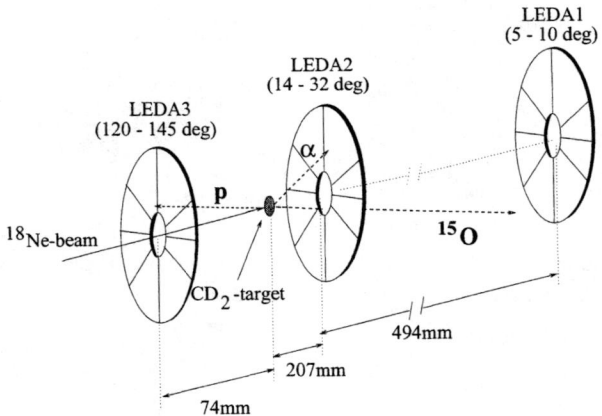

FIGURE 1. Schematic of detector set-up.

RESULTS AND CONCLUSIONS

Figure 2 shows the energy of the recoiling proton at 127 degrees. The solid line corresponds to single events and the dashed line corresponds to triple events with the correct total energy. A Monte Carlo simulation was used to determine the detection efficiency. The α-branching ratios could then be calculated from the yields for each state, taking these efficiencies into account. Preliminary analysis of a subset of the data for the 4.6 MeV state gives a α-branching ratio of 0.28±0.13, compared with 0.25±0.04 given by Magnus [6]. A full analysis of all the available statistics is expected to produce similar precision to the Magnus results. Consequently, despite the much lower beam intensity, the higher cross-section and larger solid angle allow this radioactive beam experiment to compare favourably with the stable beam experiment. Moreover, given one to two orders of magnitude more in beam intensity and longer running time, this method should make the measurement of the α-branching ratio for the 4.033 MeV state possible.

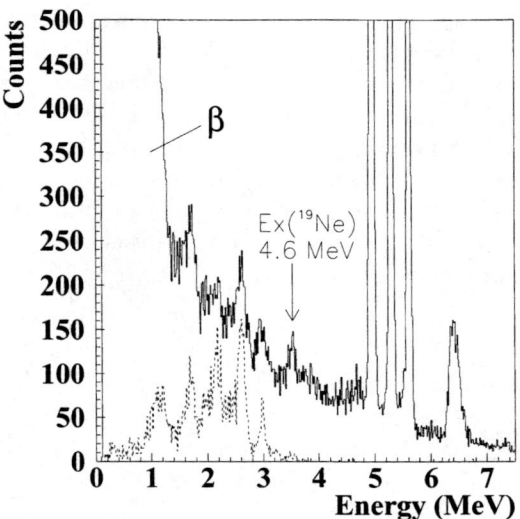

FIGURE 2. Proton energy spectrum at 127 deg. Solid line shows single events and dashed line shows triple events with a total energy cut which discriminates for the β-events seen in the singles data.

ACKNOWLEDGEMENTS

We would like to thank the technical staff in Edinburgh and Louvain-la-Neuve for their valuable contribution. This work has been made possible by funding from the UK EPSRC (Engineering and Physical Sciences Research Council) and the European Large Scale Facilities Program.

REFERENCES

1. Champagne A.E., and Wiescher M., *Annu. Rev. Nucl. Part. Sci.* **42**, 39 (1992).
2. Wallace R.K., and Woosley S.E., *Astrophys. J. Suppl.* **45**, 389 (1981).
3. Langanke K., et al., *Astrophys. J.* **301**, 629 (1986).
4. de Oliviera F., et al., *Phy. Rev.* **C55**, 3149 (1997).
5. d'Auria J., et al., *Astrophys. J.* **301**, 629 (1986).
6. Magnus P.V., et al., *Nucl. Phys.* **A506**, 332 (1990).
7. Davinson, T., et al. *Nucl. Instrum. Methods* **A288**, 245 (1990).

Measurement of Low-lying States in ^{40}Sc.

V. Y. Hansper[1], A. E. Champagne[2], C. Iliadis[2], S. E. Hale[2] and D. C. Powell[2]

[1] Institut for Fysik og Astronomi, Aarhus Universitet, Ny Munkegade, DK-8000 Aarhus C, Danmark
[2] University of North Carolina at Chapel Hill and Triangle Universities Nuclear Laboratory, c/o P.O. Box 90308, Durham, North Carolina, 27008-0308, U.S.A.

Abstract. In explosive hydrogen burning nucleosynthesis material is processed via the proton capture sequence ^{39}Ca(p, γ)^{40}Sc(p, γ)^{41}Ti. It has been predicted that the isotope ^{39}Ca represents a waiting point for a continuous reaction flow. Therefore, its reaction rate is of interest. The ^{39}Ca(p, γ)^{40}Sc reaction rate is determined by three resonances corresponding to the 2nd, 3rd and 4th excited states of ^{40}Sc. Improved nuclear structure information of the low-lying levels of ^{40}Sc is necessary to reduce uncertainties in the reaction rate of ^{39}Ca(p, γ)^{40}Sc. Results from a current measurement of ^{40}Ca(^{3}He, t)^{40}Sc at TUNL indicate that the 4th excited state is a doublet and further investigation is warranted.

INTRODUCTION

Explosive hydrogen burning is an important nucleosynthesis process as well as the power source for events such as novae and x-ray bursts. Certain isotopes have been predicted to be waiting points for the flow of nucleosynthesis via the Hot CNO cycle and rp-process. [7], [8] Since flow impedances affect the time evolution of the outburst and could lead to observable spectral features, it has been suggested that the measurement of (p, γ) cross sections of capture reactions at waiting point nuclei could be important [5]; many of these require radioactive ion beams for direct measurement. The work described herein describes initial experimental results of such measurements. The focus here is on the ^{39}Ca(p, γ)^{40}Sc reaction. At temperatures above 3 x 10^8 K, the flow of nucleosynthesis through A = 40 proceeds via the proton capture sequence

$$^{39}\text{Ca(p, } \gamma)^{40}\text{Sc(p, } \gamma)^{41}\text{Ti.}$$

It has been suggested that the isotope ^{39}Ca represents a waiting point for a continuous rp-process reaction flow. [8] Therefore the competition between (p, γ) and β^+ decay will determine the flux to heavier nuclei. As the Q-value for this reaction

CP495, *Experimental Nuclear Physics in Europe*, edited by B. Rubio et al.
© 1999 American Institute of Physics 1-56396-907-6/99/$15.00

is low, Q = 0.5391 MeV, its reaction rate is determined by three resonances corresponding to the 2nd, 3rd and 4th excited states of ^{40}Sc. [8] A previous measurement suggests that the 4th excited state could be a triplet state. [6] An experimental investigation of the nuclear structure of the low-lying levels of ^{40}Sc could resolve the issue of the multiplicity of the 4th excited state and thereby help reduce uncertainties in the reaction rate of ^{39}Ca(p, γ)^{40}Sc. Consequently we have measured the ^{40}Ca(^{3}He, t)^{40}Sc reaction to address these concerns.

EXPERIMENTAL DETAILS.

The ^{40}Ca(^{3}He, t)^{40}Sc reaction was measured using 26 MeV ^{3}He beams from the TUNL FN Van de Graaff accelerator. A target composed of 47μg/cm^2 of natural Ca on a 22μg/cm^2 C backing was used for the measurement. Reaction particles were detected at three angles (5°, 10°and 15°) at the focal plane of the Enge split-pole spectrometer using a multiwire avalanche counter [2]. The field settings for the spectrometer were selected so that the states of interest and those from calibration reactions could be measured at the same magnetic field.

The beam energy and spectrometer angle were measured by an intercomparison of states populated in the ^{27}Al(^{3}He, t) and ^{27}Al(^{3}He, d) reactions at 5°, 10°and 15°. The beam energy was determined to be 26.066 ± 0.004 MeV. There was no discernable error in angle.

The energy dispersion of the spectra was measured by means of (^{3}He, p) reactions on targets of ^{12}C, ^{27}Al and ^{28}Si leading to well-known states in ^{14}N, ^{29}Si and ^{30}P. (Figure 1.) As both protons and tritons are bent by the same field and through the same ρ, the triton energy can be given in terms of the proton energy by

$$E_t = (m_t^2 + (2m_p E_p + E_p^2))^{1/2} - m_t \tag{1}$$

where the masses are the rest mass energies. The error in the calibration was found to be ±1 keV. Since this ignores differential energy loss in the target, excitation energies in ^{40}Sc are quoted relative to the ground state.

RESULTS.

The majority of the data was taken at 10°, and the interesting features were also confirmed from data at 15°. Peaks from tritons which left ^{40}Sc in the ground state and the first three excited states were very distinct. Some of the higher states were, however, contaminated by ^{16}O in the target at 10°. Surprisingly, the 1st excited state is slightly higher in energy than quoted in the literature. [1] There was evidence of a doublet near E_x = 1.67 MeV, and was found to be about 32 keV higher than the 4th excited state. Two new peaks have also been identified above the 5th excited state. One is quite prominent and corresponds to an excitation energy of 1.925 MeV. The other has less statistics and sits on the low energy tail

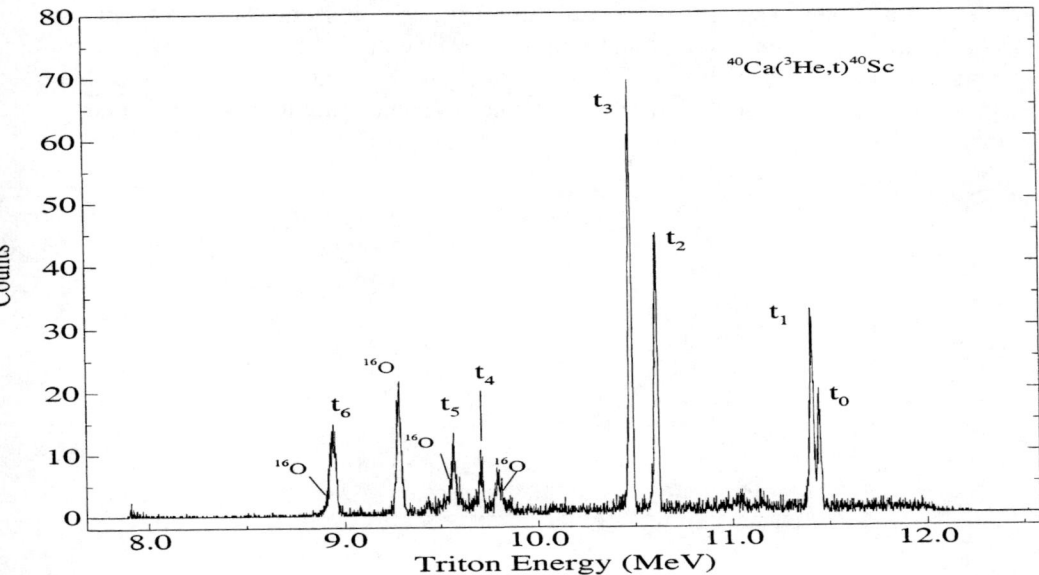

FIGURE 1. ^{40}Ca(^{3}He, t)^{40}Sc spectrum at 10°calibrated as described in the text. The subscripts for the triton peaks indicate the excited state of the ^{40}Sc nucleus. The oxygen contaminant has also been identified.

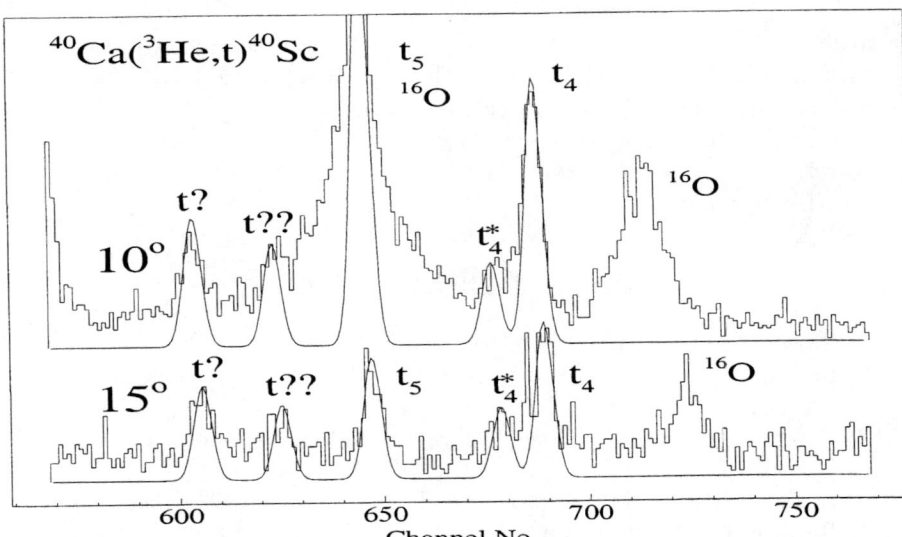

FIGURE 2. Magnified view of the region for tritons going to the 4th and 5th excited states of ^{40}Sc. The solid lines represent fits to the data, and indicate where the doublet (labelled t_4^*) appears to be. The other two new states labelled t? (1.925 MeV) and t?? (1.871 MeV) are indicated.

of the contaminant peak. It was readily fitted with a similar line shape to the data at both 10°and 15°, and correponds to a level in ^{40}Sc at 1.871 keV. Figure 2 gives a magnified view of the region which covers these newly identified peaks and Table 1 gives a list of results obtained from this measurement compared with previously tabulated data.

Table 1.

Energy Level [1]	Energy Level (This work)
33.6 ± 1.5	35.1 ± 1.6
772. ± 2	772.2 ± 1.6
892. ± 2	895. ± 2
1667. ± 4	1671.6 ± 1.9
***	1703.2 ± 2.2
1799. ± 4	1796.1 ± 2.4
***	1871 ± 6
***	1925 ± 11
2366. ± 4	2376. ± 5

* all energies in keV.

CONCLUSIONS

We have measured the level structure of ^{40}Sc by means of the ^{40}Ca(^{3}He, t)^{40}Sc reaction. Evidence of a multiplet state at $E_x = 1.67$ MeV and new levels at $E_x = 1.925$ and 1.871 MeV are indicated, though further experimental investigation is required to confirm their origin in ^{40}Sc. Some shifts in energy levels have been observed, in particular for the 1st excited state. The current results appear consistent with the level structure used in a recent calculation of the reaction rate by Iliadis et al. [4], and thus would support their conclusion that ^{39}Ca is not a waiting point for the rp-process. Confirmation of this requires reaction rate calculations for ^{39}Ca(p,γ)^{40}Sc with the new data.

REFERENCES

1. Endt P. M., 1998 Nuc. Phys. A633, 1
2. Hansper V. Y., Hale S. E. & Champagne A. E., 1997, AIP Conference Proceedings, 392, 1031
3. Herndl H., Görres J., Wiescher M., Brown B. A. & Van Wormer L., 1995, Phys. Rev. C 52, 1078
4. Iliadis C., Endt P. M., Prantos N. & Thompson W. J., 1998, ApJ in print
5. Rembges F., Freiburghaus C., Rauscher T., Thielemann F-K., Schatz & H., Wiescher M., 1997, ApJ 484, 412
6. Shulz N., Alford W. P. & Jamshili A., 1971, Nuc. Phys. A162, 349
7. Van Wormer L., Görres J., Iliadis C. & Wiescher M., 1994, ApJ 432, 326
8. Wiescher M. & Gorres J., 1989, ApJ 346, 1041

Halflives of rp-Process Waiting Point Nuclei

E. Wefers [a], T. Faestermann [a], M. Münch [a], R. Schneider [a], A. Stolz [a],

K. Sümmerer [b], J. Friese [a], H. Geissel [b], M. Hellström [c], P. Kienle [a], H.-J. Körner [a],

G. Münzenberg [b], C. Schlegel [b], P. Thirolf [d], H. Weick [b], and K. Zeitelhack [a]

[a] Technische Universität München, 85748 Garching, Germany, [b] Gesellschaft für
Schwerionenforschung mbH, 64291 Darmstadt, Germany, [c] Lund University, 22100 Lund,
Sweden and [d] Ludwig-Maximilians-Universität, 85748 Garching, Germany

Abstract. The fragment separator at GSI, Darmstadt, has been used to produce
and separate very proton rich nuclei in the ^{100}Sn region. By fragmentation of a ^{112}Sn
beam at 1 A·GeV we produced nuclei along the rp-process path between ^{77}Y and ^{98}In.
By implanting these ions into a silicon detector stack we were able to determine their
halflives. Preliminary data are presented.

Neutron deficient nuclei near ^{100}Sn have been produced by fragmentation of a
1 A·GeV ^{112}Sn beam in a beryllium target. The fragments were separated in the
0° magnetic spectrometer FRS at GSI, Darmstadt, and identified with detector
systems, tracking the ion position through the FRS and measuring the energy loss
and the time of flight. The unambiguously identified ions (see fig. 1) were stopped
in an implantation detector. Among others we identified two ^{76}Y and one ^{78}Zr ions.

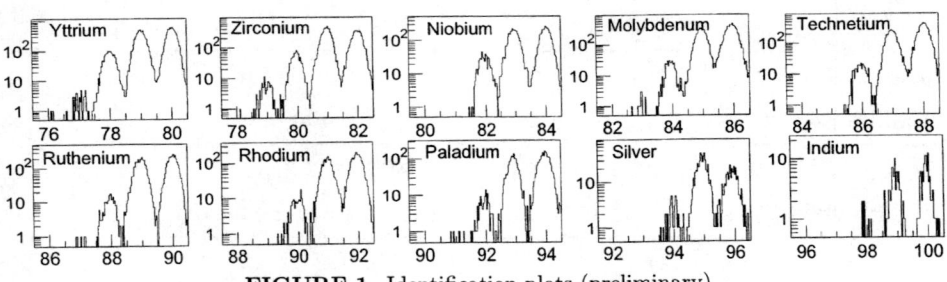

FIGURE 1. Identification plots (preliminary).

The implantation zone of our detector consists of four two-sided Si-strip detectors
($64 \times 25 \times 0.5 \, \text{mm}^3$) with a strip pitch of 0.5 mm on both sides (x and y direction),
mounted in a compact stack. By correlating the implantation position of an identified ion with its decay position we were able to suppress the background effectively
and to determine the halflives of the implanted ions. Around the implanation zone
two 10 mm thick stacks of Si detectors were mounted for the β-energy measurement.
For gamma spectroscopy about 25% of the solid angle was covered by a segmented
Clover germanium detector.

CP495, *Experimental Nuclear Physics in Europe*, edited by B. Rubio et al.
© 1999 American Institute of Physics 1-56396-907-6/99/$15.00

One purpose of this experiment was to measure the Gamov-Teller-Strength and the γ spectrum in the decay of ^{100}Sn, ^{102}Sn and ^{98}Cd. During a 50 hrs run only one event could be clearly identified as ^{100}Sn, for ^{102}Sn and ^{98}Cd we observed some 1000 decays respectively. These data are still being analyzed.

The nuclei with even Z near the proton dripline below ^{100}Sn are waiting points in the rp-process and their halflives determine the flux towards heavier nuclei and the respective isotopic abundances. We measured for the first time the halflives of these nuclei from ^{80}Zr up to ^{92,93}Pd.

On the other hand the rp-process flux is determined by the proton dripline. Therefore we investigated the possibly proton-unstable nuclei ^{77}Y and ^{81}Nb. ^{77}Y ions could be identified and we observed their decay with a short halflife, consistent with a superallowed Fermi beta decay. From the non-observation of ^{81}Nb we deduce a halflife shorter than 200 ns, considering the flight time through the fragment separator.

We also investigated the six cases of odd-odd $N = Z$ nuclei between ^{78}Y and ^{98}In, which are the heaviest nuclei where one can hope to study pure Fermi beta transitions. In a very preliminary analysis of our data we are able to show that even ^{90}Rh and ^{94}Ag are Fermi beta emitters. For the three lighter members of this series ^{78}Y to ^{86}Tc a low lying T=1, 0^+ state decaying with a superallowed Fermi transition was recently detected at GANIL [1]. Our new data on these nuclei are in good agreement.

Finally we measured the halflives of so far unknown Rh and Tc isotopes near the $N = Z$-line. First (preliminary) results and their statistical errors are shown in the table below. Up to now only a part of the data is analyzed.

Isotope	^{77}Y	^{78}Y	^{80}Zr	^{82}Nb	^{84}Mo	^{85}Mo
Halflife	53^{+29}_{-14} ms	67^{+18}_{-12} ms	$5.7^{+0.9}_{-0.7}$ s	44^{+8}_{-6} ms	$3.6^{+1.0}_{-0.7}$ s	$6.3^{+1.3}_{-1.0}$ s
Isotope	^{86}Tc	^{87}Tc	^{88}Ru	^{89}Ru	^{90}Rh	^{91}Rh
Halflife	45^{+10}_{-7} ms	$1.9^{+0.2}_{-0.2}$ s	$1.1^{+0.4}_{-0.2}$ s	$1.6^{+0.3}_{-0.2}$ s	44^{+60}_{-16} ms	$1.8^{+0.3}_{-0.3}$ s
Isotope	^{92}Rh	^{93}Rh	^{92}Pd	^{93}Pd	^{94}Ag	^{98}In
Halflife	$2.9^{+1.5}_{-0.8}$ s	$5.7^{+1.3}_{-0.9}$ s	$0.7^{+0.4}_{-0.2}$ s	$9.3^{+2.5}_{-1.7}$ s	38^{+38}_{-13} ms	to be analyzed

Acknowledgments

We would like to thank J. Gerl and R. Simon for providing us the γ-detectors, and the FRS group for their help during our beamtime.

This work was supported by BMBF (06TM872 TPI) and SFB 375.

REFERENCES

1. C. Longour et al., Phys. Rev. Lett. **81** (1998) 3337

Cross section measurements of (p,γ)-reactions relevant to p-process. [1]

S. Harissopulos[1], P. Tsagari[1], E. Skreti[1], G. Souliotis[1],
P. Demetriou[1], T. Paradellis[1], J.W. Hammer[2], R. Kunz[2],
C. Angulo[3], S. Goriely[4], T. Rauscher[5].

[1] *Inst. of Nuclear Physics, NCSR "Demokritos", 153.10 Aghia Paraskevi, Athens, Greece.*
[2] *IfS, Universität Stuttgart, Allmandring 3, D-70569 Stuttgart, Germany.*
[3] *ULB, Physique Nucleaire Theorique, CP-229, B-1050 Brussels, Belgium.*
[4] *ULB, Inst. d' Astronomie et d' Astrophysique, CP-226, B-1050 Brussels, Belgium.*
[5] *Dept. of Physics and Astronomy, Univ. of Basel, CH-4056 Basel, Switzerland.*

Abstract. The cross sections of the $^{93}\text{Nb}(p,\gamma)^{94}\text{Mo}$ and $^{89}\text{Y}(p,\gamma)^{90}\text{Zr}$ reactions have been measured over the proton energy range 1.4–4.8 MeV and 1.6–2.4 MeV respectively. The results are compared with the predictions of statistical model codes.

INTRODUCTION

The main nuclear physics uncertainties that affect the modeling of the so-called *p-process* concern the involved reaction rates of the proton or the α-particle captures by more or less neutron-deficient nuclides, as well as the relevant photodis-integrations. Except for very few cases [1–6], no experimental data are available in the mass region A≥70. Hence, all extended network calculations performed to simulate p-process nucleosynthesis had to rely almost completely on theoretical reaction rates based on the Hauser-Feshbach model (HFM) predictions. Until now and despite the uncertain nuclear physics input all p-process models are capable of reproducing p-nuclei abundances within a factor of 3 (see e.g. in ref. [7]). However, all these models do have problems in describing the light p-nuclei with A≤100: The relatively large abundances of ^{92}Mo, ^{94}Mo, ^{96}Ru and ^{98}Ru are severely underpredicted. On the other hand, the abundances of the lighter p-nuclei ^{74}Se, ^{78}Kr, and ^{84}Sr are systematically overpredicted.

These difficulties may suggest that certain aspects of p-process modeling have to be subject to further investigation. In any case, a more rigorous treatment of the p-process definitely requires a considerable improvement of the nuclear physics data set. This fact has motivated the present work.

[1] Work supported by NATO Collab. Research Grants Programme, Grant CRG 961086

CP495, *Experimental Nuclear Physics in Europe*, edited by B. Rubio et al.
© 1999 American Institute of Physics 1-56396-907-6/99/$15.00

EXPERIMENTAL PROCEDURES

The measurements reported in the present work have been carried out at the 4 MV Dynamitron accelerator of the University of Stuttgart, as well as at the 5.5 MV VdG Tandem accelerator of NCSR "Demokritos", Athens. Experimental details concerning the ^{93}Nb(p,γ)^{94}Mo reaction are given in ref. [5]. The ^{89}Y(p,γ)^{90}Zr reaction has been studied by using a 103.4 μg/cm^2 thick target, which has been fabricated by evaporating metallic ^{89}Y on a Ta-backing. By using the setup used of Stuttgart, which is described also in [5], the angular distributions of all the γ-transitions of interest have been measured at each beam energy, which has been variied from 1.6 to 2.4 MeV with a step of 25 keV. Spectra have been additionally taken at E$_p$=3 MeV. The beam current was about 40 μA. The target was water-cooled during the whole experiment. In Athens, spectra at E$_p$=2.4, 2.3, and 2.2 MeV have been measured with very good statistics, in order to clarify certain aspects of the decay pattern of the produced ^{90}Zr nucleus. Hereby, one Ge detector of 80% relative efficiency was placed at 55° to the beam axis and at a distance of about 3 cm from the target. The beam current was 2-4 μA. All detectors used in Athens as well as in Stuttgart have been calibrated by means of the 992 keV resonance of the ^{27}Al(p,γ)^{28}Al reaction. A typical γ-spectrum is shown in fig. 1, where the direct feeding to the ground, first and second excited state is labeled as

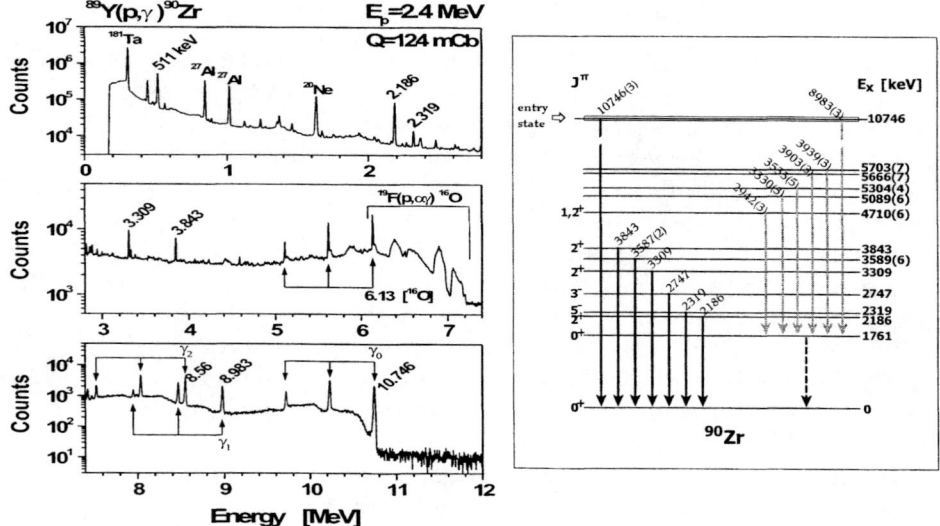

FIGURE 1. Left: γ-spectrum of the ^{89}Y(p,γ)^{90}Zr reaction. Right: Partial level scheme of ^{90}Zr.

γ_0, γ_1 and γ_2 respectively. The corresponding arrows indicate the respective first and second escape peaks. Furthermore, the strongest γ-rays depolulating excited states in the ^{90}Zr nucleus are labeled with the respective energy (in MeV). The

γ-rays arising from the $^{19}F(p,\alpha\gamma)^{16}O$ reaction or other contaminating γ-lines are indicated accordingly.

DATA ANALYSIS AND RESULTS

In order to obtain the cross section of a (p,γ) reaction, the absolute yield of all γ-transitions feeding into the ground state of the produced nucleus has to be determined. This task requires a very detailed spectra analysis in order to identify all γ-rays that populate the ground state. In the case of the $^{93}Nb(p,\gamma)^{94}Mo$ reaction this analysis supports the picture in which all capture events proceed mostly ($\gg 99\%$) through the first 2^+ state in ^{94}Mo since no γ_0-transition has been observed in the measured spectra [5]. Hence, the cross section of this reaction has been derived from the absolute yield of the $2^+_1 \rightarrow 0^+_1$ γ-ray. In the case of the $^{89}Y(p,\gamma)^{90}Zr$ reaction, the ground state of ^{90}Zr is fed by several γ-transitions. These are shown in fig. 1. with black arrows. It has to be emphasized that, the first excited state of ^{90}Zr is feeding into the ground state by means of an E0-transition shown in fig. 1 with a dashed arrow. Hence, the cross section of the $^{89}Y(p,\gamma)^{90}Zr$ reaction has been extracted from the absolute yield of all observed γ-transitions populating the ground state or the first excited level of ^{90}Zr. The latter γ-transitions are shown in fig. 1 with grey arrows.

The cross sections measured in the present work are plotted in fig. 2 as solid circles. In the case of the $^{89}Y(p,\gamma)^{90}Zr$ reaction, the predictions of the statistical

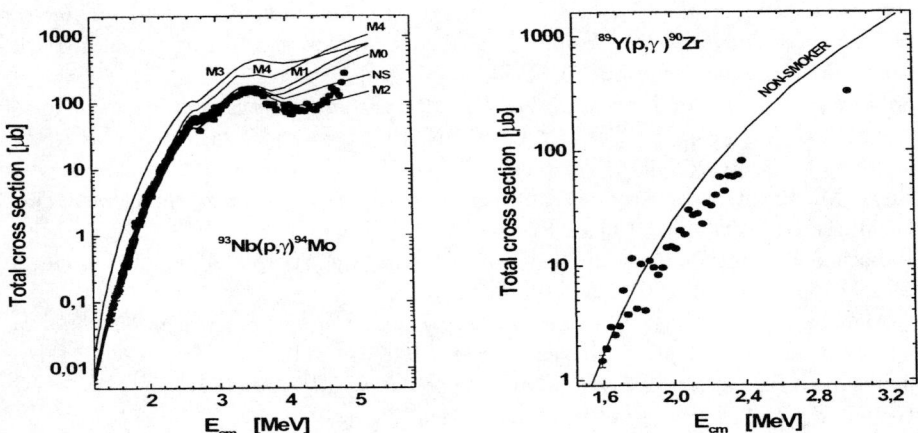

FIGURE 2. Total cross section (solid circles) measured for the $^{93}Nb(p,\gamma)^{94}Mo$ reaction (left), and $^{89}Y(p,\gamma)^{90}Zr$ reaction (right). The results of the statistical model codes NON-SMOKER and MOST are shown as solid curves. (see text).

model code NON-SMOKER [8] deviate significantly (20%–40%) from the data at energies E_p >2 MeV. In the case of the ^{93}Nb(p,γ)^{94}Mo reaction the predictions of this code indicated by the solid curve "NS" are in very good agreement with the data. The other solid curves (M0, M1, M2, M3, M4) are the results obtained from the statistical model code MOST [9] under different assumptions on the nuclear level density (NLD) and the nucleon-nucleus potential (nN-potential). Hereby, curves M0, M1 and M2 have been obtained assuming the *same* nN-potential – that of ref. [10]– and *different* NLDs, namely those of ref. [11], [12] and [13] respectively. All these curves reproduce the data quite well. This, however, does not apply for curves M3 and M4, which were both derived with the same NLD reported in ref. [11], but with different nN-potential, namely that of ref. [14] and [15] respectively. In conclusion, the NLD's used for the ^{93}Nb(p,γ)^{94}Mo reaction seem all to be rather appropriate for statistical model calculations. On the other hand, the theoretical results are rather sensitive to the adopted nN-potential. In any case, further (p,γ) cross sections measurements are necessary to improve the global parametrization of the statistical model used for astrophysical purposes.

REFERENCES

1. Laird C. *et al.*, *Phys. Rev.* **C35**, 1265 (1987).
2. Sauter T. and Käppeler F., *Phys. Rev.* **C55**, 3127 (1997).
3. Bork J., Schatz H., Käppeler F., Rauscher T., *Phys. Rev.* **C58**, 524 (1998).
4. Murphy A. S. *et al.*, in *Proceedings of the international symposium on nuclear astrophysics: Nuclei in the Cosmos V*, eds. N. Prantzos and S. Harissopulos, Paris: Editions Frontières, 1999, pp. 447-450.
5. Harissopulos S. *et al.*, in *Proceedings of the international symposium on nuclear astrophysics: Nuclei in the Cosmos V*, eds. N. Prantzos and S. Harissopulos, Paris: Editions Frontières, 1999, pp. 455-458.
6. Somorjai E. *et al.*, in *Proceedings of the international symposium on nuclear astrophysics: Nuclei in the Cosmos V*, eds. N. Prantzos and S. Harissopulos, Paris: Editions Frontières, 1999, pp. 459-462.
7. Rayet M., Prantzos N, Arnould. M., in *Origin and distribution of the elements*, ed. G. Mathews, World Scientific, 1987, p. 625
8. Rauscher T., and Thielemann F.-K., in *Atomic and Nuclear Astrophysics*, ed. A. Mezaccappa, Bristol: IOP, 1998, pp. 519-523
9. Goriely S., in *Proceedings of Conference on Nuclear Data for Science and Technology*, eds. G. Reffo *et al.*, Italian Physical Society, 1997, p. 811.
10. Jeukenne J.P., Lejeune A., Mahaux C., *Phys. Rev.* **C16**, 80 (1977).
11. Goriely S. *et al.*, *Nucl. Phys.* **A605**, 28 (1996)
12. Thielemann F.-K., Arnould M., Truran J. in *Advances in Nuclear Astrophysics*, eds. E Vangioni-Flam *et al.*, Gif-sur-Yvette: Editions Frontières, 1986, p. 525.
13. Rauscher T., Thielemann F.-K., Kratz K.-L., *Phys. Rev.* **C56**, 1613 (1997).
14. Bauge E., Delaroche J.P., Girod M., *Phys. Rev.* **C** (1998).
15. Becchetti F.D., and Greenless G.W., *Phys. Rev.* **182**, 1190 (1969).

Stopping power measurements: implications in nuclear astrophysics

Carmen Angulo, Thierry Delbar, Jean-Sebastien Graulich,
and Pierre Leleux

Institut de Physique Nucléaire, Université catholique de Louvain
Chemin du cyclotron, 2, B-1348 Louvain-la-Neuve, Belgium

Abstract. The stopping powers of C, CH_2, Al, Ni, and polyvynilchloride (PVC) for several light ions (^{9}Be, ^{11}B, ^{12}C, ^{14}N, ^{16}O, ^{19}F, ^{20}Ne) with an incident energy of 1 MeV/amu have been measured at the Louvain-la-Neuve cyclotron facility. Stopping powers are given relative to the one for 5.5 MeV ^{4}He ions with an uncertainty of less than 1%. We compare our results with two widely used semiempirical models and we discuss some implications in nuclear astrophysics studies.

INTRODUCTION

Charged particles moving through matter lose their energy by collisions with nuclei and atomic electrons [1]. The term "stopping power" usually means the differential energy loss dE per unit path length ρdx, where E is the particle energy and ρ is the density. Since dE is negative, the stopping power is defined as,

$$S_p(E) = -dE/\rho dx, \qquad (1)$$

usually given in units of $MeV/(g/cm^2)$.

Stopping power data have many applications in nuclear physics, material analysis sciences, radiology and radiation safety, and nuclear astrophysics studies. The determination of the effective beam energy inside a target and of the energy loss in a target are important sources of systematic uncertainties in many experimental studies. In order to reduce the overall systematic uncertainties, precise knowledge of the stopping power is needed.

For many ions, it does not exist any stopping power data at the lower energies so the low energy values come from an extrapolation of analytical fits of the higher energy data [2,3]. Usually, target thicknesses are obtained by using the energy loss of α particles passing through the foil. By the use of tabulated stopping power values the energy loss is translated into thickness. However, the discrepancy between stopping powers of α particles from compilations [2,3] clearly shows that we cannot rely on this method to obtain thicknesses within a few percent.

CP495, *Experimental Nuclear Physics in Europe,* edited by B. Rubio et al.
© 1999 American Institute of Physics 1-56396-907-6/99/$15.00

I EXPERIMENTAL METHOD

The measurements were performed at the Louvain-la-Neuve cyclotron facility CYCLONE. A detailed description of the beam transport system can be found elsewhere [4]; the experimental set-up and the energy calibration of the detector are described in [5]. We use the transmission technique, i. e. by obtaining the channel shift of the energy peaks coming from two consecutive measurements obtained with the same ion beam *with* and *without* the target foil. The difference between the channel of the signal peak centroid is related to the energy loss of the beam in the target ΔE. The energy calibration of the PIPS detector is obtained from the elastic scattering of the incoming projectile on a C foil. The detected energies E_f are fitted as a function of the observed channel c_f using a linear relation $E_f = a \times c_f + b$ (a and b are obtained from a least-square method). Details on the error analysis can be found in [5]. In order to give an idea of the quality of the fits, Fig. 1 shows a typical calibration curve for a ^{12}C ion beam ($\chi^2 = 0.3$ per degree of freedom). We represent the differences between the experimental energy E_{exp} and the fitted value E_f. The open (full) circles correspond to the positive (negative) angle data points. An offset angle is fitted in order to take into account a possible deviation of the beam from the mechanical axis (typically $\sim 0.2°$). We use the Time-of-Flight method to determine the absolute energy of the beam (accuracy $\sim 10^{-3}$).

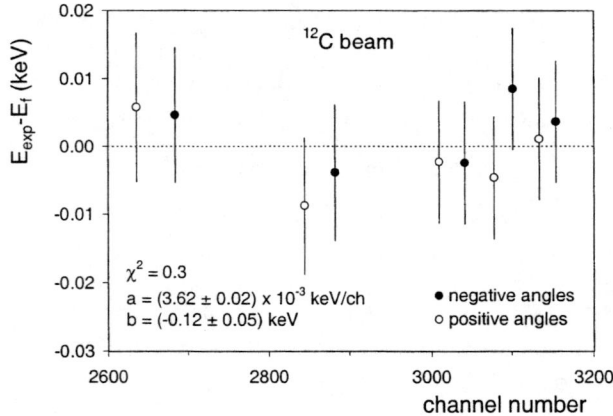

FIGURE 1. Typical calibration curve for a ^{12}C ion beam (see text).

II RESULTS AND DISCUSSION

In order to obtain stopping power data independent of the absolute target thickness, we have measured the energy loss experienced by the 5.5 MeV α particles of a very thin ^{241}Am source when passing through the same targets. In the case of a thin target, the energy loss $(\Delta E)_b$ in the target thickness δx for an ion beam is related to the corresponding energy loss for the α particles $(\Delta E)_\alpha$ in the same target by

$$R_\varepsilon = \frac{(\Delta E)_b}{(\Delta E)_\alpha} = \frac{\varepsilon_b \, \delta x}{\varepsilon_\alpha \, \delta x} = \frac{\varepsilon_b}{\varepsilon_\alpha}, \tag{2}$$

where ε_b and ε_α are the stopping power for the ion beam and the α particles, respectively. The overall uncertainty for the ratio R_ε is less than 1%.

Fig. 2 shows our results for R_ε, as well as the same ratio calculated using the predictions of refs. [2] and [3], for all ions ^{9}Be, ^{11}B, ^{12}C, ^{14}N, ^{16}O, ^{19}F, and ^{20}Ne in CH_2, C, Al, and Ni. Error bars are smaller than the size of the symbols. In the figure, NS70 refers to ref. [2], and TRIM96 to ref. [3]. There is, in general, a fair agreement with the values of ref. [2], specially for the ^{12}C, ^{14}N and ^{16}O ion beams. However, the differences between the present R_ε results and the ratio from ref. [3] are significant.

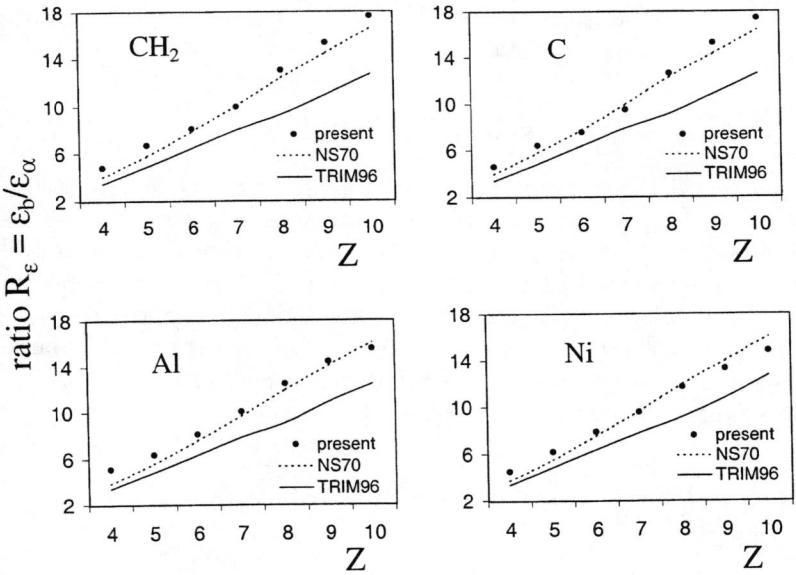

FIGURE 2. Present results of the ratio $R_\varepsilon = \varepsilon_b/\varepsilon_\alpha$ (see text).

III IMPLICATIONS IN NUCLEAR ASTROPHYSICS

One of the major problems in low-energy nuclear physics is the extrapolations from the lowest energies attained experimentally to those of most direct astrophysical relevance. Moreover, nuclear reactions studied in the laboratory are affected by electron screening [6]. Until now, all theoretical studies, based on atomic physics theory (see, for example, [7]), failed to explain the large enhancement of the cross sections at low energies. Notice that the cross sections $\sigma(E)$ are obtained from the reactions yields with the use of stopping power data. A certain error in $\varepsilon(E)$, will reflect in the same amount of error in $\sigma(E)$. Therefore, a simple estimate of screening potentials from experimental data, needs the use of accurate stopping cross sections. An independent measurement of the energy loss at such low energies is highly desirable.

In other experimental studies, the cross section is dominated by resonances. Resonance parameters are deduced from the analysis of elastic scattering and direct reaction data. Sometimes, several experiments disagree in the resulting values of such parameters. Usually, the target thickness is chosen by using tabulated stopping power values of α particles, as explained in the introduction. A typical case, is the reaction $^{18}F(p,\alpha)^{15}O$, where the results of [8] and [9] for the resonance energy (at about 650 keV) of a $3/2^+$ state in ^{19}Ne differ by 14 keV. In [8], the energy of the state and its total width are obtained from the α-spectra using the stopping powers given in ref. [2]. In [9], the stopping power values of ref. [3] is used for choosing the target thickness. A comparison of the results is not, therefore, straightforward and depends on the stopping power data.

IV   CONCLUSIONS

We have measured the stopping powers of a series of target materials for ^9Be, ^{11}B, ^{12}C, ^{14}N, ^{16}O, ^{19}F, and ^{20}Ne ion beams at an incident energy of 1 MeV/amu. In order to avoid the uncertainty related to the absolute target thickness, we give stopping powers relative to the stopping power of the ^4He ions from a α calibration source with an error of less than 1%. The differences between the present results and the values of compilations [2,3] are significant. One of the origins of large differences between data sets of reaction cross sections in nuclear physics and nuclear astrophysics studies may be the use of inappropriate values of stopping powers.

REFERENCES

1. Ziegler J.F., *The Stopping Power and Ranges of Ion in Matter*, New York: Pergamon Press, 1977.
2. Northcliffe L.C., and Schilling R.F., At. Data Nucl. Data Tables **A7**, 233 (1970).
3. Ziegler J.F., *The Stopping and Ranges of Ions in Solids*, New York: Pergamon Press, 1985, vol. 1, and TRIM96.
4. Belery P., Delbar Th., and Gregoire, Gh., Nucl. Instr. Meth. **179**, 1 (1980).
5. Angulo C., Delbar Th., Graulich J.-S., and Leleux P., to be published.
6. Assenbaum H.J., Langanke K., Rolfs C., Z. Phys. **A327**, 461 (1987).
7. Bracci L. *et al.*, Nucl. Phys. **A513**, 316 (1990).
8. Coszach R. *et al.*, Phys. Lett. **B353**, 184 (1995).
9. Rehm K.E. *et al.*, Phys. Rev. **C53**, 1950 (1996).

The program LUNA2 at the Gran Sasso

THE LUNA COLLABORATION

R. Bonetti[1], C. Broggini[2], L. Campajola[3], P. Corvisiero[4],
A. D'Alessandro[5], M. Dessalvi[4], A. D'Onofrio[3], A. Fubini[6],
G. Gervino[7], U. Greife[8], A. Guglielmetti[1], C. Gustavino[5],
M. Junker[5], C. Marino[7], P.Prati[4], V. Roca[3], C. Rolfs[8],
M. Romano[3], F. Schiemann[8], F. Strieder[8], F. Terrasi[3],
H.P. Trautvetter[8] and S. Zavatarelli[4].

1) Universita' di Milano, Dipartimento di Fisica and INFN Milano
2) INFN Padova
3) Universita' di Napoli, Dipartimento di Fisica and INFN Napoli
4) Universita' di Genova and INFN Genova
5) Laboratori Nazionali del Gran Sasso
6) ENEA Frascati and INFN Torino
7) Universita' di Torino, Dipartimento di Fisica and INFN Torino
8) Institut fur Experimentalphysik III, Rhur-Universitat Bochum

Abstract. The scientific program of nuclear astrophysics at the INFN underground laboratories of Gran Sasso for the next five years (LUNA phase II or LUNA2) is presented and discussed.

After the successful end of the first research program of nuclear astrophysics in the INFN Gran Sasso Laboratories, where we showed that it is possible to measure the nuclear cross section down to the thermal energy of the star nucleosynthesis using the modern techniques of low background measurements in a underground laboratory[1], a second phase has been approved and scheduled. In the next year a new electrostatic accelerator facility will be installed in the Gran Sasso tunnel, with a maximum proton energy of 400 keV. The accelerator and the beam transport and focussing set-up has been designed to reach the luminosity of 500 µA on the target in the energy range of 50 -200 keV with a long term stability better than $1 \ 10^{-4}$. The energy upper limit of 400 keV has been chosen in order to have a good overlapping region to the existing nuclear astrophysics data and to carry on energy callibrations as well as solid target stoichiometry tests exploiting standard nuclear resonances. A new gas target and new detection systems

CP495, *Experimental Nuclear Physics in Europe,* edited by B. Rubio et al.
© 1999 American Institute of Physics 1-56396-907-6/99/$15.00

for heavy charged particles and for gamma rays will be developped. A special attention must be spent for the detectors: some of the cross section of interest are of the order of 100 fbarn or lower, so it is crucial to reach a high detection efficency and a very low background as good as to measure one event per day in the energy region of interest. To detect charged massive particles, according to the successful experience of the $^3He(^3He,2p)^4He$ measurements[1], we plan to develop new very large area, low intrinsic background silicon telescopes using wafers of thickness larger than 1.2 mm. In order to detect the gamma rays emerging from nuclear fusion reactions two kind of detectors are under investigations: a high purity HPGe and a BGO summing crystal. The know-how on germanium detector system with a very low background has been developed at the Gran Sasso Laboratories in the neutrinoless double beta decay experiments and in the local Low Activity Counting Rate Facility. A HPGe allows a very good energy resolution and an efficient suppression of the beam induced background, but, on the other side, germanium detectors have a quite low intrinsic efficency at photon energy of few MeV and they cover only a small solid angle fractions. This makes it difficult to identify all the different gamma ray lines induced by the nuclear reaction studied and to evaluate without great uncertainties the solar rates. A summing crystal made from BGO can cover a solid angle of nearly 4π, adding up all the photons of one cascade in the sum-peak. While some spectroscopy informations will be lost, the stellar reaction rate could be quite easily extracted without any additional assumption. The first reactions that we plan to study, starting in the autumn of 2000, are $^3He(\alpha,\gamma)^7Be$, a key reaction of the pp-chain, and $^{14}N(p, \gamma)^{15}O$ for the fundamental CNO-cycles. Both of these reactions have never been studied in or even near their solar Gamow peaks, and both are critical for solar neutrino puzzle. In addition $^{14}N(p, \gamma)^{15}O$ is also one of the cross section data needed to determine the theoretical scenario used to constrain both the age and the distance of the oldest stellar system in our galaxy, namely the Globular Cluster. We plan to be able to measure this reaction at least as down as 80 keV, that means a counting rate of one event/day (the Gamow peak is at 26 keV, but this figure is not reachable because the cross section is dropping exponentially with the energy, the last direct measurements available stop at 205 keV[2]).

REFERENCES

[1] C. Arpesella et al. Phys. Lett. B389(1996)452
[2] U. Schroder et al. Nucl. Phys. A467(240)1987

HIGH ENERGY COLLISIONS

Nuclear Bremsstrahlung, a tool to study the free and in-medium NN interaction

Herbert Löhner

KVI Groningen, 9747 AA Groningen, The Netherlands

for the TAPS and SALAD collaboration [1]

Abstract. At the AGOR cyclotron of the KVI a series of bremsstrahlung measurements has been carried out with proton and α beams. The elementary bremsstrahlung process for real and virtual photons has been studied in the pp system. Coherent bremsstrahlung has been found in the $\alpha + p$ system. A strong quenching of bremsstrahlung is observed in the low-energy regime of the photon spectrum in p+nucleus reactions.

INTRODUCTION

In collisions between nucleons bremsstrahlung can be emitted due to the rapid change in velocity if at least one of the nucleons is charged. In nuclear collisions the individual nucleon collisions contribute to the bremsstrahlung cross section. Dynamical nuclear reaction models include bremsstrahlung production in the incoherent quasi-free collision limit. However, in the dense nuclear medium, and in particular in the presence of multiple scattering, the quasi-free limit is a strong approximation. Various medium effects have been discussed in the literature, but quantitative predictions have not been given. Coherent bremsstrahlung may become important in reactions where quasi-free processes are suppressed. At the AGOR cyclotron of the KVI bremsstrahlung has been measured with proton and α beams using the combination of the highly segmented photon spectrometer (TAPS) with the small-angle large-acceptance detector (SALAD) for charged-particles. TAPS was used in 'block mode' employing six blocks with 64 BaF_2 crystals each in the horizontal plane around the target.

---

[1] The author is indebted to his colleagues from KVI Groningen, The Netherlands; GANIL Caen, France; GSI Darmstadt, Germany; Univ. Giessen, Germany; NPI Řež u Prahy, Czech Republic; IFIC Valencia, Spain.

REAL AND VIRTUAL NN BREMSSTRAHLUNG

NN bremsstrahlung is the most fundamental process for the study of off-shell contributions to the NN-interaction. High precision cross sections (1% accuracy) and analyzing powers have been measured [1] for $p + p$ bremsstrahlung with 190 MeV polarized protons. In the same experiment, exclusive differential cross sections of virtual bremsstrahlung have been measured for the first time. The data are compared with state-of-the-art microscopic calculations [2] and with relativistic gauge-invariant calculations in the soft-photon approximation (SPA) [3]. In fig. 1 the measured cross section and analyzing power for the process $\vec{p} + p \rightarrow p_1 + p_2 + \gamma$ are shown as function of one of the proton angles. One of the proton angles is fixed at $16° \pm 2°$ while the photon angle (on the same side as θ_1) was kept at $145° \pm 5°$. The cross section is consistent with a previous measurement at 200 MeV. The cross sections are in good agreement with the SPA calculation (lower curve), but not with a potential-model calculation (dotted curve) and a fully relativistic microscopic calculation – with and without meson-exchange currents and Δ contributions. The analyzing power is neither described by the SPA nor the microscopic calculation. Better NN phase-shifts may improve the microscopic calculation, but obviously an improved modelling of the off-shell NN-interaction is needed.

More detailed information about the NN interaction is obtained through the study of e^+e^- pairs resulting from the virtual bremsstrahlung process. A virtual photon can be longitudinally polarized and, therefore, can probe electromagnetic currents propagating in the direction of the photon. The small cross sections could be measured [4] almost background-free due to the overdetermined kinematics and the high granularity of TAPS. The differential cross sections in fig. 2 have a sys-

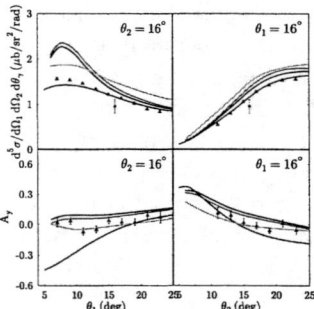

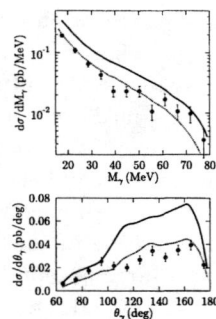

FIGURE 1. Coplanar $pp\gamma$ cross section and analyzing power measured at 190 MeV incident energy at a photon angle of 145°. A data point from literature measured previously with lower accuracy at 200 MeV is shown as a filled square. The different theoretical calculations are explained in the text.

FIGURE 2. Virtual-photon cross section as a function of the invariant mass (top) and the laboratory photon angle for invariant masses between 15 and 80 MeV (bottom). The solid and dotted lines are the results of a microscopic model and an SPA calculation, respectively. Statistical errors are shown.

390

tematic uncertainty of 15% and are compared to two calculations obeying the same cuts in energy and invariant mass as applied to the measured data. For both the real and virtual bremsstrahlung the SPA calculation [5] gives in general a better description of the data. The state-of-the-art microscopic calculation [2] overestimates the measured differential cross sections. The observed large discrepancy points to the need of modifications in off-shell dynamics.

COHERENT AND IN-MEDIUM BREMSSTRAHLUNG

Coherent bremsstrahlung may become important in reactions where quasi-free processes are suppressed. Because of the strong binding of the α-particle, the most energetic photons ($E_\gamma > 22$ MeV) can only be produced coherently in the $\alpha + p$ system at 50 MeV/nucleon [6]. Photon-particle coincidences allow to extend the coherent bremsstrahlung spectrum to low photon energies with $E_\gamma \geq 10$. For high-energy photons coherent bremsstrahlung can be associated with direct capture to the two lowest states of the unbound ^{5}Li. The low-energy photon spectrum (fig. 3) has a 'classical' $1/E_\gamma$ shape. The data have been compared with the potential model [7] and the SPA calculation [8], respectively. The calculations reproduce qualitatively the main features of the data.

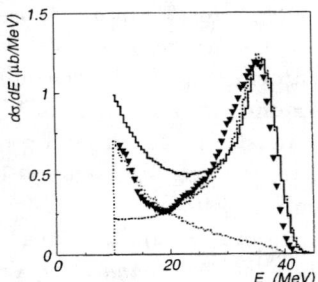 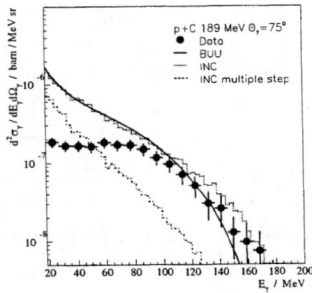

FIGURE 3. Coherent bremsstrahlung spectrum in the c.m. system integrated over 4π. The dash-dotted line represents the 'classical' bremsstrahlung spectrum, while the solid and dotted histograms indicate calculations with the potential model and the soft photon approximation, respectively.

FIGURE 4. Inclusive photon spectrum for 190 MeV p+C at a photon lab. angle of $75°\pm5°$, compared with results form the modified INC model (dotted histogram) and BUU (solid line). The dashed histogram shows the multiple-step contribution from the INC model.

In 190 MeV proton-nucleus reactions the influence of the nuclear medium on the photon production has been studied [9] by analyzing coincidences of photons with leading protons. The most intriguing effect is expected when the individual bremsstrahlung contributions can not be added incoherently. Intra-nuclear cascade (INC) calculations have been modified to include the NN bremsstrahlung process on basis of SPA calculations together with a kinematical correct treatment of the production process in subsequent scatterings. Fig. 4 shows the inclusive photon spectrum for 190 MeV p+C at a lab. angle of $75° \pm 5°$, compared with results from the modified INC model and BUU calculations. Most notable is the large discrepancy of the models with the experimental data in the soft part of the photon spectrum. Below $E_\gamma \approx 80$ MeV the models agree very well with each other and exhibit the characteristic soft photon $(1/E_\gamma)$ dependence. Exclusive photon spectra gated on protons ($E_{proton} > 30$ MeV) reveal a purely exponential slope distinct from the inclusive data. The modified INC result overestimates the data by a factor of 5. We conclude that the suppression of the soft-photon cross section is caused by a reduced photon probability in subsequent scattering processes. This appears to indicate that important medium effects are missing in the calculation.

In conclusion, we note the following: Cross sections and analyzing powers for real bremsstrahlung have been measured with high accuracy in the pp system. Virtual bremsstrahlung has been measured for the first time. For both the real and virtual bremsstrahlung the SPA calculation gives in general a better description of the data than the state-of-the-art microscopic calculation. The observed large discrepancy with the microscopic model indicates important contributions from higher-order terms which have not been considered yet. In the $\alpha+p$ system coherent bremsstrahlung has been observed. Model calculations qualitatively reproduce the main features of the data but do not reproduce the relative magnitude of the low and high energy part of the spectrum. A strong quenching of bremsstrahlung is observed in the low-energy regime of the photon spectrum in p+nucleus reactions, which seems to indicate that important medium effects are missing in the calculation and need to be understood better. Extended experiments, in particular for virtual bremsstrahlung and real bremsstrahlung studies in transfer reactions, will soon be carried out at the KVI with the highly granular and spherical symmetric Plastic Ball detector in combination with the SALAD detection system.

The effort of the AGOR team in providing high-quality beam is gratefully acknowledged. This work was supported in part by FOM, the Netherlands, by IN2P3 and CEA, France, BMBF and DFG, Germany, DGICYT and the Generalitat Valencia, Spain, by GACR, Czech Republic, and by the European Union HCM network contract.

REFERENCES

1. Huisman H. et al., accepted for publication in *Phys. Rev. Lett.*
2. Martinus G.H., Scholten O., Tjon J.A., *Phys. Rev.* **C 56**, 2945 (1997).
3. Liou M.K., Timmermans R.G.E., Gibson B.F., *Phys. Rev.* **C 54**, 1574 (1996).
4. Messchendorp J.G. et al., *Phys. Rev. Lett.* **82**, 2649 (1999).
5. Korchin A.Yu., and Scholten O., *Nucl. Phys.* **A 581**, 493 (1995).
6. Hoefman M., PhD thesis, Univ. Groningen 1999, to be published.
7. Baye D. et al., *Nucl. Phys.* **A 550**, 250 (1992).
8. Yan D. et al., *Phys. Rev.* **C 45**, 331 (1992).
9. van Goethem M.J., PhD thesis, Univ. Groningen 1999, to be published.

Thermal hard-photons from hot fragmenting nuclei

D.G. d'Enterria[*†] and G. Martínez[†]

for the TAPS collaboration

[*] *Grup de Física de les Radiacions, Universitat Autònoma de Barcelona, 08193 Bellaterra, Catalonia (Spain)*
[†] *Subatech, Ecole des Mines de Nantes, BP 20722, 44307 Nantes Cedex 3, France*

Abstract. Hard-photon ($E_\gamma \geq 30$ MeV) production has been studied in four different heavy-ion reactions (^{36}Ar+^{197}Au,^{107}Ag,^{58}Ni,^{12}C at $60A$ MeV) with TAPS electromagnetic calorimeter, in coincidence with LCPs ($Z{\leq}2$) and IMFs ($3 \leq Z \leq 20$) detected by two phoswich multidetectors. Aside from the main known emission mechanism, pn bremsstrahlung during the first high-density stage, the existence of a component of lower energy bremsstrahlung photons emitted in secondary pn collisions from a thermalizing source is demonstrated. Thermal hard photons arise as a promising probe of the thermodynamical properties of nuclear systems undergoing multifragmentation.

INTRODUCTION

Subtreshold particle production is generally considered to be a sensitive probe of the dynamics of intermediate-energy (20-100A MeV) heavy-ion collisions, as well as of the thermodynamical properties of the hot and compressed nuclear systems produced in such reactions [1]. Among the experimental probes, photons are preeminent observables because they are not absorbed and do not suffer any Coulomb or nuclear final-state interaction with the surrounding hadronic medium, giving a true image of the emission source. The production of hard photons ($E_\gamma \geq 30$ MeV) in nucleus-nucleus reactions in the intermediate-energy domain has been widely studied experimentally and theoretically the last 14 years. All experimental observations (energy spectra, angular distributions, source velocities and cross sections), as well as transport model calculations, are consistent with the assumption that they mainly originate from the incoherent summation of individual first-chance proton-neutron bremsstrahlung collisions within the participant zone [1,2]. Recent experimental results [3–5] of the TAPS collaboration, however, deviate from the pure first-chance emission scenario and point out to the existence of a softer photon contribution ("thermal hard-photons") issuing from secondary pn collisions in a later stage of the reaction when part of the available energy has been thermalised.

CP495, *Experimental Nuclear Physics in Europe,* edited by B. Rubio et al.
© 1999 American Institute of Physics 1-56396-907-6/99/$15.00

To confirm this interpretation and to try to extract informations concerning the thermodynamical state of the system at stages of the reaction where fragment formation is supposed to take place, two experimental campaigns of the TAPS collaboration have been carried out in 1997 and 1998 at the KVI and GANIL facilities, coupling, for the first time, a photon spectrometer with 2 different particle multidetectors covering more than 80% of 4π. Here we report some preliminary results of hard-photon and fragment production in ^{36}Ar+^{197}Au,^{107}Ag,^{58}Ni,^{12}C at $60A$ MeV.

PRELIMINARY RESULTS

More than $1.3 \cdot 10^6$ hard photons out of $180 \cdot 10^6$ recorded events were collected for the four mentioned systems using the experimental setup shown in fig. 1 [6]. The existence of an enhanced production of low energy hard-photons in the region between 30 and 60 MeV is clearly seen in the spectra, fig.2, for the heaviest (^{197}Au) system but not for the lightest (^{12}C) one. The energy spectra in the NN cm frame obtained for the four reactions after pion substraction have been, accordingly, fitted to the sum of two exponentials corresponding to the direct and thermal components:

$$\frac{d\sigma}{dE_\gamma} = K_{dir}exp(E_\gamma/E_0^{dir}) + K_{ther}exp(E_\gamma/E_0^{ther}) \tag{1}$$

The contribution of thermal hard-photons to the total hard-photon yield decreases for decreasing target mass as can be seen from the ratio I_{ther}/I_{tot} listed in

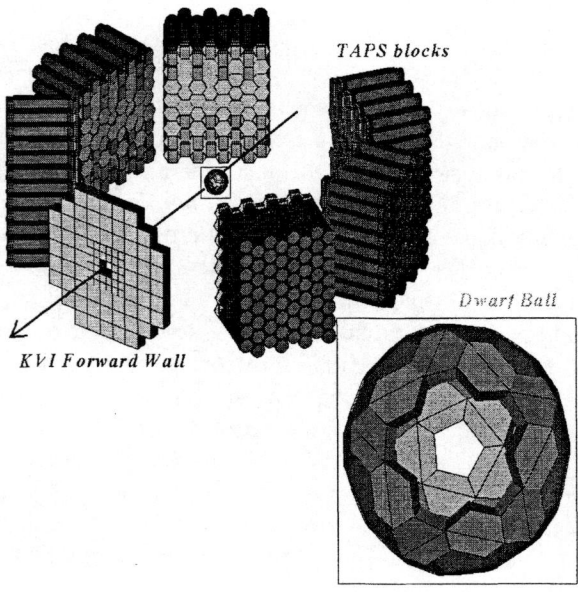

FIGURE 1. GEANT layout of TAPS, Dwarf Ball and Forward Wall multidetectors.

Table 1. Thermal hard-photon production is, thus, a process dependent on the total size of the system. In the small ^{36}Ar+^{12}C projectile-target combination, the resulting zone of participant nuclear matter does not have the necessary volume to achieve sufficient compression and/or thermalization. For that system, direct bremsstrahlung clearly dominates photon emission. The values obtained for the slope parameters of the direct component follow the known dependence with the bombarding energy in the lab as expected for first-chance NN collisions. The thermal hard-photon slopes, on the other hand, scale roughly with the available energy in the nucleus-nucleus center-of-mass, suggesting that they are emitted at a later stage of the reaction when the incident kinetic energy has been dissipated, at least partially, in the AA system into internal degrees of freedom.

TABLE 1. Measured ratios of thermal to total intensities, direct and thermal slopes and Coulomb-corrected energies in the lab and in the AA cm frames for the four studied reactions

System	I_{ther}/I_{tot}	E_0^{dir} (MeV)	E_0^{ther} (MeV)	E_{Cc}^{lab} (AMeV)	E_{Cc}^{AA} (AMeV)
^{36}Ar+^{197}Au	0.18 ± 0.02	19.2 ± 0.3	6.2 ± 0.4	55.5	7.3
^{36}Ar+^{107}Ag	0.17 ± 0.02	19.3 ± 0.3	6.5 ± 0.4	57.0	10.7
^{36}Ar+^{58}Ni	0.15 ± 0.02	19.5 ± 0.3	8.1 ± 0.4	58.1	13.7
^{36}Ar+^{12}C	0.00 ± 0.02	17.5 ± 0.3	0.0 ± 0.4	59.5	11.1

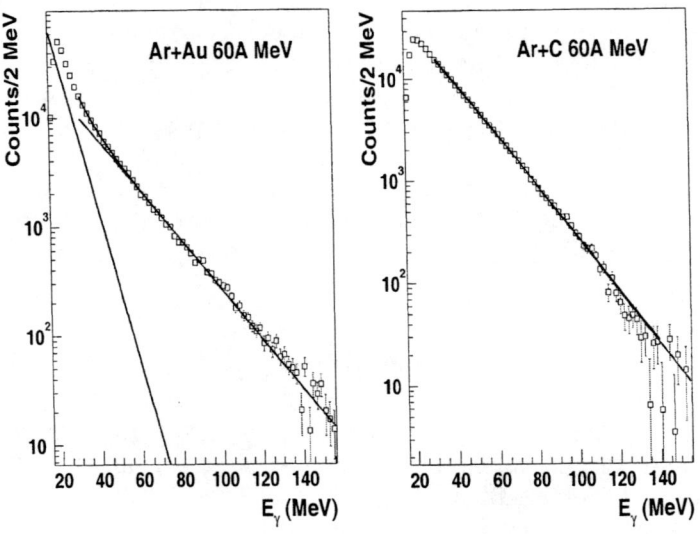

FIGURE 2. Hard-photon energy spectra of the ^{36}Ar+^{197}Au, (left) and ^{36}Ar+^{12}C (right). The ^{36}Ar+^{197}Au spectrum has been fitted to the sum of a direct and a thermal exponential (eq. (1)).

This interpretation is confirmed by the study of the Doppler-shifted lab angular distributions. For photons between 30 and 60 MeV and for the heavy systems, we obtain a source velocity $\beta_S \approx 0.14$, whereas for $E_\gamma \geq 60 MeV$, $\beta_S \approx \beta_{NN} = 0.17$. Finally, selecting specific centrality channels by means of the measured LCP multiplicity, we have been able to study the yield of different types of photons (GDR, thermal+direct and "pure" direct) for the ^{36}Ar+^{197}Au reaction as a function of impact-parameter. Thermal and direct bremsstrahlung photons are produced in very similar reaction channels that are clearly different to those that lead to GDR photon emission, suggesting that the elementary mechanism governing their production is basically the same, namely pn bremsstrahlung [6].

SUMMARY

Hard-photon production has been measured in coincidence with most of the reaction products in four different systems (^{36}Ar+^{197}Au,^{107}Ag,^{58}Ni,^{12}C at 60 AMeV). The first results demonstrate the existence of a thermal component showing up as a deviation of the pure exponential spectra in the region 30-60 MeV. The thermal photon yield is a system-size dependent process accounting for almost 20% of the total hard-photon yield for the heaviest targets (Au and Ag) and being negligible for the ^{36}Ar+^{12}C system. Thermal slopes scale with the AA center-of-mass energy, whereas direct slopes scale with the available energy in early NN collisions. Source velocity analysis are consistent with a later emission process for the thermal component, and hard-photon yields versus LCP multiplicities confirm their elementary origin different than that of photons coming from the decay of collective GDR excitations. Thermal slopes become, thus, a good candidate variable for the determination of the temperature of the system after the first pre-equilibrium phase of the reaction giving information of the thermodynamical state of the system at moments just before when fragment formation is supposed to take place.

The data presented here are the result of a common effort of the TAPS collaboration. We want specially to thank the work of L. Aphecetche, M.J. Mora, M. Hoefman and M.J. van Goethem.

REFERENCES

1. W. Cassing, V. Metag, U. Mosel and K. Niita, *Phys. Rep.* **188**, 363 (1990).
2. H. Nifenecker and J. Pinston, *Annu. Rev. Nucl. Part. Sci.* **40**, 113 (1990).
3. G. Martínez, *Phys. Lett.* **B349**, 23 (1995).
4. F.M. Marqués, *Phys. Lett.* **B349**, 30 (1995).
5. A. Schubert, *Phys.Rev. Lett.* **72**, 1608 (1994).
6. D.G. d'Enterria, *Proceedings XXXVII Int. Winter Meeting on Nuc. Phys.*, Bormio, ed. I.Iori, Univ. Milano, 1999, pp. 268-283.

Proton-Proton Interferometry in Ni+Al Collisions at 45 AMeV

V. Bellini, G. Riera, M.L. Sperduto, C.M. Sutera, M. Urrata

and the CHIC Collaboration

Dipartimento di Fisica dell'Università, Catania, Italy
INFN – LNS, Catania, Italy
INFN – Sez. Catania, Catania, Italy

Nuclear intensity interferometry (1-2) is a powerful method to translate correlation functions for pairs of fermions (p-p, p-n, n-n) into space-time information concerning nuclear reactions. However both quantum interference effects and final-state interactions obscure such a space-time information (3-5); so in order to get rid of mutual Coulomb interaction and of Pauli exclusion contribution it is required simultaneous measurement of p-p, p-n and n-n correlations with precise measurements of two momentum vectors at small relative momentum.

We performed an interferometry experiment at the C.S.-L.N.S. (Catania, Italy), where p-p, n-n and p-n correlation functions were measured simultaneously in 45 AMeV Ni+Al induced reaction.

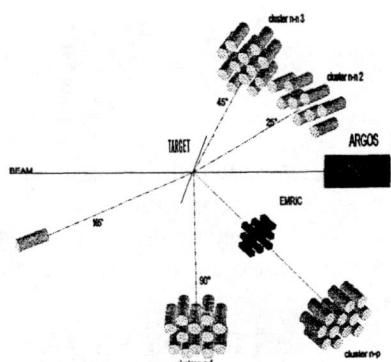

Figure 1 Experimental setup

CP495, *Experimental Nuclear Physics in Europe*, edited by B. Rubio et al.
© 1999 American Institute of Physics 1-56396-907-6/99/$15.00

The experimental setup consisted of 13 CsI (proton detectors), arranged in spherical symmetry, 70 cm. far from the target, at 45 degrees and of 48 liquid scintillators (neutron detectors) arranged in 4 clusters of 12 at 45 and 90 degrees. One cluster of liquid scintillators was positioned behind holes existing in the CsI array in order to guarantee the minimum angular separation between the detected n-p pairs. A forward wall of 32 high-resolution phoswich detectors, made of a thin (up to 700 μm) plastic scintillator optically coupled to a 10 cm. thick BaF_2, was added to characterize the event. The calibration of CsI detectors was performed by placing a two-step Si telescope as close as possible to the single CsI detector and using the Range-Energy relations for p, d, t, ^{3}He and ^{4}He, separately. Furthermore, it was implemented a Geant simulation program including a very detailed description of the experimental setup, in order to match the experimental data.

We analyzed the proton-proton correlation data and we extracted preliminary results about the radius and life-time of the emitting source, showed in the pictures below.

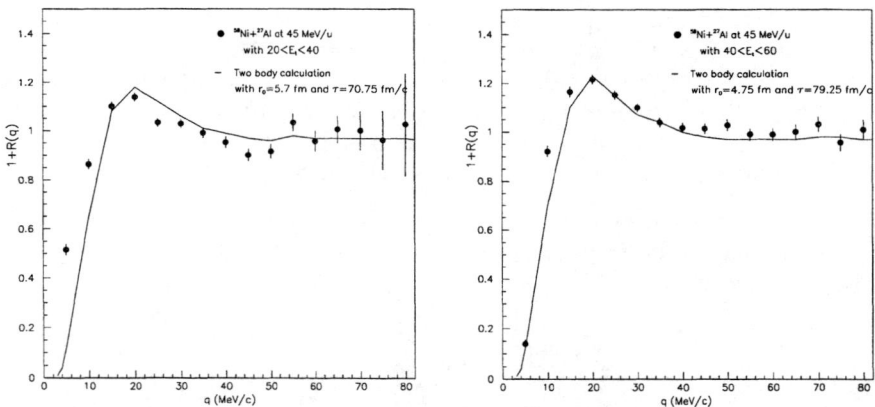

Figure 2 Correlation functions gated with different pair total energies

Our data confirm previous results already known in the literature (6), and strongly call for more exclusive experiments in order to identify the reaction mechanism which produced the correlated pair. This should be accomplished when all the relevant information from our experiment will be available.

REFERENCES

1. Koonin, S.E., *Phys. Rev.* **B70,** 43 (1977)
2. Ghetti R., *N.I.M.* **A335,** 156 (1993)
3. Dunnweber W., *Phys. Rev. Lett.* **65** 297-300 (1990)
4. Jakobsson B., *Phys. Rev.* **C44** 1238-12141 (1991)
5. Colonna N., *Phys. Rev. Lett.* **75** 4190-4193 (1995)
6. Pochodzalla J., *Phys. Rev.* **C35** 1695-1719 (1987)

Dielectron Production in Heavy Ions Collisions:the HADES Experiment

J. A. Garzón, H. Alvarez-Pol, I. Durán, C. Fernández, B. Fuentes,
R. Lorenzo, M. Sánchez, A. Vázquez-Cardesín
By the HADES Collaboration

Dept. Particle Physics, University of Santiago de Compostela, E-15706 Santiago, Spain

HADES (High Acceptance Di-Electron Spectrometer) [1] is a Collaboration of 20 institutions from 9 European countries. The spectrometer is being built at the heavy ion accelerator facility UNILAC-SIS of the GSI (Gesellschaft für Schwerionenforschung, Darmstadt, Germany), with the main purpose of studying the production of dilepton pairs in nucleus-nucleus collisions at energies of 1 AGeV.

The HADES spectrometer has been designed to provide high geometrical acceptance, high granularity and a high count rate capability to study the production of light vector mesons ρ, ω and ϕ with an invariant mass resolution less than 1% via their dileptonic decays. The data obtained will provide valuable information about the properties of such mesons when they are produced in a high density nuclear medium, and about the dilepton continuum in the small mass region. The results will allow to test QCD calculations and hadronic models that predict modifications of hadrons properties when they are produced in dense nuclear matter [2–6]. From the point of view of QCD, these modifications could come from the restoration of the chiral symmetry at high temperatures and densities.

The HADES spectrometer has an hexagonal symmetry around the beam line. A fast, hadron blind Ring Imaging Cherenkov (RICH) counter placed around a segmented target is used for electron identification. Four planes of Multi-wire Drift Chambers (MDC) together with a 6-coil superconducting magnet form the magnetic spectrometer for charged particle momentum measurement. A set of electromagnetic Pre-shower detectors (at polar angles below 45°) and a Time-Of-Flight (TOF) wall constitute the Multiplicity and Electron Trigger Array (META). The META will provide a trigger on central heavy ion collisions and additional fast lepton identification. The spectrometer covers polar angles $18° < \theta < 85°$, thus providing a geometrical acceptance of about 40° over the full azimuth.

The group of the University of Santiago de Compostela collaborating with HADES has assumed the following responsibilities:

- **Reconstruction program.** An innovative feature of the HADES software is that all reconstruction and analysis programs are being developed in an Object Oriented Programming way, using C++ as programming language and ROOT

CP495, *Experimental Nuclear Physics in Europe*, edited by B. Rubio et al.
© 1999 American Institute of Physics 1-56396-907-6/99/$15.00

[7], a CERN program, as tool. The framework HYDRA (Hades sYstem for Data Reduction and Analysis) is based on a fundamental class (Hades), and contains classes that hold data, classes to manage input/output of data, classes to perform tasks and classes handling reconstruction parameters, allowing the flexibility and modularity required by the experiment.

- **Tracking Software.** In this framework, the program to reconstruct the tracks of the particles in the MDC system is being written. This program has three main steps: First, straight track pieces (segments) are found in each one of the two lever arms of the spectrometer (before and after the magnet). After that, right combinations of two track segments are selected, one from each Lever Arm, via a Principal Component Analysis (PCA) method. Finally, the momentum of the track is fitted using a training sample of simulated tracks.

- **MDC Position Monitoring System.** The alignment system monitorizes the relative position of the MDC2 and MDC3 chambers. Our RASNIK setup consists of two light emisors with coded masks located on the frame of each MDC3. Both images are focused onto an image sensor by two lenses simmetrically disposed on each MDC2.The setup was developed and tested in Santiago. It was stable in a long term test for angles up to 25°, and the resolution appears to be nearly angle independent, reaching values around $0.2 \mu m$ in the plane perpendicular to the light (XY), and $20 \mu m$ along the light direction (Z).

- **Temperature Monitoring System.** In such a big experiment many parameters are temperature dependent. In order to monitorize it we have developed an scalable system able to manage up to 512 probes, spread over the detector cave, with a single DAQ channel. That system is composed of a central module and up to three satellites. The central module is linked to a master work-station via Ethernet, having a Com-Server unit that provides the buffering required for the asynchronous reading and the communication protocols.

The relevant characteristics are good linearity and long term stability, with a resolution of $0.2°C$ C and an overall accuracy of $0.5°C$.

REFERENCES

1. Salabura, P., et al. *Nucl. Phys. B* **44**, 701-707 (1995).
2. Metag, V., *Nucl. Phys. A* **638**, 45c-56c (1998).
3. Engels, J., et al., *Nucl. Phys. B* **435**, 295-310 (1995).
4. Hatsuda, T., and Lee, S.H., *Phys. Rev. C* **46** R34-R38 (1992).
5. Klingl, F., and Weise, W., *Nucl. Phys. A* **606**, 329-338 (1996).
6. Rapp, R. et al., *Nucl. Phys. A* **617**, 472-495 (1997).
7. Brun, R. and Rademakers, F., "ROOT - An Object Oriented Data Analysis Framework", in *Proceedings AIHENP'96 Workshop, Lausanne, Sep. 1996, Nucl. Inst. & Meth. in Phys. Res. A* **389**, 81-86 (1997). See also http://root.cern.ch/.

Results of the commissioning of the Pion Beam Factory at SIS/GSI

M. Ardid, N. Yahlali, J. Díaz, M. Álvarez
for the **Pion Collaboration**[1]

IFIC (Centre Mixte Universitat de València - CSIC),
C/ Dr. Moliner 50, E-46100 Burjassot, València, Spain.

Abstract. We describe briefly the Pion Beam Factory recently installed at SIS/GSI and present the main results of the commissioning performed during September 1998.

INTRODUCTION

A pion beam factory partially funded by the RTD programme of the CEE has been recently installed at the heavy-ion synchrotron SIS at GSI (Darmstadt). The pion beam facility has been designed to deliver π^- and π^+ beams with momenta ranging from 0.5 GeV/c to 2.8 GeV/c to the detector systems in the SIS target hall (HADES, FOPI, KaoS, LAND, ALADIN, and TAPS). This will allow a broad and relevant physics program, mainly related to the study of hadron properties in nuclear matter and baryon spectroscopy.

The identification of pions among other secondary particles is achieved by time of flight measurement with a system of three hodoscopes H1, H2 and H3 which can easily be installed in any beam line of the SIS. Each of the three hodoscopes is composed of 16 detector modules of BC404 scintillating plastic rods coupled to plexiglas light guides and read out at both ends by Hamamatsu R3478 photomultipliers.

The commissioning of the facility has been done in the HADES beam line which is 33 m long. Protons of 1.6 and 3.5 GeV and ^{12}C ions of 1.7 and 2 A GeV have been used as primary beams to irradiate beryllium production targets of 2, 4, 6 and 18.4 gr/cm² thickness. The beam line has been set to different momenta ranging from 0.4 GeV/c to 2.8 GeV/c and for negative and positive polarity of the magnets.

---

[1] Pion Collaboration: GSI Darmstadt (Germany), GANIL (France), Giessen Universität (Germany), Lund University (Sweden), IFIC València (Spain).

CP495, *Experimental Nuclear Physics in Europe,* edited by B. Rubio et al.
© 1999 American Institute of Physics 1-56396-907-6/99/$15.00

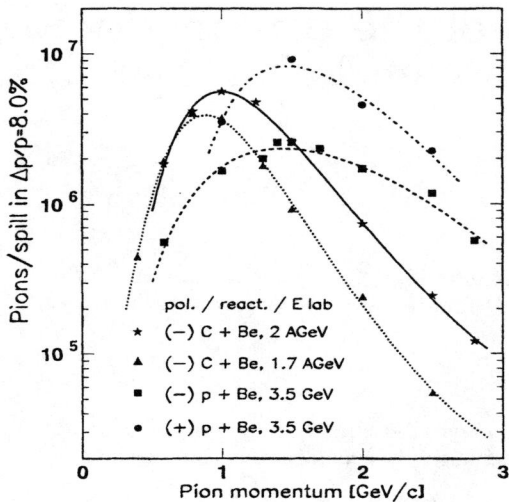

FIGURE 1. Yield of pions expected at the HADES target position for carbon and proton beams on Be(18.4 g/cm^2) target, as a function of the beam line momentum setting and polarity.

RESULTS AND CONCLUSIONS

The secondary particle time of flight and momentum have been measured with respective resolutions of 100 ps and 0.5 % within the beam line acceptance.

Separation of pions from protons has been achieved over the full momentum range, whereas the separation of pions from electrons (or positrons) is only possible at momenta lower than 1 GeV/c. Pions and muons cannot be separated. The pion intensities expected at the Hades target position, if extrapolated to the space charge limit of SIS and assuming 100 % extraction efficiency, are shown in Fig. 1. The carbon primary beam at 2 A GeV is the beam of choice for producing π at low momenta, while proton beam at 3.5 GeV is best suited for hard π production.

Another important factor for the pion yield is the density and the thickness of the production target. With a carbon beam, the pion yield increases with the thickness of the beryllium target and saturates at 18.4 g/cm^2, whereas for proton beam, saturation is not reached. Targets of higher density are presently being designed in order to optimize the pion production. They will be tested at GSI in the near future with different primary beams and in particular with the 4.7 GeV proton beam recently developed.

Quasielastic Scattering from Relativistic Bound Nucleons: R_{TL} response

J.A. Caballero[a,b], E. Moya de Guerra[b], J.M. Udías[c,b], J.E. Amaro[d]
and T.W. Donnelly[e]

[a] Dpto. de Física Atómica, Molecular y Nuclear, Universidad de Sevilla, Apdo. 1065, E-41080 Sevilla, Spain
[b] Instituto de Estructura de la Materia, CSIC Serrano 123, E-28006 Madrid, Spain
[c] Dpto. de Física Atómica, Molecular y Nuclear, Universidad Complutense de Madrid, E-28040 Madrid, Spain
[d] Departamento de Física Moderna, Universidad de Granada, E-18071 Granada, Spain
[e] Center for Theoretical Physics, Laboratory for Nuclear Science and Department of Physics, Massachusetts Institute of Technology, Cambridge, MA 02139, USA

Abstract. Predictions of relativistic calculations for electron induced knock-out from the $p_{1/2}$ and $p_{3/2}$ shells in ^{16}O are presented. Results for differential cross-section, TL response function and left-right asymmetry are compared to recent $(e, e'p)$ data at $Q^2 = 0.8 \ (GeV/c)^2$ taken at TJNAF. We show that the trend of the fully relativistic results is closely followed by the experimental data, pointing to the importance of both kinematical and dynamical relativistic effects in the nucleonic current.

In some recent works [1,2] we have studied the effect on the response functions of the relativistic dynamics of the bound nucleons. We have found that the TL response is particularly sensitive to the negative energy components of the relativistic bound nucleon wave function. At present, a certain degree of controversy surrounds the TL response measured in exclusive quasielastic electron scattering experiments for the least bound orbitals in several nuclei [3,4].

Recently, the coincidence cross section and the structure functions have been obtained for proton knockout off ^{16}O from the $1p_{1/2}$ and $1p_{3/2}$ orbits under perpendicular kinematics [5]. The data were taken at Jefferson Laboratory (TJNAF), using a nominal beam energy of 2445 MeV. We have calculated cross sections, TL responses and left–right asymmetries and have compared them with the experimental data. Different approximations have been considered:

1) Fully relativistic calculations using various current operators and gauges. We can divide the differences of this fully relativistic approach and the usual nonrelativistic ones in two categories: a) effects due to the fully relativistic kinematics, i.e., effects independent of the kind of the dynamics introduced in the model to

CP495, *Experimental Nuclear Physics in Europe*, edited by B. Rubio et al.

consider the nuclear interactions, and b) differences coming from the fact that the wave functions in the relativistic approach are solutions of Dirac-like equations, while in the nonrelativistic approach one uses Schrödinger like equations. These effects depend on the particular dynamics of each model. Within these dynamical relativistic effects there exist basically two different approaches that can be considered: b-1) differences due to the replacement of the upper component of Dirac equation's solution by a Schrödinger equation's solution, that have been long ago identified in the so-called Darwin term, and whose influence in $(e, e'p)$ observables is well known; and b-2) differences due to the removal of the lower component from the relativistic amplitude.

We have studied in deep those effects a), b-1) and b-2), as well as the fully relativistic results involving 4-spinors and negative energy component. Our calculations compared with the data on asymmetry and TL response strongly indicate the crucial role played by dynamical effects of relativity in the description of electron-nucleus scattering reactions. In order to clarify this point we have done a systematic study comparing our fully relativistic and projected calculations with non-relativistic analyses of the mechanism of the reaction. The results obtained show clearly that the experimental data measured for the left-right asymmetry can be reproduced better as far as one includes relativistic aspects in the description of the reaction mechanism.

In conclusion, experimental data show a clear preference for the relativistic results. Unfortunately, there are still few data and large error bars at some p-missing. We encourage further experimental effort to obtain data at more values of the momenta, higher p, with smaller error bars, as we face this unique opportunity to settle the issue of the importance of relativistic dynamical effects.

REFERENCES

1. J.A. Caballero, T.W. Donnelly, E. Moya de Guerra and J.M. Udías, Nucl. Phys. **A632** (1998) 323.
2. J.A. Caballero, T.W. Donnelly, E. Moya de Guerra and J.M. Udías, Nucl. Phys. **A643** (1998) 189.
3. L. Chinitz *et al.*, Phys. Rev. Lett. **67** (1991) 568.
4. G.M. Spaltro *et al.*, Phys. Rev. **C48** (1993) 2385.
5. J. Gao, Ph.D. Thesis, Massachusetts Institute of Technology (1999). T Saha *et al.* TJNAF proposal 89-003 (1989).

APPLICATIONS

Accelerator Mass Spectrometry: Analysing our World Atom by Atom

Walter Kutschera

Vienna Environmental Research Accelerator, Institut für Radiumforschung und Kernphysik, University of Vienna, Währinger Strasse 17, A-1090 Vienna, Austria

Abstract. In this paper an attempt is made to convey the fascination of accelerator mass spectrometry (AMS), a tool of unprecedented sensitivity for analysing our world atom by atom. As it is impossible to touch all the various fields where AMS is being applied, on the one hand a broad review of its possibilities is presented, and on the other hand some examples familiar to the author are selected to demonstrate specific applications. For those interested to get more information about the field an extensive list of references is added.

INTRODUCTION

Some 2400 years ago Democritus of Abdera (430-370 BC) introduced the atom by stating "nothing exists except atoms and empty space, everything else is opinion." It is interesting to speculate what he meant by "opinion". In modern terms we may identify this with our theoretical models to describe the material world, based essentially on the notion that matter consists of atoms. Accelerator mass spectrometry (AMS) literally analyses our world atom by atom, and thus provides one of the essential ingredients for the "Democretian" world. As our knowledge about the distribution of atoms in matter increases, one may hope to form a better "opinion" about the world as a whole.

Among the various methods to study our world in this analytical manner, mass spectrometry (MS) plays a special role. It allows one to use the "isotope language", i.e. the subtle shifts in stable isotope ratios within an element, which are due to a small but finite mass dependence in essentially all physical and chemical processes. The steadily increasing precision and sensitivity of MS leads to a corresponding refinement of the stable isotope language. In addition to the stable isotopes, there exist long-lived radioisotopes in nature, which can also be measured by mass spectrometry. They exhibit both a mass dependent and a time dependent abundance shift. Whereas the former can be assessed from adjacent stable isotope ratios, the latter allows one to determine the age of a sample if certain boundary conditions are well known (e.g. the initial ^{14}C abundance for radiocarbon dating). However, to study long-lived radionuclides at natural levels by mass spectrometry, the measurement of extremely small isotope ratios (10^{-10} to 10^{-16}) is required. The use of an accelerator as an integral part of the mass spectrometric system provides the necessary discriminatory power to identify these minute isotope signals.

CP495, *Experimental Nuclear Physics in Europe*, edited by B. Rubio et al.

Accelerator mass spectrometry (AMS) is a relatively new addition to the arsenal of mass-analytical methods. It evolved from nuclear physics laboratories some twenty years ago (1-4) when it was realized that long-lived radionuclides - in particular ^{14}C - can be measured at natural levels by counting atoms directly. It had been noted earlier (5) that during a typical beta decay measurement of ^{14}C lasting two days, only about one out of a million ^{14}C atoms decays (the half-life of ^{14}C is 5730 years). With a ^{14}C/^{12}C isotopic ratio of 1.2×10^{-12} in modern carbon, one needs a few grams of carbon to obtain enough decays in two days for a statistical uncertainty of 0.5% (corresponding to an age uncertainty of 40 years). In contrast, with a modern AMS facility one can easily obtain counting rates of 50 ^{14}C ions/sec using only one milligram of carbon in the ion source. This high counting rate leads to a counting statistics of 0.5% in 15 minutes. In practical terms, the amount of sample material needed is reduced by at least a factor of 1000 and the measuring time by a factor of 100 as compared to decay counting. Such an enormous gain in detection sensitivity ($\sim 10^5$) is similar to the gain in light gathering capability of a very large astronomical telescope, e.g. the one on Mt. Palomar with an opening aperture of $\sim$5 m, as compared to the unequipped eye, which has an opening aperture of about 5 mm.

Over the years, AMS has developed into an analytic tool of great versatility, with applications in almost every field of science where the measurement of minute traces of long-lived radioisotope is of interest (6-11). So far, the following long-lived radionuclides have been measured with AMS (the most used ones are marked in bold): ^{3}H ($t_{1/2}$ = 12.33 y), ^{10}**Be** (1.5×10^6 y), ^{14}**C** (5.73×10^3 y), ^{26}**Al** (7.1×10^5 y), ^{32}Si (135 y), ^{36}**Cl** (3.01×10^5 y), ^{39}Ar (268 y), ^{41}**Ca** (1.04×10^5 y), ^{44}Ti (59.4 y), ^{53}Mn (3.7×10^6 y), ^{55}Fe (2.73 y), ^{60}Fe (1.5×10^6 y), ^{59}Ni (9.2×10^4 y), ^{63}Ni (100 y), ^{79}Se (1.1×10^6 y), ^{81}Kr (2.3×10^5 y), ^{90}Sr (28.8 y), ^{99}Tc (2.11×10^5 y), ^{126}Sn (2.35×10^5 y), ^{129}**I** (1.7×10^7 y), ^{205}Pb (1.5×10^7 y), ^{236}U (2.34×10^7 y), ^{237}Np (2.14×10^6 y), ^{239}Pu (2.41×10^4 y), ^{240}Pu (6.54×10^3 y), ^{242}Pu (3.73×10^5 y), and ^{244}Pu (8.1×10^7 y).

Table 1 gives a summary of fields were AMS measurements are performed. In this table our environment is divided into seven "spheres", each constituting a major domain on Earth and beyond. Measurements of long-lived radionuclides provide important clues for the understanding of chemical and physical processes within each sphere. Even more important, interactions between the spheres in the past and in the presence can also be studied by AMS. Information gathered in this way will be the basis for extrapolating into our future on Earth (although any of these extrapolations have to be treated with utmost care as to their reliability of firm predictions).

Table 1. Overview of AMS applications in the environment at large

Sphere	Areas of interest where AMS measurements of long-lived radionuclides and stable trace isotopes can be performed; the respective radionuclides utilized are given in parenthesis.
Atmosphere	Production and distribution of cosmogenic and anthropogenic radionuclides (^{3}H, ^{7}Be, ^{10}Be, ^{14}C, ^{26}Al, ^{32}Si, ^{36}Cl, ^{39}Ar, ^{81}Kr, ^{85}Kr, ^{129}I) Study of trace gases: CO_2, CO, OH, O_3, CH_4 (^{7}Be, ^{10}Be, ^{14}C) Transport and origin of aerosols (^{14}C)
Biosphere	Dating in archaeology and other fields (^{14}C, ^{41}Ca) ^{14}C calibration studies in tree rings, corals and sediments (^{14}C) Studies in forensic medicine through bomb-peak dating (^{14}C) In-vivo tracer studies in plants, animals and humans (^{14}C, ^{26}Al, ^{41}Ca, ^{79}Se, ^{99}Tc, ^{129}I)
Hydrosphere	Dating of groundwater (^{14}C, ^{36}Cl, ^{39}Ar, ^{81}Kr, ^{129}I) Global ocean circulation pattern (^{14}C, ^{39}Ar, ^{99}Tc, ^{129}I) Paleoclimatic studies in ocean sediments
Cryosphere	Paleoclimatic studies in ice cores of glaciers and polar ice sheets (^{10}Be, ^{14}C, ^{32}Si, ^{36}Cl, ^{39}Ar, ^{81}Kr) Variation of cosmic ray intensity with time (^{10}Be, ^{36}Cl) Bomb-peak identification (^{36}Cl, ^{41}Ca, ^{129}I)
Lithosphere	Exposure dating and erosion of surface rocks (^{10}Be, ^{14}C, ^{26}Al, ^{36}Cl) Paleoclimatic studies in loess (^{10}Be) Tectonic plate subduction studies through volcanos (^{10}Be) Platinum group elements in minerals (stable trace isotopes)
Cosmosphere	Cosmic ray record in meteorites and lunar materials (^{10}Be, ^{14}C, ^{26}Al, ^{36}Cl, ^{41}Ca, ^{44}Ti, ^{59}Ni, ^{60}Fe, ^{107}Pd, ^{129}I); life on Mars ? (^{14}C) Evidence for supernova occurrence through extinct and life radionuclides in meteorites and manganese crusts (^{10}Be, ^{26}Al, ^{36}Cl, ^{41}Ca, ^{60}Fe, ^{107}Pd, ^{146}Sm, ^{244}Pu) Geochemical solar neutrino detection (^{98}Tc, ^{205}Pb) Search for exotic particles (superheavy elements, fractionally charged particles, strange matter)
Technosphere	Releases from nuclear industry (^{14}C, ^{36}Cl, ^{85}Kr, ^{90}Sr, ^{99}Tc, ^{126}Sn, ^{129}I) Half-life measurements (^{32}Si, ^{41}Ca, ^{44}Ti, ^{60}Fe, ^{79}Se, ^{126}Sn) Temperature measurement of fusion plasma (^{26}Al) Neutron flux of the Hiroshima bomb (^{36}Cl, ^{41}Ca, ^{63}Ni) Nuclear safeguards (^{233}U, ^{236}U, ^{237}Np, ^{239}Pu, ^{240}Pu, ^{242}Pu,) Trace elements in semiconductor materials (stable trace isotopes)

THE BASIC PRINCIPLE OF AMS

Measuring cosmogenic radionuclides at natural levels by mass spectrometry means to be capable of measuring radioisotope-to-stable isotope ratios in the range from 10^{-10} to 10^{-16}. For actual applications these extremely small isotope ratios have to be measured with a precision of 0.5% for ^{14}C dating purposes, and to a few percent for other radionuclides. This requires to solve three analytical problems: i) separation from interfering stable atomic isobars (e.g. from ^{14}N for the detection of ^{14}C), ii) separation from interfering stable molecular isobars (e.g. from ^{13}CH and $^{12}CH_2$ for the detection of ^{14}C), and iii) separation from stable isotopes (e.g. from ^{12}C and ^{13}C for the detection of ^{14}C). As it turns out, tandem accelerators offer by far the best conditions for AMS measurements. In particular, the most important long-lived radionuclide in nature, ^{14}C, can be measured with relative ease at tandem accelerators.

In Figure 1 the schematic of a modern AMS facility is shown, the Vienna Environmental Research Accelerator (12,13), which is based on a 3-MV Pelletron tandem accelerator. Since ^{14}N does not form negative ions (1), the otherwise (2) overwhelming background from ^{14}N is completely absent in tandem-based AMS measurements. However, the negative ion spectrum in Figure 2a measured before the entrance into the tandem accelerator shows a very large background of molecular ions of mass 14, which completely masks the ^{14}C signal. An important step in the consecutive acceleration in the tandem is therefore the stripping process in the terminal, which dissociates the $^{13}CH^-$ and $^{12}CH_2^-$ molecules very effectively when sufficient electrons are stripped off. With the high energy analysing magnet set to select $^{14}C^{3+}$ ions, one observes the energy spectrum shown in Figure 2b. Although the residual ^{12}C and ^{13}C peaks are greatly reduced in intensity, there is still a large number of background peaks which happen to have the same magnetic rigidity as $^{14}C^{3+}$. However, Figure 2c shows that the $^{14}C^{3+}$ ions can be cleanly selected by sending the mix of ions in figure 2b through a Wien filter (see fig. 1) set to the velocity of $^{14}C^{3+}$. In order to determine $^{14}C/^{12}C$ ratios with high precision (~0.5%), $^{12}C^-$, $^{13}C^-$, and $^{14}C^-$ are injected in fast sequence (10 times per second) into the tandem. The stable isotopes, $^{12}C^{3+}$ and $^{13}C^{3+}$, are measured via their respective ion currents in Faraday cups located at the exit of the high-energy analysing magnet, and $^{14}C^{3+}$ ions are counted in a Si detector at the end of the beam line (see Fig. 1). For details of $^{14}C/^{12}C$ ratio measurements at VERA the reader is referred to references (13-15).

For twenty years it was believed that $^{14}C^{3+}$ ions (or a higher charge state) must be selected to break up the molecules for sure. It is interesting to note that the "dogma" of stripping to at least the 3+ charge state for obtaining a clean ^{14}C signal was only recently revised, although indications for a deviation were reported much earlier (16). Using a sufficiently thick stripper it is possible to destroy the molecules in the 2+ charge state at about 1 MeV (17-19). Even 1+ stripping looked feasible for obtaining a reasonable ^{14}C separation (17). The latter assumption was recently proven to work

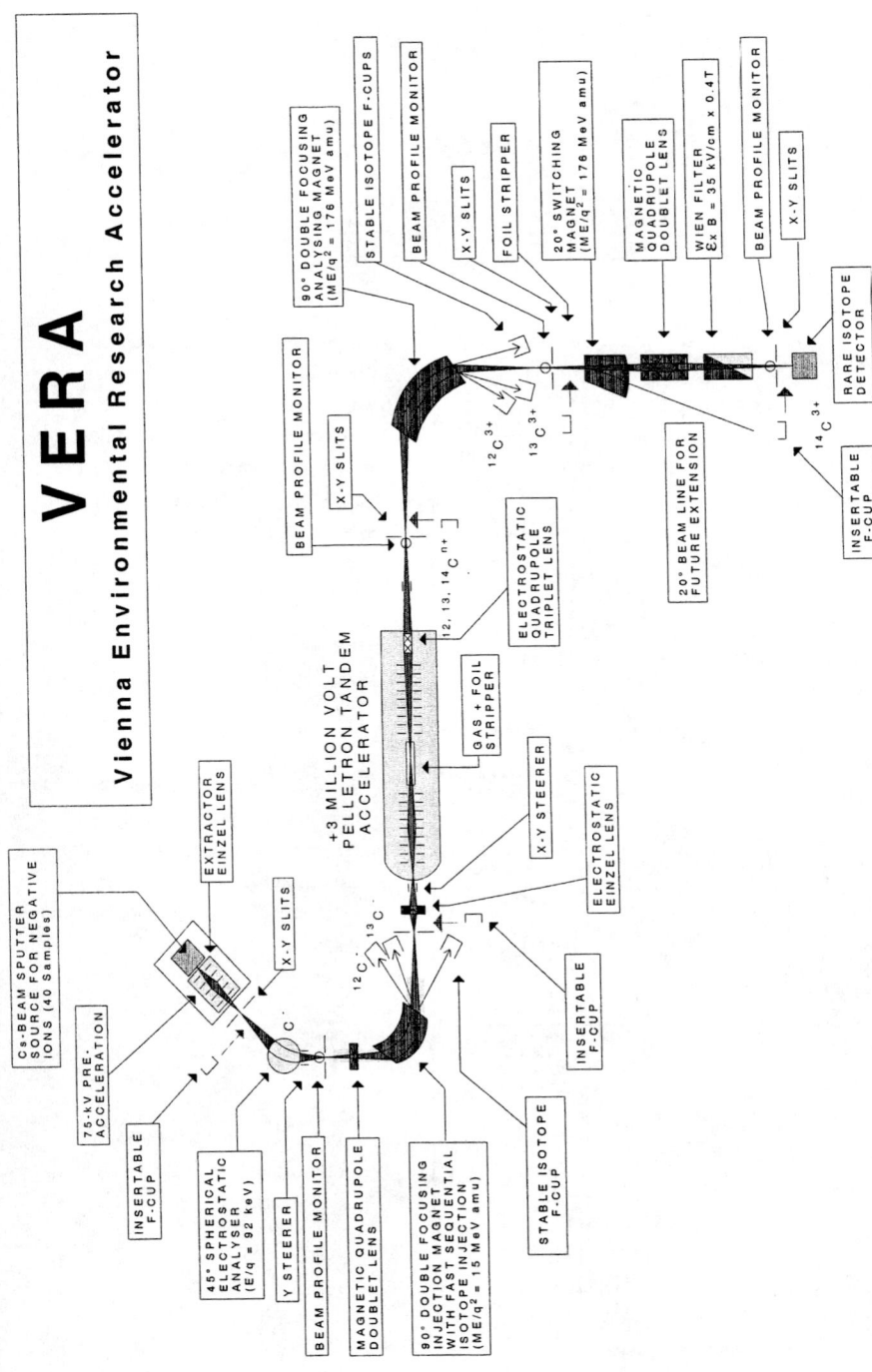

Figure 1. Schematic layout of VERA showing the essential features of the AMS system. ^{14}C measurements are typically performed at a terminal voltage of 2.7 MV

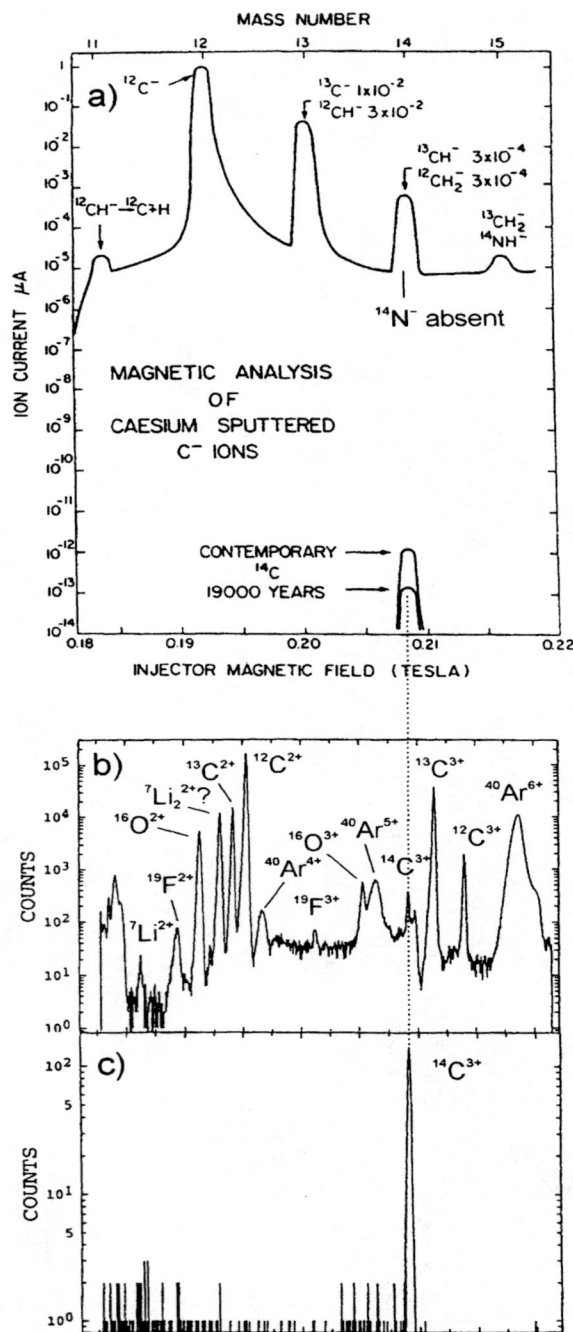

Figure 2. The three steps in the detection of ^{14}C with AMS.

a) In the negative ion mass spectrum after the ion source the ^{14}C signal is buried under an enormous background of mass-14 molecules (16). Most important however, is the absence of $^{14}N^-$ since nitrogen does not form negative ions.

b) After acceleration through the tandem and analysis in the high-energy magnet the energy spectrum measured in a Si surface barrier detector still shows many background peaks.

c) After a final analysis through a velocity filter (Wien filter) a clean ^{14}C signal emerges.

very well using a 0.5 MV Pelletron tandem in a collaborative effort of National Electrostatics Corporation (NEC) and the AMS laboratory of the ETH/PSI Zurich (20). The most surprising result was the measurement of $^{14}C/^{12}C$ ratios down to a level that corresponds to a radiocarbon age of 48,000 years. It therefore looks feasible to perform ^{14}C dating measurements with much smaller tandem accelerators than presently in use, approaching eventually the size of table top machines.

AMS Facilities World-Wide

In Table 2 a summary of AMS facilities around the world is given. It is quite obvious that AMS is almost exclusively performed with tandem accelerators. As mentioned above, AMS originally developed at accelerators in nuclear physics laboratories (1-4). A few years after the initiation of AMS, the first generation of small dedicated AMS facilities (Tandetrons) appeared on the market (21). Eventually, a second and third generation of these machines (3 MV terminal voltage) were developed, and became the workhorses for ^{14}C measurements. Recently, several new AMS facilities based on 5-MV Pelletron tandems were established. Parallel to this development a number of medium-size nuclear physics tandem accelerators were upgraded for AMS measurements. Sometimes, these tandem accelerators were shipped around the world to be re-assembled as dedicated AMS facilities in their new location: e.g. the EN tandem from Oxford in England went to Peking University in China, the FN tandem from Rutgers University in the USA went to ANSTO at Sydney in Australia, the EN tandem from Canberra in Australia went to Lower Hutt in New Zealand, the FN tandem from Washington University went to Livermore both in the USA, and the FN tandem from McMaster University in Canada went to Buenos Aires in Argentina.

The 50 facilities listed in Table 2 measure an estimated total of well over 100,000 samples per year, approximately 90% of it for ^{14}C. Although ^{14}C is by far the most ustilized radionuclide with AMS, many others are gradually increasing in importance (see Table 1). For small tandem facilities (TV $\leq$ 3 MV) there are three radionuclides which can be measuredwith relative ease, because the interfering stable isobars do not form negative ions: ^{14}C ($^{14}N^-$ absent), ^{26}Al ($^{26}Mg^-$ absent), ^{129}I ($^{129}Xe^-$ absent). In addition, beyond ^{209}Bi there is no stable nuclide in nature and thus the probablity of stable isobar interference is greatly reduced. This means that actinides and transuranic elements are also accessible to small AMS facilities (provided the analysing magnets are strong enough to bend such heavy ions). For a radionuclide where there is no primary suppression of the stable isobar, higher energies are very useful to separate the two by utilizing the difference of energy loss in matter (due to their difference in nuclear charge Z). However, for the light radionuclide ^{10}Be, the Z-difference between ^{10}Be and the strongly interfering stable isobar ^{10}B is large enough to allow a separation even at small AMS facilities. It is foreseeable that eventually all radionuclides with a half-life longer than approximately 100 years will be utilized by AMS.

413

Table2. Facilities for Accelerator Mass Spectrometry (1999)

Continent	Country	Accelerator	Location
North America	Canada	2.5 MV Tandetron	University of Toronto, Toronto
	USA	ATLAS Linac	Argonne National Laboratory, Chicago [a]
		2.5 MV Tandetron	University of Arizona, Tucson
		9.5 MV FN Tandem	Lawrence Livermore Nat. Lab, Livermore
		1 MV Pelletron	Lawrence Livermore Nat. Lab, Livermore [b]
		1.5 MV RFQ	Lawrence Livermore Nat. Lab, Livermore [b]
		K1200 Cyclotron	Michigan State Univ., East Lansing [a]
		3 MV Pelletron	Naval Research Lab, Washington D. C. [a]
		3 MV Pelletron	University of North Texas, Denton
		9 MV FN Tandem	Purdue University, West Lafayette
		2.5 MV Tandetron	Woodshole Oceanographic Institution
Europe	Austria	3 MV Pelletron	University of Vienna, Vienna
	Denmark	6 MV EN Tandem	University of Aarhus, Aarhus
	England	2.5 MV Tandetron	University of Oxford, Oxford
		5 MV Pelletron	University of York, Sand Hutton [b],
	France	2.5 MV Tandetron	Nat. Sci. Research Center, Gif-sur-Yvette,
	Germany	6 MV EN Tandem	Univ. of Erlangen-Nuernberg, Erlangen
		3 MV Tandetron	University of Kiel, Kiel
		14 MV MP Tandem	Tech. Univ. & Univ. of Munich, Garching
		3 MV Tandem	Forschungszentrum Rossendorf [a]
	Israel	14 MV Pelletron	Weizmann Institute of Science, Rehovot
	Italy	3 MV Tandem	University of Napels [a]
	Netherlands	3 MV Tandetron	University of Groningen, Groningen
		6 MV EN Tandem	University of Utrecht, Utrecht
	Sweden	6 MV EN Tandem	University of Uppsala, Uppsala
		3 MV Pelletron	University of Lund, Lund
	Switzerland	6 MV EN Tandem	Swiss Federal Inst. of Technology Zürich
		0.5 MV Pelletron	Swiss Federal Inst. of Technology Zürich [b]
Asia	China	14 MV MP Tandem	Chinese Inst. of Atomic Energy, Bejing
		6 MV EN Tandem	Peking University, Beijing
		6 MV Tandem	Shanghai Inst. of Nucl. Res., Shanghai
		Mini Cyclotron	Shanghai Inst. of Nucl. Res., Shanghai [b]
	Japan	3 MV Tandetron	Japan Atomic Energy Res. Inst., Mutsu [b]
		5 MV Pelletron	Japan Nucl. Cycle Develop. Inst., Toki [b]
		8 MV Pelletron	Kyoto University, Kyoto [a]
		10 MV Tandem	Kyushu University, Fukuoka [a]
		2.5 MV Tandetron	Nagoya University, Nagoya
		3 MV Tandetron	Nagoya University, Nagoya [b]
		5 MV Pelletron	University of Tokyo, Tokyo
		5 MV Pelletron	Nat. Inst. for Environ. Studies, Tsukuba
		12 MV Pelletron	University of Tsukuba [a]
	India	3 MV Pelletron	Institute of Physics, Bhubaneswar [a]
	Korea	3 MV Tandetron	Seoul National University, Seoul [b]
Australia & New Zealand	Australia	8 MV FN Tandem	Nucl. Sci. and Technol. Organ., Sydney
		2.5 MV Tandetron	Comm. Sci. and Industr. Res. Organ., Sydney
		14 MV Pelletron	Australian National University, Canberra
	New Zealand	6 MV EN Tandem	Inst. of Geolog.& Nucl. Sci., Lower Hutt
South America	Argentina	20 MV Pelletron	Nat. Atomic. Energy Comm., Buenos Aires [a]
		8 MV FN Tandem	Nucl. Regul. Authority, Buenos Aires [b]
	Brazil	9 MV Pelletron	University of Sao Paulo, Sao Paulo [a]

a) AMS development at existing accelerators.   b) New dedicated AMS facilities in test operation.

SOME "RANDOM" SAMPLES OF AMS APPLICATIONS

Applications of AMS are numerous, and Table 1 indicates that they penetrate essentially every domain of our environment. In the following a few selected examples have been chosen, which are neither representative nor complete. They mainly reflect the current interests of the author. For a more complete information of the field the reader is referred to several review articles and monographs on AMS (6-11), and to the proceedings of the AMS conferences (e.g. 22, 23).

The Iceman "Ötzi"

In 1991, two mountain hikers accidentally discovered the well preserved body of a man emerging from a thawing glacier at a high-altitude (3210 m) mountain pass of the Ötztal Alps near the Austrian-Italian border (24). ^{14}C measurements with AMS determined that "Ötzi", as the Iceman was nicknamed, died about 5200 years ago (25).

Figure 3 summarizes recent results of ^{14}C measurements for equipment of the Iceman, performed at three different AMS laboratories (26). Most of the data agree well with the general time range when the Iceman lived consistent with the original age determination from tissue and bone material of the Iceman himself (25). However, there are two notable exceptions. One object is about 1500 years older, corresponding to the period of mesolithic dwelling sites discovered in high altitudes of the valleys sloping north and south of the finding place. Even more surprising is the much younger date measured on a wooden artifact made of green alder. This piece of wood shows clear working traces and reveals that it was probably brought to this site by prehistoric men during the Hallstatt Period. Both of these findings indicate that the mountain pass where the Iceman perished had also been visited at other times.

The Iceman is clearly a unique object as it is the only fully preserved body from neolithic times. In addition, most of the archaeological information about this time comes from grave sites, whereas the Iceman represents a person taken out from full life. The rich outfit of the Iceman found together with him also gives important information on the neolithic life, and has led to numerous speculations about the origin of "Ötzi". Currently, it is believed that he most likely arrived from the south through one of the valleys leading up to the mountain pass. Whether he was a shepherd or had some other profession (e.g. a farmer from the valley or a mining worker) cannot be decided at this time. Many scientific investigations have been performed on the Iceman and his equipment, and some of them will soon be available in a forthcoming monograph (27). As indicated by the two exceptional dates in Figure 3, the finding place of the Iceman may have been visited also at other times, indicating other warm periods. Through ^{14}C dating of botanically identified species extracted from sediments at the Iceman finding place one hopes to find clues about climatic variations in the past.

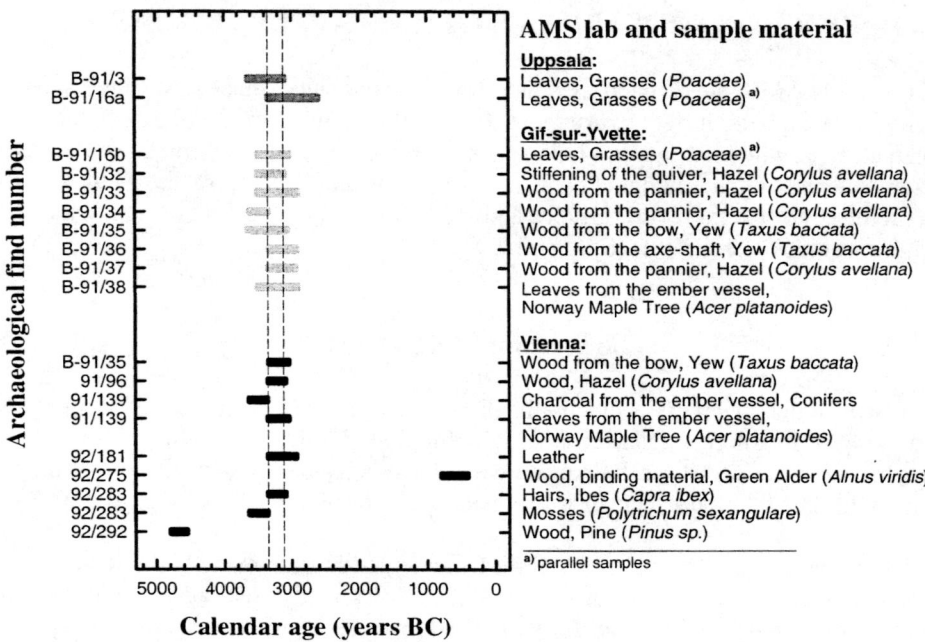

Figure 3. Summary of radiocarbon dating results for samples from the Iceman's equipment, measured at three different AMS laboratories (26). Horizontal bars indicate 2σ ranges for the calibrated dates (95% confidence). The dashed vertical lines show the 2σ range obtained from tissue and bone samples measured at the AMS facility in Zurich (25).

The Great Ocean Conveyor Belt

CO_2 is highly soluble in water and the volume of the oceans is very large. As a consequence, much more CO_2 is stored in the oceans than in the atmosphere. Correspondingly, the oceans also store the largest fraction of ^{14}C on Earth (~93%). Through continuous exchange with the atmosphere the ocean surface water approximately reflects the atmospheric $^{14}C/^{12}C$ ratio, whereas the deep ocean water shows a reduced ratio depending on how long it had been disconnected from exchange with the surface. A detailed study of apparent age differences between surface and deep water around the globe revealed the global circulation pattern shown in Figure 4.

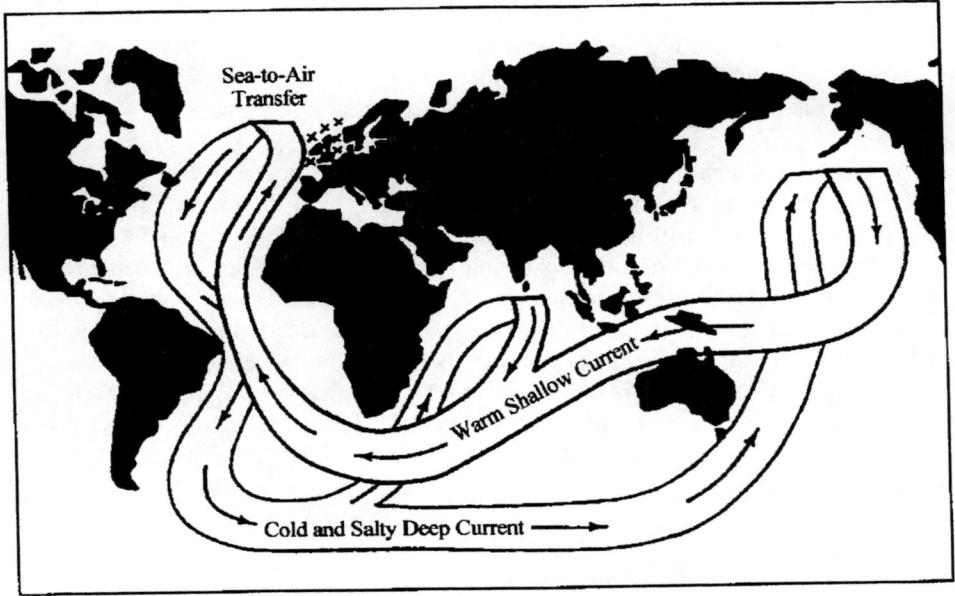

Figure 4. The "Great Ocean Conveyor Belt" describing in a simplified way the global ocean circulation (28). This pattern emerges from measuring the difference between $^{14}C/^{12}C$ ratios in surface water and deep-sea water, which increases as the deep-water ages on its way from the North Atlantic to the Pacific Ocean. The area for ^{129}I releases from reprocessing plants in England and France is marked with crosses (x).

According to this model (28), warm and shallow surface currents transport heat from aequatorial regions to the North Atlantic. After giving off heat and moisture, the cooler water becomes more salty and more dense and sinks into the deep ocean. It then returns as a cold and salty current to close the "conveyor belt" loop. The detailed global pattern is more complex (29) than indicated in Figure 4. In order to get a more complete understanding of the ocean circulation, a dedicated AMS facility was set up at the Woods Hole Oceanographic Institution (30, 31). In a few years approximately 14,000 ocean water samples will have been measured within the World Ocean Circulation Experiment (WOCE), supplying a wealth of information on global ^{14}C and $\delta^{13}C$ distributions (the δ notation indicates a measure of stable isotope ratios).

There is growing evidence from $\delta^{18}O$ fluctuations observed in ice cores that abrupt temperature changes happened in the past, possibly caused by switching the conveyor belt on and off (32, 33). An interesting new approach to oceanographic tracing was suggested a few years ago (34) utilising the long-lived fission product ^{129}I ($t_{1/2} = 1.7 \times 10^7$ y) accumulated in spent nuclear fuel rods. Since approximately 1000 kg of ^{129}I were released into the ocean from the two nuclear fuel reprocessing plants in

417

La Hague (F) and Sellafield (UK) over the past 20 years or so, a strong "point" source both in space and time (see Figure 4) was inadvertently set free. What first seemed to be only a menace, now turns out to be a valuable tracer for Oceanography (34). The importance of this tracer can be assessed by comparing the total number of ^{129}I atoms released (5×10^{27}) with the total volume of the world oceans (1.3×10^{21} litre). Since the natural pre-bomb ^{129}I concentration in ocean water is in the order of 10^5 atoms/litre H_2O (35), even complete mixing of the released material with the ocean will leave a strong ^{129}I signature per litre of water. With AMS it is quite possible to make a $^{129}I/^{127}I$ ratio measurement with the 4 million fissiogenic ^{129}I atoms in one liter of ocean water. This example illustrates the severe influence of man on the natural atom budget of a specific radionuclide. However, it should also be noticed that due to the very long half-life of ^{129}I, the total radioactivity (175 Ci) does not constitute a radiation hazard, and that the radioactivity of one liter of ocean water after complete mixing is only a single beta-decay in six years! Very recently ^{99}Tc ($t_{1/2} = 2.11 \times 10^5$ y), a fission product also partly released into the environment, may become another anthropogenic tracer measured with AMS.

Bomb-peak dating

Starting around 1950, the period of intense nuclear weapons testing in the atmosphere increased ^{14}C considerably, because the neutrons released in the explosions converted ^{14}N into ^{14}C in the same way as cosmic-ray secondary neutrons do (n + ^{14}N → ^{14}C + p). At the time of the Nuclear Test Ban Treaty in 1963, the atmospheric ^{14}C had doubled. Since then the ^{14}C redistributed into the other reservoirs (ocean and biosphere), leading to a gradual decrease of the excess atmospheric ^{14}C concentration. At present the rate of decrease of atmospheric ^{14}C is about 1% per year. This is considerably faster than the decrease by radioactive decay (1% in 83 years), and thus allows for "dating" of much shorter time periods. Figure 5 illustrates how this can be used for determining the recent time of death of humans, which is of interest in forensic medicine (36). The comparison of the ^{14}C concentration in the collagen and in the lipid fraction of bone clearly shows an affect of the longer residence time of ^{14}C in collagen. The more mobile lipids resemble the atmospheric ^{14}C concentration at the time of death within a few years and are thus reasonably well suited for this purpose.

It is possible that a detailed understanding of the residence time of ^{14}C in collagen may eventually allow an improvement of dating fossil bones. Provided that one can somehow determine the age of a bone itself (e.g. from anatomical evidence), an age-dependent correction of the measured ^{14}C age for the collagen used for dating may be applied.

Bomb-peak dating of 30-year-old man

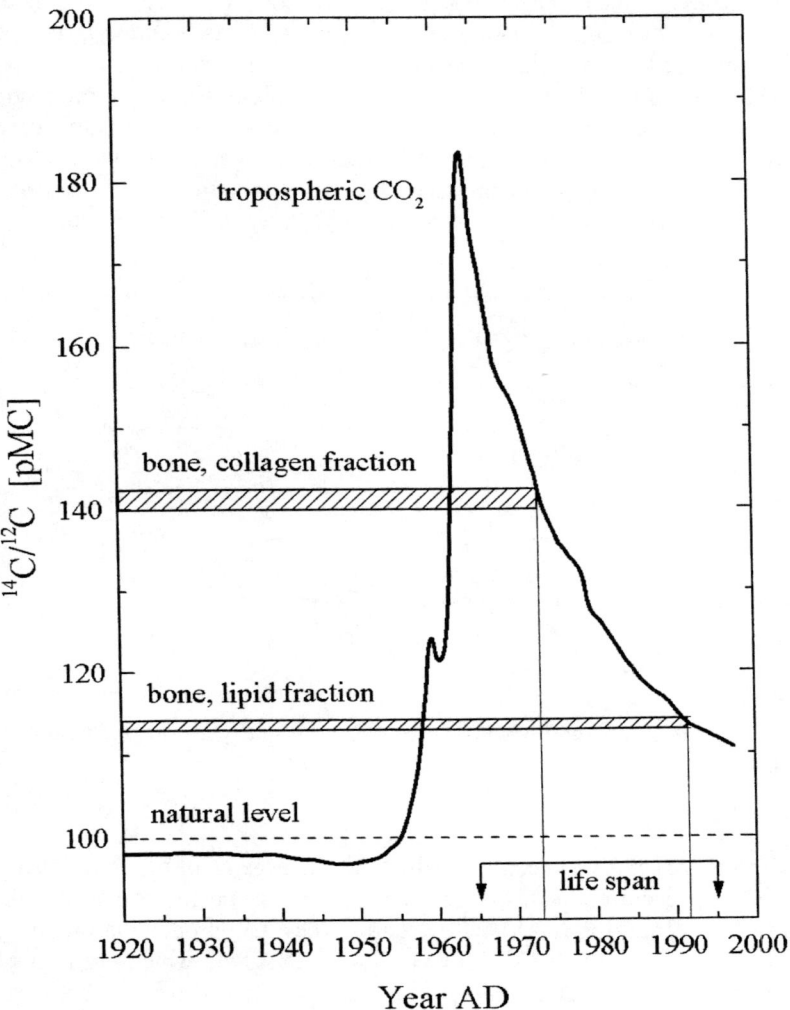

Figure 5. Measured $^{14}C/^{12}C$ ratios in tropospheric CO_2 given in percent Modern Carbon (pMC). Before 1950, the dilution with ^{14}C-free CO_2 from fossil-fuel burning (Suess effect) reduces the $^{14}C/^{12}C$ ratio below natural level. The strong increase after 1955 is due to atmospheric nuclear testing. The 2σ ranges for the $^{14}C/^{12}C$ ratios measured in collagen and lipid fractions of a long bone from a 30-year old man of known life span are indicated (36).

419

Atmospheric ^{14}CO

When ^{14}C is formed through the nuclear reaction of cosmic-ray secondary neutrons with ^{14}N, the "hot" (electronically excited) atom quickly reacts with oxygen in the air to form ^{14}C monoxide (Fig. 6). After a mean residence time of several months, most of the ^{14}CO is oxidized to CO_2 by reacting with the OH radical. While the production of ^{14}CO is essentially constant over time periods in the order of months or years (constancy of cosmic ray intensity), the OH radical concentration varies with the intensity of sunlight, because it is produced by the photolytic dissociation of ozon followed by the reaction of an excited oxygen atom with water vapor (Fig. 6).

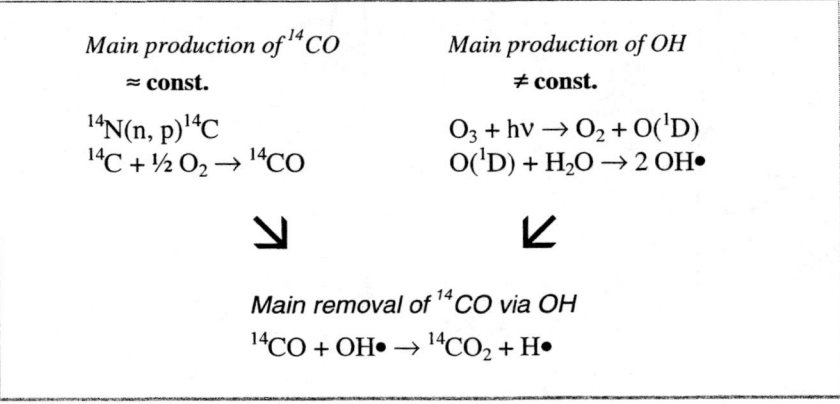

Figure 6. Principle of the coupling between ^{14}CO and OH in the atmosphere adapted from Rom et al. (37).

The extremely reactive OH molecule is called the "detergent" of the atmosphere (38). Measuring ^{14}CO concentration in the atmosphere allows one to determine indirectly the OH-radical concentration (provided one knows the ^{14}CO production rate in air and the chemical reaction rate of ^{14}CO with OH). Such a program, which has been started with great effort in the pre-AMS era (39), can now be pursued on a global basis (40) since the minute concentrations of ^{14}CO (6-30 molecules per cm^3 air) require small-sample AMS. This program headed by the Max Planck Institute for Chemistry in Mainz aims towards the complete isotopic specification of CO ($^{14}C/^{12}C$, $\delta^{13}C$, $\delta^{17}O$, $\delta^{18}O$). Eventually this will lead to a wealth of information on trace gas chemistry in the atmosphere, important for modeling the great atmospheric "reactor". Results of AMS measurements of ^{14}CO concentrations in air collected during a two-year period at a high-altitude station in the Austrian Alps are shown in Figure 7. The depletion of ^{14}CO in summer is evident . It is due to the higher OH radical concentration since there is more sunshine available for producing OH (see Fig. 6).

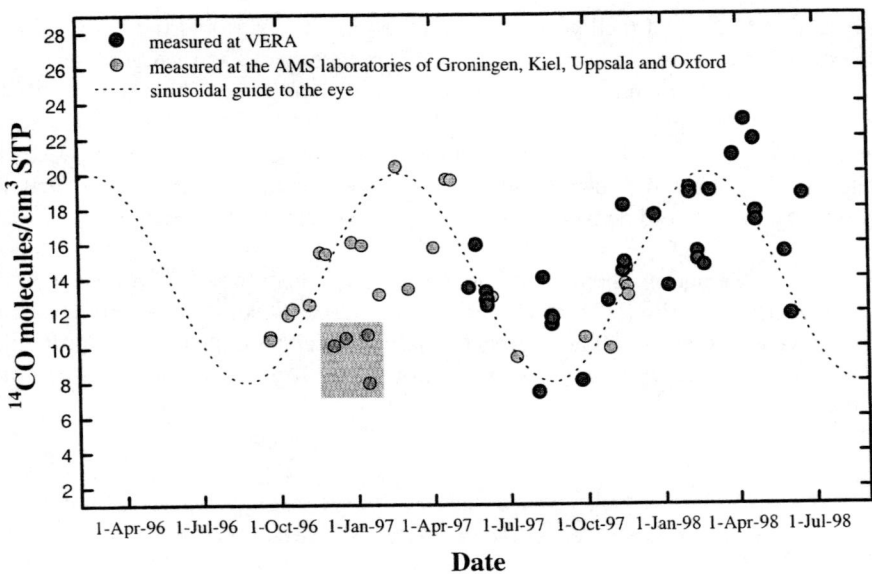

Figure 7. Atmospheric ^{14}CO data obtained from air samples taken at Mt. Sonnblick (3106 m) in 1996-98. The main feature (see dashed line) of this first detailed 2-year record for the whole temperate northern hemisphere can be explained by the OH radical sink. Since the overall precision has been determined to be 1/4 ^{14}CO molecule per cm^3 of air, the scatter of the data reflects the general complexity of atmospheric conditions. In particular, the low points in January 1997 are probably due to transport of air from southern regions (verified by backward trajectory calculations), where more sunshine and higher H_2O content leads to a depletion of ^{14}CO (37, 41).

Dating with ^{41}Ca

The long half-life of ^{41}Ca ($t_{1/2} = 1.04 \times 10^5$ y) and its presence in bone inspired attempts (42) to use it for archaeological dating in analogy to radiocarbon, but with the prospect to go back much further in time. Unfortunately, these attempts failed primarily because ^{41}Ca is distributed unevenly in the environment. The cosmogenic production happens through the capture of secondary thermal neutrons on ^{40}Ca at the surface of the earth. Without forming a chemically long-lived gas like ^{14}CO$_2$ that provides a well-mixed reservoir, it is perhaps not surprising that ^{41}Ca/^{40}Ca ratios were found to vary by more than an order of magnitude in modern samples (43, 44). Clearly these conditions will probably not allow one to establish a globally valid calibration curve such as the tree-ring calibration for ^{14}C. It is, however, intriguing to speculate about an absolute dating technique for ^{41}Ca, i. e. a method where both the parent (^{41}Ca) and the daughter (^{41}K) are measured, similar to the well-known ^{40}K-^{40}Ar dating method. In general it looks like a hopeless task to identify the minute amount of radiogenic ^{41}K in the presence of an overwhelming background of stable ^{41}K. However, the recoil energy of ^{41}K after the electron-capture decay of ^{41}Ca is very small (~2.3 eV). It may therefore be possible to separate radiogenic from non-radiogenic ^{41}K in the relevant bone material (calcium phosphate) by some yet to be found sophisticated crystallographic method.

In addition to ^{41}Ca dating, there are various useful applications of ^{41}Ca measurements in extraterrestrial materials such as lunar rocks and meteorites (45). Last not least, very interesting venues open to ^{41}Ca tracing of the calcium metabolism in living humans (46, 47). The extremely low radiation dose allows long-time experiments with humans never thought feasible before the invention of AMS.

Exposure Dating with ^{10}Be and ^{26}Al

Primary cosmic rays produce various secondary particles through nuclear reactions in the atmosphere. At the surface of the Earth the secondary particle flux is dominated by neutrons and muons. These particles react with the nuclei in the surface of the rocks producing long-lived radionuclides. If the rock surface was shielded long enough (e.g. by ice or water coverage) before the exposure to cosmic rays started, then the build-up of radionuclides with a known production rate can be used to obtain the time of exposure (48). A particularly useful mineral for this purpose is quartz (SiO_2), since it is abundant in rocks, exists naturally in chemically pure form, and allows the production of two important radionuclides: ^{10}Be and ^{26}Al. Typical production rates in quartz at sea level and high latitudes (>60° N) are (49): 6 ^{10}Be atoms per gram of SiO_2 per year, and 37 ^{26}Al atoms per gram SiO_2 per year. At an altitude of 3000 m both values increase by a factor of 10.

An interesting application of this method was the determination of the age of the Meteor Crater in Arizona, the largest known crater on Earth with associated meteorites (Canyon Diablo). From *in-situ* production of ^{10}Be and ^{26}Al a lower bound on the crater age of 49,000 ± 1,700 years was determined (50). Another impressive application of exposure dating was recently reported by earth scientists from Los Angeles, who gained information on the incision rate of the Indus river in the northwestern Himalayas (51). For the exposure dating the abandoned river bed was used and the exposure time was calculated from the measured ^{10}Be and ^{26}Al concentration in quartz and the known production rates. The incision rates then follow from these exposure times and the height of the abandoned river bed above the current surface of the river. Very high incision rates in the order of a few millimeter per year were found, apparently counterbalanced by a strong bedrock uplift of the Himalayas, resulting in an equilibrium state of this rapidly deforming region.

AMS with Positive Ions: the Detection of ^{39}Ar and ^{81}Kr

In contrast to the applications described so far, noble gas radionuclides cannot be detected at tandem accelerators because they do not form negative ions. Therefore AMS with positive ions has been initiated (52) to develop techniques for the detection of ^{39}Ar ($t_{1/2}$ = 268 y) and ^{81}Kr ($t_{1/2}$ = 230,000 y). The extremely low isotope ratio of ^{39}Ar/^{40}Ar in the atmosphere (8×10^{-16}) prevented any AMS application of this radionuclide so far. However, it would be very useful as an oceanographic tracer and thus constitutes a strong challenge for future AMS developments.

Sometimes very big accelerators are necessary to measure particular radionuclides. This was the case in developing an AMS method for measuring cosmogenic ^{81}Kr ($t_{1/2}$ = 230,000 yr) in the atmosphere (53, 54), and in ground water (55). The experiments were performed at a positive ion machine, the K1200 superconducting cyclotron at Michigan State University. In order to get rid of the stable isobar ^{81}Br, which strongly interferes with ^{81}Kr ($\Delta M/M$ = 3.7×10^{-6}), 17+ ions from the superconducting ECR source were accelerated to an energy of 45 MeV/nucleon (3.65 GeV). At this high energy, 80% of the ^{81}Kr ions can be fully stripped to the 36+ charge state and separated in a magnetic spectrometer from fully stripped ^{81}Br, which can only acquire a maximum charge of 35+. In order to measure small Kr gas samples (~0.4 cm^3 STP), a special gas handling system was developed (53, 54) and a comparison with pre- and post-nuclear krypton was performed (54). Since no difference between the two Kr sources was found, a first ^{81}Kr dating of groundwater from the Great Artesian Basin in Australia, the largest Artesian groundwater system in the world, was attempted (55).

Table 3. Summary of ^{81}Kr/Kr isotopic ratios and deduced ages for groundwater samples of the Great Artesian Basin.

Sample	^{81}Kr/Kr (10^{-13})	Mean residence time (10^3 y)
Raspberry Creek	2.63 ± 0.32	225 ± 42
Oodnadatta	1.78 ± 0.26	354 ± 50
Duck Hole	2.19 ± 0.28	287 ± 38
Watson Creek	1.54 ± 0.22	402 ± 51
Atmospheric Kr	5.20 ± 0.40	0

Four samples of 16,000 l of groundwater each were degassed in the field and the extracted gas (320,000 cm^3/sample) were subjected to a rigorous separation procedure at the University of Bern. This resulted in 0.5 cm^3 Kr per sample containing approximately 3 million ^{81}Kr atoms. Typically, 60 to 100 ^{81}Kr^{36+} ions could be counted in the final detection system, resulting in an overall efficiency of ~2x10^{-5} (atoms detected/atoms in the sample). In Table 3, results for the measured groundwater ages are listed.

Although the overall efficiency of ^{81}Kr detection with AMS is a factor of 1000 lower than the one typical achieved for ^{14}C measurements, it was possible to obtain a definite result for very old groundwater samples. Clearly, a substantial improvement in efficiency would be desirable to start "routine" measurements for groundwater samples.

CONCLUSION

Hopefully this brief review of the methodology and applications of AMS was able to convey the feeling for the enormous breadth of information one can gather with AMS measurements. Combined with high-precision stable isotope measurements, this constitutes the "isotope language", which may allow us one day to disentangle even very complex processes in the environment. Since we can reasonably expect that the power of both AMS and stable isotope MS will increase with time, a bright future for these fields lies ahead of us.

REFERENCES

1. Purser, K. H., Liebert, R. B., Litherland, A. E., Beukens, R. P., Gove, H. E., Bennet, C. L., Clover, H. R., and Sondheim, W. E., "An attempt to detect stable N⁻ ions from a sputter ion source and some implications of the results for the design of tandems for ultrasensitive carbon analysis", *Revue de Physique Appliquée* **12**, 1487-1492 (1977).

2. Muller, R. A., "Radioisotope dating with a cyclotron", *Science* **196**, 489-494 (1977).

3. Nelson, D. E., Korteling, R. G. and Stott, W. R., "Carbon-14: Direct detection at natural concentrations", *Science* **198**, 507-508 (1977).

4. Bennet, C. L., Beukens, R. P., Clover, M. R., Gove, H. E., Liebert, R. B., Litherland, A. E., Purser, K. H.,and Sondheim, W., "Radiocarbon dating using electrostatic accelerators: negative ions provide the key", *Science*, **198**, 508-510.

5.   Oeschger, H., Houtermans, J., Loosli, H., and Wahlen, M., "The constancy of cosmic radiation from isotope studies in meteorites and on the Earth", in *12th Nobel Symposium on Radiocarbon Variations and Absolute Chronology* , ed. Olssen, I. U., New York: John Wiley & Sons, 1970, pp. 471-498.

6.   Litherland, A. E., "Ultrasensitive mass spectrometry with accelerators", *Ann. Rev. Nucl. Part. Sci.* **30**, 437-473 (1980).

7.   Elmore, D. and Philiips, F. M., "Accelerator mass spectrometry for measurement of long-lived isotopes", *Science*, **236**, 543-550 (1987).

8.   Kutschera, W. and Paul, M., "Accelerator mass spectrometry in nuclear physics and astrophysics", *Ann. Rev. Nucl. Part. Sci.*, **40**, 411-438 (1990).

9.   Finkel, R. C. and Suter M., "AMS in the earth sciences: techniques and applications", *Advances in Anal. Geochem.* **1**, 1-114 (1993).

10.   Tuniz, C., Bird, J. R., Fink, D., and Herzog, G. F., *Accelerator mass spectrometry: ultrasensitive analysis for global science*, Boca Raton: CRC Press, 1998, pp. 1-371.

11.   Gove H. E., *From Hiroshima to the Iceman: the development and applications of accelerator mass spectrometry*, Bistol and Philadelphia: Institute of Physics Publishing, 1999, pp. 1-226.

12.   Kutschera, W., Collon, P., Friedmann, H., Golser, R., Hille., P., Priller, A., Rom, W., Steier, P., Tagesen, S., Wallner, A., Wild, E. and Winkler, G., "VERA: A new AMS facility in Vienna", *Nucl. Instr. Meth.* B **123**, 47-50 (1997).

13.   Priller, A., Golser, R., Hille, P., Kutschera, W., Rom, W., Steier, P., Wallner, A. and Wild, E., "First performance tests of VERA", *Nucl. Instr. Meth.* B **123**, 193-198 (1997).

14.   Rom, W., Golser, R., Kutschera, W., Priller, A., Steier, P. and Wild, E., "Systematic investigations of ^{14}C measurements at the Vienna Environmental Research Accelerator", *Radiocarbon*, **40/1**, 255-263 (1998).

15.   Wild, E., Golser, R., Hille, P., Kutschera, W., Priller A., Puchegger, S., Rom, W. and Steier, P., "First ^{14}C results from archaeological and forensic studies at the Vienna Environmental Research Accelerator", *Radiocarbon*, **40/1**, 273-281 (1998).

16.   Litherland, A. E., "Accelerator mass spectrometry", *Nucl. Instr. Meth.* B **5**, 100-108 (1984).

17.   Suter, M., Jacob, St. and Synal, H. A., "AMS of ^{14}C at low energies", *Nucl. Instr. Meth.* B **123**, 148-152 (1997).

18.   Mous, D. J. W., Purser, K. H., Fokker, W., van den Broek, R. and Koopmans, R. B., "A compact ^{14}C isotope ratio mass spectrometer for biomedical applications", *Nucl. Instr. Meth.* B **123**, 153-158 (1997).

19.   Mous, D. J. W., Fokker, W., van den Broek, R. and Koopmans, R. B., "An ion source for the HVEE ^{14}C isotope ratio mass spectrometer for biomedical applications", *Radiocarbon* **40/1**, 283-288 (1998).

20. Suter, M., Huber, R.., Jacob, S. A. W., Synal, H. A. and Schroeder, J. B., "A new small accelerator for radiocarbon dating ", Proceedings of the International Conference on the Application of Accelerators in Research and Industry, Denton, Texas, 1988, to be published.

21. Purser, K. H., Schneider, R. J., J. Dobbs, J. McG., and Post, R., "A preliminary description of a dedicated commercial ultra-sensitive mass spectrometer for direct atom counting of [14]C", Proc. Sympcsium on Accelerator Mass Spectrometry, *Argonne National Laboratory Report ANL/PHY-81-1*, 431-462 (1981).

22. Jull, A. J., Beck, J. W., and Burr, G. S., eds., Proceedings of the 7th International Conference on Accelerator Mass Spectrometry, Tucson, Arizona, USA, 20-24 May 1996, *Nucl. Instr. Meth. B* **123**, 1-612 (1997).

23. Kutschera, W., Golser, R., Priller, A., and Strohmaier, B., eds., Proceedings of the 8th International Conference on Accelerator Mass Spectrometry, Vienna, Austria, 6-10 September 1999, to be published in *Nucl. Instr. Meth B*.

24. Spindler, K., "The Iceman's last weeks", *Nucl. Instr. Meth. B* **92**, 274-281 (1994).

25. Bonani, G., Ivy, S., Hajdas, I., Niklaus, Th. R., and Suter, M., "AMS ^{14}C age determination of tissue, bone and grass samples from the Ötztal Iceman", *Radiocarbon* **36/2** (1994) 247-250.

26. Rom, W., Golser, R., Kutschera, W., Priller, A., Steier, P., and Wild, E. M., "AMS ^{14}C dating of equipment from the Iceman and of spruce logs from the prehistoric salt mines of Hallstatt", *Radiocarbon* **41/2** (1999), in press.

27. Bortenschlager S. and Oeggl K., eds., *The Iceman and his natural environment*, Wien/New York: Springer-Verlag, to be published

28. Broecker, W. S., "The great ocean conveyor", *Oceanography* **4/2**, 79-89 (1991).

29. Broecker, W. S., "Chaotic climate", Scientific American (November 1995) 44-50.

30. Von Reden, K. F., McNichol, A. P., Peden, J. C., Elder, K. L., Gagnon, A. G., and Schneider, R. J., "AMS measurements of the ^{14}C distribution in the Pacific Ocean", *Nucl. Instr. Meth. B* **123**, 438-442 (1997).

31. Von Reden, K. F., Peden, J. C., Schneider, R. J., Bellino, M., Donoghue, J., Elder, K. L., Gagnon, A. G., Long P., McNichol, A. P., Morin T., Stuart, D.,and Hayes, J. M., "High-precision measurements of ^{14}C as a circulation tracer in the Pacific, Indian, and Southern Oceans with accelerator mass spectrometry (AMS)", Proceedings of the 8th International Conference on Heavy Ion Accelerator Technology, Argonne Natinal Laboratory, Illinois, USA, 5.9 October 1998, *AIP Conf. Proc.* **473**, 410-421 (1999).

32. Broecker, W.S:, "Thermohaline circulation, the Achilles heel of our climate system: will man-made CO_2 upset the current balance?", *Science* **278**, 1582-1588 (1997).

33. Barber, D. C., Dyke, A., Hillaire-Marcel, C., Jennings, A. E., Andrews, J. T., Kerwin, M. W., Bilodeau, G., McNeely, R., Southon, J., Morehead, M. D., and Gagnon, J.-M., "Forcing of the cold event of 8,200 years ago by catastrophic drainage of Laurentide lakes", *Nature* **400**, 344-348 (1999).

34. Yiou, Y., Raisbeck, G.M., Zhou, Z. Q., and Kilius, L. R., "^{129}I from nuclear fuel reprocessing; potential as an oceanographic tracer", *Nucl. Instr. Meth. B* **92**, 436-439 (1994).

35. Fabryka-Martin, J., Bentley, H., Elmore, D., and Airey, P. L., "Natural iodine-129 as an environmental tracer", *Geochim. Cosmochim. Acta* **49**, 337-347 (1985).

36. Wild, E., Golser, R., Hille, P., Kutschera, W., Priller, A., Puchegger, S., Rom, W., and Steier, P., "First ^{14}C results from archaeological and forensic studies at the Vienna Environmental Research Accelerator", *Radiocarbon* **40/1**, 273-281 (1998).

37. Rom, W., Brenninkmeijer, C. A. M., Bräunlich, M., Golser, R., Mandl, M., Kaiser, A., Kutschera, W., Priller, A., Puchegger S., Röckmann, Th., and Steier, P., "The "CO-OH-Europe" project and measurements of ^{14}C monoxide concentrations in air from the high-altitude observatory Sonnblick (3106 m) in the Austrian Alps", *Proceedings of the International Workshop on Frontiers in Accelerator Mass Spectrometry*, 6-8 January 1999, National Institute of Environmental Science, Tsukuba, and National Museum of Japanes History, Sakura, Japan, pp. 228-243.

38. Graedel, Th. E. and Crutzen, P., *Chemie der Atmosphäre: Bedeutung für Klima und Umwelt*, Heidelberg: Spektrum Akademischer Verlag, 1994, pp. 1-511.

39. Volz, A., Ehalt, D. H., and Derwent, R. G., "Seasonal and latitudinal variation of ^{14}CO and the tropospheric concentration of OH radicals", *J. Geophys. Res.* **86**, 5163-5171 (1981).

40. Brenninkmeijer, C. A. M., "Measurement of the abundance of ^{14}CO in the atmosphere and the ^{13}C/^{12}C and ^{18}O/^{16}O ratio of atmospheric CO with applications in New Zealand and Antarctica", *J. Geophys. Res.* **98**, 10,595-10,614 (1993).

41. Rom, W., Brenninkmeijer, C. A. M., Bräunlich, M., Golser, R., Mandl, M., Kaiser, A., Kutschera, W., Priller, A., Puchegger, S., Röckmann, Th., and Steier, P., A detailed 2-year record of atmospheric ^{14}CO in the temperate northern hemisphere, Proceedings of the Conference on Ion Beam Analysis (IBA-14)/ European Conference on the Application of Accelerators in Research and Technology (ECAART-6), 26-30 July 1999, Dresden, Germany, to be published in *Nucl. Instr. Meth. B*.

42. Henning, W., Bell, W. A., Billquist, P. J., Glagola, B. G., Kutschera, W., Liu, Z., Lucas, H. F., Paul, M., Rehm, K. E., and Yntema, J. L., "Calcium-41 concentration in terrestrial materials: prospects for dating of pleistocene samples", Science **236**, 725-727 (1987).

43. Kutschera, W., Ahmad, I., Billquist, P. J., Glagola, B. G., Furer, K., Pardo, R. C., Paul, M., Rehm, K. E., Slota, P. J., Taylor, R. E., and Yntema ,J. L., "Studies towards a method for radiocalcium dating of bones", *Radiocarbon* **31/3**, 311-323 (1989).

44. Kutschera, W., "Accelerator mass spectrometry: a versatile tool for research", *Nucl. Instr. Meth, B* **50**, 252-261 (1990).

45. Fink, D., "^{41}Ca: past, present and future", *Nucl. Instr. Meth. B* **52**, 572-582 (1990).

46. Johnson, R. R., Berkovits, D., Boaretto, E., Gelbart, Z., Ghelberg, S., Meirav, O., Paul, M., Prior, J., Sossi, V., and Venczel, E., "Calcium resorption from bone in a human studied by ^{41}Ca tracing", *Nucl. Instr. Meth. B* **92**, 483-488 (19994).

47.   Freeman, S. P. H. T., King, J. C., Vieira, N. E., Woodhouse, L. R., and Yergey, A. L., "Human calcium metabolism including bone resorption measured with ^{41}Ca", *Nucl. Instr. Meth. B* **123**, 266-270 (1997).

48.   Lal, D., "Cosmic ray labeling of erosion surfaces: *in situ* nuclide production rates and erosion models", *Earth Planet. Sci. Lett.* **104**, 424-439 (1991).

49.   Nishiizumi, K., Winterer, E. L., Kohl, C. P., Klein, J., Middleton, R., Lal, D., and Arnold, J. R., "Cosmic ray production rates of ^{10}Be and ^{26}Al in quartz from glacially polished rocks", *J. Geophys. Res.* **94**, 17,907-17,915 (1989).

50.   Nishiizumi, K., Kohl, C. P., Shoemaker, E. M., Arnold, J. R., Klein, J., Fink, D., and Middleton, R., "*In situ* ^{10}Be-^{26}Al exposure ages at Meteor Crater, Arizona", *Geochim. Cosmochim, Acta* **55**, 2699-2703 (1991).

51.   Burbank, D. W., Leland, J., Fielding, E., Anderson, R. S., Brozovic, N., Reid, M. R., and Duncan, Ch., "Bedrock incision, rock uplift and threshold hillslopes in the northwestern Himalayas", *Nature* **379**, 505-510 (1996).

52.   Kutschera, W., Paul, M., Ahmad, I., Antaya, T. A., Billquist, P. J., Glagola, B. G., Harkewicz, R., Hellstrom, M., Morrissey, D. J., Pardo, R. C., Rehm, K. E., Sherrill, B. M., and Steiner, M, "Long-lived noble gas radionuclides", *Nucl. Instr. Meth. B* **92**, 241-248 (1994).

53.   Collon, P., Antaya, T. A., Davids, B., Fauerbach, M., Harkewicz, R., Hellstrom, M., Kutschera, W., Morrissey, D. J., Pardo, R. C., Paul, M., Sherrill, B. M., and Steiner, M., "Measurement of ^{81}Kr in the atmosphere ", *Nucl. Instr. Meth. B* **123**, 122-127 (1997)

54.   Collon, P., Cole, D., Davids, B., Fauerbach, M., Harkewicz, R., Kutschera, W., Morrissey, D. J., Pardo, R., Paul, M., Sherrill, B. M. and Steiner, M., "Measurement of the long-lived radionuclide 81Kr in pre-nulcear and present-day atmospheric krypton", *Radiochimica Acta* **85**, 13-19 (1999).

55.   Collon, P., Kutschera, W., Loosli, H. H, Lehmann, B. E., Purtschert, R., Love, A., Sampson, L., Anthony, D., Cole, D., Davids, B., Morrissey, D. J., Sherrill, B. M., Steiner, M., Pardo, R. C., and Paul, M., "^{81}Kr in the Great Artesian Basin, Australia: a new method for dating yery old groundwaters", to be published.

Application of Multidimensional Spectrum Analysis for Analytical Chemistry

Yuichi Hatsukawa, Takehito Hayakawa, Yosuke Toh, Nobuo Shinohara
and Masumi Oshima

Japan Atomic Energy Research Institute,

Tokai, Ibaraki 319-1195, Japan

Abstract. Feasibility of application of the multidimensional γ ray spectroscopy for analytical chemistry was examined. Two reference igneous rock (JP-1, JB-1a) samples issued by the Geological Survey of Japan (GSJ) were irradiated at a research reactor with thermal neutrons, and γ rays from the radioisotopes produced by neutron capture reactions were measured using a γ–ray detector array. Simultaneously 27 elements were observed with no chemical separation.

Multidimensional spectroscopy methods are widely used in the field of the nuclear spectroscopic studies, and have produced many successful results. In order to scrutinize the ability of multidimensional gamma ray method for analytical chemistry, minor elements in reference rock samples, each element of which is already determined by various methods, were measured. In the case of neutron activation analysis of geo-chemical specimen, measurements of gamma rays from minor elements are strongly interfered by the gamma rays from ^{24}Na and ^{56}Mn. So usually chemical separation processes are required to get rid of the sodium and manganese for determination of the minor elements in geo-chemical samples. In this study, we tried to get rid of influence of the gamma rays from ^{24}Na and ^{56}Mn, and measured the minor elements in the reference rock samples by using multidimensional gamma ray spectrum.

Two reference igneous rock samples (JP-1(peridotite), JB-1a(basalt)) issued by the Geological Survey of Japan (GSJ) were used in this study. Compilation data for major, minor and trace elements in GSJ reference samples of JP-1 and JB-1a are complied in the reference [1]. Neutron irradiation of the samples was performed at JRR-4 Research Reactor of the Japan Atomic Energy Research Institute(JAERI). The rock samples were irradiated for 10 min. Multiple γ-rays from the radioisotopes produced by neutron capture reactions were measured using an array[2] of 12 germanium detectors with BGO Compton suppressors, GEMINI, for about 2 days for JP-1 and 4 days JB-1a. The γ–γ coincidence relationships were derived from a 4k x 4k matrix. Simultaneously 27 elements were observed without chemical separation.

CP495, *Experimental Nuclear Physics in Europe*, edited by B. Rubio et al.

Only 4 ppb of Eu contained in JP-1 was also detected (see in Fig.1). Although there are no previous data for contents of bromine in the JP-1 and JB-1a samples, bromine in the both samples were detected for the first time in this study (see in Fig.2). The contents of bromine are determined as 0.7 ppm for JB-1a and 1.1 ppm for JP-1, respectively.

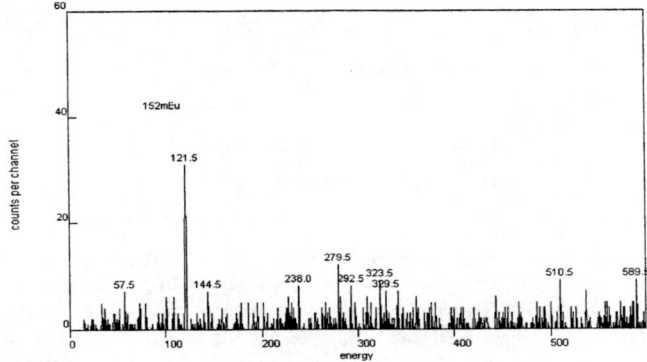

Figure 1 A gated spectrum coincidence with 842 keV γ-ray from JP-1 sample. The 841.6 keV-121.8 keV γ-ray sequence from ^{152m}Eu (9.3 h) was found.

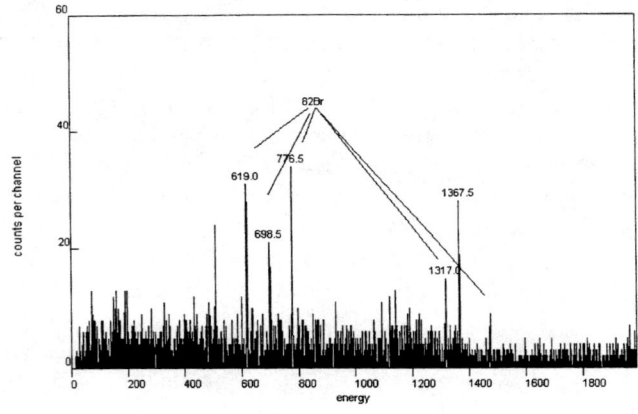

Figure 2   A gated spectrum coincidence with 554 keV γ-ray from JB-1a sample.   The 554.3 keV-1317.5 keV-776.5 keV and 554.3 keV -619.1 keV-698.3 keV-776.5 keV γ-ray sequences from ^{82}Br(35.3 h) were found.

REFERENCES

1. Imai, N., Terunuma, S., Itoh, S., and Ando, A., Geochemical Journal, **29**,   pp.91-95 (1995).

2. Furuno, K., et al., Nucl. Instr. and Meth. **A 421** 211(1999).

Ion Beam Analysis Techniques in Interdisciplinary Applications

Miguel A. Respaldiza[*,†], Francisco J. Ager[*,†]

*Depto. de Física Atómica, Molecular y Nuclear, Universidad de Sevilla, SPAIN
†Centro Nacional de Aceleradores, Av. Thomas A. Edison , Sevilla E-41092 SPAIN

Abstract. The ion beam analysis techniques emerge in the last years as one of the main applications of electrostatic accelerators. A short summary of the most used IBA techniques will be given as well as some examples of applications in interdisciplinary sciences.

INTRODUCTION

In the last decades, the small electrostatic accelerators became out of range for most of the experimental nuclear research. However, far from becoming obsolete or useless, they turned out to be essential for applied research in very fascinating and interesting fields. Among these, it is worth mentioning the materials modification by using high energy ion beams (ion implantation, ion beam mixing), the dating using the so-called Accelerator Mass Spectrometry (AMS), the studies of radiation damage and synthesis of new materials, and the ion beam analysis techniques.

Nowadays the Ion Beam Analysis (IBA) techniques are well-established analytical methods with very attractive features: high sensitivity, non-destructive character, multielemental analysis, relatively low cost, high depth resolution, etc. In fact, in the last years, many electrostatic accelerators providing ion beams in the MeV/uma range have been set-up all around the world to perform IBA techniques. Besides, a continuously increasing scientific activity using these techniques is developing in many interdisciplinary fields.

A summary of the most used IBA techniques, PIXE (Proton Induced X-Ray Emission), RBS (Rutherford Backscattering Spectrometry) and NRA (Nuclear Reaction Analysis), and their applications will be done. Some examples of recent applications in Material Science, Medicine, Biology, Environmental research, Art and Archaeometry will also be given. Finally, a short presentation of the first Spanish IBA facility, recently installed in Seville and based in a 3 MV Pelletron tandem accelerator, will be made.

CP495, *Experimental Nuclear Physics in Europe*, edited by B. Rubio et al.
© 1999 American Institute of Physics 1-56396-907-6/99/$15.00

ION BEAM ANALYSIS (IBA) TECHNIQUES

Rutherford Backscattering Spectrometry (RBS)

This technique is based on the elastic collision of an incident charged projectile with the nucleus of an atom (1). The recoil energy depends both on the mass of the target nucleus and on its depth within the sample matrix. Therefore, RBS can, in principle, carry out major element analysis and depth profiling.

Ion backscattering spectrometry has been used for analytical purposes for more that three decades. A traditional RBS experiment is performed with ^{4}He or H ion projectiles in the Coulomb scattering energy region around 2 MeV.

The analytical capability of the technique lies in its mass sensitivity, which it is based on the energy exchange between the elastically colliding particles. The energetics of the elastic collision may be calculated precisely when the masses of the colliding partners and the collision geometry are known. Conservation of energy and momentum in the interaction gives rise to constant ratios K (kinematic factors) of the incident and exit energies.

Depth analysis is based in the knowledge of the energy loss of ions in the target material. The physical mechanisms of the energy loss involve interactions between the incident particle, target electrons and target nuclei and they are well characterized by the stopping cross-sections.

Particle Induced X-ray Emission (PIXE)

Particle induced X-ray emission (PIXE) is a method in which X-ray emission is used for elemental analysis (2). High energy H or He ions normally create vacancies in inner atomic shells, with kinetic energies in the range of the MeV. Major features of PIXE are its multielemental character (in principle, all elements from B to U can be measured), high sensitivity (absolute detection limits down to 1 pg and relative detection limits down to ppm), smooth variation of relative detection limits with atomic number, the ability to analyze small amounts of material (1 mg or less), speed of analysis (1-10 min per sample) and the possibility of automation.

PIXE is generally considered as a non-destructive technique. Sample preparation, if any, is relatively simple. In principle, any dry sample that can be fitted into a vacuum chamber can be analyzed "as is". It is also possible to extract an ion beam out of a chamber (in vacuum) into atmosphere in the so-called "external beam mode" that enables analysis of bulky samples of delicate and valuable specimens.

The use of a focused ion beam as an excitation source provides a microanalytical tool with unique potential: standard PIXE typically uses a beam size of the order of 10 mm diameter. Micro-PIXE is performed using a beam probe which can be as small as 0.1 • m and the absolute sensitivity can be extremely high (10^{-17} g).

Nuclear Reaction Analysis (NRA)

When a beam of MeV energy ions impinges on a sample, it causes a variety of interactions, as we have already seen. Some particles may induce nuclear reactions, and when the emitted radiation is detected, it can provide information that is not available by other techniques. However, there is no simple correlation of the nuclear excitation response with A or Z and, consequently, it is more realistic to appreciate NRA as element specific. Even though this may appear a limitation if our quest was for multi-elemental capability, in fact NRA can claim that there is a nuclear reaction appropriate for chemical analysis for virtually every element of the Periodic Table.

In particular, NRA plays an important role in the analysis of the lightest nuclides, which are inaccessible to PIXE. In some cases, relatively narrow resonances exist that allow high-resolution depth profiling. They can be exploited as a very precise nondestructive method for determining the elemental concentrations in the depth of the sample (3).

IBA APPLICATIONS

Material Science: Improvement of the Oxidation of Stainless Steels

It is well known that the addition of small amounts of some elements, called reactive elements, to Cr_2O_3 or Al_2O_3 forming alloys greatly enhances the resistance of the alloy to high temperature oxidation (4,5). This effect, known as Reactive Element Effect (REE), is not well understood yet, and a definitive theory that accounts for it has not been established at present. This effect can be produced by a variety of reactive element addition methods such as microalloying, dispersion of oxides and surface coating with a thin layer of the reactive element oxide (6). Among these alternatives, ion implantation has been chosen as a way of obtaining surface or subsurface alloys with the desired composition (7).

A detailed study of the composition and evolution with time of oxide scales formed onto lanthanum-coated AISI-304 stainless steel specimens by means a modified CVD deposition method (PYROSOL) at 900 °C in air, was done with the help of ion beam analysis techniques (3). RBS and NRA of chromium and manganese were used for profiling the oxide scale formed on top of lanthanum deposited AISI-304 steel specimens. Complementary data were obtained by means of other analytical techniques: scanning electron microscopy (SEM), energy dispersive X-ray analysis (EDX) and thermogravimetric measurements (TG).

By means of RBS analysis using 1.6 MeV H^+, distinct information about the amount of oxygen incorporated in the samples can be obtained (Fig. 1(a)). The oxygen signal is clearly visible on the top of the continuous RBS spectrum from the steel matrix (mostly iron). For a more detailed depth profiling of the elements involved in the oxidation and diffusion processes, 1.6 MeV He$^+$ RBS analysis was performed. In this case, lanthanum can be readily profiled in the first few hundred nanometers, since its signal is well separated from that of iron, chromium or manganese. However, we cannot profile the latter three because their signals are overlapping. It is feasible to distinguish Cr from Mn+Fe only at the surface, and from there on a slow variation in the composition until the bulk is reached can be assumed. Some of the computer simulations of the spectra with RUMP (8) have been performed in this way. Those species are better profiled by NRA. Knowing the Cr and Mn profiles from NRA and introducing them into the simulations, an approximation of the Fe profile can be obtained.

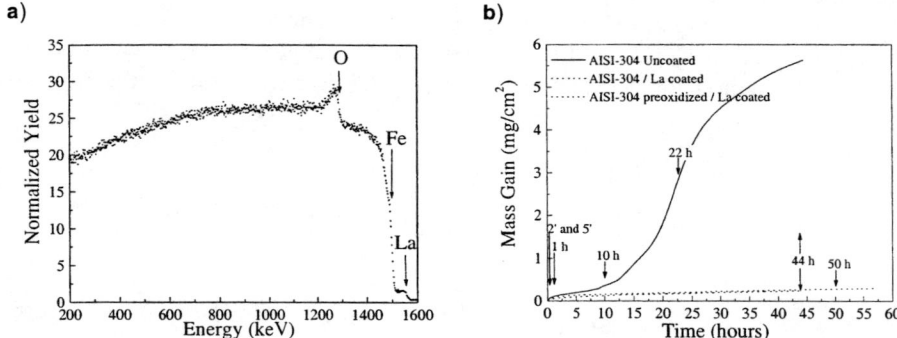

FIGURE 1. (a) RBS spectra taken with 1.6 MeV H$^+$ of a sample coated with lanthanum and oxidized for 60 minutes at 900 °C. (b) Thermogravimetric measurements indicating the mass gain of the three sets of samples with the oxidation time. Arrows indicate the oxidation times at which samples were analysed. 22 hours is only for the uncoated sample, and 50 hours only for the non-preoxidised coated sample.

It is clear form the thermogravimetric measurements, that the addition of lanthanum clearly improves the oxidation resistance at 900 °C of the AISI-304 stainless steels (Fig. 1(b)). After 10 hours, the uncoated sample shows no longer a refractory behaviour, whereas the coated samples keep it during all the studied period.

In conclusion, IBA techniques are very useful for understanding the processes going on during the oxidation of stainless steels. They allowed us to develop a low cost method that may result in homogeneous coating with industrial application.

Arts and Archaeometry: Tartesic Gold Jewelry (Ébora Treasure)

Gold jewelry artifacts of Tartesic origin (700-500 BC) have been studied using external beams for quantitative PIXE (9). Collimators along the incident proton beam (2.7 MeV) allowed the artifacts to be irradiated in narrow regions down to 350 μm in diameter. Special attention has been paid to the procedure of soldering in various narrow regions of the bindings of filigrees, twisted wires, narrow strips and granulations, on finely decorated items.

By using reference materials and thick target PIXE programs including all matrix effects (X-ray cross sections, attenuation coefficients, and secondary fluorescence), relative concentrations of Au, Ag and Cu in various regions have been determined. The results seem to indicate that solderings have been made by local fusion and brazing. No procedure of solid state diffusion bonding like in Etruscan jewelry has been identified. Qualitative results using imaging and EDX induced by a very narrow electron beam (SEM) give complementary information to the quantitative determinations obtained from the PIXE measurements.

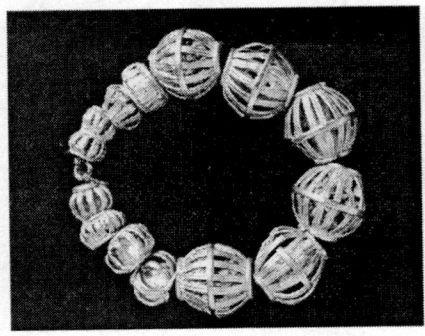 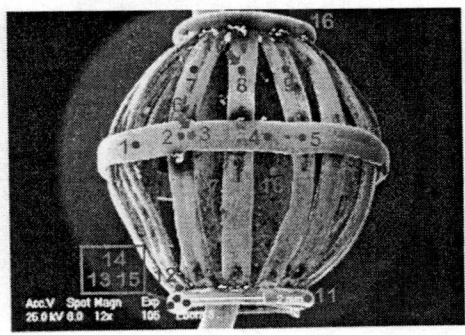

FIGURE 2. Left: necklace of the Ébora treasure (700-500 B.C.), conserved at the Archaeological Museum of Sevilla. Right: bead of the Ébora necklace made by strips attached on a hollow cylinder, showing the impact points.

The sample for this study is from the "Ébora" treasure (Fig. 2) and it was found with a large group of Oriental jewels, during agricultural work in the "Sanlúcar de Barrameda" area, a village near Cádiz (Spain).

The compositions of the alloy in various impact regions on the jewel are closely distributed around 84.5% Au, 8% Ag and 7.5% Cu. A closer look shows that impacts outside solders contain more copper than in almost all soldering regions. The lower Cu concentrations at solders may be understood if we think that some forging procedure has been used to bind strips together. No additional material is used: parts of the item to be bound are simply locally heated at a temperature slightly higher than the solidus temperature. The metal is partially fused and copper, the less noble metal in the alloy, is selectively eliminated by forming oxides that are lost during the final procedure of brushing. At point 3, both copper and silver are slightly more abundant than anywhere and we may think that a brazing procedure has been used to bind both ends of the central circular ring. The central ring is then soldered by a brazing procedure, the strips on the bottom and upper rings are bound by forging.

Biomedicine: "Ötztaler Ice-Man" Scalp Hair Analysis

Hair is a body tissue which survives for a long period after death. Because hair grows continuously through the life and the trace elements content depends on the underlying body chemistry at the time of growth, in principle, chronological changes can be followed. PIXE analysis is well suited to this type of analysis because of the sensitivity to trace elements.

Hair analysis was used in the investigation of a number of ancient bodies. In a recent study by Prof. G.W. Grime and co-workers (10), the Oxford Scanning Proton Microprobe (SPM) was used to analyze the scalp hair of the ice body dated around 3000 BC discovered in 1992 in a high alpine pass in the Ötzaler Alps on the Austrian-Italian border.

Fig. 3 shows diametric line profiles of S, Cu and As in a hair. These results show clearly the difference between the surface concentration of Cu and the internal concentration of As. One possible explanation is that the Ötzaler iceman was directly or indirectly associated with the production of copper artifacts (As is a common constituent of copper ores).

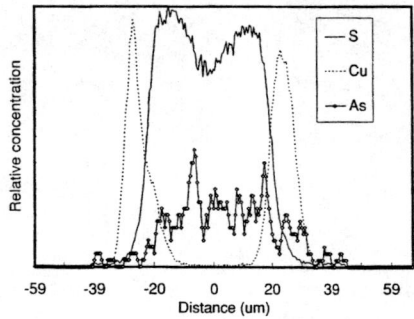

FIGURE 3. Distribution of S, Cu and As across the diameter of a scalp hair from the Ötzaler ice body. The hair was embedded in resin and polished transversely to give a flat surface which was analysed using the 1 • m beam of the Oxford SPM scanned in a line along a diameter of the hair.

Environmental Sciences: Analysis of the Huelva Estuary

Near the town of Huelva (SW of Spain), along the estuary formed by the confluence of the Tinto and Odiel rivers, there has been for 30-40 years, a large industrial complex. Some of the factories discharge directly or indirectly their waste to those rivers, producing a clear anthropogenic contamination.

The environmental impact of the industrial complex in the sediments of the river basins can be inferred from the determination of their elemental concentrations. In this sense, it is well known that an appropriate technique to determine elemental concentrations is thick-target PIXE (TTPIXE), due to its high sensitivity (ppm), its multielemental character and the short collection time.

TTPIXE technique was applied to sediment samples collected in the estuary (Fig. 4), aiming not only to determine the contamination and its origin, but also to study of the influence of tides in the distribution of the heavy metal deposition. For this purpose, the concentrations of about 20 elements were obtained (11).

The Al, Si, K and Ti concentrations were in general uniform and were in agreement with the average values expected in sediments. Concerning the P results, the obtained values showed the existence of a clear contamination over all the studied area, having its origin in the industrial activity of the FORET fertilizer factory. The P contamination in the rivulet Estero Domingo Rubio, which is free of dumps in its margins, showed that the extension of the contamination along the estuary is due to the tidal action. The Cu, Zn, As and Pb concentrations were also of anthropogenic origin, and they turned out to be very high in the entire estuary. In the case of Fe, it was clear that the waste of the fertilizer factory contains very low proportion of this metal, while extremely high values were obtained in the sample taken near the RTM metal factory. The S and Ca concentrations were high or moderate in the Odiel and Tinto rivers, due to discharges of the fertilizer factory to the Odiel or the uncontrolled deposition of phosphogypsum from the piles in the Tinto, but the concentrations are only moderate to low in the rest of the samples.

It is clear how through the application of the TTPIXE technique to environmental samples, the contamination produced for a big industrial complex can be evaluated, and even how the contribution of the different possible sources to general contamination can be discerned.

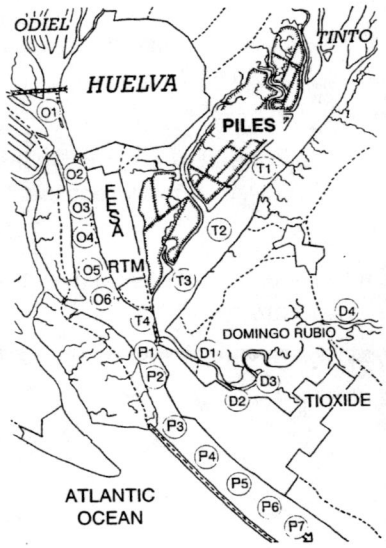

FIGURE 4. Map showing the estuary formed by the Tinto and Odiel rivers near Huelva. It shows the fertilizer (FORET), metal (RTM) and paint (TIOXIDE) factories, the phosphogypsum piles, the Estero Domingo Rubio rivulet and the sample names sited in the collection points.

THE SPANISH NATIONAL ACCELERATORS CENTER

The Centro Nacional de Aceleradores (CNA) is born as a joint enterprise between the University of Seville, the Junta de Andalucía (Andalusian's regional government) and the Consejo Superior de Investigaciones Científicas (Spanish Council for Scientific Research) for the creation of the first Spanish ion beam facility. Our Center is intended as a multidisciplinary research center, being its primary aim the material modification and analysis by making use of ion beam based techniques.

The National nature of our Center gives all Spanish researches, from either private or public institutions, the possibility of incorporating research proposals in a wide variety of fields: Material Science, Medicine, Biology, Geology, Environmental Sciences, Art and Archaeometry.

The main characteristics of our facility include a 3 MV Pelletron accelerator from NEC, two ion sources and seven possible beam lines. The performance tests were done during 1998 few months after the arrival of the machine. Since then three beam lines have been installed and are already operational.

At the 0° port of the switcher magnet a simple chamber was initially installed for calibrations and tests and now a multipurpose scattering chamber substitutes the former; this chamber has been designed to carry out RBS, PIXE, NRA and PIGME experiments simultaneously.

A commercial Charles-Evans scattering chamber is located in the +30° line; this line is primarily devoted to channeling analysis of crystalline samples (RBS/Channeling).

At -15° there is an ion microprobe of the Oxford Microbeam Ltd., focused on studies in biomedicine, material science and environmental analysis by means of • - PIXE, RBS and NRA.

The fourth line is currently under installation and will be devoted to an external beam for studies in Art and Archaeometry.

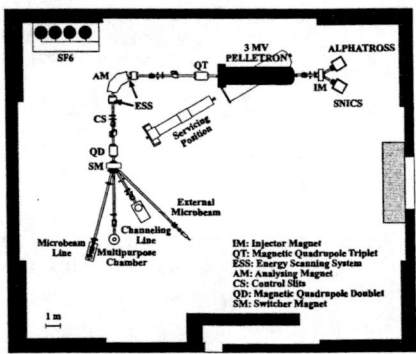

FIGURE 5. Schematic layout of the acceleration system and experimental beam lines.

ACKNOWLEDGMENTS

The authors would like to acknowledge Prof. G.W. Grime and Prof. P. Mandó for their involvement in the preparation of this contribution to the ENPE 99 conference, and also M.A. Ontalba and J.E. Martin for their help with the preparation of the paper.

REFERENCES

1. Chu, W.K., Meyer, J.W., and Nicolet, M.A., *Backscattering Spectrometry*, New York: Academic Press, 1978.

2. Johanson, S.A.E. and Campbell, J.L,*"PIXE, A novel technique for elemental analysis"*, Chichester: J. Wiley, 1988

3. Ager, F.J., Respaldiza, M.A., Paúl, A., Odriozola, J.A., Da Silva, M.F., Soares, J.C., Nuclear Inst. and Meth. B 136-138, 1045 (1998).

4. Stringer, J., Mater. Sci. Eng., A120, 129 (1989).

5. Bonnet, G., Larpin, J.P., and Colson, J.C., Solid State Ionics, 51, 11 (1992).

6. Konno, H., Tokita, M., and Furusaki, A., Electrochim. Acta, 37, 2421 (1992).

7. Laursen, T., Claphan, L., Whitton, J.L., and Jackman, J.A., Nucl. Instr. and Meth. B, 59/60, 768 (1991).

8. Doolitle, L.R., Nucl. Instr. and Meth. B9, 344 (1985).

9. Ontalba-Salamanca, M.A., Demortier, G., Fernández Gómez, F., Coquay, P., Ruvalcaba, J.L., Respaldiza, M.A., Nuclear Inst. and Meth. B 136-138, 851 (1998).

10. Grime, G.W., "IBA Applications in biogenic Materials in Archaeology" in *Applications of ion Beam Analysis Techniques to Arts and Archaoemetry,* edited by M.A. Respaldiza and J. Gómez-Camacho, Sevilla: Universidad de Sevilla, 1997, pp. 145-157

11. Martín,J.E., Respaldiza, M.A., García-Tenorio, R., Bolívar, J.P., Gómez-Camacho, J., Nuclear Inst. and Meth. B 109-110, 506 (1996).

A 5MV Tandetron to Universidad Autonoma de Madrid.

Olof Tengblad*,
for the project committee and Universidad Autonoma de Madrid†

*Instituto de Estrucutura de la Materia, CSIC, Serrano 113, E-28006 Madrid, Spain
† El Centro de Analisis de Materiales, Universidad Autonoma de Madrid, Ciudad Universitaria
de Canto Blanco, E-28049 Madrid, Spain.

Abstract. A 5MV Tandetron accelerator is being projected for the Centre of Material Analysis of the Universidad Autonoma de Madrid. The accelerator will be dedicated to Material Science but it meant to be open to all fields of science and industry that can profit from this kind of installations. Estimated construction time and delivery of the accelerator implies that the first experiments can be performed in the spring 2001.

With help of European community money a new accelerator is being projected for the Centre of Material Analysis of the Universidad Autonoma de Madrid. The project is in the state that the machine has been ordered in july 1999; a 5 MV TANDETRON from High Voltage Engineering [1], the Netherlands. A new building will be erected for the accelerator including space for a mechanical and a electronic workshop, together with a meeting room and a few offices for experimentalists. The time-plan is such that the building construction should start in september 1999 and be finished in time for the delivery of the accelerator in december year 2000. After 3 month of installations and test activities the center should be ready for real research activity in the spring year 2001.

The accelerator specifications are presented in the table below.

CP495, *Experimental Nuclear Physics in Europe*, edited by B. Rubio et al.

Accelerator specifications;

5 MV TANDETRON
High Voltage Engineering Europa B.V.
Amsterdamseweg 63, 3812 RR Amersfoort, The Netherlands

Terminal Voltage   0,2 - 5 MV
Terminal voltage stability   400 V_{pp}
Terminal voltage ripple   200 V_{pp} guaranteed, (50 V_{pp} design aim)
Beam current from HV supply   1,0 mA
Deflection angles $0^{0}, \pm 10^{0}, \pm 30^{0}, \pm 45^{0}$
Beam currents in the 30^{0} beam-line 2m from HE magnet in a 3 x 3 mm^{2} spot:

Ion	Source	Terminal voltage (MV)	Charge Z	Beam Energy (MeV)	Beam Current (μA)
^{1}H	Duoplasma	1,25	1	2,5	15,0
^{2}D	Duoplasma	0,2	1	0,4	5,0
^{2}D	Duoplasma	0,3	1	0,6	8,0
^{4}He	Duoplasma	1,0	2	3	1,0
7Li	Sputter	2,0	2	6,0	1,5
^{14}N	Sputter	4,0	3	14,2	15,0
^{15}N	Duoplasma	1,8	3	6,8	1,0
^{28}Si	Sputter	4,7	5	28,0	25,0
^{28}Si	Sputter	3,0	3	12,0	125,0
^{166}Er	Sputter	5,0	3	19,6	1,0
^{197}Au	Sputter	5,0	3	20,0	20,0
^{4}He	Duoplasma	1,0	2	3,0	10 pnA *
^{28}Si	Sputter	4,0	6	28,0	10 pnA *

* in the 10^{0} degree beam line and in a rectangle defined by a slit of 2 x 0.5 mm^{2}
** X-ray level at 100 cm from the tank wall with a 0,5 μA He beam at
   maximum terminal voltage < 2 μSv/h

REFERENCES

1. High Voltage Engineering Europa B.V. Amsterdamseweg 63, 3812 RR Amersfoort, The Netherlands.

Charge Changing Cross Sections and Stopping Powers in Very Thin Layers for Ions of Frozen Charge State

A. Blažević*, H.G. Bohlen* and W. von Oertzen*

*Hahn-Meitner-Institut, Glienicker Str. 100, D-14109 Berlin, Germany

Abstract. A method is developed to measure charge changing cross sections and charge state state dependent energy losses. It is relevant for stopping powers in very thin layers, where the charge state distribution has not yet reached a dynamical equilibrium. First results are reported for neon ions at 2 MeV/u in thin carbon foils.

The specific energy loss, dE/dx, of heavy ions penetrating into very thin layers is subjected to changes, because the ions after their acceleration to the final energy are usually in a charge state lower than charge equilibrium. Data on charge changing cross sections and the dependence of the specific energy loss on frozen charge states are needed for a prediction of the energy loss. Measurements have been performed

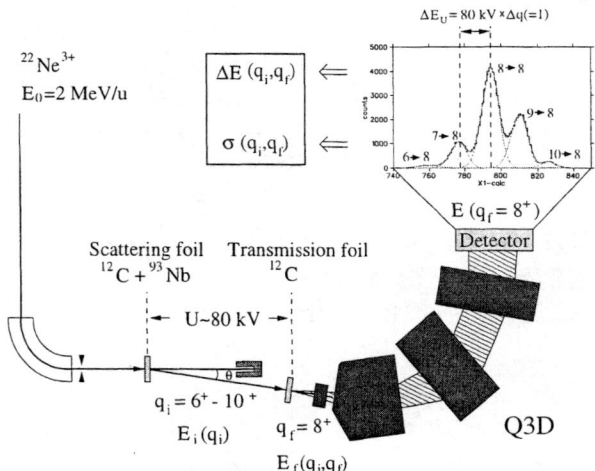

FIGURE 1. Setup at the Q3D spectrograph for energy loss and charge exchange measurements using a high voltage U to separate different charge states q_i incident on the transmission foil.

CP495, *Experimental Nuclear Physics in Europe*, edited by B. Rubio et al.

at the Q3D-spectrograph at HMI, the setup is shown in Fig. 1 for the case of an incident beam of $^{22}Ne^{3+}$ at 2 MeV/u. An initial charge state distribution is created in a thin scattering foil. The second (transmission) foil is the material of interest, where charge changing processes and energy loss are studied. By varying its thickness we can investigate the transition from non-equilibrium to equilibrium conditions. The Q3D-spectrograph selects only one outgoing charge state q_f of ^{22}Ne. Different incoming charge states q_i are distinguished by applying a high voltage between the two foils, an energy difference $\Delta E_U = U \times (q_i - q_f)$ appears in the focal plane. In this way separate peaks are produced for the different incident charges q_i as indicated in Fig. 1 for $q_f = 8^+$. The energy loss is calibrated by measuring the peak position without transmission foil. Selecting different q_f by an appropriate magnetic field and unfolding the five separated peaks (Fig. 1) related to the initial charge states 6^+ - 10^+, we obtain a matrix of yields for charge changing processes and energy loss $\Delta E(q_i, q_f, d)$. Fig. 2 (left panel) shows the dependence

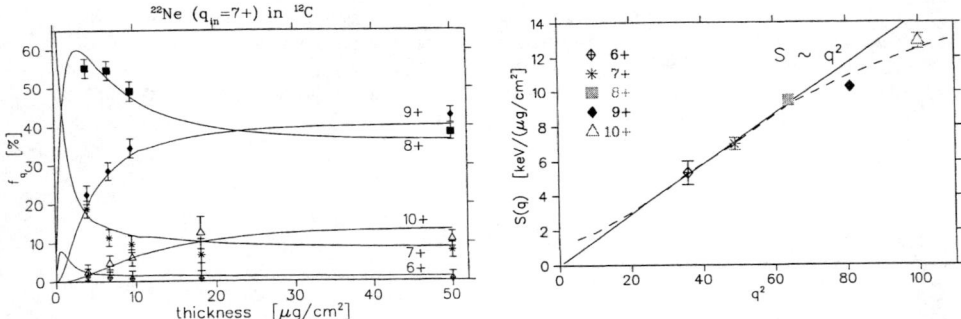

FIGURE 2. Left panel: Dependence of the charge state distribution on the layer thickness. Right panel: Stopping power S(q) for frozen charge states of Ne in carbon as a function of q^2.

of the yields of final charge states, q_f, on the thickness of the transmission foil for $q_i = 7^+$, the equilibrium is reached at a thickness larger then 30 $\mu g/cm^2$.

The data are compared with calculations for charge changing processes, where cross sections for electron-loss and -capture enter. We calculated such cross sections for radiative and non-radiative capture, ionisation, excitation and Auger decay with the code ETACHA [J.P. Rozet et al., Nucl. Instr. and Meth. B107 (1996) 67] and solved the rate equation for the charge state distributions. The theoretical description agrees well with the data for all q_i (6^+ - 10^+), results for 7^+ are shown as solid lines in Fig. 2. Energy losses have been analysed for $q_i = q_f$. In the theoretical calculation we take into account charge changing processes within the foil by Monte-Carlo simulation using the microscopic cross sections, which describe the charge state distributions. Stopping powers for frozen charge states S(q) could be extracted in this way, they are shown in Fig. 2 (right panel) as a function of q^2. There is an obvious deviation from the S(q) $\sim q^2$ scaling law (solid line) for the high charge states 9^+ and 10^+. The dashed line shows a possible realistic dependence.

Tumor Therapy with Heavy Charged Particles

Hans Blattmann

Division of Radiation Medicine, Paul Scherrer Institute, 5232 Villigen PSI, Switzerland

Abstract. Nuclear science has contributed significantly to the development of tumor therapy with heavy charged particles. Interest evolved for neutron therapies in the forties because of the increased radiobiological effectiveness (RBE) compared to photon irradiation. The development of more powerful proton and heavy ion accelerators with higher energies or higher intensities, made new particles for radiation therapy available. Pions, protons, light ions, from helium up to silicon were studied in view of precision dose delivery and increased RBE. Without the parallel development of new diagnostic techniques such as computer tomography (CT) and positron emission tomography (PET) the rapid development would not have been possible. Heavy-charged particle therapy has now come into a consolidation phase. Hospital-based facilities are built by industry, and research institutes focus on refinements in dose delivery and treatment planning, as well as systems for monitoring dose delivery and for dose distribution verification.

HISTORY

The motivation for the use of heavy particles in radiotherapy were their advantages - compared to contemporary photon therapy - in relative biological effectiveness (RBE) in tumor cells, compared to normal cells, better dose distributions or both. Neutrons were the first hadrons to become available at appropriate energies, and thus penetration depth, and at the same time with an adequate dose rate. Neutron therapy began more than 60 years ago shortly after the discovery of the neutron. These early studies were made possible by the availability of the first cyclotrons developed by E.O. Lawrence at Berkeley, California. The first clinical trial of neutron therapy resulted in too bad late sequelae to be continued (1). The trial itself suffered from disadvantages such as inadequate neutron dosimetry and a lack of understanding of the effects of fractionated treatments on the relative biological effect of the beam. A new program was started at the Hammersmith Hospital, London with the first patient treatments in 1966 and with very encouraging results (2). This lead to many more neutron therapy facilities in the world and to clinical trials demonstrating a clear advantage of densely ionizing radiation at least for selected medical indications.

Charged hadrons were expected to improve the spatial dose distribution. It was again Berkeley, 1954, which pioneered the use of protons in radiotherapy. During the first

CP495, *Experimental Nuclear Physics in Europe*, edited by B. Rubio et al.

treatments, small volumes were irradiated by geometrically overlapping individual proton beams in the plateau region of their depth dose profile, thereby neglecting the well defined range of the heavy charged particles (3). Uppsala, Sweden, 1957–1976; Harvard, MA, USA, since 1961; Moscow and Dubna, Russia, since 1967 were other pioneering centers for proton therapy (4).

Charged particle therapy experienced a remarkable thrust in development when CT scanners became available. Projects for radiotherapy with protons, light ions, and pions were simultaneously started to achieve higher doses to the target tissue while sparing normal critical structures adjacent to it. Each of the projects had a slightly different justification for the choice of the particle. While for protons clearly the physical dose distribution argument was most important, pions were expected to offer primarily a favorable biological dose distribution, as the biologically more efficient dose would be concentrated to the target volume. To produce pions with a reasonable dose rate, an accelerator with a high proton current and an energy of at least 400 MeV is necessary. Only three facilities world wide were able to offer this. Pions were tested from 1974 to 1982 at LAMPF, Los Alamos, USA; from 1980 to 1993 at SIN (now PSI) Villigen, Switzerland; and from 1979 to 1994 at TRUMF, Vancouver, Canada, with an overall total of close to 1000 patients. At all three centers the therapy was discontinued because of high costs and non-convincing clinical results.

The light ions offer a combination of an increased linear energy transfer (LET) in the target volume, a low LET in the entrance channel, and excellent properties for shaping dose distributions. Heavy ions from helium to silicon were available from 1975 to 1992 at the BEVALAC in Berkeley in 1994 this was followed by the carbon ion program at the HIMAC at Chiba, Japan. Since 1997 patients are also treated with carbon ions at GSI, Darmstadt, Germany.

All of the charged particle projects were based on accelerators built, and most of them already in operation, for physics research. It was not until 1990 that the first hospital based proton accelerator in Loma Linda came into operation and which has treated more than 4000 patients since. The only heavy ion facility dedicated to radiotherapy is HIMAC where close to 500 patients have been treated since 1994.

Further hospital based facilities for protons and heavy ions are under construction or in planning today.

MEDICAL CONSIDERATIONS

Physical dose distribution

The goal of radiotherapy of tumors is to give a high enough dose to the tumor tissue, while sparing the normal tissue to avoid complication. This goal can be approached by physical selectivity, i.e. by optimizing the spatial distribution of the dose, thus limiting the high dose to the tumor tissue. This is usually difficult because of the intrusion of tumor cells into the adjacent normal tissues. The target volume is defined as the volume of visible tumor surrounded by a safety margin which includes those areas of predicted expansion of the tumor. The physical characteristics of the radiation source, depth dose or lateral dose fall-off may also limit the potential to spare normal tissues.

Because of the sharp increase of complication probability with dose, distributing the dose to normal tissue over a larger volume by selecting multiple treatment ports is frequently necessary. It is, thereby, necessary to avoid radiosensitive tissues. A gantry system, allowing an isocentric treatment from several different angles is, therefore, essential. Heavy charged particles with their characteristic dose maximum close to the end of their range, and with no or little dose beyond that depth, have an advantage compared to photons or neutrons with their exponential dose fall-off with depth.

Biological parameters

Biological parameters can also be used to optimize treatment results, either by selecting dose fractionation schemes which spare normal tissues better than tumor cells, or by choosing radiations with a higher ionization density and, therefore a higher RBE. This is of special importance for poorly oxygenated tumor cells which are up to three times more resistant to photon irradiations.

While densely ionizing particles are more efficient for the treatment of radioresistant tumors they usually also have a higher RBE for late tissue complications.

Beam delivery systems

In photon radiotherapy, isocentric beam delivery systems have improved dose distributions since beams have become available with sufficient penetration to be superimposed from different angles in the target volume. Without change in position the patient is treated from different ports. The entrance and exit dose is, thereby, distributed over a larger volume, avoiding sensitive normal structures, which permits higher doses to the target volume without overexposing sensitive organs.

Even though heavy charged particles have intrinsically favorable dose distributions, the flexibility of using several different entrance ports is essential in order not to jeopardize the potential advantages.

Comparison of heavy charged particles with state-of-the-art photon therapy

Any new radiation therapy modality will be compared with the best available technique from conventional radiotherapy equipment. This may be, from linear accelerators with multi-leaf collimators and inverse planning, or brachytherapy or the Gamma-knife. Today's demand for evidence based medicine require proven advantages in terms of uncomplicated tumor control combined and cost efficiency. Photon radiotherapy is still under improvement. Due to more powerful computer techniques, combinations of therapy modalities to further improve results are also feasible.

Areas where photon and heavy charged particles compete are radiosurgery, i.e. irradiation of small lesions, usually in the head, with multi-port geometry in a single treatment. The advantage of heavy charged particles for these treatments is the superior lateral dose fall-off.

A second area of competition is conformal radiotherapy. Physical dose distribution calculations for the various techniques are compared but, more and more, the need for comparing biologically efficient dose distributions arises (5). The biological models available are not verified due to inadequate data. This is a handicap, especially if very sophisticated dose distributions are used resulting in partial volume irradiation of more than one critical normal tissue or if the radiation sources compared have different biological efficiencies. This is particularly true if large volumes of tissue are at risk, receiving a dose which is lower than the tolerance dose, and for which biological data are missing. At some distance from the target volume, the dose from charged particles may be a factor of two or even lower than from photons. Careful interpretation of the results is therefore essential.

As developments proceeds toward precision radiotherapy, quality assurance is important to guarantee that the advantageous dose distributions calculated by the treatment planning program correspond to the dose distribution applied to the patient, in each of frequently more than thirty irradiations.

RESEARCH ACTIVITIES

A major contribution to the development and application of proton radiotherapy came from the group at the Harvard Cyclotron together with the Massachusetts General Hospital in Boston. With a 160 MeV horizontal proton beam, precision radiotherapy has been performed since many years for lesions mainly at the base of the skull. Ocular melanoma and radiosurgery for arterio-venous malformations were other indications.

Today treatment of patients with heavy charged particles is being transferred more and more to hospitals. No clinical facility comparable to the proton therapy centers of Loma Linda or the North East Proton Center in Boston is planned in Europe at the present time but studies are underway. An international scientific collaboration for a "proton and ion medical accelerator study" (PIMMS) with participation of MedAustron (Vienna, Austria), GSI (Darmstadt Germany), CERN (Geneva, Switzerland), KFA (Jülich, Germany) and TERA (Milan, Italy) has come up with a design for a combined proton and light ion facility which could form the base for clinical facilities in Vienna, Heidelberg and northern Italy.

Other developments, on a smaller scale, aim at optimizing gantry design or beam delivery systems. Worldwide, an increasing interest in dynamic beam delivery systems is observed which will increase flexibility in shaping dose distributions in three dimensions for conformal therapy and make full use of the potential of charged particles. Two dimensional spot scanning for low energy protons has been developed at Chiba since the late seventies (6). PSI and GSI have developed three dimensional scanning for protons and for carbon ions for deep seated lesions. The spot scanning at PSI uses magnetic scanning for the fastest scan direction and a mechanical range shifter and patient movement for the other scan directions (7). GSI scans magnetically in two directions and uses energy variation to control depth of penetration of the particles (8). For both projects, the beam size is much smaller than used previously at PSI and TRIUMF who developed beam scanning techniques for pions. This results in a significantly improved shaping of the dose distribution but also leads to a higher sensitivity to motion of the target volume.

The achieved higher flexibility creates a need for fast measurement of dose at a large numbers of points. Several groups of detector specialists and physicists are contributing already to this field.

Dynamic dose application, combined with the sharp dose fall-off laterally and distally of the charged particle beams, gives rise to dose errors due to motion of the patient. Real time measurement of the position of the tumor during therapy, and synchronizing the irradiation with this motion, will be important.

OUTLOOK

Clinical research

Clinical studies are underway in the Proton Radiation Oncology Group (PROG). The goal of this group is to provide a centralized research base for clinical trials employing proton therapy. They will also study improvement of the control of primary and regional malignant disease, and the pattern of failure and foster the design and implementation of clinical protocols.

Due to the fact that heavy charged particle radiotherapy not only has the potential of precise high-dose volumes shaped in three dimensions, conformal to the target volume, but also give a substantially lower integral dose to normal tissues outside the target volume, it may prove especially advantageous for combination of radiotherapy with adjuvant systemic therapy for disseminated cancer cells. Even though the dose at some distance from the target volume is usually adjusted to be well below tolerance, for combination with other therapies, the sparing typical for heavy charged particle therapy, could turn out to be an important progress. New imaging techniques which visualize biological conditions of cells or tissues will be instrumental to exploit this potential.

Plans worldwide

Hospital-based proton facilities are planned in the US and Japan. In the United States, besides the almost completed new facility in Boston, plans for five proton centers for a private health organization have been reported. In Japan, several centers are planned or under construction, one of them at Tsukuba where proton radiooncology has a long tradition at the accelerator of KEK.

In Europe some of the ongoing projects, i.e. Clatterbridge, Darmstadt, Louvain la Neuve, Nice, Orsay, Uppsala, Villigen and recently Berlin, intend to expand and new facilities are underway at Catania and Bratislava (Table 1).

Even when all the presently planned facilities are in operation, the number of patients treated by heavy charged particles will be small compared to the total number of patients undergoing radiotherapy. The relative importance of proton and carbon-ion therapy in the long term will mainly depend on the costs for routine therapy and on political decisions in the different participating countries.

TABLE 1. Proposed NEW FACILITIES for PROTON & ION BEAM THERAPY - July 1999
(from ref. 9)

INSTITUTION	PLACE	TYPE	1ST RX	COMMENTS
INFN-LNS, Catania	Italy	p	1999	70 MeV; 1 room, fixed horiz. beam
NPTC (Harvard)	MA USA	p	2000	at MGH; 230 MeV cyclotron; 2 gantries + 2 horiz
Hyogo	Japan	p, ion	2001	2 gantries; 2 horiz; 1 vert; 1 45 deg;under construction
NAC, Faure	South Africa	p	2001	new treatment room with beam line 30° off vertical.
Tsukuba	Japan	p	2001	270 MeV;2 gantries;2 fixed (research);under construction
Wakasa Bay	Japan		2002	multipurpose accelerator; building completed mid 1998
Bratislava	Slovakia	p, ion	2003	72 MeV cyclotron; p; ions; +BNCT, isot prod.
IMP, Lanzhou	PR China	C-Ar ion	2003	C-ion from 100MeV/u at HIRFL expand to 900 MeV/u at CSR; clin. treat; biol. research; no gantry; shifted patients

REFERENCES

1. Stone, R.S.: The American Journal of Roentgenology and Radium Therapy, **59**, 771-785, 1948
2. Catterall, M.: Cancer. **334**, 91-95, 1974
3. Larsson,B., Proton Therapy: Review of the Clinical Results. Proceedings of the EULIMA Workshop on the potential value of light ion beam Therapy. Ed. P.Chauvel, A.Wambersie, Nice November 3-5, 139-164, 1988, EUR 12165 EN
4. Raju MR, Radiat Res **145**, 391-407, 1996
5. Lomax AJ; Bortfeld T; Goitein G; Debus J; Dykstra C; Tercier PA; Coucke PA; Mirimanoff RO. Radiotherapy and Oncology. **51**, 257-271, 1999
6. Kanai T., Kawachi K., Kumamoto Y., Med Phys **7**,:365-9, 1980
7. Pedroni E, Bacher R, Blattmann H, Bohringer T, Coray A, Lomax A, Lin S, Munkel G, Scheib S, Schneider U, et al Med Phys **22**,:37-53, 1995
8. Haberer Th., Becher W., Schardt D., and Kraft G., Instr. Meth. Phys. Res. **A330**, 296-305, 1993
9. Sisterson J. Particles, Newsletter for proton, light ion and heavy charged particle radiotherapy, No. 24, Massachusetts General Hospital, Boston., July 1999

On the determination of the in-air spatial spread of clinical electron beams

María A. Aragón[1], Antonio M. Lallena[1], Javier González[2], Manuel Vilches[2] and Juan C. Zapata[2]

[1]*Departamento de Física Moderna, Universidad de Granada, E-18071 Granada, Spain*
[2]*Servicio de Radiofísica, Hospital Universitario "San Cecilio", Avda. Dr. Olóriz, 16, E-18012 Granada, Spain*

Abstract. The method of the penumbra, that of the reconstruction of the gaussian profile and that of the background are used to determine the in-air spatial spread of clinical electron beams. The method of the background appears to be the most accurate. Besides, the failures of the Fermi-Eyges theory in describing the beam transport are shown.

The measurement of the spatial spread of clinical electron beams crossing a medium is important to determine the corresponding dose distributions. In this work we are interested in measuring this parameter in air. To do that we have considered electron beams generated by a Siemens Mevatron KDS of various nominal energies; we have collimated them at the center using the adjustable collimators of the LINAC, and we have measure dose profiles at different distances to the lower edge of the jaws. For this setup, the pencil beam approach based on the Fermi-Eyges multiple scattering theory [1] predicts, below the collimator, dose profiles given by

$$D(z, x, y) = \frac{1}{2} D_\infty(z) \operatorname{erfc} \left[\frac{x_{\text{cent}}(z) - x}{\sqrt{2}\sigma_x(z)} \right]. \tag{1}$$

Here we compare three methods to determine σ_x. The first one, MP, is based on the previous measurement of the penumbra [1,2] and consists in performing a linear regression of the doses measured for some x points (15 in our case) around x_{cent}. A second approach, MG, [3] measures σ_x taking the derivative of the dose data (1) numerically (with a three point formula) and then fitting to the obtained profiles a Gaussian like function. The last approach, MB, [4] assumes the obvious presence of a background (which is due to second order effects) in the profiles and fits directly a function of the type (1) extended by adding to it a function $B(z)$. The results obtained for the spatial spread with the three methods are shown in Fig. 1. For MP and MB the centroids x_{cent} are fixed to the experimental values obtained as described in Ref. [4].

CP495, *Experimental Nuclear Physics in Europe*, edited by B. Rubio et al.
© 1999 American Institute of Physics 1-56396-907-6/99/$15.00

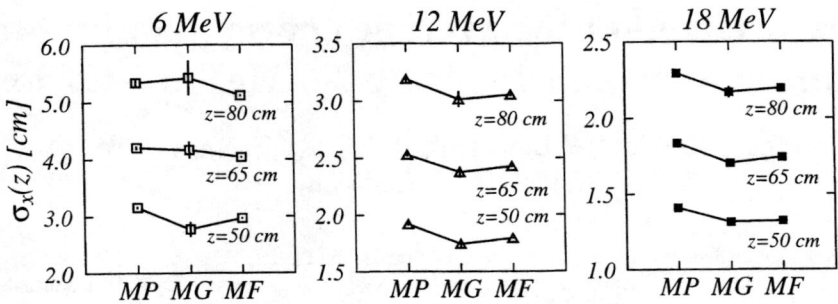

FIGURE 1. The spatial spreads found for the three methods considered.

On the other hand, the theory predicts for σ_x^2/z^2 a linear behaviour with z, the slope being the linear scattering power $T(E)$. The corresponding linear regressions produce the results quoted in Table 1. The $T(E)$ reported in [2] for the mean energies of our beams are also included. The discrepancies of these values with the results of our analysis are apparent, mainly for MP and MB. This puts some doubts on the validity of the theoretical description based on the pencil beam within the Fermi-Eyges theory. The method of the background appears to be the most accurate.

TABLE 1. Values of $T(E)$ obtained with the three methods considered compared to those of Ref. [2].

Energy [MeV]	$\overline{E}_0$ [MeV]	$T(\overline{E}_0)$ (Ref. [2]) $[\times 10^{-5}$ rad^2 cm$^{-1}]$	$T(E)$ $[\times 10^{-5}$ rad^2 cm$^{-1}]$		
			MP	MG	MB
6	5.3	24.983	10.4 ± 2.3	27.2 ± 10.6	11.4 ± 0.7
8	7.2	14.663	6.8 ± 1.3	8.4 ± 4.2	8.5 ± 0.2
12	11.0	7.015	2.4 ± 0.7	4.9 ± 1.6	3.8 ± 0.2
15	13.9	4.670	1.8 ± 0.4	1.8 ± 1.0	3.3 ± 0.1
18	16.9	3.324	0.8 ± 0.3	0.9 ± 0.7	1.0 ± 0.1

REFERENCES

1. Hogstrom K.R., Mills M.D., and Almond P.R., *Phys. Med. Biol.* **26**, 445 (1981); Huizenga H. and Storchi P.R.M., *Phys. Med. Biol.* **32**, 1011 (1987); Sandison G.A. and Huda W., *Med. Phys.* **15**, 498 (1988).
2. ICRU Report 35, International Commission on Radiation Units and Measurements, Bethesda, U.S.A., 1984.
3. Werner B.L., Khan F.M., and Deibel F.C., *Med. Phys.* **9**, 180 (1982).
4. Vilches M., Zapata J.C., Guirado D., Fernández D., Burgos D. and Lallena A.M., *Med. Phys.* **26**, 550 (1999).

Measurement of the average energy required to produce an ion pair in air by 7.6 MeV/u $^{12}C^{6+}$ ions

J.Rodriguez-Cossio[1,2], C.Brusasco[2], D.Schardt[2], B.Voss[2], U.Weber[2], F.Flesch[3], W.Heinrich[3]

1) Universidad de Santiago, Experimental Group of Nuclear and Particle Physics, E-15706 Santiago, Spain. 2) Gesellschaft für Schwerionenforschung / Biophysik, Planckstrasse 1, D-64291 Darmstadt, Germany. 3) Universität-GH Siegen, Fachbereich Physik, D-57068 Siegen, Germany.

Abstract. The average energy required to produce an ion pair (W value) in air by carbon ions must be known with high precision for the ion-chamber dosimetry performed at the GSI tumor treatment unit. As there is a lack of data in the energy range of interest for the therapy we performed a measurement in air with 7.6 MeV/u $^{12}C^{6+}$-ions. The experimental method is described and the preliminary result of $w_E = (34.1 \pm 0.4)$ eV is discussed.

In radiation therapy, an accurate measurement of the deposited dose (the energy absorbed per unit mass) is essential to verify that the target volume receives the prescribed dose and that the dose release distribution is in accordance with the treatment plan. When using ionization chambers (IC) as dosimeters, the dose deposited in the detector cavity is obtained from the collection of the charge created in the gas, with the W value as multiplication factor, being this the average energy required to produce an ion pair in the gas. A differential form of W is defined in order to take into consideration the energy dependence as $w_E = dE/dN$, where dE is the mean energy deposited by a charged particle of energy E in traversing an absorber of thickness dx, and dN is the mean number of ion pairs produced when dE is completely dissipated in the gas. The main contribution to the error in the absolute dose determination actually comes from the uncertainty of the W value.

The value of W in a gas is the result of a complex chain of processes involving mainly excitation and ionization. In the interaction of heavy ions with matter, a theoretical approach to the calculation of the ratio of the cross sections of ionization over excitation is extremely complex. It involves the cross sections for each kind of excitation, the total cross section for ionization and the energy distribution of the secondary particles in each ionizing collision. The experimental measurements, even though difficult to perform, are still by far the most important and accurate source of W values (Fig.1).

In a measurement of the differential W value, three quantities have to be determined independently: the energy E of the incident ions, the mean energy ΔE deposited by one ion when traversing the gas gap and the mean number of ion pairs produced in the gas gap by one ion when dissipating ΔE. This last quantity can be obtained as the ratio of the number of primary ions N_{ions} and the chamber charge output Q, integrated during a certain interval.

According to this, the differential W-value can be expressed as: $w_E = \Delta E \cdot (N_{ions}/Q) \cdot e$

CP495, *Experimental Nuclear Physics in Europe*, edited by B. Rubio et al.
© 1999 American Institute of Physics 1-56396-907-6/99/$15.00

We performed a measurement with 7.6 MeV/u $^{12}C^{6+}$-ions in air. Using a parallel plate IC with very thin windows (0.5 μm thick) and 2.94 cm of gas gap. The initial energy E of the ions entering the gas gap was obtained with a time-of-flight technique, and the energy lost in the vacuum window and in the air was calculated to obtain the mean value of the ions in the gas gap. The energy loss ΔE in the gas gap (Fig 2, upper plot) was obtained as the difference between the energy deposited in a Si detector (500 μm thick) when placed directly after the IC and when placed after only the IC foils. Carbon ions of three different energies (11.5, 8.6 and 5.9 MeV/u) were used for the energy calibration in vacuum. For the ratio N_{ions} /Q (Fig.2, lower plot), the charge output Q from the IC was integrated by an electrometer, while the corresponding number of ions traversing the detector, N_{ions}, was recorded simultaneously by a CR39 track-etch detector [4] placed behind the IC. The small leakage current was regulary checked between the measurements.

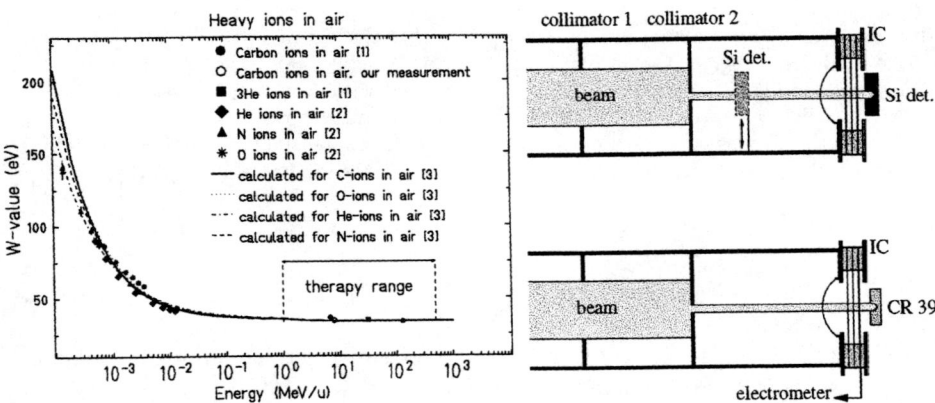

FIGURE 1. Compilation of the reported experimental values for heavy ions in air, showing the lack of data available in the energy range of interest for therapy.

FIGURE 2. Setup used for the measurement of the energy loss, and the ratio N_{ions} /Q .

With this method, we found for the W value of 7.6 MeV/u $^{12}C^{6+}$-ions in air the preliminary value: w_E = (34.1 ± 0.4) eV. The result is in good agreement with the parametric formula of Dennis [3], but is somewhat lower than the value of (36.2 ± 1) eV for 6.7 MeV/u $^{12}C^{6+}$-ions reported by Kanai et al [1] .

In principle, the same technique can be applied at higher ion energies to measure the variation of w in the energy range used for therapy.

REFERENCES

[1] Kanai et al. , Rad. Res. 135, 293-301 (1993)
[2] Huber et al., Rad. Res. 101, 237-251 (1985)
[3] J.A. Dennis, Phys. Med. Biol., Vol. 18. No. 3, 379-395 (1973)
[4] J.Dreute et al., Nucl. Tracks 12 (1-6), 261 (1986)

Production of epithermal neutron beams for BNCT

P. Colangelo, N. Colonna, P. Santorelli, V. Variale, V. Paticchio, G. Maggipinto

Istituto Nazione Fisica Nucleare, Sez. di Bari and Dip. Fisica, Università, V. Amendola 173, 70126 Bari, Italy

Abstract. Boron Neutron Capture Therapy, a promising modality for the treatment of malignant tumors, relies on the use of neutron beams of suitable energy and intensity. For deep-seated tumors, simulations indicate that the optimal neutron energy is in the epithermal region, and in particular between 1 and 10 keV. Therapeutic neutron beams of high spectral purity could be produced with low-energy accelerators, through a suitable neutron producing reaction. In this talk we present an overview of some recently investigated reactions for the production of intense epithermal neutron beams for BNCT, and their potential use towards the setup of an hospital-based BNCT facility.

INTRODUCTION

Boron Neutron Capture Therapy (BNCT) is a promising two-component modality for the treatment of some malignant tumors against which ordinary methods are largely ineffective. In particular, this kind of hadrotherapy is currently being considered for the treatment of Glioblastoma Multiforme (GBM), an aggressive form of brain cancer, as well as of other types of brain, liver and skin cancer [1]. The method is based on the radiation damage produced by high LET particles emitted following neutron capture by ^{10}B. To obtain a high tumor control probability with minimal collateral effects on healthy tissues, large concentrations of ^{10}B have to be selectively accumulated in the tumor cells, by means of specific boronated compounds. The patient is then irradiated with neutron beams of energy and intensity suitably chosen so that a maximum density of thermal neutron is reached in the proximity of the tumor region.

Promising results in the treatment of various tumor diseases by BNCT justify increasing efforts to improve the two components of the therapy, the ^{10}B carriers and the neutron beams. In particular, while clinical trials are currently undergoing at nuclear reactors, recent developments in the accelerator technology are making realistic the possibility of setting up high-quality, safe and cost-effective

CP495, *Experimental Nuclear Physics in Europe*, edited by B. Rubio et al.

epithermal neutron sources for BNCT based on accelerators (see for example ref. [2]). Compared to nuclear reactors, accelerator-based sources may allow to produce epithermal neutron beams with higher spectral purity, that is with energy distribution narrowly centered around the optimal value required for the treatment of deep-seated tumor. Furthermore, the advantages of accelerator sources may enable the installation of BNCT facilities in metropolitan areas and even in hospitals.

In this talk we report on a systematic search of neutron producing reactions, to be used in conjunction with a high-current accelerator as an epithermal neutron source for BNCT. The needs and the main features of the reactions are presented and discussed with regards to their potential use for the setup of an hospital-based BNCT facility.

EPITHERMAL NEUTRON BEAMS FOR BNCT

Analysis of the dose deposition in tumor and in normal tissues shows that the therapeutic effect would be optimized by using a monoenergetic neutron beam with the energy chosen on the basis of the tumor depth [3]. The optimal neutron energy for the treatment of deep-seated tumors has been investigated by means of realistic simulations of the dose produced by the neutron beam. In the simulations, performed with the Geant/Micap package, a realistic description of the geometry and composition of the head were used, with a tumor region located at 5 cm depth inside the brain. Concentrations of 10 and 43 parts-per-million of ^{10}B in blood and tumor respectively were assumed (these values are typical of BPA, a ^{10}B-carriers currently employed in clinical trials performed at research reactors).

Figure 1a shows the therapeutic gain (TG) as a function of neutron energy. TG is defined as the ratio between the dose released to the tumor and the maximum dose to normal tissues. The results here shown indicate that the optimal neutron energy, for the tumor location and ^{10}B concentrations considered, is between 1 and 10 keV. Neutrons of lower energy thermalize at depths smaller than the tumor location, while for higher energies the recoiling protons lead to a sharp increase of the dose released to the normal tissues, in particular to the skin and at the brain surface. The depth-dose profile in brain is shown for the close-to-ideal case of a 4 keV neutron beam in figure 1b. The total dose (symbols) is shown together with the partial doses (dashed curves) produced by the various reactions induced by neutrons in biological tissues in presence of ^{10}B (10 ppm), that is n-p elastic scattering, the ^{1}H(n,γ)^{2}H, the ^{14}N(n,p)^{14}C and the ^{10}B(n,α)^{7}Li capture reactions.

Although in practice monoenergetic neutron beams of few keV cannot be produced, one should try to approach such a condition to optimize the therapeutic effect. In this respect, an important role is played by the optimization of the material and geometry of the beam shaping assembly, used to moderate to the useful energy neutrons produced in a nuclear reactor. However, a large improvement in the quality of the epithermal neutron beam could be achieved by using accelerator-based neutron sources, provided that an appropriate proton- or deuteron-induced

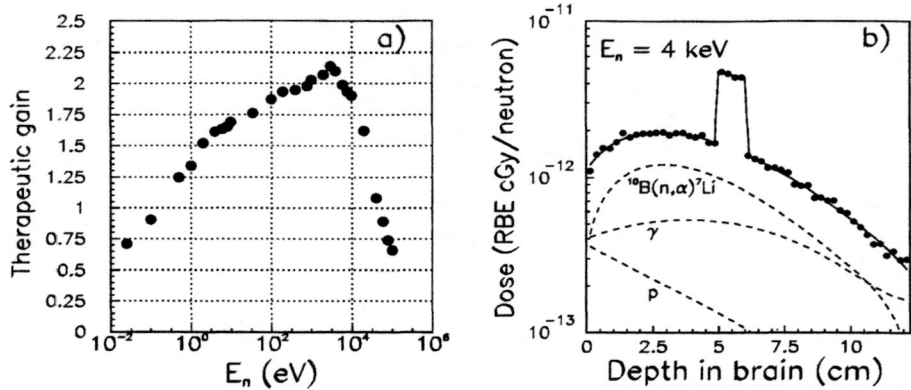

FIGURE 1. a): Ratio between the BNCT dose released to tumor and the maximum dose to normal tissues as a function of neutron energy. b) Depth-dose distribution in the brain for a neutron beam of 4 keV energy. For more details see ref. [4].

reaction is used for neutron production. The choice of the beam energy and of the target is mainly dictated by the need of a high yield of low-energy neutrons ($E_n < 1$ MeV), with a small contamination of high-energy neutron and γ-rays. Furthermore, the size and cost of the accelerator could be minimized by choosing a low energy for the primary beam, while the target should present good mechanical and thermal properties. Finally, a stable residue should be produced in the reaction, to reduce safety problems associated with storage and disposal of used targets.

NEUTRON-PRODUCING REACTIONS

A quick survey of some neutron producing reactions reveals that the conditions mentioned above are not easily met by a unique reaction, and in fact are often conflicting. This is the case, for example, of the need of high neutron yield and that of a low-energy primary proton or deuteron beam. In order to search for a reaction that fulfills the best compromise between the various needs, it is important to collect data on several neutron producing reactions, in terms of neutron yields, energy and angular distributions. To this aim, a systematic study of neutron production for BNCT has recently been started by the SOLONE group, of INFN, Italy. In particular, (p,n) and (d,n) reactions at energies around 2 MeV have been measured at the 88" Cyclotron of Lawrence Berkeley National Laboratories, in collaboration with groups from the Nuclear Science and Life Science Divisions of LBNL [4]. Liquid scintillator cells were used in these measurements to detect neutrons of $E_n > 100$ keV. Studies of near-threshold reactions are instead being performed at INFN Laboratori Nazionali Legnaro, Italy, by means of ^{3}He spherical detector

TABLE 1. Neutron producing reactions of potential interest for BNCT

Reaction	E_{in} (MeV)	Tot. Yield (n/μC)	$<E_n>$ at $0°$ (MeV)	Frac. of neut. at $0°$ with $E_n >1$ MeV
^{7}Li(p,n)^{7}Be	2.5	$9.8 \cdot 10^8$	0.6	0
^{9}Be(p,n)^{9}B	2.5	$3.9 \cdot 10^7$	0.4	0
^{9}Be(d,n)^{10}B	1.5	$3.3 \cdot 10^8$	1.66	50 %
^{13}C(d,n)^{14}N	1.5	$1.9 \cdot 10^8$	1.08	30 %
^{7}Li(p,n)^{7}Be	1.95	$6. \cdot 10^7$	0.06	0

(LND 27036), that allow to measure neutrons of energy < 100 keV.

In table 1, the results of the reactions studied so far are summarized. Although a final decision on the applicability of the different reactions to BNCT requires complete simulations of the moderation process, some general conclusions can already be drawn from the features of the reactions here presented. The well known ^{7}Li(p,n)^{7}Be reaction at proton energies around 2.5 MeV is very convenient in terms of neutron energy and yield, but its use is complicated by the low melting point of the Li target (170°) and the production of radioactive ^{7}Be residue. The ^{9}Be(p,n)^{9}B reaction is also characterized by the emission of low energy neutrons, but with a considerably lower yield. For the ^{9}Be(d,n)^{10}B reaction at $E_d=1.5$ MeV, recently proposed as an alternative to the previous two reactions, our measurements revealed the presence of a large contamination of high energy neutrons, which makes this reaction not suitable for neutron production for BNCT. On the contrary, the relatively large yield and low contamination of high energy neutrons make the ^{13}C(d,n)^{14}N reaction at $E_d=1.5$ MeV potentially interesting for an hospital-based facility, thanks also to the good thermal properties of the C target and the particularly low-energy of the primary beam, which would result in a relatively simple and inexpensive accelerator. The high current required by this reaction (~ 100 mA) may be within reach of the high-intensity accelerator technology currently being developed. Finally, a class of convenient neutron sources may be represented by the near-threshold reactions. In this case, the low neutron yield is compensated by the very small neutron energy, which requires little or no moderation to obtain a therapeutic beam. In particular, considering the drawbacks of the Li as target, an interesting reaction could be the Be(p,n) at $E_p \sim 2.2$ MeV. This reaction, currently being investigated, may turn out to be useful in conjunction with a 30 mA RFQ accelerator, like the one presently being developed at Laboratori Nazionali Legnaro.

REFERENCES

1. Barth R.F., Soloway A.H. and Brugger R.M., *Cancer Investigations* **14**, 534 (1996).
2. Chu, W.T., *Proceedings of the 7th Intern. Symposium on Neutron Capture Therapy for Cancer*, Zurich, Sept. 1996.
3. Wallace S.A et al., *Phys. Med. Biol.* **39**, 897 (1994).
4. Colonna N. et al., *Med. Phys.* **26**, 793 (1999).

Spallation reactions: a tool for RNB production and a neutron source for nuclear waste transmutation

J. Benlliure[a], P. Armbruster[b], M. Bernas[c], A. Boudard[d], E. Casarejos[a], S. Czajkowski[e], T. Enqvist[b], F. Farget,[c] R. Legrain[d], S. Leray[d], M. Pravikoff[e], K.-H. Schmidt[b], C. Stéphan[c], J. Taieb[b,c], L. Tassan-Got[c], C. Volant[d]

[a] Univ. of Santiago de Compostela, [b] GSI Darmstadt [c] IPN Orsay [d] SPhN Saclay [e] CENBG Bordeaux

Abstract. A large experimental program was initiated at GSI to study in detail the spallation reactions. The use of the inverse kinematics allows to determine the production cross section and recoil momentum of the spallation residues with high accuracy. The comparison of the experimental data with model calculation gives valuable information about the reaction mechanism and the application of these reactions to RNB production and to the problematic of nuclear waste transmutation.

INTRODUCTION

Spallation reactions recently gained new interest because of their application for RNB production [1] and to the problematic of nuclear waste transmutation where they constitute an optimum neutron source [2]. A common point to these two applications is the residue production in these reactions. While in the first case the reaction residues constitute the possible nuclei that can be used as RNB, in the second one they will determine the radioactive pollution and material damages in the spallation source.

The experimental information available until now about the production of residues in spallation reaction was mainly obtained by mean of spectroscopic techniques where the spallation residues are identified after β-decay and only few shielded isotopes can be identified as direct spallation residues. Similar situation concerns the theoretical description of these reactions. Most of the existing simulation codes are able to describe qualitatively these reactions but gives rather poor quantitative predictions.

In order to improve the experimental knowledge of spallation reactions, a program was started at GSI to measure the reaction kinematics and the residue cross sections down to 0.1 mb with an accuracy of 10% for a series of different projectile-target combinations. In addition, big efforts are being done to improve the theoretical description of this process.

CP495, *Experimental Nuclear Physics in Europe*, edited by B. Rubio et al.

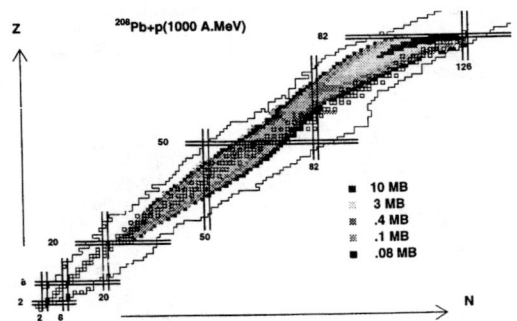

FIGURE 1. *Measured isotopic cross sections of the residues produced in the reaction* $^{208}Pb(1\ A$
GeV) + p.

EXPERIMENTAL METHOD AND RESULTS

The experiments at GSI are performed in inverse kinematics with relativistic
energies. The advantage of using inverse kinematics is the forward focusing of
the reaction products which then can be identified in-flight with a recoil separator
prior to their β-decay. The experiments have been performed at the GSI fragment
separator [3]. The primary beams of ^{197}Au, ^{208}Pb and ^{238}U with an energy of 1 A
GeV impinged on a liquid hydrogen or deuteron target which were mounted at the
entrance of the FRS.

The acrhomatic spectrometer FRS [3] equipped with an energy degrader, two position sensitive scintillators and a multisample ionisation chamber allow to identify
in charge and mass number the reactions products with a resolution $\Delta A/A \approx 400$.
More details about the experiments and the data analysis can be found in Ref.
[4–6].

In Fig. 1 we present in a chart of the nuclides all the residues measured in the
reaction ^{208}Pb+p. More than 800 different spallation residues were identified in
this reaction. As can be seen in this figure, the spallation residues populate two
different regions of the chart of the nuclide. The upper region correspond to the
spallation-evaporation residues which populate the so call evaporation-residue cor-
ridor. The second region populates medium-mass residues produced in spallation-
fission reactions. Then, both reactions mechanism,fission and evaporation, should
be considered to describe the production of spallation residues.

MODEL CALCULATIONS

In the frame of this work we developed a Monte-Carlo model calculation to describe these reactions. These reactions are a quite complex process that can be
explained in two steps. First the projectile interacts with the target. As a consequence of this fast interaction the target-like nucleus will lose some nucleons and

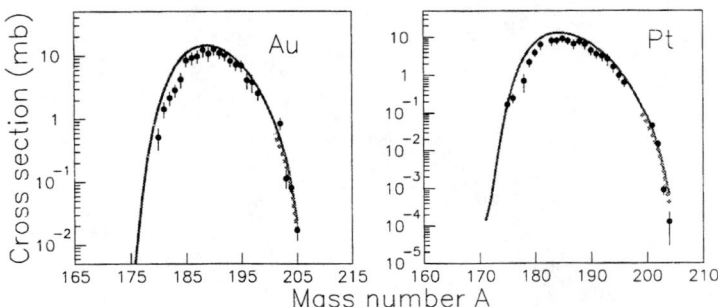

FIGURE 2. *Isotopic production cross sections of Au and Pt isotopes (evaporation residues)*
obtained in the reaction $^{208}Pb(1\ A.GeV)+Cu$ (circles) [12] compared with our model calculations.

will be excitate. This prefragment will deexcitate in a second step by evaporating particles or fissioning. This two processes determine the nature of the final spallation residues.

In the model the first interaction between the projectile and the target can be described by means of an intranuclear-casdace model (in the case of light projectiles) [8] or by an abrasion model (for heavy-ion reactions) [7]. In the second stage the particle evaporation is described by using the Weisskopf formalism with an improved description of the level densities [9]. The limitation of these calculations is that the computation times are so long that we can not afford to calculate isotope production of very neutron-rich nuclei which have cross sections lower than 1 μb. To overcome this limitation we developed also an analytical model describing the production of very neutron-rich nuclei in fragmentation-evaporation reaction. A full description of this model can be find in Ref. [10].

The fission decay width is described according to the transition-state method of Bohr and Wheeler but including the effect of the nuclear matter dissipation [9]. We introduced also a semiempirical description of the fission-fragment properties taking into account the influence of nuclear structure on the fission process. The model is supposed to be able to predict the magnitudes and widths of the different fission channels (symmetric and asymmetric), the even-odd fluctuations and the neutron-to-proton ratio of the fission-fragment distributions as a function of nuclear charge, mass and excitation energy of the fissioning nucleus in a global way. A more detailed description of the model can be found in ref. [11].

A complete calculation with this model allows us to predict the complex nuclide distribution resulting from spallation or peripheral relativistic heavy-ion collisions. In figures 2 and 3 we compare the predictions obtained with our model with existing experimental data for both evaporation and fission residues. As can be seen the agreement between the data and the calculations is better than a factor of two. This comparison allows us to be confident on the predictive power of this model to be used for investigating the possibilities of production of RNB [1]

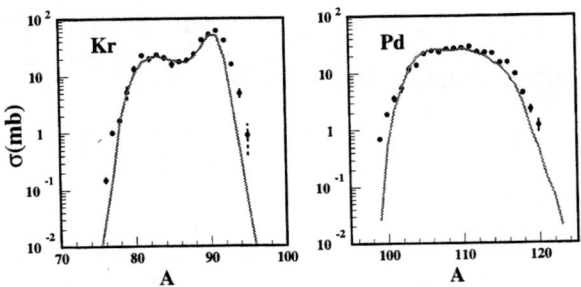

FIGURE 3. *Isotopic production cross sections of Kr and Pd isotopes (fission residues) obtained in the reaction* $^{238}U(1\ A.GeV)+Pb$ *(circles) [13] compared with our model calculations.*

I CONCLUSION.

The investigation of spallation and fragmentation reactions in inverse kinematics is a powerful tool which gives new insight for the description of these reaction mechanisms. The large amount of experimental information obtained with this technique is very important for the design of future neutron sources for radioactive nuclear waste transmutation, but also for studies on RNB production. This experimental data also allow to improve considerably the model description of these reactions.

REFERENCES

1. J. Benlliure et al., Proc. of the XXXVII Int. Winter Meeting on Nuclear Physics, Bormio, Italy, Feb. 1999
2. C.D. Bowman et al, Nucl. Instr. Methods A 320 (1992) 336
3. H. Geissel et al., Nucl. Instr. Methods B 70 (1992) 286
4. L. Tassan-Got et al., Proc. of the Int. Conference on the Physics of Nuclear Science and Technology, Island Mariott, Long Island, USA, Oct. 1998
5. W. Wlazlo et al., Proc. of the 3rd Int. Conference on Accelerator Driven Transmutation Technologies and Applications, Praha, Czech Republic, 1999
6. E. Casarejos et al., Proc. of this conference
7. J.-J. Gaimard, K.-H. Schmidt, Nucl. Phys. A 531 (1991) 709
8. J. Cugnon et al., Nucl. Phys. A 620 (1997) 475
9. A.V. Ignatyuk et al., Nucl. Phys. A 593 (1995) 519
10. J. Benlliure et al., to be published in Nucl. Phys. A
11. J. Benlliure et al., Nucl. Phys. A 628 (1998) 458
12. M. de Jong et al., Nucl. Phys. A 613 (1997) 435
13. T. Enqvist et al., to be published in Nucl. Phys. A

Transmutation in ADS and needs for nuclear data, *with an introduction to the n-TOF at CERN*

E. González, M. Embid, R. Fernández, J. García and D. Villamarín

FACET project, Department of Nuclear Fission, Centro de Investigaciones Energéticas, Medioambientales y Tecnológicas (CIEMAT), Av. Complutense 22, 28040 Madrid, Spain. Project supported by Empresa Nacional de Residuos Radiactivod, S.A. (ENRESA), Spain.

Abstract. Transmutation can help in the nuclear waste problem by reducing seriously the life and amount of the most dangerous isotopes (radiotoxicity, heat, packing volume and neutron multiplication reductions). ADS are one of the best technologies for nuclear waste transmutation at large scale. Although enough information is available to prepare conceptual designs and make assessments on their performance, a large R&D campaign is required to obtain the precision data required to optimize the detailed engineering design and refine our expectations calculations on waste reduction by the different transmutation strategies being proposed. In particular a large R&D effort is required in nuclear physics, where fundamental differential measurements and integral verification experiments are required. In this sense, the PS213 n-TOF at CERN PS (at Switzerland) will become one of the largest installations to perform the fundamental differential measurements and a wide international collaboration has been setup to perform the cross section measuring campaign. Similarly, the MUSE and several other experiments taking place and in preparation in Europe, USA and Japan will provide the integral verification.

TRANSMUTATION AND ACCELERATOR DRIVEN SYSTEMS

Nuclear wastes are one of the main concern about the energy production by nuclear fission. Managing these wastes is a major problem for which no solution, free of technical uncertainties and of severe difficulties for public acceptance, has been found. Transmutation is considered in the last years as a viable option to reduce the nuclear waste problem and in this way complement the radioactive waste repositories. The present objective for the transmutation technology is to convert the long lived isotopes contained in the nuclear wastes to other isotopes with shorter half-life that finally will decay to stable isotopes.

The projected radioactive waste world-wide by year 2010 is 300000 tons of spent fuel, containing: 3000 tons of Plutonium, 140 tons of Neptunium, 120 tons of Americium and higher actinides and 10000 tons of Fission Fragments (FF) from which about 400 tons are very long lived (^{99}Tc, ^{135}Cs, ^{129}I). There are long lived isotopes both in the activated materials and in the fission fragments, but the most urgent to be transmuted are the long lived actinides. The main reasons are: first they represent the main contribution to the radiotoxicity inventory after 1000 years, second they contain a large amount of energy than can be released by fission and finally they constitute the largest proliferation hazard in the nuclear waste.

CP495, *Experimental Nuclear Physics in Europe*, edited by B. Rubio et al.
© 1999 American Institute of Physics 1-56396-907-6/99/$15.00

Transmutation of different isotopes is achieved by different physical processes. For fissile isotopes like ^{239}Pu, the main transuranic component of the nuclear wastes, direct fission converts these isotopes on FF, of much shorter average half-life. For the not so easily fissionable actinides, as most minor actinides (MA), one or several captures are required before the fission can finally transmute the actinide on FF. These two methods produce large amount of energy and some extra neutrons during the transmutation process. Finally neutron absorption, mainly by neutron capture, can be used to convert long lived FF on others isotopes of much shorter half-life. All these transmutation processes require large amounts of neutrons, that in addition must be either epithermal (En: 0.1 eV – 1 KeV) or very fast (En > 1 MeV), in some cases. The most efficient methodology to produce these neutrons is the use of nuclear multiplying assemblies, either critical (reactors) or subcritical with an external neutron source. In the second option the combination of one accelerator plus one spallation target appears as the most efficient choice for the external neutron source. This last type of systems, that combine a subcritical nuclear device with an accelerator and a target, are known as Accelerator Driven Systems (ADS).

When comparing the two possibilities, The ADS concept provides a very important *flexibility* that make more *feasible* the actinide transmutation. The most important flexibility aspects are: the choice of the safety operation margin (subcriticality level vs. the low β_{eff} of Pu and MA in a critical reactor), flexibility in the relation of neutron multiplication and total power or the corresponding transmutation performance of the device and flexibility on the neutron economy. This last element is the fundamental key that allows to consider: new fuel cycles (use of Pu and MA as fuel, use of Th based fuels, ...), the possibility to use impossible combinations of fuels and spectrum (like thermal Th reactors), the possibility to include absorbing targets (as in the case of FF being transmuted in targets) and the possibility to simplify the present fuel cycle (no fuel enrichment, ...). In addition the general flexibility will open the possibility of progressive licensing, because in the ADS very high power density can be reached even with a small fraction of the total fuel.

NEEDS FOR NUCLEAR DATA

At present there is no prototype of ADS and only very few installations that allow to verify the basic concepts and the achievable performance of different ADS designs. In parallel with the rich experimental activity on those few installations, a very large effort is being developed world wide to better understand and to optimize the ADS designs using detailed and very sophisticated computer simulations of these devices. These simulations require precise and complicated algorithms able to describe in detail the 3D structure of the ADS core, the neutron transport inside the ADS, their moderation, the induced nuclear reactions and the consequent isotopic evolution of the materials at the different positions of the ADS. Suitable algorithms and codes start to be available and more will come, however none of these codes can be more precise than the nuclear databases used on the simulation. The international community studying the transmutation process in ADS has recognized an important lack of data of enough quality to allow optimization of detailed designs of ADS. Examples of the codes for ADS studies, the type of simulations performed with these codes, the results of these simulations and the uncertainties introduced in the simulations by the lack of precise nuclear data, can be found in some publications and presentations to recent international conferences from CIEMAT (1-5) and other groups.

The main physical processes taking place in one ADS devoted to actinide transmutation and requiring new or better nuclear data can be classified on four groups. First, the processes in the spallation target, where better data is needed for the proton cross section for nuclear interaction, (p,xn+X), the neutron multiplicity of the spallation process and the energy and angular distribution of produced neutrons as well as the nature and amounts of the produced residual nuclei. In addition the cross section of processes with very fast neutrons (fast fission, (n,xn),...) are relevant for this element of the ADS. Second, the processes producing the neutron multiplication and actinide transmutation in the core. In this case the fission, elastic, capture, (n,2n), inelastic and total differential cross sections for all actinides with half live larger than few days together with the average number of neutrons per fission and the fission yields for fast neutrons on the different actinides are the most urgent needs. In addition a better resolution and precision at intermediate and fast neutron energies will be very interesting. Third, the processes for the transmutation of long-lived FF, LLFF, in special targets. For this purpose the elastic, capture, total and absorption differential cross sections for all the energy regions (thermal, epithermal and fast neutrons) and with good precision and high energy resolution should be measured for the LLFF: (^{99}Tc, ^{129}I, ^{135}Cs, ^{93}Zr,...). And fourth, the processes in the coolant, diffusing media and structural materials that require investigation on the elastic, absorption, inelastic, (n,xn) and total differential cross sections for all the energy regions for these materials (Pb, Bi, Fe, ...).

Different types of experiments are required to obtain all this nuclear information required by the ADS application to transmutation. In particular, specific experiments will be required to obtain the spallation and fission yields (like some experiments already going on at GSI, Germany), and differential experiments, based on the time of flight (TOF) techniques and alike, should be used to obtain precision determination of the different nuclear cross section. On the other hand, integral experiments with zero power, small and full scale, ADS-equivalent multiplying systems and fast critical assemblies (like the MUSE experiments already going on at Cadarache, France) are needed to fine tuning the cross sections and other parameters related to criticality and to validate the concepts, models and the data coherency at full ADS level.

THE PS213 N-TOF EXPERIMENT AT CERN

A highly relevant, recently approved, experiment for the measurement of basic nuclear data relevant to transmutation and ADS is the CERN PS213 (6). This experiment proposes to build a large neutron TOF facility using the PS accelerator of CERN. The PS can deliver 2-3×10^{13} 24 GeV protons per pulse, for pulses of ~4 ns

RMS at a repetition rate of 0.1-1 Hz. The protons will be sent to a lead target to produce the most instantaneously intense neutron beam shortly available. An existing 230 m long tunnel will host the time of flight vacuum pipe and instrumentation. The strongest points of PS213 will be this very high instantaneous neutron beam intensity and the high energy resolution. These properties should allow performing the very difficult measurements of radioactive samples, low cross section reactions, high resolution in a wide energy range and the coverage of the high-energy part.

PS213 was approved the 15th of April 1999 by the CERN Research Board and the first beam is expected by April 2000. A large international collaboration has been set up to develop the experimental program of the approved facility. This collaboration is preparing a proposal to the 5th Framework program of the European Union and further agreements with international organizations, IAEA and NEA/OCDE, for evaluation and dissemination of the measured cross section.

In addition to the cross sections relevant to nuclear waste transmutation and ADS, PS213 will include cross section measurements interesting for astrophysics and basic nuclear physics. For this purpose the collaboration is planning to develop and deploy a large set of detectors including: detectors for neutron flux determination and monitoring (^{3}He ionization and scintillation fast detectors, ^{6}Li based detectors, fast NE213 liquid scintillators and BF_3 and CeF_3 for time-energy calibration), fission detectors: (parallel plate avalanche counters, PPAC, and parallel plate induction chamber, PPIC) and neutron capture detectors (total absorption 4π segmented crystal ball, probably of BaF_2, total absorption 4π segmented liquid noble gas γ detectors, and C_6D_6 and Moxon-Rae detectors for the experiments in the year 2000).

REFERENCES

1. Fernández, R. et al., *Performance on actinide transmutation of Lead-Thorium based ADS*, and Embid, M. et al., *Systematic uncertainties on Montecarlo simulation of lead based ADS*, contributions to the 5th International Information exchange meeting on actinide and fission product partitioning and transmutation MOL (Belgium). 1998.

2. García-Sanz, J. M. et al., *Isotopic Composition Simulation of the Sequence of Discharges from a Thorium TRU's, Lead Cooled, ADS*, contribution to ADTTA 99, Prague (Czech Republic), 1999.

3. González, E. Et al., *EVOLCODE: ADS combined neutronics and isotopic evolution simulation system*, Accepted to be presented in MC'99 Conference, to be held in Madrid, September 1999.

4. Díez, S. et al., *Simulation of the TARC neutron flux measurements (0.1 eV to 10 KeV)*, CIEMAT DFN/TR-01/II-98, 1998. And Díez, S. et al., *Simulation of the* ^{99}*Tc capture rate measurements performed in the TARC experiment*, CIEMAT DFN/TR-01/II-98, 1998.

5. Embid, M. et al., *Sensitivity Study for the Keff of a Lead based ADS on the cross sections evaluations: ENDFB-6.4 and JENDL-3.2*, CIEMAT Technical Report **846**, 1998.

6. Abramovich S. et al., *European collaboration for high-resolution measurements of neutron cross sections between 1 eV and 250 MeV, The TOF Collaboration*, CERN/SPSC99-8 and CERN/SPSC/P310, 1999.

Comparison between calculations and experimental results on spallation reactions of interest for hybrid systems

C. Volant, A. Boudard, R. Legrain, S. Leray and W. Wlazlo

DAPNIA/SPhN CEA/Saclay F-91191 Gif-sur-Yvette Cedex, France

Abstract. Data on proton induced spallation reactions concerning the production of neutrons obtained at SATURNE and of spallation residues at GSI have been compared with two-step models including different Intra-Nuclear Cascades (INC) followed by different evaporation codes. Results and improvements of the codes are discussed.

In the design of Accelerator Driven Systems, it is necessary to precisely predict the energy spectrum and angular distribution of the neutrons produced in the spallation target. This is important for the optimization of the target geometry in terms of useful neutron production and spatial distribution of the neutron flux. The spallation residue production has also to be well understood since these nuclei contribute to the radioactivity or/and could cause structural material embrittlement.

The spallation processes are generally described in two steps: the incident light particle first interacts with the target nucleons through successive nucleon-nucleon hard collisions modeled by the so called Intra Nuclear Cascade (INC), leading to the emission of high energy nucleons. At the end of this stage, the remaining nucleus is left with excitation energy and then de-excites either by evaporation or by fission. The reliability of the available calculation code systems is not yet sufficient to assess all the parameters required to design spallation targets. In this contribution, results of codes are confronted to new data and the physical ingredients inside the models are discussed.

As an exemple, fig. 1 presents neutron double-differential cross-sections in Pb(p,xn)X reactions at 1200 MeV measured at SATURNE [1]. The histograms are numerical calculations performed with the TIERCE [2] code system developed at Bruyères-le-Châtel. Within TIERCE two different Intra-Nuclear Cascade models followed by the same evaporation-fission model from Dresner-Atchison [3] have been used: Bertini [4] (solid line) and Cugnon [5] (dotted line) models. At 0°, the more realistic parametrization of the $NN \to N\Delta$ reaction angular distribution [6], introduced in the Cugnon code, solves the problem of the pathologic behaviour

CP495, *Experimental Nuclear Physics in Europe*, edited by B. Rubio et al.
© 1999 American Institute of Physics 1-56396-907-6/99/$15.00

obtained with the Bertini cascade. At higher energy, both models are in quite good accordance with the data between 10° and 85° but Bertini underpredicts the backward angles. Whatever the angle, calculations with the Bertini cascade over-estimates the production cross-sections below 20 MeV while the Cugnon model generally leads to a much better agreement.

The averaged neutron multiplicities per reaction have been deduced from the SATURNE neutron cross-sections and compared with results obtained with TIERCE-Bertini and Cugnon codes for the different studied targets: Al, Fe, Zr, W, Pb and Th at three incident energies: 0.8, 1.2 and 1.6 GeV. From this comparison, it can be stated that the Cugnon INC+evaporation is systematicaly closer to the experiment and is well suited to reproduce the new neutron measurements.

Remaining difficulties with the Cugnon code are illustrated on fig. 2. The description of events with low number of collisions although improved by using new n-p scattering parametrization [6], is still poor. This is mainly visible at forward angles and high outgoing energies. However, the contribution of this part is rather small compared to the total neutron production. Possible improvements are discussed in [5], particularly the introduction of a surface diffuseness is foreseen.

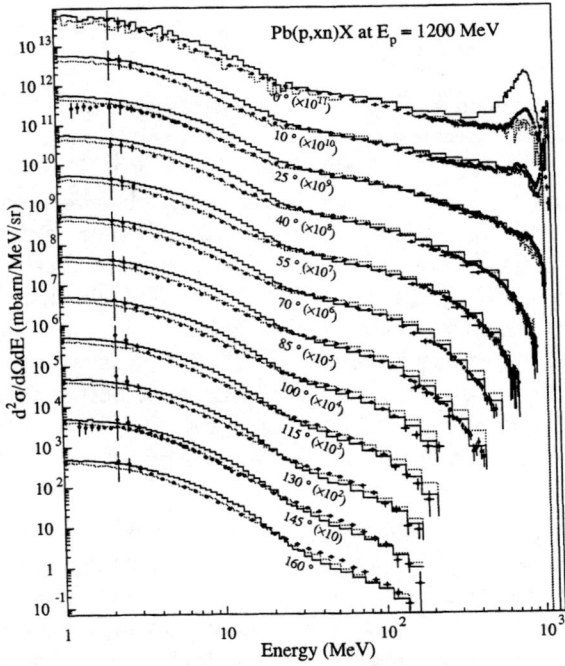

FIGURE 1. *Neutron production double-differential cross-sections measured in proton induced reactions on a 2 cm thick Pb target at 1200 MeV [1]. The histograms represent TIERCE calculations [2] using Bertini [4] (full line) or Cugnon [5] (dotted line) cascade model followed by the same evaporation model.*

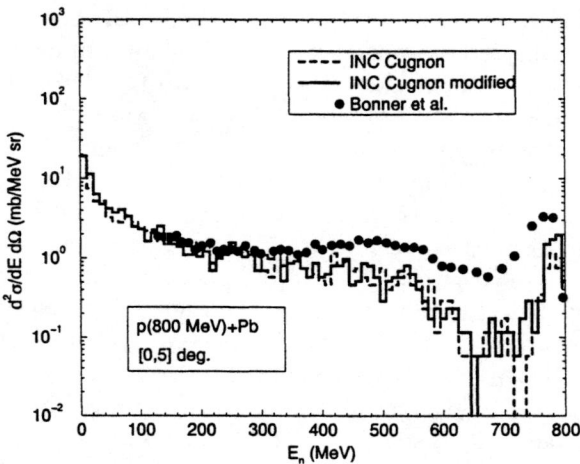

FIGURE 2. *Influence of modification of elastic n-p cross-sections (solid curve). The dashed curve is obtained with the old parametrization.*

The fact that the Cugnon INC model gives a better agreement with the data than the Bertini one, is likely because it leads to lower excitation energies at the end of the INC stage. Since this model includes explicitely the time development of the cascade, a stopping time to switch towards evaporation has been phenomenologicaly parametrized [5]. Variation of this time has weak influence on neutron results but can affect strongly the predictions for residue production as illustrated on the upper part of fig. 3 for the coupling of Cugnon INC to the GSI evaporation code [7]. Here a better agreement with the fragmentation data obtained at GSI [8] is obtained when shortening the stopping time (higher excitation energies). However, some modifications inside the evaporation code, lead to a better agreement with the data when using the nominal time (see bottom of fig. 3). The fission bump, not well reproduced, could be a mean to infer the nuclear viscosity [7].

These ambiguities between the respective role of INC and evaporation could be studied in more details in forthcoming exclusive experiments [9]. The correct description of the evaporation and fission products in spallation reactions is still a large challenge as well as the description of the light composite particle production, not correctly described in the present models. Inclusion of a preequilibrium stage between the INC and evaporation steps or the use of a percolation procedure following the INC stage need to be explored [5] to account for the emission of high energy composite particles.

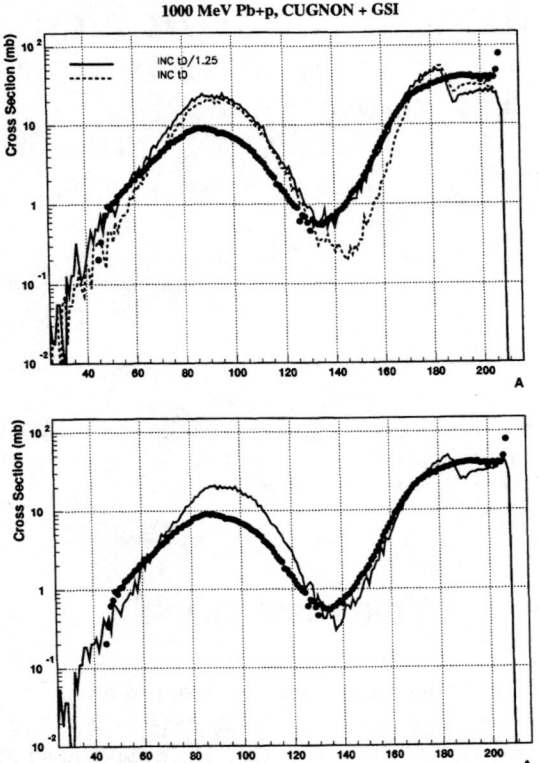

FIGURE 3. *Comparison between measured residual mass production and calculations for the system Pb + p at 1 GeV/A [8]. Upper part : two different stopping times of INC coupled with GSI evaporation code [7]. Bottom part : nominal time t_0 and modified evaporation code.*

REFERENCES

1. X. Ledoux *et al.*, Phys. Rev. Lett. 82 (1999) 4412.
2. O. Bersillon, 2nd Int. Conf. on Accelerator Driven Transmutation Technologies, Kalmar, Sweden, June 3-7, 1996.
3. L. W. Dresner, Oak Ridge Report ORNL-TM-196 (1962); F.Atchison, Jül-Conf-34, paper II, 17 (1980).
4. H. W. Bertini, Phys. Rev. 131 (1963) 1801.
5. J. Cugnon, C. Volant and S. Vuillier, Nucl. Phys. A620 (1997) 475.
6. J. Cugnon et al., Phys. Rev. C56 (1997) 2431.
7. J. Benlliure et al., Nucl. Phys. A 628 (1998) 458.
8. W. Wlazlo et al., Proceedings of the XXXVII International Winter Meeting on Nuclear Physics, Bormio Italy, January 1999.
9. COSY proposal N^0.73.1, Spokesmen A. Boudard and D. Filges.

Experimental Study of Neutron Production in Proton Reactions with Heavy Targets

Vladimír Wagner*, Andrej Kugler*, Czabo Filip* and Petr Kovář†

*Nuclear Physics Institute ASCR, 25068 Řež, Czech Republic
† Škoda, 31600 Plzeň, Czech Republic

Abstract. The spatial distributions of neutrons produced in reactions of 1.5 GeV protons with a 60 cm long tungsten target and with a 50 cm long lead target were determined using the activation method. Both data sets are compared with complex Monte Carlo calculations.

INTRODUCTION

Radioactive waste management is nowadays one of the main problems of nuclear power. Some of the produced radioactive isotopes have a very long lifetime, potentially representing long term radiation hazards. Effective transmutations of these isotopes into the short lived or even stable ones needs continuous neutron fluxes with very high intensity in the 10^{16} n/cm^2 range [1]. Spallation reactions induced by high energy ($\sim 1 GeV$) and intense ($\sim 100 mA$) proton beam in heavy target are proposed as currently technologically achievable source of neutrons in Accelerator Driven Transmutation Technologies (ADTT) projects, see [1]. Detail experimental studies of neutron production in proton induced reactions will allow verification of currently available simulation codes used in design of ADTT.

One of important aspects of such experimental studies is thoroughly knowledge of neutron field around the thick production target. Results of thick targets irradiation[1] with relativistic protons carried out at Laboratory of High Energy (LHE) of Joint Institute of Nuclear Physics (JINR) at Dubna (Russia) are given bellow, as well as their comparison with computer based simulations carried out at Nuclear Physics Institute at Řež.

---

[1] Collaboration with M.I. Krivopustov C.A. Novikov, P.A. Rukoyatkin, Ts. Tumendelger, D. Chulten (LHE JINR) and J. Adam, V.S. Pronskikh, V.I. Stegailov, V.M. Tsoupko-Sitnikov (LNP JINR).

CP495, *Experimental Nuclear Physics in Europe*, edited by B. Rubio et al.
© 1999 American Institute of Physics 1-56396-907-6/99/$15.00

EXPERIMENT

Tungsten and lead targets were irradiated by proton beam with energy 1.5 GeV accelerated by synchrophasotron at the LHE of JINR Dubna. Intensity of produced neutrons was measured by activation method using ^{197}Au and ^{28}Al foils.

The cylindrical tungsten target had thickness 60 cm and diameter 2 cm. The proton beam was collimated to diameter of about 1.8 cm in the front of target. Defocusation changed beam profile from the circle to the ellipse on the distance 40 cm from the front of target with axes of about 1.4 cm and 2.6 cm. The range of 1.5 GeV protons in tungsten is 54.5 cm, hence they were completely stopped in tungsten target. The target was surrounded by 14 thin Au and Al foils which were fixed at surface of the special holder, thin-wall (0.4 cm) Al tube with outer diameter of 3.6 cm.

The cylindrical lead target had thickness 50 cm and diameter 9.6 cm. Uncollimated beam had diameter 3.0 cm. In this case protons were not stopped in the target. The target was placed into the hole in the box with dimension about 1 × 1 × 1 m made from boron plastic. The Au and Al foils were placed on the surface of lead target.

Interaction of neutrons within Au and Al foils leads to the reactions with main characteristics given in Table 1. Corresponding cross sections as function of neutron energy were deduced from literature [2] for energy up to 40 MeV and calculated using LAHET code for energy higher than 40 MeV. The isotope ^{198}Au is produced mainly by low energy neutrons, cross section is very high (1.3-2.1 barn) between 5-10 keV, it fells down approximately linearly in log-log scale from value of 1 barn at 15 keV to value of 0.01 barn at 4 MeV. Production of ^{24}Na, ^{196}Au and ^{194}Au starts from the neutron energy of 6.5 MeV, 8 MeV and 28 MeV, respectively. Hence, foils of ^{197}Au and ^{27}Al can be used as energy threshold detectors of neutrons.

TABLE 1. Interaction of neutrons within Au and Al foils.

Reaction	Halftime [hours]	Gamma lines [keV]
^{27}Al(n,α)^{24}Na	14.96	1368.6, 2754.0
^{197}Au(n,γ)^{198}Au	64.68	411.8
^{197}Au(n,2n)^{196}Au	148.39	333.0, 355.7, 426.0
^{197}Au(n,4n)^{194}Au	39.5	328.5

Gamma decay of isotopes produced in foils were measured using two HPGe detectors to check possible error in their detection efficiency. Corresponding gamma spectra were analyzed using PC-code DEIMOS, which carries out gaussian fit of gamma peaks. Obtained areas bellow peak were corrected to accidental summing of photon cascade. Resulting number of produced radioactive nuclei obtained from both detectors did not differ more than 5 %.

Number of protons hitting the tungsten target was 1.86 × 10^{12}. It was deduced from the activation of aluminium foil placed in front of target. Cross section of ^{27}Al(p,X)^{24}Na reaction for 1.5 GeV protons is 10.0 mbarn [3]. The contribution

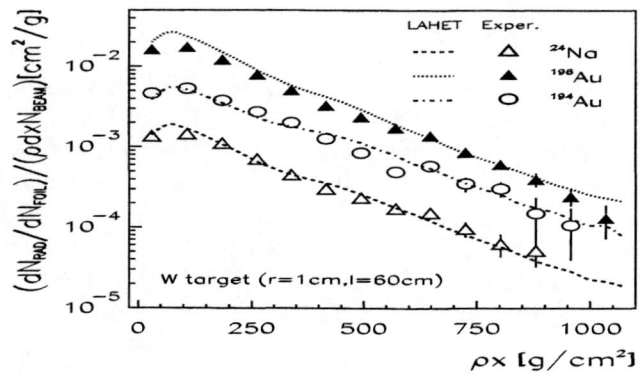

FIGURE 1. Number of produced radioactive nuclei per one proton hitting tungsten target and one foil nucleus as function of thickness of tungsten to be passed by primary protons to achieve corresponding foil position. Symbols correspond to experimental data, curves to results of simulation, see text.

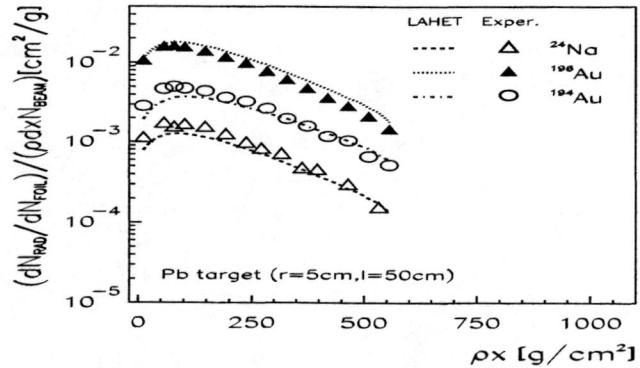

FIGURE 2. The same as on Figure 1 but for lead target.

due to neutrons emitted backwards from the target was estimated using LAHET based simulation. The error of beam dose is estimated to be around 20%. Number of protons hitting the lead target was 42.2×10^{12}. This number was reported to us by operator staff as deduced from the beam current integrator. Its error is estimated to be around 30%.

The obtained spatial distributions of numbers of radioactive nuclei produced per one proton and one foil nucleus are presented at Figure 1-3. Only statistical errors are shown. For low energy neutrons the intensity of neutron field surrounding lead target is more than two orders of magnitude higher in comparison with tungsten target, see Figure 3. Detailed LAHET simulations, see bellow, revealed that the source of these low energy neutrons is scattering of high energy neutrons produced in target within huge "moderator", which surrounded lead target.

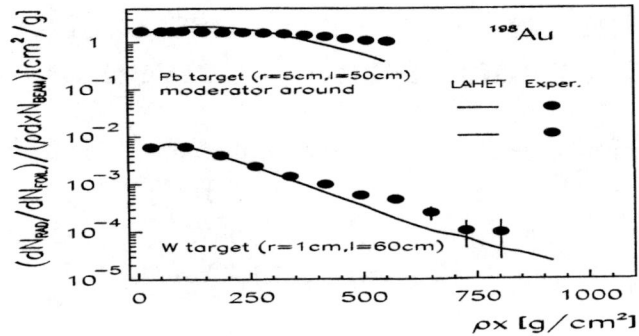

FIGURE 3. The same as on Figure 1 but for ^{198}Au and lead (upper part) or tungsten (lower part) target.

SIMULATIONS

Simulations were carried out using LAHET (Bertini)+MCNP4B code combination. Energy spectra of neutrons obtained by simulation for each foil positions were convoluted with corresponding cross sections.

Diameter of beam was close to the diameter of tungsten target. In this case, emission of secondary protons due to spallation reaction as well as scattering out of primary proton beam has to be taken into account. Scattering of secondary particles in aluminium holder of foils and the defocusation of beam were simulated, too. The proton energy spectra for each foil positions were convoluted with corresponding cross sections. In the case of lead target, influence of the moderator box around this target was simulated. This effect influenced only low energy part of neutron spectra. The number of protons for foil positions was negligible because diameter of lead target was much larger than diameter of beam.

CONCLUSIONS

We have studied neutron production on the thick tungsten and lead targets. Detailed simulation by LAHET(Bertini)+MCNP4B lead not only to good description of spatial distribution shape but also to absolute values compatible within 30 % with experimental data.

REFERENCES

1. Bowman C.D., et al, *Nucl. Instr. and Meth. A* **320**, 336 (1992).
2. T-2 Nuclear Information Service, *ENDF/B-VI Neutron Data*
   http://t2.lanl.gov/cgi-bin/nuclides/endind
3. Brandt, R., et al, *Phys. Rev. C* **45**, 1194 (1992).

Spallation-reaction cross sections relevant for accelerator-driven systems

E. Casarejos[a1], J. Benlliure[a], P. Armbruster[b], M. Bernas[c],
A. Boudard[d], S. Czajkowski[e], T. Enqvist[b], F. Farget[c], R. Legrain[d],
S. Leray[d], B. Mustapha[b], M. Pravikoff[e], K.-H. Schmidt[b],
C. Stephan[c], J. Taieb[b], L. Tassan-Got[c], C. Volant[d], W. Wlazlo[d]

[a] *Univ. of Santiago de Compostela,* [b] *GSI Darmstadt,* [c] *IPN Orsay,* [d] *SPhN Saclay,* [e] *CENBG Bordeaux*

Abstract. A new experimental method has been developed at GSI to study spallation in inverse kinematics. The use of the in-flight spectrometer FRS, allows the isotopic identification of the rection residues in a very short time ($\approx$200 ns). The method and some results for (^{238}U+p) and (^{238}U+d) reactions at 1 A·GeV are shown.

Since the discovery of spallation reactions in the field of astrophysics, they gained recently in interest due to the technical application to the nuclear waste problem [1] and the radio-nuclear beam production [2]. The knowledge of the residues production in spallation reactions is essential for those purposes. It is also stated [3] that the understanding of the process is not enough nowadays for the technical application.

The experiments are performed in inverse kinematics on the FRagment Separator (FRS) the in-flight spectrometer at GSI [4]. The isotopic identification is based in the **magnetic rigidity** of the nuclides: $B\rho = \frac{c \cdot u}{e} \frac{A}{Q} \beta \gamma$ with the atomic mass u, the relativistic parameters β and γ, the elementary charge u and the speed of light c. The scintillators at the focal planes of the FRS, measure the Time-of-Flight and the positions: $\beta\gamma$ is determined as well as the Bρ of the particle. Some details of the analysis are found at [6].

We isolate the **different nuclear charges** (see Figure 1) by using the energy losses at the DEGRADER [5] and two ionization chambers. The contributions from secondary reactions and charge-changing proceses are taken away in this process. The spectrometer resolution in Bρ ($\frac{\Delta B\rho}{B\rho} \approx 10^{-4}$) allows a good resolution in charge ($\frac{charge(Q)}{FWHM(Q)} \approx 200$) and mass ($\frac{mass(A)}{FWHM(A)} \approx 400$). (See Figure 1)

---

[1] Corresponding authors: Experimental Group of Nuclear and Particle Physics, Facultad de Física, University of Santiago de Compostela, E-15706 Spain

CP495, *Experimental Nuclear Physics in Europe,* edited by B. Rubio et al.
© 1999 American Institute of Physics 1-56396-907-6/99/$15.00

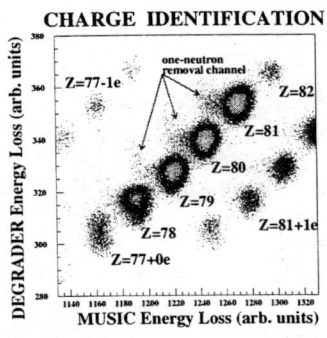

CHARGE IDENTIFICATION

Z=77-1e
one-neutron
removal channel
Z=82
Z=81
Z=80
Z=79
Z=81+1e
Z=78
Z=77+0e

DEGRADER Energy Loss (arb. units)

MUSIC Energy Loss (arb. units)

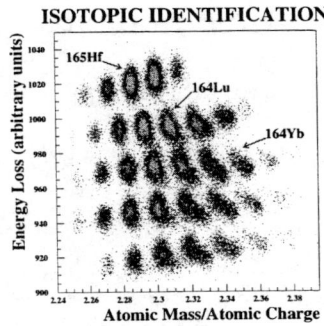

ISOTOPIC IDENTIFICATION

165Hf
164Lu
164Yb

Energy Loss (arbitrary units)

Atomic Mass/Atomic Charge

FIGURE 1. On the left, a typical pattern for elements identification. On the right, a group of isotopes of Z=72,71,70,...

More than **450 different isotopes** have been already identified for $(^{238}U+p)$ and $(^{238}U+d)$ at 1 A·GeV. The comparison between these new data and the simulation codes [7] allow to improve our knowledge on the spallation process. Also the kinematic information is studied [8,9]. (See Figure 2)

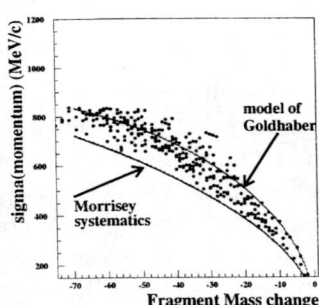

sigma(momentum) (MeV/c)

model of
Goldhaber

Morrisey
systematics

Fragment Mass change

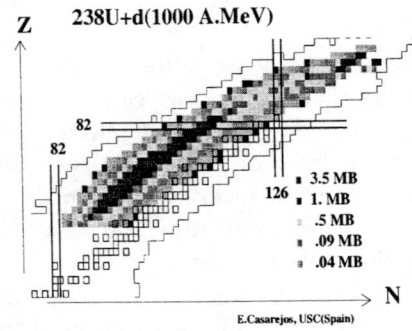

Z

238U+d(1000 A.MeV)

82

82

126

- 3.5 MB
- 1. MB
- .5 MB
- .09 MB
- .04 MB

N

E.Casarejos, USC(Spain)

FIGURE 2. On the left, kinematic information on momenta, of interest in material applications. On the right, a chart of niclides with the measured cross-sections for $^{238}U+d$ at 1A·GeV.

REFERENCES

1. Bowman C.D., *Annu.Rev.Nucl.Part.Sci.* **48**, 505 (1998)
2. Munzenberg G., *Nuc. Inst. and Meth. B* **70**, 265 (1992)
3. Michel R., Nagel P., *"International codes and model ..."*, OECD (1997)
4. Geissel H., et al., *Nuc. Inst. and Meth. B* **70**, 286 (1992)
5. Schmidt K.H., et al., *Nuc. Instr. and Meth. A* **260**, 287 (1987)
6. Tassan-Got L. et al., Proc. of the Int. Conf. on the Physics of Nuclear Science and Technology, Island Mariott, Long Island, USA, Oct.1998
7. Benllire et al., Proc. of this conference
8. Goldhaber A.S., Phys. Lett. **B53**, 306 (1974)
9. Morrisey D.J., Phys. Rev. **C39**, 460 (1989)

Infrared Synchrotron-Accelerator Diagnostic

Anatoly A. Maltsev and Mikhail A. Maltsev

Joint Institute for Nuclear Research, Dubna 141980, Russia

Abstract

Research and development in the technology of accelerator physics experiments using infrared methods, measuring and information systems for passive, nondestructive diagnostics of charged-particle bunches or beams and for nondestructive studies of fast radiation processes are discussed.

The quality of a modern accelerator experiment depends directly on precise and strict satisfaction of the requirements imposed on the accelerated beam. The problem of monitoring and checking the beam during acceleration (accumulation) is a very pressing problem today for all accelerators and storage rings, and it is especially important for the new designs. It requires the development of different methods of diagnostics, including elaboration of existing methods, as well as the development of data acquistion and processing (DAP) systems for implementing the methods in practice.

The objectives of this work are as follows:

* We present the methods and systems of nondisruptive diagnostics and studies of charged-particle (electron, electron-ion, and proton) beams based on the use of their magnetic-bremsstrahlung and/or synchrotron radiation (SR) in a wide spectral range, from the ultraviolet to the far long-wave infrared (IR) region (1).

* We draw attention to the great diversity of problems, both in accelerator experiments, for example the study of the coherence of SR or of the coherent processes at colliders (2) and in other, sometimes quite unrelated fields, for example for experimental research on the problem of transmutation of nuclear waste (3).

It is mainly the visible part of the SR spectrum that is used for diagnostics. Judging from the number of papers, the IR region has still not been given the attention that it undoubtedly merits. This can be explained, to some extent, by difficulties with the optical and detection systems used in the IR part of the spectrum. Nonetheless, the IR region of the SR spectrum is undoubtedly of interest for the designer of diagnostic methods and systems, since it makes it possible to expand substantially the possibilities of SR utilization for scientific and applied purposes.

The IR SR diagnostic methods and systems were based on the results of spectral calculations and the latest achievements of IR technology (optical materials and radiation detectors). The results obtained have shown that both the diagnostic methods and the apparatus are highly efficient and reliable. The technical characteristics of the integral and position-sensitive processing DAP systems, which employ SR (predominantly in the IR region) as the carrier of information about the

CP495, *Experimental Nuclear Physics in Europe,* edited by B. Rubio et al.
© 1999 American Institute of Physics 1-56396-907-6/99/$15.00

beam parameters in the process of beam acceleration, meet all requirements and conditions of an accelerator experiment. They permit measurement and monitoring of the instantaneous (within 1 μs) values of the main parameters and characteristics of a beam, while providing the possibility of detecting radiation in the spectral region 0.3-45 μm, which is much larger than the spectral region employed mainly for this purpose in modern accelerator practice (0.3-1.1 μm). This made it possible to solve the problems arising in an accelerator experiment of providing measurements and monitoring the instantaneous values of the following basic parameters and characteristics of a beam: particle number (current), geometric parameters (dimensions), position in the space of the accelerator, particle distribution in space and time, particle energy, and others (4).

The methods developed and the DAP systems built in order to implement them have been found to be the only possibility of obtaining valuable scientific information about the capability and possibilities of electron ring bunches to trap, confine, and accelerate heavy ions. Unfortunately, the collective method of accelerating heavy ions by electron rings has been found to be unrealizable under present conditions because of the impossibility of forming high-density electron ring bunches (4). However, a special feature of the low-energy electron (and/or electron-ion) ring accelerator (EIRA) of JINR is the large current of the electron ring, which reaches a kiloampere (the beam currents in synchrotrons and storage rings are usually less than an ampere). In addition, the fact that the intensity of the long-wavelength part of the SR spectrum depends weakly on the electron energy ensures that the EIRA is superior to other types of accelerator as a source of SR in the IR region.

The detection systems possess a high sensitivity and good stability against the action of pulsed electromagnetic and radiation fields of the accelerator. This makes it possible to place the detecting units close to the magnetic units and systems of the accelerator without distorting in so doing the configuration of the magnetic fields. The measurement systems are simple to operate and service. They are highly reliable, and their parameters and characteristics remain unchanged over long periods of time, despite the quite severe operating conditions.

REFERENCES

1. Mal'tsev, A. A., and Mal'tsev, M. A., *Atom. Energy,* 80 **3** 190-197 (1996).
2. Kuraev, E.A. and Mal'tsev, A.A., Preprint JINR P9-97-242. Dubna, 1997. p 6.
3. Arkhipov, V.A., et al, *Accelerator Driven System Based on Plutonium Subcritical Reactor and 660 MeV Phasotron*, presented of the Conference on Experimental Nuclear Physics in Europe, Seville, Spain, June 21-26, 1999.
4. Mal'tsev, A. A., and Mal'tsev, M. A., *Techn. Phys.,* 42 **4** 378-384 (1997).
5. Dolbilov, G.V., Mal'tsev, A.A., Sarantsev, V.P. et al, *Synchrotron Radiation Characteristics Electron-Ion Ring of KUTI-20 and it's Perspective*, in Proc. of the Tenth All-Union Sumposium on Charged-Particle Accelerators, JINR, Dubna, 1987, D9-87-105, v.1, pp. 390-393.

Accelerator Driven System Based on Plutonium Subcritical Reactor and 660 MeV Phasotron

V.A. Arkhipov[1], V.S. Barashenkov[1], V.S. Buttsev[1], D. Chultem[1], S.Yu. Dudarev[2], V.I. Furman[1], W. Gudowski[3], J. Janczyszyn[4], A.A. Maltsev[1], L.M. Onischenko[1], G.N. Pogodajev[1], A. Polanski[5], Yu.P. Popov[1], I.V. Puzynin[1], A.N. Sissakian[1], S. Taczanowski[5]

[1]Joint Institute for Nuclear Research, Dubna, Russia; [2]MR-NEC, Moscow, Russia; [3]Royal Institute of Technology, Stockholm, Sweden; [4]Soltan Institute for Nuclear Studies, Swierk, Poland; [5]University of Mining and Metallurgy, Cracow, Poland

Abstract

The proposal presents a PLUTONIUM BASED ENERGY AMPLIFIER TESTING CONCEPT which employs a plutonium subcritical assembly and a 660 MeV proton accelerator. operating in the the JINR (Dubna, Russia). To make the present conceptual design of the Plutonium Energy Amplifier we have chosen a nominal unit capacity of 20 kW (thermal). This corresponds to the multiplication coefficient keff between 0.94 and 0.95 and the energetic gain about 20.

Increased social demands regarding the safety of nuclear power have inspired scientists to search for a higher level of safety by reducing inter alia the risk and consequences of reactivity excursions and the hazard of nuclear waste. Thus the objective: safety including non-proliferation has drawn world wide attention to the idea of accelerator-driven subcritical (ADS) systems.

The main advantage of subcritical system from the safety viewpoint is its distance from the absolutely forbidden area of super prompt criticality, which is about one order of magnitude larger than that of critical systems. This is particularly important when transmuting minor actinides distinct by minute delayed neutron fraction, making the safety margin too narrow. Thus a lowering of the k_{eff} at the desired system energetic gain is the target of the system design.

The non-proliferation issue is connected with Pu and especially with the W-Pu (weapon). The well known idea of W-Pu denaturation by blending it with ^{238}Pu, offering the fastest way of it removing yet leaving the possibility of its use as a fuel open, needs further consideration and intensive investigation. The features of an optimum subcritical unit such as its structure, material composition and distribution, neutron spectrum, material trajectories, etc. - require thorough research.

It should be emphasized that the future industrial scale of the accelerator-driven transmutation technology and its economic determinants demand a very reliable knowledge of the processes and data while the technological complexity

CP495, *Experimental Nuclear Physics in Europe,* edited by B. Rubio et al.

gives rise to a need for detailed computer simulations. Thus, a confrontation of the results of calculations with the experimental data is a general target of the present research project. Two types of ADS are expected to be investigated: Actinide Waste Transmuting System and Plutonium Transmuting System.

The following measurements on the test assembly are planned: energetic gain and its variation for different target material compositions, system neutron multiplication k_{eff} and its variation, neutron generation, neutron spectra, benchmark of various reaction rates and their spatial distributions etc.

We plan to investigate the kinetics of the processes in the Pu-subcritical assembly by the proton-neutron flash in the target inside the Pu-zone from 650-660 MeV protons (1). As we know, these processes have not been investigated experimentally up to now. At the same time this question is very important from the viewpoint of the safety of a new generation of power stations based on the subcritical accelerator-reactor system. One of the interesting questions is the stability of the neutron multiplication coefficient value for the subcritical assembly at the different intensity of the proton beam. This problem is of special importance for the checking the safety multiplication coefficient value for the subcritical system of future real subcritical reactors for power production and for the burning of radioactive waste.

Main Parameters of the System

Phasotron Data

- max. average beam power	1.2 kW
- max. proton energy	660 MeV
- proton energy spread	6 MeV
- max. average beam intensity	$2 \cdot 10^{13}$ p/s
- number of protons per pulse	$0.8 \cdot 10^{11}$
- pulse rate	250 Hz
- pulse length (fwhm)	20 μs
Pulse microstructure:	
- bunch length	10 ns
- interval between bunches	70 ns
- number of bunches per pulse (approx.)	300

Subcritical Assembly Data

- multiplication coefficient - k_{eff}	< 0.95
- max. power of the ADS assembly	20 kW
- forced air cooling	

The installation designed is intended for:

* research in the electronuclear system's dynamics, in particular, methods for measurement and monitoring of the value of k_{eff} and its fluctuations;

* research in the efficiency of the electronuclear technology for utilization of the weapon-grade and technical plutonium;

* obtaining data for designing a full-scale plutonium transmuter by means of optimization of the system's parameters depending on the selection of a reflector, a slower-down and other components of the system;

* measurements of integral cross-sections (radiation capture and fission) of fragmentary and actinide isotopes in various neutron spectra depending on the type of reflector, slower-down and other components of the system;

* studying some possibilities of increase in effectiveness of the electronuclear installations by way of sectioning the breeding assembly (by introducing valve layers);

* obtaining data to correct the physical model and software support by means of comparison with computations on neutron yield and leaking out.

As the main parameters of the accelerator, the beam channel and the assembly have been considered above, we are presenting only the programme of experimental research that consists of the following two parts:

* Research into the characteristics and parameters of the electronuclear installation itself (power of the beam, generation and spectrum of neutrons, heat production per unit of beam power, yield of radioactive isotopes).

* Researches which are carried out with the help of the installation.

As to the second stage of the programme, the detailed description will be presented until the creation of the installation is over. as a result of discussion on the matter at a special JINR seminar on the "Programme of experimental research on the electro-nuclear installation "PLUTON" (obtaining of nuclear data required for designing the full-scale industrial electronuclear system for energy production and nuclear power radioactive waste transmutation; optimization of unit power; sizes and design of the future energy systems and transmuters; researches in material properties connected to the creation of a new type of nuclear energy systems, such as electronuclear ones).

In order to perform this programme, the following measuring systems have been developed:

* neutron spectrometer on time of slowing down in a leaden column;

* gamma spectrometry complex with a pneumomail for the express analysis of the radioactive isotopes formed in the field of electronuclear neutrons;

* the information-measuring detection systems consist of monitoring system for measuring the proton beam intensity, precision and automated infrared integrated and position sensitive detectors for measuring the teat production rate in the assembly, and calorimetric system on the basis of metal thermometers of resistance and sensitive micro thermocouples;

* operation modes automatic control system of the experimental installations with the modules for accumulation and processing of physical information.

In order to measure the spectra of the neutrons leaving the plutonium assembly and the shielding, it is offered to use a method of spectrometry on time of slowing down the neutrons in a leaden column. The fast neutrons generated in a primary target from tungsten by a 660 MeV proton beam of the JINR phasotron, interacting with plutonium are bred as a result of fission. These neutrons, passing through a steel cladding, get in the leaden column placed in tungsten shielding. For the slowing down time in lead, the spectrum of neutrons taking off the limits of the steel shielding of the assembly is measured. The time of slowing down neutrons is inversely proportional to $\sqrt{E}_n$, is measured by detectors of boron, lithium and isotopes of uranium and plutonium. The 660 MeV extracted proton beam from the phasotron will be controlled with the help of external start-up. The phasotron operation in the mode of external start-up is provided by a package of impulses which are emitted by the switching system from the pulse generator of the accelerator start-up and allows one to select the needed time interval. The time interval is determined by the slowing down time of the neutrons having the least energy in the spectrum.

To investigate the neutron spectrum in the electronuclear installation, to irradiate and measure the samples, in neutron reactions, a pneumomail channel and a stationary γ-spectrometer with a AccuSpec/B card and an integrated signal processor will be installed. The γ-spectrometer with high efficiency and precise resolution is expected to be used for studying the effective cross-section of transmutation of long-lived radioactive isotopes in the field of electronuclear neutrons. The results of these studies are of special interest for creation in the future of an industrial scale ecological transmutor.

The automated infrared information-measuring detection system is intended for measurement of heat production in the plutonium assembly. Infrared sensors will be used as detectors. They possess high sensitivity and possibility of performing a study of space and temporary distribution of heat production in the assembly. In addition to the infrared sensors for calorimetry, other thermometric gauges and detectors such as microthermocouples, thermoresisters, etc., can be also used. However, the infrared methods are selected only from the viewpoint of convenience when working in high radiation conditions. The jamming-protected precision device for measuring the thermal radiation intensity is intended for registration of spectra and measurement of the radiation intensity in the infrared spectrum under the conditions of strong electromagnetic and radiation jamming. It can be used for measurement of one-dimensional structures and the sizes of objects under study such as emitters and beams of radiation.

REFERENCE

1. Barashenkov, V.S., Puzynin, I.V., Sissakian, A.N., et al, Report R2-98-74, JINR, Dubna, 1998.

Fundamental information in nuclear data

S.I.Sukhoruchkin

Petersburg Nuclear Physics Institute, 188350 Gatchina, Russia

Suggested by S.Devons possibility that a nucleon structure might influence on fine nuclear effects (which would be difficult to observe in high-energy physics) was checked by comparison of correlations in both nuclear and particle mass data ("tuning effect") [1]. General character of such influence was shown in analysis of few-nucleon effects on different nuclear shells. Systematic correlations in excitations (E^*) and in differences of binding energies (ΔE_B) were found by applying of such specific methods of detection as common plot of E^* (Fig.1a) or spacing distributions $D_{ij} = E_i^* - E_j^*$ in near-magic nuclei (Fig.1b) and plot of ΔE_B in certain types of nuclei (Fig.2-3 [1]). For study of noticed earlier stable character of mass/energy intervals close to $n \times m_e$ ($m_e = 511$ keV) and $D_o = m_n - m_p = 1293$ keV data files were collected.

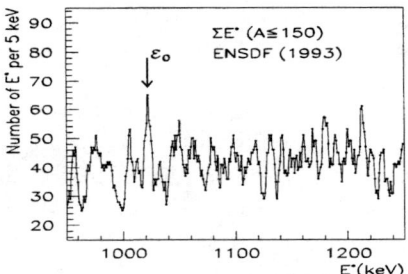 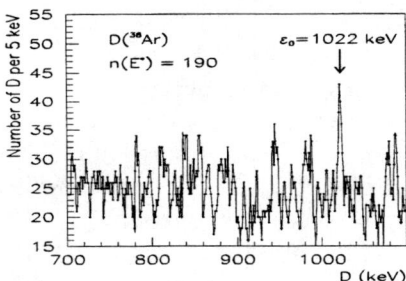

FIGURE 1. *Left*: Sum distribution of excitation energies E^*) in all nuclei with $A \leq 150$. *Right*: Spacing distribution in all levels of ^{38}Ar. Groupings take place at parameter $\varepsilon_o = 2m_e$ $= 1021.8(2)$ keV derived from spin-flip splitting of ^{10}B-levels with $J^\pi = 0^+$(T=1) and 1^+(T=0).

Empirical relations in particle masses and energies of nuclear states [1,2] were considered from Y.Nambu point of view [3] on recent Standard Model (SM):

"a) When we discover new phenomena which we do not understand, the first thing to do is to collect data and try to find some empirical regularities among them, b) one next tries to build concrete models, c) finally there emerges a real theory ... Standard Model qualifies as such a theory. But ... it is theoretically unsatisfactory because a) the unification of forces is only partially realized; and b) there are two many input parameters, especially concerning the masses, which are not explained. The nature can be at the same time more complicated than we think, and simpler in a way we do not know yet. ... I assume that, with Standard Model, we have reached the third stage of a three-stage cycle, and we are now at

CP495, *Experimental Nuclear Physics in Europe*, edited by B. Rubio et al.
© 1999 American Institute of Physics 1-56396-907-6/99/$15.00

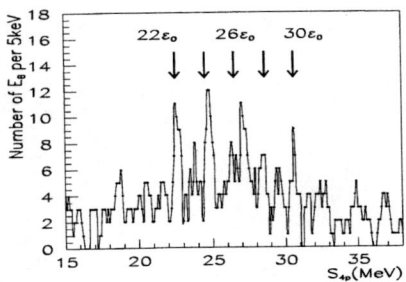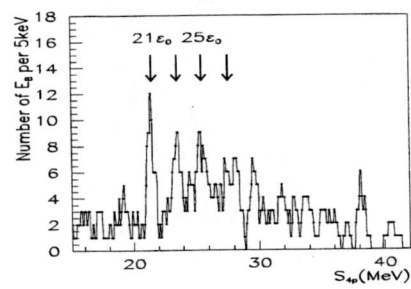

FIGURE 2. *Left*: Sum distribution of nuclear binding energy differences $S_{4p} = \Delta E_B$ in N-even nuclei (Z=50-82) with $\Delta Z=4$. *Right*: The same in N-odd nuclei (Z=50-82) [6,9].

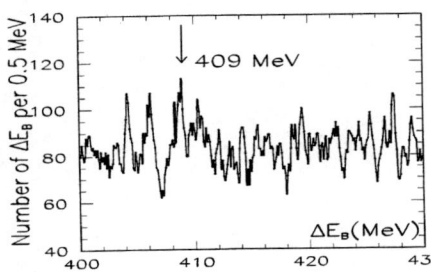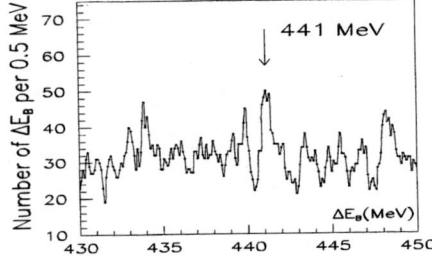

FIGURE 3. Sum distribution of ΔE_B in odd-odd (*left*) and even-even (*right*) nuclei.

the door step of a new cycle. The mass problem is already an early signal for it".

The proximity of electromagnetic mass difference of pion to $9m_e$ [1] permits to notice 1) that small difference between them is close to SFERC (Scaling Factor equal to Electrodynamics Radiative Correction $= \alpha/2\pi$) and 2) masses of light quarks and electron (SM-parameters) are of comparable magnitude. Ratio SFERC involves two other SM-parameters (M_Z, m_μ) as well as parameter of baryon quark interaction $\Delta M_\Delta = (m_\Delta - m_N)/2 = 147$ MeV ($m_e/3\Delta M_\Delta = m_\mu/M_Z = \alpha/2\pi$). Distinguished character of ΔE_B - intervals proximate to integers of $9m_e$ (n=8-12 and n=32 for 147.1 MeV [1]) is supplemented by shown in Fig.3 groupings at 3×147 MeV=441 MeV (right, close to estimation of constituent quark mass) and 409 MeV=441 MeV- $32\varepsilon_o$ (left, close to mass differences in mesons $m_{\eta'} - m_\eta = m_{\eta^-} m_{\pi^\pm} = 409$ MeV). The similarity of intervals in nuclear and particle mass data opens a possibility to evaluate fundamental information from observed "tuning effects".

REFERENCES

1. Sukhoruchkin S.I., *J. Phys.G: Nucl. Part. Phys.* **25**, 921 (1999) and references therein.
2. Frosch, R., *Nuovo Cim.* **A104**, 913 (1991).
3. Nambu Y., *Nucl. Phys.* **A629**, 3c (1998).

CLOSING REMARKS

Concluding Remarks – A Personal View

W Gelletly

School of Physics and Chemistry, Department of Physics
University of Surrey, Guildford, UK   GU2 5XH

I.   Introduction

I have been asked by the organisers to make a few remarks at the conclusion of this first International Conference on Nuclear Physics to be held in Spain. It will not be a summary of the meeting; instead I shall try to give you a few simple messages to carry away from this very enjoyable and successful meeting.

At the Opening Ceremony on Monday, I said that we are very lucky to work in a subject which is vibrant, full of intellectual vigour and a prolific source of applications to everyday problems and other scientific disciplines. If anyone had any doubts about this they would have been dispelled by what we heard here in Sevilla.

My first message comes from Claus Rolfs' talk. He reminded us that

> *"Nuclear Physics has been a great success! We can be very proud of this."*

He is quite right! Even if we confine ourselves to what we heard here you need think only of the following:

- The dramatic discoveries of elements 114[1], 116 and 118[1] reported here by S Hofmann; discoveries built on years of patient and steady work at GSI. These experiments confirm that the fabled Island of Superheavy nuclei exists and that we have reached the shallow waters near its shores.
- The long series of experiments[2] on ground state proton emitters, reported and summarised here by P J Woods, which have helped to map out the proton drip-line.
- The first studies[3] of gamma rays from trans-Fermium nuclei which were reported here by H Kankaanpää. From the level spacing in the ground state band of $^{254}_{102}$No the experimenters deduce $\beta=0.27(2)$ in good agreement with the value of $\beta=0.246$ predicted from microscopic-macroscopic calculations[4].

CP495, *Experimental Nuclear Physics in Europe,* edited by B. Rubio et al.
© 1999 American Institute of Physics 1-56396-907-6/99/$15.00

- The many studies of gamma rays from structures populated with less than 10^{-5} of the total cross-section in fusion-evaporation reactions carried out with EUROBALL and GAMMASPHERE and reported by S Lunardi and C J Lister.
- The remarkable result reported by C Rolfs of the measurement, with a small accelerator in the Gran Sasso Underground Laboratory, of a cross-section of only $10fb$ for the $^3He + ^3He$ reaction at stellar energies.

These and many other examples show just how much we have improved the sensitivity of our measurements. This improvement shows no signs of slackening in pace and is a key feature in our success in trying to understand the structure of the complex, many-body quantal system which constitutes the atomic nucleus.

II.   The Future and Applications of Nuclear Physics

The fruits of this success and of our continuing improvements in sensitivity eventually pay off in terms of applications even although there may be a long time lag. The organisers were at great pains to make sure that we got this message and nowhere was it clearer than in the opening talk by W Kutschera. He showed us that Accelerator Mass Spectroscopy, where we count small numbers of atoms, can be used to study almost every aspect of our environment from the build up of CO_2 in the atmosphere to determining whether artesian wells in Australia are being mined of water or are being replenished. Given the tendency of administrators everywhere, including Spain I am sure, to think that the only worthwhile science is that which produces a short-term pay-off the organisers did well, in their insistence on a leavening of talks on some key applications, to remind us not only that applications are important but that the most successful applications often come decades after the basic science which nurtures them.   Any country which has politicians or administrators with the vision to see that science requires long term investment and that it takes patience to await the rewards will be blessed indeed.

Turning to the future of our subject we heard much to be excited about. Guerreau told us about new sources of radioactive beams and about upgrades to GSI, RIKEN, MSU and GANIL; All of which represent a considerable advance. In this context, E Roeckl (Darmstadt) reminded us of Message No.2 namely,

"We already have radioactive beams!"

There is a tendency for many of us to talk about radioactive beams as if they were in the future, with the implication that when they arrive there will be a dramatic change in what we can do. This is a profound mistake. We have had a wide range of beams of radioactive nuclei at ~1 keV/u for 30 years or so at CERN-ISOLDE. We have excellent beams from fragmentation in a number of laboratories. It is only in terms of high intensity, high quality beams of precise energy at the Coulomb barrier that we have a serious gap. To date we have only a few places providing a few such beams using the ISOL method. A number of ISOL-based facilities are under construction or are about to start, notably SPIRAL and REX-ISOLDE, but it will be some time before we have good provision in this area. It is an area of high promise but it will not transform Nuclear Physics overnight.

III.   The Vital Need for Nuclear Physics

Message No.3 is one we need to address seriously every time we make a grant application. It is the answer to the question

"Why should we study nuclear physics?"

It is the favourite question of every administrator, particle physicist, etc. It would be wise for us to take it seriously since it is asked in the context of the jostling, scientific market place where there are large fish ready to gobble us up.

It may seem obvious to us that understanding 99.8% of the matter in our World is important and that our subject is full of vigour but the people who ask this question are clearly looking for something more. We have seen here in Sevilla, because the organisers went to great lengths to ensure that there was a range of plenary talks on applications, just how wide ranging the applications of Nuclear Physics are. Nuclear Physics has not only created whole new areas of science but, quite literally, has changed our lives. Amongst its offspring we find reactor- and spallation-based neutron sources, synchrotron radiation sources, particle physics, materials modification by implantation, carbon dating and much more. Nuclear power, nuclear weapons, radiation therapy and various, non-invasive medical imaging techniques all stem from basic research in Nuclear Physics and have changed the course of our lives. Much of this was touched on in the Conference.

The answer to the question in Message No. 3 is really that in modern societies there is a vital need first for the ideas and applications of nuclear physics and secondly for manpower trained in Nuclear Physics. Two further examples will suffice to make this point. Firstly in Western Europe one sixth of the population will have radiation therapy at some time in their lives. It needs trained manpower to maintain and improve the network of clinical systems needed to deliver the therapy. Secondly we still generate a large fraction of our electrical power in nuclear power stations and this is likely to be true in the long term. J A Rubio (CERN) discussed this latter point in his talk on "Energy production and Transmutation of Nuclear Waste". In my view he understated the need for future nuclear generating capacity. Consider the facts. Firstly everything points to a considerable increase in the generating capacity required in the next century. Forecasting such demand is notoriously difficult but even if the industrialised countries can keep their power consumption at present levels there will be a steadily rising demand from developing countries. Secondly there is now a consensus that global warming is real. If one examines the total production of CO_2 over the last two hundred and fifty years then the dramatic increase due to our activities is abundantly clear. The result is a steady rise in the ten year global average temperature over the same period.

How can we stem the tide? Clearly burning coal and oil must be phased out as rapidly as possible. Quite apart from the CO_2 their use produces it is wilfully foolish to burn the feedstocks for so many materials. One can hope to use as much in the way of renewable sources as possible but even on an optimistic scenario this seems unlikely to fill the gap. This leaves nuclear power as the only option unless there is an unexpected breakthrough elsewhere. This scenario does not demand the instant building of many nuclear stations. However existing reactors start to come off-line in ~2010. Even with the reactors planned at present this timescale is only extended to 2015-2020. Since the lead times are long we will be forced to make decisions on new nuclear stations in 5 years or so. To do so and decommission present nuclear power plants will need trained manpower. We also need such trained manpower to study and carry out research on waste transmutation, accelerator driven reactor systems etc. Our governments will regret it if they neglect our subject. The manpower required may not be there when it is needed. Our politicians must have the vision to see beyond the next election, or the one after that, and put in place the means to allow us to build nuclear power stations if and when they are needed. If they are to deliver their promises at successive environmental summits it is not an issue they can shirk[5].

IV.   Leadership in Nuclear Physics

We cannot leave it all to the politicians however. We also have a duty to educate the public; to explain the risks inherent in all industrial processes and give them the information required to make rational decisions about their environment. They are constantly assaulted by fears about global warming on the one hand and nuclear discharges from reprocessing plants on the other. One moment journalists tell them that renewable energy sources are the answer and the next that wind farms are a hideous eyesore. It is high time for reasoned discussion on these issues not hysteria. We must play our part in informing and illuminating the debate.

This leads me to Message No.4 which is to all nuclear physicists. It is:

"Hang together or be hanged separately."

It is important that nuclear physicists co-operate to create the large, cutting edge facilities we need but it is also vital to nurture and support University groups for it is in these groups we nourish the students, the raw material of the future. By its nature

Nuclear Physics needs many different kinds of facility. As a result local loyalties to facilities tend to divide our community. It is vital that we see beyond the petty rivalries. We need strong leadership from NuPECC and other such bodies to ensure that all the key facilities are built and are well funded. They must also support and foster university groups. At the same time this means that we cannot defend every facility. In this respect Particle Physics is a shining example. Despite many internal disagreements particle physicists usually present a single face to the World in terms of their demands on the public purse. We would do well to emulate this aspect of their behaviour.

V.   Nuclear Physics in Spain

Turning now to my final topic I recall that when Berta Rubio told me that she was thinking of a Conference in Spain to draw the World's attention to Spanish Nuclear Physics, I thought it was a good idea. When she told me that she and Manolo Lozano would be the organisers and it would be held in this very attractive City of Sevilla, I thought it was an even better idea. Since it was one of their aims I have asked myself "What have we learned of Spanish Nuclear Physics?"

Firstly we were able to confirm our view that there is much excellent theoretical work carried out by a number of groups. During the Conference we heard about the work of our Chairperson's (E Moya de Guerra) group from P Sarriguren (Madrid). Their work is varied but here it concerned deformed Hartree-Fock calculations of β-decay strength distributions and their dependence on nuclear deformation. We also heard of the work of the Madrid-Strasbourg collaboration on full Shell Model calculations from A Poves. We were also reminded of the sizeable group of young people here in Sevilla led by M Lozano and C Dasso. Secondly it was heartening to learn from Respaldiza that the small accelerator in Sevilla, to be dedicated to applications, is about to begin operations. In addition O Tengblad in his poster informed us that a 5MV Van de Graaff devoted to applications is under construction in Madrid. We also heard from E Gonzalez (CIEMAT, Madrid) of Spanish involvement in research on Accelerator Driven Systems for transmutation and power generation at CERN. J-L Tain also took a little time out from his presentation of the work of the Valencia group on beta decay to mention a new PET project for clinical studies in Valencia. Unfortunately this last project is not yet properly funded. Thirdly the most striking comment I heard came from M Lozano's opening statement on Monday when he told us that in Nuclear Physics there are four theorists for every experimenter in Spain. Thus the experimental effort is small in terms of numbers of people but of high quality. Rubio and Tain (Valencia) have done excellent work on beta decay studies and, as we heard from the latter, they have been able to observe clearly the Gamow-Teller resonance in certain rare-earth decays. Borge and Tengblad (Madrid) have carried out a wide range of beta decay and reaction studies of quality at ISOLDE and GANIL. Fourthly we could not fail to be impressed by excellent talks by Benlliure, Gadea, Martel, Cortina-Gil, Marques, Angulo and Gonzalez and poster presentations by Cano-Ott, Garrido and others. There is no shortage of young Spaniards of talent in our field.

Thus my last Message (No.5) is addressed to the Spanish authorities. It is in four parts, namely:

- *Continue and strengthen your support for Theoretical Nuclear Physics. It is of international quality and deserves support.*
- *Strengthen your support for applications of Nuclear Physics. You will find it profitable in the long run.*
- *Support for applications will be of little use however, unless you support research in basic nuclear physics. Applied Physics without basic science to underpin it is like fish without water. It cannot thrive without the nourishing medium of basic science where the ideas come from and the techniques are pushed to the limit. As an urgent priority you need to expand and properly support activity in experimental nuclear physics.*
- *Your first priority in this regard and the first important step should be to create posts for your talented young people.*

In this context it has been announced recently[6] that the Spanish government intends to increase considerably the fraction of Spain's GDP devoted to Science. An increased investment in Nuclear Physics would be very profitable in the long run.

Finally as we leave Sevilla flushed with enthusiasm after this first Nuclear Physics Conference in Spain, I say thank you to our hosts, Berta Rubio and Manolo Lozano. Their choice of speakers was exemplary and they were rewarded with a high standard of presentations followed by lively discussions. I hope that Berta and Manolo feel rewarded for their efforts by our participation and our enjoyment. On behalf of all the participants I think I can say that we hope to see Spanish Nuclear Physics flourish. Official Spain has been slow compared with most countries to invite foreigners to join in national panels and committees[6]. In my view it is a mistake. Informed criticism from outsiders is a vital and healthy part of the system in other European countries. When this change comes, as I am sure it will, I can say that we are ready to help in this way. Do not hesitate to ask us.

REFERENCES

1. Yu, T., *Nature* **400**, 242 (1999); Ninov, V. et al, *Phys.Rev.Letters* (submitted)
2. For example, Woods, P.J. and Davids, C.N., *Rev.Nucl.Part.Sci.* **47**, 541 (1997)
3. Reiter, P. et al, *Phys.Rev.Letters* **82**, 509 (1999); Kankaapaa, H. et al, *Eur.Phys.J. A* (in press)
4. Moller, P. et al, *At.Dat.Nucl.Dat.Tab*, **59**, 185 (1995)
5. *Nuclear Energy – the future climate*, Report of the Royal Society and Royal Academy of Engineering, Roy.Soc.Document, **11/99** (June 1999)
6. Bosch, X., *Nature* **400**, 393 (1999)

◻ SCIENTIFIC PROGRAMME

ENPE 99

Monday, June 21

Chairperson: Manuel Lozano (Sevilla, Spain)

9.30 Welcome

10:15 Accelerator Mass Spectrometry: analysing our world atom by atom
Walter Kutschera, VERA (Austria)

11:00 COFFEE BREAK

11:30 From the nuclear physics microcosm to the astrophysics macrocosm: a
challenge for the next millennium
Marcel Arnould, Universite Libre de Bruxelles (Belgium)

12:15 Tumor therapy with heavy charged particles
Hans Blattmann, Paul Scherrer Institut (Switzerland)

13:00 TIME FOR LUNCH

15:00 Oral Contributions of Sessions:
           Exotic Nuclei I
           Nuclear Astrophysics

ORAL CONTRIBUTIONS - EXOTIC NUCLEI I

Chairperson: Mª José García Borge (Madrid, Spain)

15:00 *C. Monsanglant*   First results using a new technology for measuring
                            masses of very short-lived nuclides with very high
                            accuracy: the MISTRAL program at ISOLDE

15:20 *F. P. Hessberger*   Nuclear structure investigations of neutron deficient
                            transactinide nuclei

15:40   *A. Türler*   Identification of Heavy and Superheavy Nuclides
                      Using Chemical Separator Systems

16:00   *L. M. Fraile*   B(E1) rates in ^{220}Ra and traces of octupole correlations at
                      the upper border of the A=225 island

16:20   COFFEE BREAK

16:50   *P. T. Greenlees* Observation of Excited States in ^{220}U: Evidence for
                      Octupole Deformation

17:10   *A. Andreyev*   Fine Structure in the Alpha Decay of ^{193m,g}Po, ^{195m,g}Po
                      and ^{197m}Po Nuclei Identifying Intruder States in Odd-
                      Mass Lead Isotopes

17:30   *J. C. Batchelder* Proton radioactivity studies at HRIBF

17:50   *S. Ichikawa*   ß-decay half-lives of new neutron-rich isotopes of elements
                      from Pm to Tb

ORAL CONTRIBUTIONS - NUCLEAR ASTROPHYSICS

Chairperson: William Gelletly (Surrey, England)

15:00   *M. Gai*   Measurements of the ^{7}Be (p,γ) ^{8}B With Radioactive
                      Beams, and the ^{8}B Solar Neutrino Flux.

15:20   *M. Hass*   A new Measurement of the ^{7}Be(p,γ)^{8}B Cross- Section with
                      an Implanted ^{7}Be Target.

15:40   *R. E. Tribble*   Indirect determination of the ^{7}Be(p,γ)^{8}B astrophysical S-
                      factor from reactions induced by a ^{7}Be radioactive beam

16:00   *A. M. Laird*   A study of the ^{15}O(α, γ) reaction via the α-decay of ^{19}Ne*

16:20   COFFEE BREAK

16:50   *V. Y. Hansper* Measurement of low-lying nuclear levels in ^{40}Sc

17:10   *S. Harissopulos* Cross section measurements of (p,gamma) reactions
                      relevant to the nucleosynthetic p-process

17:30   *C. Angulo*   Stopping power measurements: implications in nuclear
                      astrophysics

Tuesday, June 22

9:30 Status of the present radioactive beam facilities and perspectives for second generation installations
Daniel Guerreau, GANIL (France)

10:15 Nuclear structure physics at GSI - Challenges and perspectives
Gottfried Muenzenberg, GSI (Germany)

11:00 COFFEE BREAK & POSTER SESSION

12:00 Quests in experimental nuclear astrophysics
Claus Rolfs, Institut für Experimentalphysik III (Germany)

12:45 Light dripline nuclei
Björn Jonson, Chalmers University of Technology (Sweden)

13:30 TIME FOR LUNCH

ORAL CONTRIBUTIONS - EXOTIC NUCLEI II

Chairperson: Rauno Julin (Jyväskylä, Finland)

15:00 *F. Le Blanc* Large Deformation Change in Iridium Isotopes from Laser Spectroscopy

15:20 *P. Sarriguren* A deformed Approach to β^+ Decay of Proton Rich Nuclei

15:40 *P. Dessagne* Gamow-Teller and Fermi decay of N=Z nuclei above A=70

16:00 *E. Roeckl* B$^+$ Decay of very Proton-Rich Nuclei, a Topic of Particular Nuclear-Physics and Astrophysics Interest

16:20 *P. Baumann* The ^{34,35}Al beta decay to ^{34,35}Si

16:40 COFFEE BREAK

17:10 *O. Tengblad* Beta decay asymmetry in mirror nuclei: A=9

17:30 *F. M. Marques* Probing the halo with intensity interferometry

17:50 *D. Cortina-Gil* Nuclear structure studies via one-nucleon removal reactions at the FRS

18:10   *R. Raabe*   Elastic 2n-transfer in the ⁴He(⁶He,⁶He) ⁴He scattering

Wednesday, June 23

Chairperson: Ramon Wyss (Stockholm, Sweeden)

9:30   Shell Model calculations of medium - light nuclei
        Alfredo Pores, Universidad Autónoma de Madrid (Spain)

10:15   New results from Euroball
        Santo Lunardi, Universita Degli Studi di Padora (Italy)

11:00   COFFEE BREAK

11:30   Study of Nuclei Far from Stability at Intermediate Energies: Present
        Status and Future
        Marek Lewitowicz, GANIL (France)

12:15   Shell structure of neutron-rich lignt nuclei: New Vista
        Faisal Azaiez, Institut de Physique Nucleaire d'Orsay (France)

13:00   TIME FOR LUNCH

ORAL CONTRIBUTIONS - GAMMA RAY SPECTROSCOPY I

Chairperson: Peter Kleinheinz (Valencia, Spain)

15:00   *F. Brandolini*   Collective phenomena at the middle of the 1f₇/₂ shell

15:20   *T. Ishii*   Core-excited states in the doubly magic ⁶⁸Ni and its neighbor ⁶⁹Cu

15:40   *A. Gadea*   Collective excitations in N≈Z nuclei studied with GASP
                        and EUROBALL

16:00   *C. Borcan*   First identification of excited states in the N=Z nucleus ⁷⁰Br

16:20   *A. Algora*   Shape changes induced by quasiparticle alignment in ⁷⁴Kr

16:40   COFFEE BREAK

17:10   *H. Schnare*   Magnetic Rotation in the odd-odd nuclei ^{82,84}Rb

17:30   *C. Fransen*   Investigation of mixed-symmetry states in ^{94}Mo

17:50   *J. Gizon*   Terminating high-spin bands in the ^{98,99,100}Ru nuclei: new
information on the $2d_{5/2}$-$1g_{7/2}$ energy spacing

18:10   *M. Górska*   Nuclear stucture in the region of ^{100}Sn at N≈Z

ORAL CONTRIBUTIONS - NUCLEAR REACTIONS AND RADIOACTIVE BEAMS I

Chairperson: Cosimo Signorini (Padova, Italy)

15:00   *R. C. Johnson* Few-body theories of reactions involving weakly
bound nuclei

15:20   *H. G. Bohlen* Structure Studies of ^{11}Be and ^{12}Be: Observation of
Molecular Rotational Bands

16:40   *G. Tagliente*   Two-neutron interferometry in low- and intermediate-
energy heavy-ion reactions

16:00   *S. Tudisco*   Evidence of preequilibrium γ-ray emission in heavy
ion collisions at intermediate incident energies

16:20   *N. Alahari*   Radioactive-Beam Studies of Single-Particle
Structure: Measurements of Spectroscopic Factors
and Angular-Momentum Assignments

16:40   COFFEE BREAK

17:10   *R. Sagaidak*   Study of complete fusion reactions leading to the
production of heavy and superheavy nuclei

17:30   *H. Löhner*   Nuclear Bremsstrahlung a tool to study the free and
in-medium NN interaction

17:50   *D. G. d'Enterria* Thermal hard photons from hot fragmenting nuclei

18:10   *H. Geissel*   Experiments with Stored Exotic Nuclei at
Relativistic Energies

Thursday, June 24

Chairperson: Ikuko Hamamoto (Lund, Sweeden) *

9:30   Microscopic structure of isospin and spin-isospin modes in charge-
exchange reactions
Muhsin Harakeh, KVI (Netherlands)

10:15   Beta-strength measurements in nuclei
Jose Luis Tain, IFIC (Spain)

11:00   COFFEE BREAK & POSTER SESSION

12:00   Exotic beams in ion traps
Georg Bollen, EP - ISOLDE (Switzerland)

12:45   Energy production and transmutation of nuclear waste
Juan Antonio Rubio, CERN (Switzerland)

Afternoon free

21:00   *Social activity: boat cruise down the River Guadalquivir*

Friday, June 25

Chairperson: José M. Arias (Sevilla, Spain)

9:30   Towards the Quark-Gluon Plasma: experiments with Ultra-Relativistic
Nuclear Collisions
Peter Braun - Munzinger, GSI - KP1 (Germany)

10:15   Ion beam analysis of materials
Miguel A. Respaldiza, Universidad de Sevilla (Spain)

11:00   COFFEE BREAK

11:30   Research on exotic nuclei - Overview
Juha Äystö, University of Jyväskylä (Finland)

12:15   The Production and Study of Neutron-rich Nuclei
John Durell, University of Manchester (United Kingdom)

13:00   TIME FOR LUNCH

15:00   Oral Contributions of Sessions:
         Gamma Ray Spectroscopy II
         Nuclear Reactions and Radioactive Beams II
         Applications

ORAL CONTRIBUTIONS - GAMMA RAY SPECTROSCOPY II

Chairperson: Giacomo de Angelis (Legnaro, Italy)

15:00   *D. Jenkins*   Confirmation of Magnetic Rotation in the A~110 region

15:20   *B. Million*   The γ-Continuum in ^{114}Te: Mass Dependence of Rotational
                           Damping

15:40   *H. C. Scraggs*   DSAM Lifetime Measurements in ^{119}Xe

16:00   *R. Chapman*   Signature-dependent Electromagnetic Transition
                           Rates in the $\pi h_{11/2}$ Rotational Sequence of $^{167-3}$Ta

16:20   *D. M. Cullen*   Ten Quasiparticle K-isomers in ^{178}W

16:40   COFFEE BREAK

17:10   *M. Hass*   *g* factors of superdeformed and quasicontinuum states in
                           ^{193,194}Hg.

17:30   *J. Styczen*   First Observation of Excited States in ^{199}At with the
                           Recoil Filter Detector

17:50   *M. Rejmund*   High Spin Structure in Nuclei Near ^{208}Pb

18:10   *H. Kankaanpää*Excited states in the heavy nuclide ^{254}No

ORAL CONTRIBUTIONS - NUCLEAR REACTIONS AND RADIOACTIVE BEAMS II

Chairperson: Antonio Lallena (Granada, Spain)

15:00   *H. R. Jaqaman*   Quark deconfinement in the modified quark meson
                           coupling model

15:20	*U. Georg*	Radioactive Ion Beams at ISOLDE/CERN Recent Developments and Perspectives
15:40	*B. V. Pritychenko*	Intermedite-energy Coulomb excitation of ^{28,29,30,31}Na

16:20	*M. Pfützner*	New Spectroscopy of Heavy Neutron Rich Nuclei: Isomeric Studies Following Relativistic Fragmentation
16:40	COFFEE BREAK	

ORAL CONTRIBUTIONS - APPLICATIONS

Chairperson: José Luis Taín (Valencia, Spain)

17:10	*J. Benlliure*	Spallation reactions investigations at the FRS: from RNB production to the application for nuclear waste incineration
17:30	*E. González*	Transmutation in ADS and needs for nuclear data
17:50	*C. Volant*	Comparison between calculations and experimental results on spallation reactions of interest for hybrid systems
18:10	*V. Wagner*	Experimental study of neutron production in proton reactions with heavy targets
18:30	*N. Colonna*	Production of epithermal neutron beams for BNCT

Saturday, June 26

Chairperson: Elvira Moya de Guerra (Madrid, Spain)

9:30	Studies of superheavy elements - status and prospects *Sigurd Hofmann, GSI (Germany)*
10:15	Nuclear explorations beyond the Proton Drip-line *Phil Woods, University of Edinburgh (Scotland, UK)*

11:00 COFFEE BREAK

11:30 Gamma ray spectroscopy of very unstable nuclear states
Christopher J. Lister, Argonne National Laboratory (USA)

12:15 Closing remarks

12:45 Closing ceremony

 **POSTER SESSIONS**

(Not all the Posters were presented)

Posters will be displayed throughout the Conference. The schedule of the Conference contains two special sessions devoted to posters. During these sessions, authors are expected to be at their boards to present details and answer questions about their contributions.

There will be an area assigned for every session (not for every contribution). Please, note that posters can be put during the first day of the conference. Boards are 90 cm wide and 150 cm high. The title of the poster, author(s) and their affiliations should be clearly stated in large letters at the top of the display board.

The following posters were presented and accepted to the conference

EXOTIC NUCLEI

L. Axelsson — Momentum distributions from core-breakup reactions of ^{11}Be and ^{11}Li

N. Belcari — Absence of blocking effect in the N=Z odd-odd nucleus ^{62}Ga: a new signature of the n-p pairing?

U. C. Bergmann — β-Delayed Neutron Emission from ^{12,14}Be

M. J. Chromik — Evidence for Two-proton-Radioactivity of the First Excited State of ^{17}Ne

J. F. C. Cocks — Fine Structure in the Alpha Decay of ^{191m,g}Bi and ^{193m,g}Bi

B. V. Danilin	Borromean halo nuclei
P. Descouvemont	High-spin states in ^{9,10,11}Be isotopes
D. O. Eremenko	Blocking technique measurements of the induced fission time of Pa and U nuclei
L. M. Fraile	Nuclear structure of ^{231}Ra
B. Fuentes	Study of the β^+ decay of the nucleus ^{75}Rb
M. Gai	Nuclear Molecular Halo: Soft Dipole or Threshold Effect in 11Li?
E. Garrido	Structure of 11Li from breakup reactions.
M. G. Itkis	Low Energy Fission of ^{256}No, ^{270}Sg, ^{271}Hs and ^{286}112 nuclei Formed in Reactions with ^{22}Ne and ^{48}Ca Ions
V. Lapoux	The role played by the polarization potential in the elastic scattering of light exotic nuclei on protons and carbon
A. Marinov	Discovery of Long-Lived Shape Isomeric States which Decay by Strongly Retarded High-Energy Particle Radioactivity
K. Markenroth	Studies of halo nuclei using logitudinal momentum distributions
I. Martel Bravo	Beta-delayed multi-particle emision in the ^{31}Ar decay
Masljuk	New Statistical Approach for Systematization of Fragments of the Heavy Nuclei Fission
M. Oinonen	Probing the N ~ Z line via beta decay
O. Parlag	Product Yields for the Photofission of Th-232 and Np-237 in Energy Range 9.0 - 17.5 MeV
I. Piqueras	Beta Decay study of the N=Z nucleus ^{72}Kr
S. Yu. Platonov	The novel Experimental Technique for investigation of formation and decay of heavy and superheavy nuclei
E. C. Pollacco	Study of low lying resonant states in 6He by proton scattering
F. Rejmund	A new interpretation for the odd-even effect in fission fragments yields
D. Santonocito	Proton Scattering on ^{8}B Unstable Nucleus
D. Sicora	The Nuclear Structural Effects in the Heavy Nuclei Alpha-Dacay constans

D. Sicora	On the Internal Energy Distribution on the Fission Products of Heavy Neutron Deficite Nuclei
C. Signorini	Fusion Barrier Distribution in "Be+²⁰⁹Bi
I. V. Sokolyuk	Influence of Nucleus Shell Structure fpg-Shell for Excitation Metastable States in (γ,n) ReactionI.
I. V. Sokolyuk	Population of Isomeric States of Nuclei from A=100-160 Mass Range in Photoneutronic reactions
I. V. Sokolyuk	Sc Isomer Population in Nuclear Reactions
S. P. Tretyakova	Experimental Investigation of Heavy Ion Radioactivity
M. Trotta	Sub-Barrier Fusion with a Halo Nucleus
A. Wolf	Nuclear Structure of Exotic Nuclei from Magnetic Moments Measurements

GAMMA RAY SPECTROSCOPY

M. Abbas	Analytical calculation of gamma Scintillators efficiencies part II: Total efficiency for wide circular sources
M. Bassiouni	Direct mathematical calculation of the photopeak efficiency for gamma rays in cylindrical NaI(TI) detectors
M. Belleguic	In-beam gamma spectroscopy of very neutron-rich nuclei at GANIL
E. Camera	Exclusive Studies of GDR g- Decay in Hot Heavy Nuclei
D. Cano-Ott	Total Absorption Spectroscopy of ¹⁵⁰Ho 2- and ¹⁵⁰Ho 9⁺ Decay.
T. V. Chuvilskaya	About the New Representation of the γ-Ray Spectrum
J. Eberth	MINIBALL: A Progress Report
F. Everling	Systematics of Coulomb-Energy Differences of Mirror Nuclides in the 0d₃⁄₂ Subshell
G. Falconi	Collectivity in the mass A=110 studied with EUROBALL
E. Farnea	High Spin States and Forbidden E1 Decays in the ⁶⁴Ge Nucleus
J. Garcés Narro	Channel Selection Using Euroball Spectrometer
W. Gelletly	Complete spectroscopy and the ¹⁶²Dy nucleus
A. Gillibert	Inelastic scattering of Magnesium isotopes

NUCLEAR REACTIONS AND RADIOACTIVE BEAMS

505

HIGH ENERGY HEAVY ION COLLISIONS

LIST OF PARTICIPANTS

Mahmoud I. Abbas
mabbaJ@ghazi.net
CIRM National Physical Laboratory
Physics Department - Faculty of science
Alexandria University
Alexandria (Egypt)

Navin Alahari
alahari@nscl.msu.edu
Michigan State University
NSCL
MI 48824 East Lasing (USA)

Kjell Aleklett
aleklett@studsvik.uu.se
Uppsala University
Dep. of Neutron Research
611 82 Nyköping (Sweden)

Alejandro Algora Pineda
algora@lnl.infn.it
INFN - Laboratori Nazionali di Legnaro
Via Romea, 4
35020 Legnaro (Padova) (Italy)

Clara Alonso Alonso
Universidad de Sevilla
Facultad de Fisica
Departamento de FAMN
Apdo. 1065
E-41080 Sevilla (Spain)

Mª Victoria Andres Martin
Universidad de Sevilla
Facultad de Fisica
Departamento de FAMN
Apdo. 1065
E-41080 Sevilla (Spain)

Andrei Andreyev
andrei.andreyev@fys.kuleuven.ac.be
Katholieke Universiteit Leuven
Instituut voor Kern-en Stralingsfysica
Celestijnenlaan 200 D
B-3001 Leuven (Belgium)

Carmen Angulo
angulo@fynu.ucl.ac.be
Universite Catholique de Louvain
Institut de physique nucleaire
Chemin du Cyclotron 2
1348 Louvain-La-Nouve (Belgium)

Miguel Ardid Ramirez
mardid@evala3.ific.uv.es
IFIC. Institut de Fisica Corpuscular
Dr. Moliner, 50
Burjassot
46100 Valencia (Spain)

Jose M. Arias Carrasco
ariasc@cica.es
Universidad de Sevilla
Departamento de Fisica Atomica Molecular y
Nuclear
Apdo. 1065
E- 41080 Sevilla (Spain)

Marcel Arnould
marnould@astro.ulb.ac.be
Universite Libre de Bruxelles
Institut d'Astronomie et d'Astrophysique
Campus de la Plaine - CP-226
Boulevard du Triomphe
B-1050 Brussels (Belgium)

Saleh Ashrafi Saribaglou
saleh.ashrafi@ijs.si
Jozef Stefan Institute
F2 Department
Jamova 39
Ljubljana (Slovenia)

Leif Axelsson
f2bla@fy.chalmers.se
Chalmers University of Technology
Department of Physics
S-412 96 Göteborg (Sweden)

Faisal Azaiez
azaiez@ipno.in2p3.fr
Institute de Physique Nucleaire d'Orsay
IPN Orsay
F-91406 Orsay Cedex (France)

Mohamed Bassiouni
aaeng006@aa.aast.egnet.net
Arab Academy for Science & Technology
Basic & Applied Science Dept.
Faculty of Engineering & Technology
Abou Kir
Alexandria (Egypt)

Paule Baumann
Paule.Baumann@IReS.in2p3.fr
Institut de Recherches Subatomiques
IReS
Bat 24
B.P. 28
F 67037 Strasbourg Cedex 2 (France)

Uffe Christian Bergmann
ucb@ifa.au.dk
Aarhus Universitet
Institut for Fysik og Astronomi
Bygn. 520
DK-8000 Ny Munkegade (Denmark)

Sigfrido Boffi
boffi@pv.infn.it
Universita ' di Pavia
Dipartamento di Fisica Nucleare e Teorica
Via Bassi 6
I-27100 Pavia (Italy)

Juha Äystö
aysto@snafu.phys.jyu.fi
University of Jyväskylä
Department of Physics
Accelerator Laboratory
Survontie 9
FIN-40351 Jyväskylä (Finland)

Jose Barea Muñoz
Universidad de Sevilla
Facultad de Física
Departamento de FAMN
Apdo 1065
E-41080 Sevilla (Spain)

Jon Batchelder
batcheld@mail.phy.ornl.gov
Oak Ridge Associated Universities
JIHIR / ORNL
P.O Box 2008
Bld. 6008
Tn 37381-6374 Oak Ridge (USA)

Jose Benlliure
benlliur@fpddux.usc.es
University of Santiago de Compostela
Dpto de Fisica de Particulas
15706 Santiago Compostela (Spain)

Hans Blattmann
hans.blattmann@psi.ch
Paul Scherrer Institut
Radiation Medicine
CH-5232 Villigen (Switzerland)

Hans Gerhard Bohlen
bohlen@hmi.de
Hahn-Meitner-Institut Berlin
Glienicker Str. 100
D-14109 Berlin (Germany)

510

Georg Bollen
georg.bollen@cern.ch
EP - ISOLDE
CERN
CH-1211 Geneve 23 (Switzerland)

Christina Borcan
borcan@alpha.fz-rossendorf.de
Froschunszentrum Rossendorf
Bautzner Landstrasse 128 (B6)
Postfach 51 01 19
01314 Dresden (Germany)

Christian Bourgeois
bourgeoi@ipno.in2p3.fr
IPN Orsay
F 91406 Orsay Cedex (France)

Franco Brandolini
brandolini@pd.infn.it
INFN
Via Marzolo 8
I-35131 Padova (Italy)

Peter Braun-Munzinger
p.braun-munzinger@gsi.de
GSI
KP1
Planckstr. 1
D - 64291 Darmstadt (Germany)

Lothar Buchmann
lothar@triumf.ca
TRIUMF
4004 Westbrook Mall
V6T 2A3 Vancouver B.C. (Canada)

Vladimir Butsev
Joint Institute for Nuclear Research
Joliot Curie 6
Dubna
RU-141980 Moscow Region (Russia)

Juan Antonio Caballero
jacarr@cica.es
Universidad de Sevilla
Facultad de Fisica
Departamento de FAMN
Apdo 1065
E-41080 Sevilla (Spain)

Daniel Cano Ott
ott@ciemat.es
CIEMAT
Facet Group
Dept. of Nuclear Fission
Avda. Complutense 22
28040 Madrid (Spain)

Enrique Casarejos
enrique@fpddux.usc.es
Universidad de Santiago de Compostela
Dpto. de Particulas
Facultad de Fisica
E-15706 La Coruña (Spain)

Robert Chapman
rc@odin5.paisley.ac.uk
University of Paisley
Dept. Physics
High Street
Paisley Renfrewshire
PA1 2BE Scotland (United Kingdom)

Danka Chmielewska
danka@ipj.gov.pl
The Anrezej Soltan Institute for Nuclear
Studies
05-400 Swierk / Otwock (Poland)

James Cocks
jfcc@pulsar.phys.jyu.fi
University of Jyvaskyla
Department of Physics
Survontie, 9
40351 Jyvaskyla (Finland)

Nicola Colonna
nicola.colonna@ba.infn.it
Instituto Nazionale Fisica Nucleare - Sez . di
Bari
V. Amendola 173
70126 Bari (Italy)

Dolores Cortina
d.cortina@gsi.de
GSI Darmstadt
KP II
Planckstras. 1
D - 64291 Darmstadt (Germany)

Evelina Costanzo
costanzo@ct.infn.it
Dipartimento di Fisica
Corso Italia 57
95129 Catania (Italy)

Raquel Crespo
raquel@wotan.ist.utl.pt
Instituto Superior Tecnico
Departamento de Fisica
Av.Rovisco Pais
1096 Lisboa (Portugal)

David Cullen
dmc@ns.ph.liv.ac.uk
University of Liverpool
Oliver Lodge Laboratory
Department of Physics
L69 7ZE Liverpool (United Kingdom)

David d'Enterria
enterria@in2p3.fr
Universitat Autonoma de Barcelona - Subatech
Grup de Fisica de les Radiacions
08193 Bellaterra (Spain)

Giacomo De Angelis
deangelis@lnl.infn.it
Laboratori Nazionali di Legnaro
LNL INFN
Via Romea 4
I-35020 Legnaro (Italy)

Pascale Delbourgo
delbo@bruyeres.cea.fr
Commissariat a l'energie atomique CEA
DPTA / SPN (Service de Physique Nucleaire)
Centre d'Estudes de Pbruyeres-le-Chatel
BP 12
F-91680 Bryeres-le-Chatel (France)

Pierre Descouvemont
pdesc@ulb.ac.be
Universite Libre de Bruxelles
Physique Nucleaire Theorique CP 229
Campus Plaine
Boulevard du Triomphe
1050 Bruxelles (Belgium)

Philippe Dessagne
philippe.dessagne@ires.in2p3.fr
Institut de Recherches Subatomiques
BP 28
67037 Strasbourg Cedex (France)

Ignacio Duran Duran
duran@fpddux.usc.es
Universidad de Santiago de Compostela
Facultad de Fisica
15706 Santiago (Spain)

John Durell
nsd@mags.ph.man.ac.uk
University of Manchester
Department of Physics and Astronomy
M13 9PL Manchester (United Kingdom)

Friedrich Everling
f.everling@t-online.de
IPN at University of Kiel
Ringheide 24 f
D- 21149 Hamburg (Germany)

Carmen Fernandez
carmen@fpddux.usc.es
Universidad de Santiago de Compostela
Facultad de Fisica
Santiago de Compostela
15706 La Coruña (Spain)

Luis Mario Fraile Prieto
imtfp4d@fresno.csic.es
CSIC
Instituto de Estructura de la Materia
C/ Serrano, 113. bis
28006 Madrid (Spain)

Beatriz Fuentes
beatriz@fpddux.usc.es
Universidad de Santiago de Compostela
Departamento de Fisica de Particulas
Facultad de Fisica
Santiago de Compostela
15706 La Coruña (Spain)

Ettore Gadioli
gadioli@mi.infn.it
University of Milano
Dipartimento di Fisica
Via Celoria 16
20133 Milano (Italy)

Jose M. Espino Navas
Universidad de Sevilla
Facultad de Física
Departamento de FAMN
Apdo. 1065
E-41080 Sevilla (Spain)

Enrico Farnea
Enrico.Farnea@ific.uv.es
Instituto de Física Corpuscular
Avda. Dr. Moliner 50
Burjasot
46100 Valencia (Spain)

Pierpaolo Figuera
figuera@lns.infn.it
INFN Laboratori Nazionale del Sud
Via S. Sofia 44
I- 95123 Catania (Ital)

Christoph Fransen
fransen@ikp.uni-koeln.de
University of Cologne
Institut fuer Kernphysik
Zuelpicher Strasse 77
D-50937 Koeln (Germany)

Andres Gadea
andres.gadea@ific.uv.es
Instituto de Fisica Corpuscular
Avda. Dr. Moliner 50
Burjassot
46100 Valencia (Spain)

Enrica Gadioli Erba
enrica.erba@mi.infn.it
University of Milano
Departimento di Fisica
Via Celoria 16
20133 Milano (Italy)

513

Moshe Gai
gai@uconnvm.uconn.edu
University of Connecticut
Department of Physics, U46
2152 Hillsdie RD.
CT- 06269- 3046 Storrs, (USA)

Maria Jose Garcia Borge
emborge@iem.csic.es
Consejo Superior de Investigaciones Científicas
Instituto de Estructura de la Materia
Serrano 110 - 123
28006 Madrid (Spain)

Eduardo Garrido
imteg57@pinar2.csic.es
Instituto de Estructura de la Materia - CSIC
Consejo Superior de Investigaciones Cientificas
Serrano 123
28006 Madrid (Spain)

Hans Geissel
h.geissel@gsi.de
GSI
KP II
Planckstras. 1
D-64291 Darmstadt (Germany)

Gianpiero Gervino
gervino@to.infn.it
INFN - Torino
Dipartimento di Fisica Sperimentale
Via P. Giuria 1
I-10125 Torino (Italy)

Joaquin Gómez Camacho
Universidad de Sevilla
Facultad de Fisica
Departamento de FAMN
Apdo. 1065
E- 41080 Sevilla (Spain)

Joaquin Garces Narro
j.garces-narro@surrey.ac.uk
University of Surrey
Department of Physics
Guildford
GU2 5XH Surrey (United Kingdom)

Jose E. García Ramos
Universidad de Sevilla
Facultad de Fisica
Departamento de FAMN
Apdo 1065
E-41080 Sevilla (Spain)

Juan A. Garzon
hans@fpddux.usc.es
Universidad de Santiago de Compostela
Facultad de Fisica
Santiago de Compostela
15706 La Coruña (Spain)

William Gelletly
w.gelletly@surrey.ac.uk
University of Surrey
Physics Dept.
Guildford
GU2 5XH Surrey (United Kingdom)

Jean Gizon
gizon@isn.in2p3.fr
Institut des Sciences Nucleaires
53 Avenue des Martyrs
F-38026 Grenoble Cedex (France)

Enrique Miguel González-Romero
enriques@ciemat.es
CIEMAT
Dept of Nuclear Fission
Av. Complutense, 22
28040 Madrid (Spain)

514

Magdalena Gorska
m.gorska@gsi.de
GSI
Planckstr. 1
64291 Darmstadt (Germany)

Dominique Goutte
goutte@bruyeres.cea.fr
CEA-DAM
Service de Physique Nucleaire
CEA/ Bruyeres-le Chatel
F-91680 Bruyeres-le Chatel (France)

Paul Greenlees
ptg@phys.jyu-fi
University of Jyvaskyla
Physics Department
Survontie , 9
40351 Jyvaskyla (Finland)

Daniel Guerreau
guerreau@ganil.fr
GANIL
BP 5027
F - 14021 Caen Cedex 5 (France)

Ikuko Hamamoto
ikuko@matfys.lth.se
University of Lund
Department of Mathematical Physics LTH.
P.O.Box 118
S-22100 Lund (Sweden)

Vera Hansper
vyh@ifa.au.dk
Institute of Physics and Astronomy
University of Aarhus
Ny Munkegade
DK 8000 Aarhus C (Denmark)

Muhsin Harakeh
harakeh@kvi.nl
Kernfysisch Versneller Institut
University of Groningen
Zernikelaan 25
NL-9747 AA Groningen (Netherlands)

Sotiris Harissopulos
sharisop@mail.demokritos.gr
Institute of Nuclear Physics
NCSR "Demokritos"
POB 60228
Aghia Paraskevi
GR-153,10 Athens (Greece)

Michael Hass
fnhass@wicc.weizmann.ac.il
Department of Particle Physics
Weizmann Institute of Science
76100 Rehovot (Israel)

Kerttuli Helariutta
KERTTULI.HELARIUTTA@PHYS.JYU.FI
University of Jyvaskyla
Deparament of physics
P.O. BOX 35 (Y5)
40351 Jyvaskyla (Finland)

Fritz Peter Heßberger
f.p.hessberger@gsi.de
Gesellschaft fuer Schwerionenforschung mbH
Plankstrasse 1
D-64291 Darmstadt (Germany)

Sigurd Hofmann
s.hofmann@gsi.de
GSI
Kernphysik II
Postfach 110552
D-64220 Darmstadt (Germany)

515

Shin-ichi Ichikawa
sichi@popsvr.tokai.jaeri.go.jp
Japan Atomic Energy Research Institute
Advanced Sciece Research Center
Tokai-mura
319 - 1195 Ibaraki (Japan)

Tetsuro Ishii
ishii@tdmalph0.tokai.jaeri.go.jp
Japan Atomic Energy Research Institute
Advanced Science Research Center
Tokai
319-1195 Ibaraki (Japan)

Henry Jaqaman
hjaqaman@bethlehem.edu
Bethlehem University, Palestine
Department of Physics
P.O. Box 9 Bethlehem (Palestine)

David Jenkins
oej@yksc.york.ac.uk
University of York
Department of Physics
Heslington
Y01 5DD York (United Kingdom)

Ronald Johnson
R.Johnson@surrey.ac.uk
University of Surrey
Department of Physics
Guildford
GU2 5XH Surrey (United Kingdom)

Björn Jonson
bjn@fy.chalmers.se
Chalmers University of Technology
Department of Experimental Physics
S-412 96 Göteborg (Sweden)

Rauno Julin
rauno.julin@phys.jyu.fi
University of Jyväskylä
Department of Physics
P.O. BOX 35
FIN 40351 Jyväskylä (Finland)

Harri Kankaanpää
kankaanpaa@phys.jyu.fi
University of Jyväskylä
Department of physics
P.O. Box 35
SF-40351 Jyvaskyla (Finland)

Aleksandra Kelic
kelic@ireS.in2p3.fr
IReS
BP 28
23, rue de Loess
F-67037 Strasbourg-Cedex (France)

Peter Kleinheinz
peter@evalvx.ific.uv.es
IFIC
Dr. Moloner 50
Burjassot
E 46100 Valencia (Spain)

Walter Kutschera
walter.kutschera@univie.ac.at
University of Vienna
Institut fur Radiumforschung und Kernphysik
VERA Laboratory
Waehringer Strasse 17
A-1090 Vienna (Austria)

Marco La Commara
m.lacommara@gsi.de
GSI, Gesellschaft fur Schwerionenforschung
Planckstrasse, 1
D- 64291 Darmstad (Germany)

Alison Laird
aml@ph.ed.ac.uk
University of Edinburgh
Department of Physics and Astronomy
JCMB, The Kings Buildings
Mayfield Road
EH9 3JZ Edinburgh (United Kingdom)

Antonio Miguel Lallena Rojo
lallena@ugr.es
Universidad de Granada
Departamento de Física Moderna
18071 Granada (Spain)

Marie- Odile Lampert
lampert_eurisys@compuserve.com
Eurisys Mesures
BP 311
67834 Tanneries Cedex (France)

Edoardo Lanza
lanza@cica.es
Universidad de Sevilla
Facultad de Física
Departamento FAMN
Apdo 1065
E-41080 Sevilla (Spain)

Yves Le Coz
lecoz@cea.fr
CEA / SACLAY
Daphnia /SPhN
Bat 703
Saclay
91191 Gif-sur-Yvette (France)

Matti Leino
matti.leino@phys.jyu.fi
University of Jyvaskyla
Department of Physics
P.O. Box 35
Fin-40351 Jyvaskyla (Finland)

Marek Lewitowicz
lewitowicz@ganil.fr
GANIL
BP 5027
F - 14076 Caen Cedex (France)

Andrej Likar
andrej.likar@ijs.si
J. Stefan Institute
F2 Department
Jamova 39
1000 Ljubljana (Slovenia)

Axel Lindroth
axell@studsvik.uu.se
Uppsala University
Institution for Neutron Research
NFL Studsvik
S- 611 82 Nyköping (Sweden)

Christopher J. Lister
lister@anlphy.phy.anl.gov
Argonne National Laboratory
9700 South Cass Ave.
IL 60439-4843 Argonne (USA)

Herbert Loehner
loehner@kvi.nl
KVI, University of Groningen
Zernikelaan 25
NL_9747 AA Groningen (Netherlands)

Maria Jose Lopez Jimenez
mjlopez@ganil.fr
GANIL
Boulevard Hènri Becquerel
B.O. 5027
14076 Caen Cedex (France)

517

Manuel Lozano
lozano@cica.es
Universidad de Sevilla - Facultad de Física
Departamento de Fisica Atomica Molecular y
Nuclear
Apdo. 1065
41080 Sevilla (Spain)

Concettina Maiolino
maiolino@lns.infn.it
INFN-LNS
Via S. Sofia
44-95123 Catania (Italy)

Fco. Miguel Marques Moreno
marques@caelav.in2p3.fr
Laboratoire de Physique Corpusculaire de Caen
LPC-CAEN (ISMRA)
6 Bd du Marechal Juin
14050 Caen Cedex (France)

Christiane Miehe
christiane.miehe@ires.in2p3.fr
Institut de Recherches Subatomiques CNRS /
ULP
B.P. 28
F - 67037 Strasbourg Cedex (France)

Alberto Molina Caballes
Universidad de Sevilla
Facultad de Fisica
Departamento de FAMN
Apdo 1065
E-41080 Sevilla (Spain)

Antonio Moro Muñoz
Universidad de Sevilla
Facultad de Física
Departamento de FAMN
Apdo 1065
E-41080 Sevilla (Spain)

Santo Lunardi
santo.lunardi@pd.infn.it
Universita Degli Studi di Padova
Departamento di Fisica Galileo Galilei
Via F. Marzolo, 8
I - 35131 Padova (Italy)

Karin Markenroth
karin.markenroth@fy.chalmers.se
Chalmers University of Technology
Institute of Experimental Physics
Subatomic Physics
S-412 96 Göteborg (Sweden)

Paolo Maurenzig
zig@fi.infn.it
INFN - Sezione di Firenze
Largo Fermi 2
I-50125 Firenze (Italy)

Benedicte Million
million@mi.infn.it
INFN - Sez Milano
Via Celoria, 16
I- 20133 Milano (Italy)

Celine Monsanglant
monsangl@csnsm.in2p3.fr
CERN site de Meyrin
bat26-1-017 division EP/SC
CH-1211 Geneve 23 (Suisse)

Wilhelm Mueller
wilhelm.mueller@fys.kuleuven.ac.be
University of Leuven
Instituut voor kern en Stralingsfysica
Celestijnenlaan 200D
B-3001 Leuven (Belgium)

518

Gottfried Muenzenberg
g.muenzenberg@gsi.de
GSI Darmstadt
Kernphysik II
Planckstr. 1
D - 64291 Darmstadt (Germany)

Maarit Muikku
maarit.muikku@phys.jyu.fi
University of Jyvaskyla
Deparment of Physics
PL 35 (Y5)
40531 Jyvaskyla (Finland)

Thomas Nilsson
thomas.nilsson@cern.ch
CERN, ISOLDE
CERN, EP-div
CH-1211 Geneve 23 (Switzerland)

Filomena Nunes
filomena@wotan.ist.utl.pt
Universidad Fernando Pessoa
Departamento de Fisica
Av.Rovisco Pais
1069 Lisboa (Portugal)

Chris O'Leary
cdo@ns.ph.liv.ac.uk
University of Liverpool
Department of Physics
Oliver Lodge Laboratory
L69 7ZE Liverpool (United Kingdom)

Masao Ogawa
mogawa@nr.titech.ac.jp
Tokyo Institute of Technology
Research Laboratory for Nuclear Reactors
2-12-1 Ohokayama, Meguro-ku
152-8550 Tokyo (Japan)

Markuku Oinonen
markku.oinonen@cern.ch
CERN / ISOLDE
EP Division
Bat. 26-1-014
CH-1211 Geneva (Switzerland)

Masumi Oshima
oshima@jball4.tokai.jaeri.go.jp
Japan Atomic Energy Research Institute
Advanced Science Research Center
Tokai-mura, Naka-gun
319-1195 Ibaraki-ken (Japan)

Edward Paul
esp@ns.ph.liv.ac.uk
University of Liverpool
Oliver Lodge Laboratory
P. O. Box 147
L69 7ZE Liverpool (United Kingdom)

Costel Marian Petrache
petrache@pd.infn.it
INFN - Padova
Dipartimento di Fisica
Universita di Padova
Via Marzolo 8
I-35131 Padova (Italy)

Marek Pfuetzner
pfutzner@mimuw.edu.pl
Warsaw University
Institute of Experimental Physics
ul. Hoza 69
00-681 Warszawa (Poland)

Sara Pirrone
sara.pirrone@ct.infn.it
Instituto Nazionale di Fisica Nucleare Sezione
di Catania
Sezione infn Catania
Corso Italia 57
95129 Catania (Italy)

Emanuel Pollacco
lolly@phnx7.saclay.cea.fr
DSM/DAPNIA/SPhN
Bat 703
CEA Saclary
F-91191 Gif/Yvette Cedex (France)

Boris Pritychenko
Pritychenko@nscl.msu.edu
National Superconducting Cyclotron Laboratory
East Lansing
48824-1321 Michigan (USA)

Jose M. Quesada Molina
Universidad de Sevilla
Facultad de Física
Departamento FAMN
Apdo 1065
41080 Sevilla (Spain)

Subramanian Raman
raman@mail.phy.ornl.gov
Oak Ridge National Laboratory
Building 6010
Mail Stop 6354
TN 3783 Oak Ridge (USA)

Maurycy Rejmund
rejmund@csnsm.in2p3.fr
CSNSM Orsay
bat 104
91405 Orsay Campus (France)

M. A. Respaldiza
respaldiza@cica.es
Universidad de Sevilla
Centro Nacional de Aceleradores
Parque Tecnologico Cartuja' 93
Avda. Thomas A. Edison s/n Isla de la Cartuja
41092 Sevilla (Spain)

Alfredo Poves
poves@nucphys1.ft.uam.es
Universidad Autónoma de Madrid
Departamento de Física Teórica
Facultad de Ciencias CXI
Cantoblanco
28049 Madrid (Spain)

Elena Prokhorova
lena@nrsun.jinr.ru
Joint Institute for Nuclear Research
Flerov Laboratory of Nuclear Reactions
Dubna
RU-141980 Moscow Region (Russia)

Riccardo Raabe
riccardo.raabe@fys.kuleuven.ac.be
University of Leuven
Instituut Voor Kern- en Stralingsfysica
Celestijnenlaan 200 D
Haverlee
B-3001 Leuven (Belgium)

Jack Rapaport
rapaport@ohio.edu
Ohio University
Physics Department
OH 45701 Athens (USA)

Fanny Rejmund
farget@ipno.in2p3.fr
IPN
Universite Paris XI
F-91406 Orsay Cedex (France)

Eszter Retfalvi
retfalvi@sunserv.kfki.hu
Research Institute for solid State Physics
Konkoly
Thege 29- 31
H- 1121 Budapest (Hungary)

Francesca Rizzo
rizzo@ct.infn.it
Dipartimento di Fisica
Corso Italia 57
95129 Catania (Italy)

Jose Rodriguez Cossio
cossio@clri6a.gsi.de
GSI Darmstadt
Biophysik
Planckstrasse 1.
D-64291 Darmstadt (Germany)

Ernst Roeckl
e.roeckl@gsi.de
GSI Darmstadt
Planckstrasse, 1
D - 64291 Darmstadt (Germany)

Claus Rolfs
rolfs@ep3.ruhr-uni-bochum.de
Institut für Experimentalphysik III
Ruhr-Universität Bochum
Universitätsstraße 150
D-44780 Bochum (Germany)

Juan Antonio Rubio
juan.antonio.rubio@cern.ch
CERN
DSU/DO
CH-1211 Geneve 23 (Switzerland)

Berta Rubio
rubio@evala4.ific.uv.es
Instituto de Fisica Corpuscular
Avda.Dr.Moliner, 50
Burjassot
46100 Valencia (Spain)

Roman Sagaidak
sagaidak@sunvas.jinr.ru
Joint Institute for Nuclear Research
Flerov Laboratory of Nuclear Reactions
Dubna
141980 Moscow Region (Russia)

Jorge Miguel Sampaio
sampaio@alf2.cii.fc.ul.pt
Universidade de Lisboa
Centro de Fisica Nuclear
Av. Prof. Gama Pinto, 2
1699 Lisboa (Portugal)

Pedro Sarriguren
emsarri@iem.csic.es
CSIC
Instituto de Estructura de la Materia
Serrano, 123
28006 Madrid (Spain)

Harald Schnare
schnare@fz-rossendorf.de
Forschungszentrum Rossendorf
Institut fuer Kern- und Hadronenphysik
Postfach 510119
D-01314 Dresden (Germany)

Helen Claire Scraggs
hcs@ns.ph.liv.ac.uk
University of Liverpool
Department of Physics
Oliver Lodge Labs
L69 7ZE Liverpool (United Kingdom)

Cosimo Signorini
signorini@padova.infn.it
Universidad de Padova
Physics Department
Via Marzulo, 8
35131 Padova (Italy)

521

Martin Smith
mbs@juno20.paisley.ac.uk
University of Paisley
Dept. Physics
High Street
Paisley Renfrewshire
PA1 2BE Scotland (United Kingdom)

Klaus-Michael Spohr
pix@juno20.paisley.ac.uk
University of Paisley
Dept. Physics
High Street
Paisley Renfrewshire
PA1 2BE Scotland (United Kingdom)

Teruo Suehiro
suehiro@titan.tohtech.ac.jp
Tohoku Ibstitute of Technology
35-1 Yagiyama-Kasumityo
Taihaku-ku
982-8577 Sendai (Japan)

Sergey Sukhoruchkin
sergeis@hep486.pnpi.spb.ru
St. Petesburg Nuclear Physics Institute
18350 Gatchina, Leningrad (Russia)

Giuseppe Tagliente
pino.tagliente@ba.infn.it
I.N.F.N.
Via Amendola 173
70126 Bari (Italy)

Olof Tengblad
olof.tengblad@cern.ch
CSIC
Instituto de Estructura de la Materia
Serrano, 113 bis
E-28006 Madrid (Spain)

Pietro Sona
sona@fi.infn.it
I.N.F.N Sez di Fierenze
Largo E. Fermi 2
50125 Firenze (Italy)

Jan Styczen
jan.styczen@ifj.edu.pl
The Niewodniczanski Institut of Nuclear
Physics
ul. Radzikowsliego 152
30-432 Krakow (Krakow)

Ziemowid Sujkowski
sujkowsk@ipj.gov.pl
Institute for Nuclear Studies
The Andrezey Soltan
05-400 Swierk (Poland)

Concetta M. Sutera
sutera@lns.infn.it
I.N.F.N. - Catania
Corso Italia, 57
95129 Catania (Italy)

Jose Luis Tain
jose.luis.tain@.ific.uv.es
IFIC
Avda. Dr. Moliner, 50
Burjassot
46100 Valencia (Spain)

Catherine Thibault
thibault@csnsm.in2p3.fr
CSNSM -IN2P3, CNRS
Bat.104-108
F-91405 Orsay Campus (France)

Katarina Wilhemsen Rolander
wilhelmsen@physto.se
Stockholm University
Department of Physics
Box 6730
SE-113 85 Stockholm (Sweden)

Hans Juergen Wollersheim
wolle@axp602.gsi.de
GSI
Postfach 110552
D-64220 Darmstadt (Germany)

Philip Woods
pjw@ph.ed.ac.uk
University of Edimburg
Department of Physics and Astronomy
JCMB, The Kings Buildings
Mayfield Road
EH9 3JZ Edinburgh (United Kingdom)

Ramon Wyss
wyss@msi.se
Royal Institut of Tecnology
KTH-Frescati
Frescativ. 24
S-104 05 Stockholm (Sweden)

Boutami, R., 131
Bracco, A., 225
Bradfield-Smith, W., 9, 313
Braga, F., 13
Brandenburg, S., 275
Brandolini, F., 189, 193, 195, 199, 201
Broggini, C., 357, 385
Brown, B. A., 309
Bruce, A. M., 79
Brusasco, C., 452
Burjan, V., 361
Buscemi, A., 193, 199
Butler, P. A., 133, 251, 253, 260
Buttsev, V. S., 478

Cocks, J. F. C., 121, 125, 133, 249, 251, 253, 258, 260
Colangelo, P., 454
Collatz, R., 107
Colonna, N., 319, 454
Conreur, G., 59
Cortina-Gil, D., 29, 33
Corvisiero, P., 357, 385
Cottle, P. D., 51
Courtin, S., 255
Cousin, R., 59
Crawford, J., 117
Crespo, R., 295
Cullen, D. M., 79, 187, 243
Cunsolo, A., 367
Czajkowski, S., 79, 458, 474

C

Caamaño, M., 113
Caballero, J. A., 403
Cabaret, L., 117
Cabibbo, M., 313
Camera, F., 225
Cameron, J. A., 189, 221
Campajola, L., 357, 385
Campbell, G. J., 239
Campisi, M. G., 353
Cann, K. J., 133
Cano-Ott, D., 97, 107
Cappuzzello, F., 367
Cardella, G., 313, 323, 353
Carstoiu, F., 361
Casarejos, E., 458, 474
Casten, R. F., 237
Catford, W. N., 79
Catherall, R., 349
Caurier, E., 55
Cavallaro, S., 315, 353
Cavinato, M., 334
Cejpek, J., 361
Champagne, A. E., 371
Chandler, C., 79
Chapman, R., 239, 255, 258
Cherubini, S., 9, 367
Chewter, A. J., 260
Chiara, C. J., 229, 231
Chultem, D., 478
Clark, H. L., 361
Clark, R. M., 79, 221

D

D'Alessandro, A., 357, 385
Daugas, J. M., 79
Davinson, T., 9, 313, 367
de Acuña, D., 201
de Angelis, G., 189, 193, 195, 199, 201, 205, 217, 332
De Filippo, E., 319, 353
Delbar, T., 381
Demetriou, P., 377
Dendooven, P., 63
d'Enterria, D. G., 393
De Poli, M., 189, 193, 195, 199, 201
de Saint Simon, M., 59
Descouvemont, P., 9, 25
Dessagne, P., 55, 77, 79
Dessalvi, M., 357, 385
Devlin, M., 229
Díaz, J., 401
Di Pietro, A., 9, 313, 323, 367
Dönau, F., 205
Donnelly, T. W., 403
D'Onofrio, A., 332, 357, 385
Dorvaux, O., 121, 125, 133, 253, 258, 260
Døssing, T., 225
Doubre, H., 59
Dracoulis, G. D., 243
Drouart, A., 13
Drumm, P., 349
Dudarev, S. Y., 478

Walter, G., 55
Warner, D. D., 187, 237
Weber, U., 452
Wefers, E., 375
Weick, H., 327, 375
Weiszflog, M., 195
Wenander, F., 19, 27
Wheldon, C., 243
Wilhelmsen Rolander, K., 19, 27
Willis, A., 275
Wilpert, M., 303
Wilson, A. N., 243
Winkler, M., 29, 33
Wittmann, V., 107
Wlazlo, W., 466, 474
Wöhr, A., 9
Wolińska, M., 255
Wollersheim, H.-J., 260
Wollnik, H., 33, 327
Woods, P. J., 89, 313
Wu, X., 237
Wyss, R., 121, 193, 195, 199, 201

X

Xu, F., 243

Y

Yahlali, N., 401
Yeremin, A. V., 145
Yurkon, J., 309

Z

Zapata, J. C., 450
Zavatarelli, S., 357, 385
Zegers, R. G. T., 275
Zeitelhack, K., 375
Zelevinsky, V. G., 51
Zetta, L., 353
Zganjar, E. F., 109
Zhang, H. Q., 340
Zhang, Y. H., 247
Zhukov, M. V., 33
Ziębliński, M., 255
Zolnai, L., 213
Zotti, C., 340